Grass Varieties in the United States

United States Department of Agriculture

AF754537

For the United States Department of Agriculture

James Alderson
Plant Materials Specialist
Temple, Texas

W. Curtis Sharp
Plant Materials Specialist
Washington, D.C.

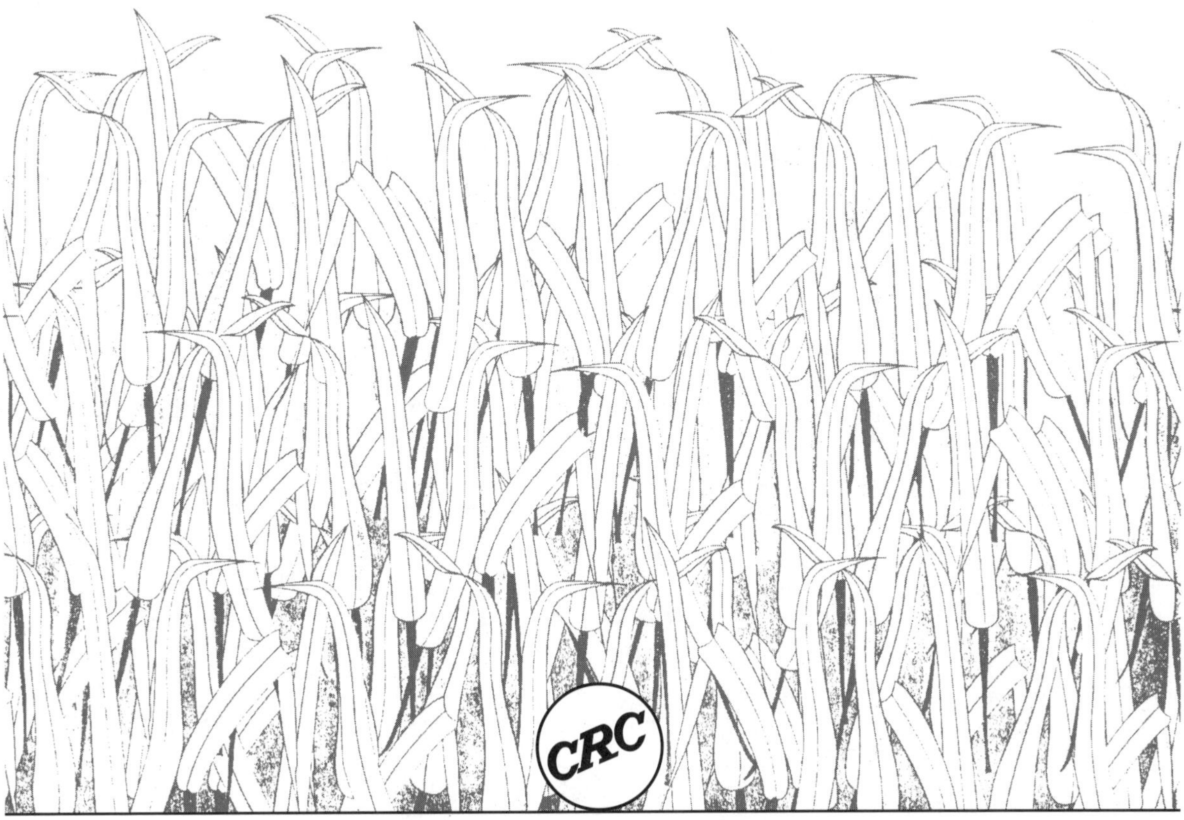

LEWIS PUBLISHERS

Boca Raton New York London Tokyo

Previously published as:
Agriculture Handbook No. 170
Revised November 1994
Soil Conservation Service
U.S. Department of Agriculture
Washington, D.C.

Library of Congress Cataloging-in-Publication Data

Catalog record is available from the Library of Congress.

This book contains information obtained from authentic and highly regarded sources. Reprinted material is quoted with permission, and sources are indicated. A wide variety of references are listed. Reasonable efforts have been made to publish reliable data and information, but the author and the publisher cannot assume responsibility for the validity of all materials or for the consequences of their use.

Neither this book nor any part may be reproduced or transmitted in any form or by any means, electronic or mechanical, including photocopying, microfilming, and recording, or by any information storage or retrieval system, without prior permission in writing from the publisher.

CRC Press, Inc.'s consent does not extend to copying for general distribution, for promotion, for creating new works, or for resale. Specific permission must be obtained in writing from CRC Press for such copying.

Direct all inquiries to CRC Press, Inc., 2000 Corporate Blvd., N.W., Boca Raton, Florida 33431.

© 1995 by CRC Press, Inc.
Lewis Publishers is an imprint of CRC Press

No claim to original U.S. Government works
International Standard Book Number 1-56670-193-7
Printed in the United States of America 1 2 3 4 5 6 7 8 9 0
Printed on acid-free paper

Preface

This handbook is a working guide to the status of named and experimental grasses available for use in the United States (U.S.). Some, no doubt, have been inadvertently missed. Several varieties that are included are currently registered in Canada, and may not currently be used in the U.S. Most of the varieties or experimental lines were developed in the U.S., while others originated in Canada or Europe.

The first issue of Grass Varieties in the United States appeared as a processed report, CR-12-58, from the Crops Research Division, Agricultural Research Service (ARS). Subsequent revisions, including the present one, were published as Agriculture Handbook No. 170 and appeared in 1959, 1965, and 1972. The format of the handbook and the basic information found in the descriptions are about the same as in previous revisions. When available, the types of information include the following:

- agency or agencies involved in developments,
- individuals who assumed leadership in selecting or increasing specific varieties, where known,
- method of breeding,
- description of variety,
- release date,
- releasing agency or agencies,
- source of breeder seed, and
- status of certified seed/stock production.

Three additional categories of information for each variety have been added:

- intended use;
- the known or anticipated area the plant is adapted to, usually presented as a combination of land resource region and plant hardiness zone; and
- the person who reviewed or prepared the information and/or the source of additional information for each entry.

Land Resource Regions and the Plant Hardiness Zone maps are located in appendix A. The adaptation information is not available for all entries or is provided only as geographic regions.

Previous editions of Agriculture Handbook 170, as well as this one, generally have not included grasses used for the production of commodity crops. Although the last edition did list a few sorghums, these have now been deleted. This edition has greatly expanded the descriptions of turf grasses.

Throughout the descriptions, the current names/acronyms of U.S. government agencies are used. Appendix B is an alphabetical list of these, as well as other abbreviations used in the handbook.

Variety registration numbers in the U.S. were assigned originally under a memorandum of understanding between the former Crops Research Division, ARS, and the American Society of Agronomy. Currently, crop registration is performed by the Crop Science Society of America. Registration is provided in Canada by the Plant Protection Division, Agriculture Canada. Registration provides a permanent record of named and distinctive characteristics of crop varieties; it does not mean that a given variety is superior in any or all respects to any other variety.

Insofar as possible, those individuals and agencies that have taken the initiative in developing specific varieties have been identified. It is recognized, however, that plant introductions, as well as some selections and gene pools, are distributed widely for evaluation purposes. Thus, it is often difficult to assign sole credit for the development of a given variety to any one individual or group. Information on the source of breeding material can often aid in identifying the origin of germplasm; it also provides partial recognition to the contribution of plant exploration and to those individuals and agencies furnishing seed and plant material. The information required to provide such appropriate recognition on identification was not available for all entries.

No attempt has been made to appraise the relative merits of varieties included in the handbook nor to verify the adaptation information provided by the preparer. Descriptions, with some exceptions, are those reported by developers. For the most part, the descriptions are based on information accumulated at originating stations. Specific attributes may not be expressed, at least not to the same degree, in all environments where the variety is adapted.

Breeder seed designation, for the purpose of the handbook, serves only to indicate the source of stock seed or planting material.

For some varieties, limited local supplies of certified seed may have been overlooked. Conversely, there may be common seed/stock of a variety sold that was in the previous revision but not included in this one. This means that breeder or other reliable germplasm of the variety is no longer available. These varieties, as well as experimental lines that fall in this same category, are listed in Appendix C, Obsolete Grass Varieties and Experimental Lines.

Descriptions are included of both named varieties and experimental lines that have been distributed for testing purposes. Insofar as reasonably acceptable descriptions are available, all varieties and experimental lines that may be distributed, sold, or easily available for use in the U.S. are included.

Common and accepted scientific names are included at the beginning of the discussion of each species. Any synonyms for the accepted name follow the accepted name. Only common names are used in the variety descriptions, unless the use of scientific names is needed for clarity. Only common names of diseases and insects are used in the descriptions. Appendix D is a list of all common names of diseases and insects used in the publication, followed by a scientific name. Because many of the original descriptions contained only common names, it is certainly possible that the pairing of the scientific name with a common name in appendix D is different from the one intended by the author of the original description.

This revision would have been impossible without the cooperation and contributions of many individuals. Richard Heizer, State Resource Conservationist, Soil Conservation Service (SCS), Stillwater, Oklahoma, provided leadership for this undertaking at its inception in April 1989. At that time, he was Plant Materials Specialist, SCS, Temple, Texas. Richard made the initial contacts for the review of entries in the current version of the handbook and for the addition of newly developed materials. Darrell Carter, Rebecca Grant, and Barbara Floyd typed the initial draft. J. Scott Peterson confirmed current accepted nomenclatures and prepared the synonymy based on Plant List of Attributes, Nomenclature, Taxonomy and Symbols, an SCS database. The nomenclature and synonymy are based on the work of John Kartesz, Biota of North America Project, University of North Carolina, Chapel Hill, North Carolina. Lara Philbert assisted with style and consistency and with editing the manuscript.

Grateful acknowledgment is made for the cooperation of scientists associated with State agricultural experiment stations, the ARS, SCS, Agriculture Canada, and private seed companies for their contribution of new varieties. To the extent this information was available, these contributions are included at the end of each variety entry. Their contributions provided the basis for this edition. Special appreciation is extended to those who provided assistance with the introduction of new species and new information on varieties in the 1972 edition.

Contents

Varieties and Experimental Lines: Descriptions

Agropyron cristatum (L.) Gaertn. - fairway crested wheatgrass

Important cool-season bunchgrass from Siberia. First distributed in 1927 in Saskatchewan, Canada, and later identified as fairway crested wheatgrass. In the U.S., crested wheatgrass was considered to be one species until 1950, when the (diploid) fairway type was identified as *A. cristatum* (L.) Gaertn., and the (tetraploid) crested typed as *A. cristatum* (L.) Gaertn. X *desertorum* (Fisch. ex Link) J.A. Schultes. At that time, many seed lots represented mixtures of the two species. Thus, in the U.S., the term fairway, used previously in Canada for a specific cultivar of *A. cristatum* (L.) Gaertn., was accepted very widely as the common name for all sources of *A. cristatum* (L.) Gaertn.. Fairway is shorter, denser, finer stemmed, and less productive than crested, but better suited for dryland lawns and general-purpose turf. Used extensively for pasture and hay in western Canada and in the northern Great Plains and intermountain region of U.S.

Ephraim

Selected at USDA-FS Shrub Sciences Laboratory, Provo, UT, and Utah Division of Wildlife Resources - USDA-FS Great Basin Field Station, Ephraim, UT. Carried as accession no. PI 109012.

Source - Ankara, Turkey.

Method of Breeding - Increase and field selection from original introduction.

Intended Use - Range rehabilitation, roadsides, critical areas.

Description - Ephraim is rhizomatous and will grow and produce adequate forage with 200 mm annual precipitation, although it does best where there is 250-360 mm. The higher the precipitation, the sooner the rhizomatous characteristics develop. Adapted to a wide range of soils, including disturbed areas and mine spoils. Salt and alkali tolerance is moderately high. It is not well adapted to silty soils having a very low water intake rate or extremely stony sites. When in pure stands, it is susceptible to the black grass bug. Ephraim is a good seed producer when standard cultural practices are followed. Seed matures fairly evenly in early August and can be harvested with a field combine. On rangeland, stands are well maintained by seed and the strong rhizomatous characteristic.

Adapted to - LRR B, D, E.

Released - 1983, by USDA-FS, ID, cooperatively with Utah Division of Wildlife Resources, SCS, and UT, AZ, and ID AESs.

Breeder Seed/Stock - Plant Materials Center, SCS, Aberdeen, ID.

Certified Seed/Stock - Available.

Preparer/Additional Information - E.D. McArthur, USDA-FS, Shrub Science Lab, 735 N. 500 E., Provo, UT 84606, (801) 377-5717; or Richard Stevens, Utah Division of Wildlife Resources, 15 S. Main, Ephraim, UT 84627, (801) 283-4441.

Agropyron cristatum (L.) Gaertn. X *desertorum* (Fisch. ex Link) J.A. Schultes - crested wheatgrass

Hycrest

ARS forage research unit at Logan, UT - Dr. Kay Asay. Carried as accession no. SCS-9028605.

Source - Central Asia.

Method of Breeding - Generated by crossing induced tetraploid *Agropyron cristatum* (L.) Gaertn. with natural tetraploid *A. desertorum* (Fisch. ex Link) J.A. Schultes. The initial crosses were made by D.R. Dewey from 1962 to 1967. The genetic base of the hybrid population was established with seven clones, each of induced tetraploid *A. cristatum* (L.) Gaertn. and *A. desertorum* (Fisch. ex Link) J.A. Schultes. Reciprocal crosses were made to insure that the cytoplasms of both species were represented in the breeding population.

Intended Use - Range and pasture seedings.

Description - Hycrest tends to be larger and more robust than the two parental species. It establishes better stands and produces significantly more forage than Nordan or Fairway, particularly during the first two years after seeding. In spaced planted trials, it produced about 20% more seed than Nordan and Fairway. Hycrest also performed significantly better than Fairway and Nordan in terms of stand establishment in the field, root development, emergence from deep plantings, and subsequent seedling vigor. It is well adapted to sagebrush and juniper vegetation sites and it also established good to excellent stands on shadscale-budsage, shadscale-desert molly, grease-wood, and Indian ricegrass sites where annual precipitation is less than 25 cm.

Adapted to - LRR B, D, E; PHZ 4, 5, 6, 7.

Released - 1984, by ARS in cooperation with Utah AES and SCS.

Breeder Seed/Stock - ARS, Logan, UT.

Certified Seed/Stock - Available in quantity.

Preparer/Additional Information - Kay H. Asay, ARS Forage Research Unit, Utah State Univ., Logan, UT 84322, (801) 750-2233.

Parkway

Source - Collection of fairway crested wheatgrass.

Method of Breeding - Several generations of recurrent selection for vigor, height, and leafiness followed by polycross progeny tests at three Saskatchewan stations. Synthetic of 16 clones distributed for testing as S-5565.

Intended Use - Hay and pasture production.

Description - Hay and seed yields 7-10% above certified fairway wheatgrass. Plants 50-75 mm taller, slightly less leafy, much greater lodging resistance than fairway. Recommended for hay, but not as turfgrass.

Adapted to - PHZ 3, 4.

Released - 1969, by Canada Department of Agriculture.

Breeder Seed/Stock - Canada Department of Agriculture, Research Station, Saskatoon, Saskatchewan.

Certified Seed/Stock - Available.

Preparer/Additional Information - Agriculture Canada, Research Station, P.O. Box 1030, Swift Current, Saskatoon S9H 3X2, (306) 773-4621.

Ruff

Developed by USDA, ARS, and Univ. of Nebraska - L.C. Newell and E.C. Leonard.

Source - Commercial lots and experiment station accessions of fairway wheatgrass collected in 1936-38.

Method of Breeding - Plants selected from seven lots of commercial fairway crested wheatgrass collected at the Nebraska station, 1936-38. Selection of spaced clones was carried through three successive generations. Seed was composited as Nebraska 3576 and increased in central latitudes in the 1940s and 1950s.

Intended Use - Recommended as a short-season forage crop in areas of low rainfall of the central plains latitudes for early spring grazing complementing grazing of summer grasses on rangelands, for plantings of roadsides, parks, and playgrounds in such areas, and in mixtures with other cool-season grasses such as western wheatgrass for such purposes, including special plantings for conservation.

Description - Ruff is an early, cool-season grass. It owes its sod-forming characteristics to its adaptation for self-seeding and to its broad-bunch plants in contrast to the narrow-bunch plants of standard crested wheatgrass. Plants are perennial, leafy, and relatively short, with fine culms which terminate in broad cristate spikes. The spikelets are several-flowered, usually awned. Seed is harvested in mid July. Seed is relatively small. The cultivar establishes easily from rapid seed germination in cool seasons of early fall or early spring. After establishment, it competes well with annual or perennial weeds or other grasses. In the central Great Plains, it produces forage yields similar to those of Nordan. It has superior resistance to root rot organisms and in a grazing trial in eastern Nebraska had significantly better persistence than Nordan.

Adapted to - LRR northern G & H, F; PHZ 3, 4, 5.

Released - 1974 by Nebraska AES and ARS cooperating.

Breeder Seed/Stock - Nebraska AES and USDA-ARS, Univ. of Nebraska, Lincoln.

Certified Seed/Stock - Not available.

Preparer/Additional Information - K.P. Vogel, ARS-USDA, 344 Keim Hall, Univ. of Nebraska or Jeff Pederson, Dept. of Agronomy, Univ. of Nebraska, Lincoln, NE 68583, (402) 472-1564, (402) 472-2811.

Agropyron desertorum (Fisch. ex Link) J.A. Schultes - standard crested wheatgrass

***Agropyron cristatum* (L.) Gaertn. ssp. *desertorum* (Fisch. ex Link) A. Love**

Major cool-season bunchgrass indigenous to eastern Russia, western Siberia, and central Asia. First successful introduction received from Turkestan in 1906. Used for pasture, hay, and erosion control primarily in northern Great Plains; important for range seeding westward to Cascade and Sierra Nevada mountains, and south to northern Arizona and New Mexico. Long lived, drought resistant; tolerates heavy grazing, but not prolonged flooding. Valuable for reseeding in areas with 230-380 mm rainfall. Starts growth in early spring and is ready to graze before native grasses.

Nordan (Reg. No. 2)

Selected at Northern Great Plains Research Laboratory, ARS, Mandan, ND - G.A. Rogler. Carried as accession no. PI 469225.

Source - Developed from selection made in old nursery at Dickinson, ND in 1937.

Method of Breeding - Single-plant selections made under open pollination for two generations; seven plants within open-pollinated progeny bulked for increase and tested as Mandan 571.

Intended Use - Range and pasture plantings.

Description - More uniform and erect than commercial, with seed more awnless and larger in size and heads more dense and compressed. Good seedling vigor and seed quality. Forage yield as good as or better than commercial.

Adapted to - LRR B, E, F, G; PHZ 2, 3, 4. Released - 1953, cooperatively by North Dakota AES, Fargo, and Plant Science Research Division, ARS.

Breeder Seed/Stock - Northern Great Plains Research Laboratory, Mandan, ND.

Certified Seed/Stock - Available.

Preparer/Additional Information - Ian M. Ray, ARS Northern Great Plains Research Laboratory, P.O. Box 459, Mandan, ND 58554, (701) 663-6445.

Summit

Source - Introduction from Western Siberian Experiment Station, Omsk, former USSR.

Method of Breeding - Rogueing of introduction to increase uniformity.

Intended Use - Forage.

Description - Fairly similar to standard crested wheatgrass strains grown in the U.S.. No attempt made to alter strain through selection, but one generation of increase was by single plants to rogue out impurities, particularly fairway-type plants. Tested as S-131. Stock designated Summit 62 resulted from mass selection of 40 plants in Summit for good seed yield, seed quality, and uniformity of plant type.

Adapted to - LRR F; PHZ 3, 4.

Released - 1953, by Canada Department of Agriculture.

Breeder Seed/Stock - Canada Department of Agriculture, Research Station, Saskatoon.

Certified Seed/Stock - Available.

Preparer/Additional Information - Agriculture Canada, Research Station, P.O. Box 1030, Swift Current, Saskatoon S9H 3X2, (306) 773-4621.

Agropyron fragile **(Roth) P. Candargy ssp.** ***sibiricum*** **(Willd.) Melderis - Siberian wheatgrass**

Agropyron cristatum **(L.) Gaertn. ssp.** ***fragile*** **(Roth) A. Love pro parte**
Agropyron sibiricum **(Willd.) Beauv.**

Cool-season bunchgrass from the former USSR. Similar to *A. desertorum* (Fisch. ex Link) J.A. Schultes in appearance and distribution, but less widely tested and used.

P-27

Selected at Plant Materials Center, SCS, Pullman, WA and Aberdeen, ID - J.L. Schwendiman and R.H. Stark. Carried as accession no. PI 108434.

Source - Original collection in 1934 from Kazakhstan; obtained from Institute of Plant Industry, St. Petersburg, Russia, by Westover-Enlow expedition.

Method of Breeding - Included in row nurseries and field-evaluation studies since 1935. Individual clones selected in 1949 by R.H. Stark are basis of present increase.

Intended Use - Range reseeding.

Description - Similar to standard crested wheatgrass, *Agropyron desertorum*, in adaptation and season of use, but differs in several important respects. Narrow, awnless heads and fine, leafy stems. Drought resistant, good seedling vigor, good seed yields. Well adapted to light, droughty soils.

Adapted to - LRR B, D; PHZ 5.

Released - 1953, cooperatively by Idaho AES, Moscow, and Plant Materials Center, SCS, Aberdeen, ID and Pullman, WA.

Breeder Seed/Stock - Plant Materials Center, SCS, Aberdeen, ID.

Certified Seed/Stock - Available.

Preparer/Additional Information - Clarence A. Kelley, Plant Materials Center, SCS, Rm. 104, Hulbert Agricultural Sciences Bldg, WSU, Pullman, WA 99164-6211, (509) 335-7376.

Agrostis canina **L. - velvet bentgrass**

Cool-season, stoloniferous grass introduced from Europe. Used on putting greens, bowling greens, and lawns in some coastal areas of Northeastern U.S.. Relatively shade tolerant, but does not thrive on poorly drained soils. Propagated from either seed or stolons.

Kingstown

Selected at Rhode Island AES, Kingston - C.R. Skogley and J.A. DeFrance.

Source - Inbred selection from Piper by H.F.A. North in 1929.

Method of Breeding - Maintained vegetatively and from seed since initial selfing, comparative testing.

Intended Use - Golf courses.

Description - Semi-brilliant dark green; excellent vigor; good texture. Good resistance to most diseases; very resistant to dollar spot.

Adapted to - New England.

Released - 1963, by Rhode Island AES.

Breeder Seed/Stock - Rhode Island AES.

Certified Seed/Stock - Available.

Preparer/Additional Information - Pickseed West Inc., 33149 Hwy 99E, Tangent, OR 97389, (503) 926-8886.

SR 7200

Source - Old turf sites in New England and Canada. Evaluated at University of Rhode Island by Dr. Richard Skogley and best four clones selected to be parental material.

Method of Breeding - The four selected clones were vegetatively established into a breeder block at Corvallis, OR and allowed to intercross. The resulting 1260 progeny (315 from each clone) planted in clonal rows in Corvallis, OR. Rogued down to approximately 100 plants based on color, leaf texture, freedom from disease, seed yield based on panicle production/floret fill, etc.

Intended Use - High quality turf primarily for golf greens in New England, Pacific Northwest. Total area of adaptation to be established.

Description - Velvets are finest textured bentgrass. This variety has improved seed yields compared to previous velvets. Original plants selected from low-maintenance sites so may reduce inputs for high-quality greens and increase area of adaptation.

Adapted to - PHZ 3, 4, 5, 6, 7, 8 NE and Pacific NW.

Released - 1993, by Seed Research of Oregon, Inc.

Breeder Seed/Stock - Seed Research of Oregon, Inc.

Certified Seed/Stock - Available in 1994.

Preparer/Additional Information - Dr. Leah A. Brilman, Seed Research of Oregon, Inc., P.O. Box 1416, Corvallis, OR 97339, (503) 757-2663.

***Agrostis capillaris* L. - colonial bentgrass**
***Agrostis tenuis* Sibthorp**

Important cool-season, stoloniferous grass from Europe. Used for golf courses, lawns, and erosion control. Well adapted in northeastern region and along northern Pacific coast. Inferior to creeping and velvet bentgrass for putting greens and other fine turf. Most of the production traces to seed collection originally from native stands found in southwestern part of State, from north of Columbia River to Olympia.

Astoria

Source - Collection made in northwestern Oregon by Engbretson and Hyslop in 1926.

Method of Breeding - Comparative testing.

Intended Use - Fine turf, putting greens.

Description - Weakly creeping; short stolons; semi-erect, slender culms. Short ligule, round to obtuse in shape, finely toothed, and often split. Panicle open with delicate form and somewhat larger than that of common colonial but cannot be readily distinguished on basis of growth habit or color. Astoria may be slightly more robust but not under all conditions. Susceptible to brown patch. Used in lawn mixtures and on fairways.

Released - Yes. Included in seed certification program in Oregon in 1926.

Breeder Seed/Stock - Not available.

Certified Seed/Stock - Available.

Preparer/Additional Information - Department of Crop and Soil Sciences, Crop Science Building 107, Oregon State Univ., Corvallis, OR 97331-3002, (503) 737-4513.

Bardot

Source - Collections from lawns, pastures, and other uncultivated areas. Dutch ecotypes.

Method of Breeding - Selection and evaluation of individual plants, clones, and families under turf conditions.

Intended Use - Cool season turfgrass. Suitable for greens, lawn mixtures.

Description - Late heading; rather dark green, narrow leaves; dense turf; moderate late flowering; prostrate turf type. Good resistance to *Fusarium nivale* and other diseases; good drought resistance; excellent winter color; suitable for lawns and fine turf. Produces a very attractive turf under low cutting and is also suitable for low maintenance.

Adapted to - PHZ 1, 2, 3, 4 - Pacific Northwest, Northeast.

Released - 1968, Agency for US and Canada, Steven J.R. Frohlich & Co., Princeton, NJ.

Breeder Seed/Stock - Barenbrug Holland.

Certified Seed/Stock - Available from Barenbrug USA.

Preparer/Additional Information - Barenbrug USA, P.O. Box 239, Tangent, OR 97389, (503) 926-5801.

Exeter

Source - Old pasture in Exeter, RI, about 1940.

Method of Breeding - Comparative testing.

Intended Use - Golf course greens and fairways, and home lawns.

Description - Similar to Astoria. Becomes green earlier in spring; holds color better in summer. Bright green, more leafy than other varieties. Very winter hardy. Some leaf spot resistance. Seems best adapted to north and east in cool-season turfgrass region. Used in lawn mixtures, fairways, lawn tennis courts, and bowling greens.

Adapted to - New England.

Released - 1963, by Rhode Island AES.

Breeder Seed/Stock - Rhode Island AES.

Certified Seed/Stock - Available.

Preparer/Additional Information - Pickseed West Inc., 33149 Hwy 99E, Tangent, OR 97389, (503) 926-8886.

Highland

Source - Collections made in southern Willamette Valley, OR, about 1930.

Method of Breeding - Comparative testing.

Intended Use - Fine turf, fairways.

Description - Astoria and common colonial bentgrass are very similar in appearance. Highland has several distinctive characteristics: bluish green, with erect robust culms; ligule longest of three types, about 1-3.5 mm, round to obtuse, finely toothed, and often split; panicles generally largest of three, pyramidal in form, with variations from Astoria to almost appearance of redtop; culms tend to be coarse and tall; culms and panicles dull, light red up to spikelets and remain so at ripening; panicles semiclosed after blooming, making it readily noticeable in fields of common colonial, which turns brown at ripening and its panicles remain open. Highland stoloniferous, slightly stronger creeper than other types. Susceptible to brown patch. Turf tends to become puffy when mowed at ordinary lawn height.

Released - Yes. Included in seed certification program in Oregon in 1934.

Breeder Seed/Stock - Oregon AES.

Certified Seed/Stock - Available.

Preparer/Additional Information - Department of Crop and Soil Sciences, Crop Science Building 107, Oregon State Univ., Corvallis, OR 97331-3002, (503) 737-4513.

SR 7100

Source - Selection from spaced planting of European variety Bardot.

Method of Breeding - One thousand-plant nursery evaluated for freedom from disease, bright green color, fine leaf texture, good panicle production and good floret fill for two years. Ninety plants allowed to interpollinate to form base population. Second generation used to verify improvement in yield.

Intended Use - Cool season turfgrass; may also be used for overseeding. Primary use for golf course fairways and lawns in the Pacific Northwest.

Description - Very fine textured colonial bentgrass with improved seed yields. Has improved color retention, seed yields and brown patch resistance when compared to parental variety Bardot. Improved turf quality in New Jersey at fairway heights.

Adapted to - PHZ 4, 5, 6, 8 NE, Pacific NW, California 8, 9, 10 SE, SW overseeding.

Released - 1993, Seed Research of Oregon, Inc.

Breeder Seed/Stock - Seed Research of Oregon, Inc., Corvallis, OR.

Certified Seed/Stock - Available.

Preparer/Additional Information - Dr. Leah A. Brilman, Seed Research of Oregon, Inc., P.O. Box 1416, Corvallis, OR 97339, (503) 757-2663.

Tracenta

Source - The Netherlands.

Method of Breeding - Plant selection began in 1954 from ecotypes from the Netherlands. Plant selection was followed by clone selection and progeny selection after a polycross. Based on six clones.

Intended Use - Turf for golf course tees and fairways.

Description - Dense, fine-leaved, dark green. Good winter hardiness.

Adapted to - PHZ 3, 4, 5, 6, 10 - Cool humid.

Released - Mommersteeg International, Vlijmen, The Netherlands.

Breeder Seed/Stock - Mommersteeg International, Vlijmen, The Netherlands.

Certified Seed/Stock - Available from Advanta Seeds West, Inc.

Preparer/Additional Information - Kenneth Hignight, Advanta Seeds West, Inc., 33725 Columbus Street S.E., P.O. Box 1496, Albany, OR 97321-0452, (503) 967-8923.

Agrostis gigantea Roth - redtop
Agrostis alba auct. non L.

Important cool-season, rhizomatous grass from Europe. Used as temporary grass in lawn seed mixtures, for pastures, erosion control, and occasionally hay. Found throughout cooler parts of U. S., especially in northeastern and north-central regions. Adapted for use on poorly drained acid soils. Less aggressive than colonial bentgrass.

9051629

Selected at the Big Flats Plant Material Center, SCS, Corning, NY. Carried as accession no. 9051629, a composite of accession nos. PI 443037 and 9046772.

Source - Accession no. 443037 was from a collection made in Orleans County, NY, in 1975 by Wendall Oaks. Accession no. 9046772 was a selection from common seed obtained from Mangelsdorf Seed Co., MO.

Method of Breeding - These two accessions were selected from 62 accessions made at the Big Flats Plant Material Center, SCS, Corning, NY. Selection criteria were for lowest incidence of leaf diseases and herbage production. Seed from the two accessions were blended and increased in bulk.

Intended Use - Erosion control, critical area revegetation, and forage production.

Description - Leafy, vigorous accession with leaves narrower than Streaker. Slightly susceptible to leaf rust.

Adapted to - LRR K, L, M, N, P, R, S, T; PHZ 4.

Released - Pending in 1994.

Breeder Seed/Stock - Big Flats Plant Materials Center, SCS, Corning, NY.

Certified Seed/Stock - Not available.

Preparer/Additional Information - P. Salon and J. Dickerson, Big Flats Plant Materials Center, SCS, RD#1 Rte 352, Box 360A, Corning, NY 14830, (607) 562-8404.

Streaker

Jacklin Seed Company, Post Falls, ID, and Lofts Seed, Inc. - A.W. Jacklin, A.D. Brede, and R.H. Hurley. Carried as accession no. P 501.2.

Source - Commercial seed from Illinois production.

Method of Breeding - Twenty-one lots of seed from Illinois production were evaluated for uniformity, vigor, seed yield, and mechanical purity and cleanliness. A production field was established and rogued thoroughly to remove any aberrant. Trueness to type tests indicated high degree of stability and uniformity.

Intended Use - Overseeding of dormant warm season grasses, low maintenance turf, reclamation, and pasture.

Description - Uniform and stable tufted perennial with fine leaves and strong rhizome system. Fine textured, high spring density turf.

Adapted to - LRR A, B, E, F, K, L, M, N, P, R, S; PHZ 5.

Released - 1982 Jacklin Seed Company, Post Falls, ID and Lofts Seed, Inc., Bound Brook, NJ.

Breeder Seed/Stock - Breeder seed is controlled by Jacklin Seed Co., Post Falls, ID.

Certified Seed/Stock - Available.

Preparer/Additional Information - Kim Peterson, Jacklin Seed Co., W. 5300 Riverbend Ave, Post Falls, ID 83854, (208) 773-7581.

Agrostis stolonifera L. - creeping bentgrass
Agrostis alba L. var. *palustris* (Huds.) Pers.
Agrostis maritima Lam.
Agrostis palustris Huds.

Cool-season, stoloniferous grass from Eurasia. Used for lawns, putting greens, and erosion control. Planted on putting greens throughout much of the U.S., especially in Northeastern, Midwestern, and Northwestern States. Two types commercially available: 1) a group of individual strains selected from established greens of South German mixed bentgrass, all propagated vegetatively; and 2) includes Penncross and Seaside strains grown from seed.

Cobra

Cooperatively developed by International Seeds Inc., Halsey, OR, and the New Jersey AES. Carried as accession no. HK.

Source - Plants collected in the northeastern U.S..

Method of Breeding - Plants screened for resistance to red leaf spot; individual plant and progeny testing of resistant plants under low soil moisture conditions was undertaken. Seven elite clones were selected to produce synthetic variety.

Intended Use - Turfgrass for golf course tees, greens, and fairways. Also recommended for bowling greens and croquet courts.

Description - Early maturing variety with excellent seed yield potential. Produces a uniform, dark green turf capable of maintaining high quality under low soil moisture, heat and other forms of environmental stress. Very good resistance to red leaf spot.

Adapted to - LRR A, B, C, E, F, G, H, J, K, L, M, N, O, P, R, S, T; PHZ 3.

Released - 1988 by International Seeds Inc., Halsey, OR.

Breeder Seed/Stock - Maintained by International Seeds Inc.

Certified Seed/Stock - Available from International Seeds, Inc. PVP #8900086.

Preparer/Additional Information - Stephen W. Johnson, International Seeds Inc., P.O. Box 168, 820 W. First St., Halsey, OR 97348, (503) 369-2251.

Emerald

Developed by W. Weibull AB, Landskrona, Sweden.

Source - Congressional (C-19) bentgrass.

Method of Breeding - Progeny of Congressional screened for turf and seed production characteristics. Single clone chosen from which to derive cultivar.

Intended Use - High quality turf for areas such as golf courses, lawn tennis courts, and bowling greens.

Description - Widely adapted, cold tolerant, moderately vigorous variety. It is very uniform and does not form the segregating patches that may appear in turfs seeded to other varieties. Exhibits very good resistance to pink snow mold and typhula blight. Thatch build-up is lower than many other cultivars.

Adapted to - LRR A, B, C, E, F, G, H, J, K, L, M, N, O, P, R, S, T, W; PHZ 3.

Released - Europe 1965 by Weibull; in the U.S., 1973, by International Seeds Inc., Halsey, OR.

Breeder Seed/Stock - W. Weibull AB, Landskrona, Sweden.

Certified Seed/Stock - Available.

Preparer/Additional Information - Stephen W. Johnson, International Seeds Inc., P.O. Box 168, 820 W. First St., Halsey, OR 97348, (503) 369-2251.

Lopez

Source - Willamette Valley Plant Breeders, Brownsville, OR.

Method of Breeding - Derived from 16 elite clones. An advance generation synthetic variety.

Intended Use - Turf, putting greens, tees, and fairways.

Description - Dark green, seed type bentgrass, fine bladed leaf, excellent disease resistance.

Adapted to - PHZ 2, 3, 4, 5, 6, 7, 8 - cool humid regions, transitional zones, cooler portions of warm humid regions.

Released - Kevin McVeigh, Willamette Plant Breeders.

Breeder Seed/Stock - Willamette Valley Plant Breeders.

Certified Seed/Stock - Available.

Preparer/Additional Information - David Lundell, Fine Lawn Research, Inc., P.O. Box 1051, Lake Oswego, OR 97034, (503) 636-2600.

Penncross (Reg. No. 1)

Selected at Pennsylvania AES, University Park - H.B. Musser.

Source - Parent strains for seed production identified under station accession numbers 10(37)4 (Pennlu creeping bentgrass), 9(38)5, and 11(38)4.

Method of Breeding - First-generation seed produced by random crossing of three vegetatively propagated clones of creeping bentgrass.

Intended Use - Fine turf and putting greens.

Description - Turf-quality records obtained over five-year period at Pennsylvania AES show Penncross significantly better in density, tolerance to disease, and rate of recovery from attacks than other commercially available seeded types. Because of general vigor, shows exceptional ability to produce better turf than other seeded bentgrasses under adverse conditions.

Adapted to - Cool temperate regions of the U.S.

Released - 1954, by Pennsylvania AES.

Breeder Seed/Stock - Parent clones maintained by Pennsylvania AES.

Certified Seed/Stock - Available.

Preparer/Additional Information - Pennsylvania State Univ., Agronomy Department, University Park, PA 16802, (814) 865-6541.

Penneagle

Developed at Pennsylvania AES, University Park - J.M. Duich.

Source - A pool of 156 vegetative bent lines, from which 21 were selcted for a crossing program.

Method of Breeding - Following turfgrass screening, combinations of seven lines were grown in Pennsylvania and Oregon and then turf tested in Pennsylvania, Ohio, Illinois, and North Carolina. Based on performance, a four-parent polycross combination was selected and further turf-tested on 26 golf course and university sites. Parents are second- and third-generation selections from Washington bent, and third-generation from Seaside and Cocoos bents.

Intended Use - Predominantly golf course fairways, as well as greens and tees.

Description - A medium-textured bent with upright growth habit and non-excessive thatching compared to Penncross bent. Good resistance to leafspot, brownpatch, and dollar spot and fair resistance to cool weather brownpatch and snowmold. Significantly differentiated from other creeping bent cultivars in mature form in vegetative and morphological characteristics and in electrophoresis isoenzyme banding patterns.

Adapted to - Cool temperate regions.

Released - 1978, PA AES, PVP No. 7900009, as certified seed only.

Breeder Seed/Stock - PA AES.

Certified Seed/Stock - Available in quantity.

Preparer/Additional Information - Pennsylvania State Univ., Agronomy Dept., University Park, PA 16802, (814) 865-6541.

Pennlinks

Developed at Pennsylvania AES, University Park - J.M. Duich.

Source - Plants collected from golf course greens and experimental lines.

Method of Breeding - Original base parental clone (#126) selected from a large segregated patch from a green at the Country Club, Shaker Heights, Ohio, in 1964 and outcrossed with nine experimental lines. The "126 line" was subjected to two cycles of reselection, backcrossed to original clone followed by further reselection. Variety consists of approximately 100 parental clones utilized to produce breeder seed. Tested on 55 golf courses prior to release.

Intended Use - Golf course greens, tees, fairways, and similar turf areas.

Description - Upright growth habit (non-graining), fine foliar texture (0.73 mm at 5/32" cut and 1.32 mm at 1/2") and minimum comparative segregation when managed as putting green turf. Differs from other bent cultivars by texture, mature form in vegetative and morphological characteristics, and electrophoresis isoenzyme banding patterns. Exhibits good heat tolerance, good leaf spot, dollar spot, and cool weather resistance, and fair snowmold and brownpatch resistance.

Adapted to - Entire U.S., except extremely warm, humid areas.

Released - 1986, PA AES, PVP No. 8700030, as certified seed only.

Breeder Seed/Stock - PA AES.

Certified Seed/Stock - Available in quantity.

Preparer/Additional Information - Pennsylvania State Univ., Agronomy Dept., University Park, PA 16802, (814) 865-6541.

Providence / SR 1019

University of Rhode Island and Seed Research of Oregon - R. Skogley and M.F. Robinson.

Source - Germplasm collected throughout northeastern North America from old turf including turf trials at the University of Rhode Island established in the 1920s and 1930s.

Method of Breeding - Polycross progeny of five clones selected in 1968 at the University of Rhode Island with selection based on superior performance and persistence in long-term tests.

Intended Use - High-quality turf for golf course greens, tees, and fairways.

Description - SR 1019 forms a turf that is very uniform in appearance and growth habit. In comparison with other bentgrass varieties, it has shown improved disease resistance, especially to dollar spot. It has shown improved winter hardiness, puttability, and reduced thatching tendency. SR 1019 has demonstrated greater tiller production and its aggressiveness is a real benefit when competing against *Poa annua*.

Adapted to - LRR A, B, C, D, E, F, G, H, I, J, K, L, M, N, O, P, R, S, T, U, W; PHZ 2, 3, 4, 5, 6, 7, 8, 9, 10 (some of 9 & 10 for overseeding).

Released - 1988, Seed Research of Oregon.

Breeder Seed/Stock - Seed Research of Oregon, Corvallis, OR.

Certified Seed/Stock - Available.

Preparer/Additional Information - Dr. Leah A. Brilman, Seed Research of Oregon, Inc., P.O. Box 1416, Corvallis, OR 97339, (800) 253-5766.

Putter

Developed by Washington State University, Puyallup, WA, and Jacklin Seed Company, Post Falls, ID - Stan Brauen, and A.D. Brede. Carried as accession no. AP-10.

Source - From 26 disease-resistant creeping bentgrasses collected from North America in 1972.

Method of Breeding - The 26 accessions were intercrossed in a nursery; selection from this polycross progeny resulted in release.

Intended Use - Golf course putting greens.

Description - Characterized in mowed turf by winter dormancy with compact spring growth; vigorous, dense, upright leaves; resistance to take-all patch.

Adapted to - LRR A, B, E, G, F, M, K, L, R, S; PHZ 5.

Released - Yes. Breeder Seed/Stock - Jacklin Seed Company, Post Falls, ID (proprietary).

Certified Seed/Stock - Available.

Preparer/Additional Information - Kim Peterson, Jacklin Seed Company, W. 5300 Riverbend Ave., Post Falls, ID 83854, (208) 773-7581.

Regent

Source - From 285 clones derived from Penncross, Emerald, and selections from old turfs in the eastern U.S. and Oregon.

Method of Breeding - Advanced generation synthetic cultivar developed by Willamette Valley Plant Breeders with Dr. Kevin McVeigh.

Intended Use - Turf.

Description - Medium dark green, fine leaf texture, superior disease resistance, persistence, dense turf.

Adapted to - All zones where creeping bentgrass is used for turf.

Released - 1991 by Barenbrug USA/Normarc.

Breeder Seed/Stock - Willamette Valley Plant Breeders and Barenbrug USA.

Certified Seed/Stock - Available from Barenbrug USA.

Preparer/Additional Information - Barenbrug USA, P.O. Box 239, Tangent, OR 97389, (503) 926-5801.

Southshore

Source - Derived from an extensive selection of plants collected from old putting greens located in New Jersey, New York, California, Oregon, and Pennsylvania.

Method of Breeding - Selections were made for persistence, aggressiveness, tolerance of environmental stresses, attractiveness, freedom from disease, upright growth, color, and fine texture.

Intended Use - Cool season turfgrass for golf course greens, tees, and fairways. Can be used alone or blended.

Description - Southshore displays a dense, upright growth habit that is superior to many other popular varieties. It exhibits a medium-fine leaf texture that produces less grain.

Adapted to - Wherever cool season grasses are adapted.

Released - Lofts Seed Inc., 1992 - First certified crop released.

Breeder Seed/Stock - Lofts Seed, Inc. Seed classes will be limited to breeder, foundation, registered and certified.

Certified Seed/Stock - Available.

Preparer/Additional Information - Marie Pompei, Lofts Seed Inc., P.O. Box 146, Bound Brook, NJ 08805, (908) 560-1590.

SR 1020

Selected at the University of Arizona and developed jointly with Seed Research of Oregon - W.R. Kneebone and M.F. Robinson.

Source - Primary plants collected from old putting greens in Arizona and germplasm introduced from Clemson University.

Method of Breeding - Polycross progenies of five clones selected for heat and drought tolerance, uniformity, and good seed production.

Intended Use - High quality turf for golf course greens, tees, and fairways.

Description - Dark green, fine textured leaves. Improved heat and drought tolerance and tolerance of reduced levels of fertility. High resistance to current races of both species of pythium blight. Susceptible to dollar spot. SR 1020 produces a uniform, closely knit turf with excellent vigor and density that does not tend to segregate into component clones.

Adapted to - LRR A, B, C, D, E, F, G, H, I, J, K, L, M, N, O, P, R, S, T, U, W; PHZ 2, 3, 4, 5, 6, 7, 8, 9, 10 (some of 9 & 10 for overseeding).

Released - 1987 by Seed Research of Oregon.

Breeder Seed/Stock - Seed Research of Oregon in Corvallis, OR.

Certified Seed/Stock - Available.

Preparer/Additional Information - Dr. Leah A. Brilman, Seed Research of Oregon, Inc., P.O. Box 1416, Corvallis, OR 97339, (800) 253-5766.

Alopecurus arundinaceus Poir. - creeping foxtail

Cool-season, sod-forming grass from Eurasia. Limited use for hay, pasture, and erosion control in moist areas of northern Great Plains, Pacific Northwest, and intermountain region. Possesses strong rhizomes, forms dense sod, and well adapted to wetland pastures and some mountain meadows.

Garrison

Selected at the Plant Materials Center, SCS, Bismarck, ND - John McDermand. Carried as accession nos. PI 436704, NDG-772.

Source - Field collection made near Max, McLean County, ND, in 1950. Information obtained from local people there indicates grass brought into area from eastern Germany or western Russia by immigrant in early days of homesteading; later, it escaped and was growing around many pothole sloughs in area.

Method of Breeding - Selected as superior to other accessions of the species and *A. pratensis* and increased at the Plant Materials Center, SCS, Bismarck, ND. It is a bulk increase of the original 1950 collection.

Intended Use - For pasture, hay, or grassed waterways in shallow potholes or wetlands that are too wet for most other grasses.

Description - Garrison does not differ significantly from other cultivars or lines of this species. It is more productive and spreads more aggressively than *A. pratensis* L. It survives periods of complete inundation in spring and shallow water during the summer. Moderate tolerance to saline and alkaline conditions. Preferred by livestock over reed canarygrass. Excellent digestibility and protein content.

Adapted to - LRR B, E, F, G, H, K, L, M, N, R, S; PHZ 2a - 5b.

Released - Released in 1963 by the Plant Materials Center, SCS, Bismarck, ND and the Wyoming AES, Laramie, WY.

Breeder Seed/Stock - Plant Materials Center, SCS, Bismarck, ND.

Certified Seed/Stock - Available.

Preparer/Additional Information - Russell Haas, SCS, P.O. Box 1458, Bismarck, ND 58502, (701) 250-4425.

Alopecurus pratensis L. - meadow foxtail

Stout perennial introduced from Eurasia; occasionally cultivated as a meadow grass; found across North America from Virginia and Missouri north and west to California; sometimes rooting at the lower nodes; flowers in June-July; generally found in moist meadows or along ditches and streams.

Dan

Source - Institute of Plant Breeding and Acclimatization in Brudzyn, Poland.

Method of Breeding - Developed from ecotypes gathered in 1939 near Wilno, Poland. Selection from this material was undertaken from 1946 to 1955 using mass selection.

Intended Use - Hay and pasture production.

Description - Heading date is variable; plant height is about 93 cm.; maturity is early; winterhardiness is good; lodging resistance is fair; and moderately resistant to rust. In trials conducted by the Ontario Forage Crops Committee, Dan ranked fourth in yield out of 12 experimental lines tested in 1984-86, although it has very similar yields to the second and third ranked lines. Since no other meadow foxtail varieties were currently registered, the four highest yielding varieties were supported for registration.

Adapted to - Recommended in Canada for use in Ontario.

Released - Paznanska Hodowla, Roslin, Sarmacka, Poland. Registered in Canada in 1987 with registration no. 2753.

Breeder Seed/Stock - Paznanska Hodowla, Roslin, Sarmacka, Poland.

Certified Seed/Stock - Available.

Preparer/Additional Information - Agriculture Canada, Food Production and Protection Branch, Plant Products Division, Ottawa, Ontario K1A 0C6. Seed is distributed by Oseco Inc., P.O. Box 219, Brampton, Ontario L6V 2L2.

Mountain

Source - Otto Pick and Sons Seeds Ltd., Blenheim, Ontario.

Method of Breeding - Mountain is a 31-clone synthetic variety developed from material selected in Ontario for rapid early spring and fall growth, seed yield, and freedom from leaf disease. Selection was conducted in 1981 and selected plants were grown in isolation to produce the parental material from which Mountain was derived.

Intended Use - Hay and pasture production.

Description - Leaves are flat and glabrous; panicles are dense, 7.6 cm.; plant height is approximately 99.7 cm.; maturity is similar to Dan; and lodging resistance is fair.

Adapted to - Recommended in Canada for use in Ontario.

Released - Otto Pick and Sons Seeds Ltd. Registered in Canada in 1989 as registration no. 3041.

Breeder Seed/Stock - Otto Pick and Sons Seeds Ltd.

Certified Seed/Stock - Available.

Preparer/Additional Information - Agriculture Canada, Food Production and Protection Branch, Plant Products Division, Ottawa, Ontario K1A 0C6. Seed is distributed by Pickseed Canada Inc., Box 126, Richmond Hill, Ontario L4C 4X9.

Ammophila arenaria (L.) Link - European beachgrass

Cool-season, rhizomatous grass from Europe. Used to control shifting dunes in coastal regions. Dies out when sand ceases to move; not suitable for permanent erosion control. Propagated vegetatively. Extensively used on Northwest coast of U.S. Varieties not available.

Ammophila breviligulata Fern. - American beachgrass

Cool-season, rhizomatous, native grass. Occurs on shores of Great Lakes, along Atlantic coast from Newfoundland to North Carolina, and sparingly on Pacific Coast. Used for initial sand-dune stabilization. Propagated vegetatively.

Cape (Reg. No. 34)

Selected at SCS Plant Materials Center, Cape May Court House, NJ - W.C. Sharp.

Source - Culms collected in 1965 from single plant growing on sand dune at Sandy Neck, Cape Cod, Barnstable County, MA.

Method of Breeding - Culms of American beachgrass from the mid-Atlantic region were evaluated at Cape May Courthouse, NJ, and in several dune sites for vigor, rate of culm multiplication, and culm size. Cape resulted from this.

Intended Use - Primary stabilization of sand dunes.

Description - Very vigorous; heavy culm producer, both in number and size of stems; leaves exceptionally broad and thick; spreads rapidly by rhizomes. Propagated by culms only.

Adapted to - LRR R, S, T; PHZ 5a - 8b.

Released - 1971, cooperatively by New Jersey AES and SCS.

Breeder Seed/Stock - SCS Plant Materials Center, Cape May Court House, NJ.

Certified Seed/Stock - Plants available. Permitted by nurseries complying with state rules for plant certification.

Preparer/Additional Information - Christopher Miller, SCS, 1370 Hamilton St., Somerset, NJ 08873, (908) 246-4110.

Hatteras

Selected at North Carolina AES, Raleigh - W.W. Woodhouse and D.S. Chamblee.

Source - A group of 18 clones, representing a wide range in plant type, selected from nursery planting in 1963.

Method of Breeding - Clones increased in nursery rows near Clayton and screened for vigor and rate of spread. Eight clones selected for testing at three sites along the coast. In 1966, four promising clones planted in large duplicate plots in dune building trial on Ocracoke Island. The clone, identified as Hatteras, selected on basis of survival and effectiveness in trapping sand.

Intended Use - Initial stabilization and revegetation of coastal sand dunes.

Description - Very stiff, large stems, wide bluish-green leaves, medium to late maturity, and light to medium seed producer. Characterized by early vigor, which is important in establishment, and by rapid spread, which permits wider spacing and rapid healing of dunes broken by storms. Well adapted to entire North Carolina coast. Could be of value elsewhere on Atlantic seaboard. Can be propagated vegetatively under nursery conditions.

Adapted to - LRR T; PHZ 8.

Released - 1969, by North Carolina AES.

Breeder Seed/Stock - North Carolina ARS.

Certified Seed/Stock - Not available.

Preparer/Additional Information - Stephen W. Broome, Dept. of Soil Science, North Carolina State Univ., Box 7619, Raleigh, NC 27695-7619, (919) 515-2643.

Andropogon gerardii Vitman - big bluestem

Important warm-season grass on relatively fertile, well-drained loam soils along eastern edge of Great Plains from North Dakota to eastern Texas. Used for pasture and hay. Has deep roots, short rhizomes, and some lateral spread.

Bison (Reg. No. CU-9)

Collected by the ARS, Northern Great Plains Research Laboratory, Mandan, ND, and developed and evaluated in cooperation with the Plant Materials Center, SCS, Bismarck, ND - George Rogler and Reed E. Barker, John McDermand, Erling T. Jacobson, and Russell J. Haas. Carried as accession no. PI 477994, NDG-4.

Source - Original plants were collected from a native stand near Price, Oliver County, ND.

Method of Breeding - Collected plants were grown for three years in comparison with 30 other accessions. Bison was selected over other accessions because of its uniform plant type with good leafiness, high plant vigor, seed yields, and winter hardiness. Its phenology, persistence, and forage yields were extensively tested by SCS in comparative field studies and on farm field plantings in ND, SD, and MN.

Intended Use - Erosion control, critical area, wildlife habitat, surface mine revegetation, range pasture, and natural area seedings.

Description - Bison is 20 days earlier in anthesis than Bonilla and 30-48 days earlier than the southern cultivars Kaw, Champ, and Pawnee. Tends to be shorter in mature height. Chromosome number 2n=6x=60.

Adapted to - LRR F, G, K; PHZ 3a - 4a.

Released - Released cooperatively in 1989 by the ARS; Plant Materials Center, SCS, Bismarck, ND; and the ND and MN AESs.

Breeder Seed/Stock - ARS, Northern Great Plains Research Laboratory, Mandan, ND.

Certified Seed/Stock - Available.

Preparer/Additional Information - Russell Haas, SCS, P.O. Box 1458, Bismarck, ND 58502, (701) 250-4425.

Bonilla (Reg. No. 0007)

Selected at the Plant Materials Center, SCS, Bismarck, ND, and developed and evaluated in cooperation with ARS, Northern Great Plains Research Laboratory, Mandan, ND - John McDermand, Erling T. Jacobson, Russell J. Haas, and Reed Barker. Carried as accession no. PI 315658, PM-SD-27.

Source - Bulk of seed collected from native stands at two sites near Bonilla, Beadle County, SD.

Method of Breeding - Selected over 16 accessions in initial evaluation studies at the Bismarck Plant Materials Center. Selection based on high seed and forage yields and winter survival. Phenology, persistence, forage yield and quality, animal performance, and wildlife potential have been extensively tested in comparative field evaluation studies and field plantings in ND, SD, and MN. Intended Use - Range and pasture seedings, wildlife habitat, natural areas, surface mine revegetation, and critical area seedings.

Description - Superior winter hardiness, persistence, and seed production ability. Forage production exceeds that of Bison and is equal to Champ and Kaw at northern latitudes. Average daily gains of yearling steers have been higher for Bonilla than Pawnee in grazing studies at Morris, MN.

Adapted to - LRR F, G, K, M; PHZ 3b - 4b.

Released - Released cooperatively in 1987 by the Plant Materials Center, SCS, Bismarck, North Dakota; ARS; and the ND, SD, and MN AESs.

Breeder Seed/Stock - ARS, Northern Great Plains Research Laboratory, Mandan, ND.

Certified Seed/Stock - Available.

Preparer/Additional Information - Russell Haas, SCS, P.O. Box 1458, Bismarck, ND 58502, (701) 250-4425.

Champ (Reg. No. 2)

Developed by USDA, ARS, and Univ. of Nebraska - L.C. Newell.

Source - Domestic collections in Nebraska by L.C. Newell, D.E. Atkinson, and R.D. Staten. Iowa introductions supplied by SCS.

Method of Breeding - Clones with good vegetative spread and large caryopses from five north-central Nebraska sandhill sources reciprocally crossed with clones from prairie sources, two of which derived from Iowa introductions and three from Pawnee County, NE. Resulting 10 progeny lines grown, and eight Synthetic 1 clones from each moved to crossing block. Seed from 80 clones used to establish small field of seeded rows (Synthetic 2 of original cross) for production of breeder seed. The name, Champ, derived from sources of germplasm: Ch-Cherry and Holt Counties, NE; AM-Ames, IA; P-Pawnee County, NE.

Intended Use - Range and pasture seedings.

Description - Moderately late maturing, but ordinarily 7-10 days earlier in seed maturity than Pawnee. Leafy; variable in awn length, culm, glume color (yellow green to purple), and foliage color (light green to glaucous gray). Seed set and seed quality superior to those of ordinary bluestem. Performed well in several forage-yield tests in Nebraska, especially on sandy and fine textured soils in area of adaptation. May be grown for seed production in central and eastern Nebraska as far north as central Platte and lower Loup and Elkhorn valley. Seed should be produced in irrigated rows. For conservation and forage use, it may be utilized in solid stands or mixtures.

Adapted to - LRR northern G & H; PHZ 4, 5.

Released - 1963, cooperatively by Nebraska AES and USDA-ARS.

Breeder Seed/Stock - Nebraska AES and USDA-ARS, Univ. of Nebraska, Lincoln.

Certified Seed/Stock - Available.

Preparer/Additional Information - K.P. Vogel, ARS-USDA, 344 Keim Hall, Univ. of Nebraska or Agricultural Research Division, Univ. of Nebraska, Lincoln, NE 68583, (402) 472-1564, (402) 472-2811.

Kaw

Selected at Kansas AES, Manhattan, KS. Carried as accession no. PI 421276.

Source - Composite of lines selected after four or more generations from progeny of 200 accessions collected in 1935 in native Flint Hills grasslands south of Manhattan, KS.

Method of Breeding - Synthetic variety consisting of several lines.

Intended Use - Range and pasture plantings.

Description - Tall and more uniformly leafy than field-run types. Medium late in maturity. Somewhat resistant to rust. Forage yields greater in plot tests than those of field-run accessions with which it has been compared. Seed yields relatively high and seed set good.

Adapted to - LRR H, M, N; PHZ 4, 5.

Released - 1950, by Kansas AES

Breeder Seed/Stock - Kansas AES.

Certified Seed/Stock - Available.

Preparer/Additional Information - Kansas AES, Agronomy Department, Manhattan, KS 66506, (913) 532-6101.

Niagara

Selected at the Big Flats Plant Materials Center, SCS, Corning, NY - W.C. Sharp. Carried as accession no. NY 1145, PI 315656.

Source - Seed collected in 1957 by Harry Porter from a native stand of big bluestem along Buffalo Creek, Elma Township, Erie County, NY.

Method of Breeding - The original accession was selected by W. Curtis Sharp from an evaluation nursery in comparison with other big bluestem accessions. Plants were then isolated for seed increase.

Intended Use - Used for warm season forage production, for revegetating droughty sites, such as sand and gravel mines, strip mines, or roadsides, and wildlife habitat improvement.

Description - The flag leaf of Niagara is slightly but significantly wider than those of Kaw, Pawnee, and Rountree. Regrowth was more rapid in late summer and early fall. Leaves had less damage from leaf spot disease than plants of Kaw and Pawnee.

Adapted to - LRR R, S, L, M, northern P & O; PHZ 4.

Released - Cooperatively in 1985 by the SCS, ARS, and the Pennsylvania AES.

Breeder Seed/Stock - Plant Materials Center, SCS, Big Flats, NY.

Certified Seed/Stock - Available.

Preparer/Additional Information - P. Salon & J. Dickerson, Big Flats Plant Materials Center, SCS, RD 1, Rt 352, Box 360A, Corning, NY 14830, (607) 562-8404.

Pawnee (Reg. No. 1)

Developed by USDA, ARS, and Univ. of Nebraska - L.C. Newell and E.C. Leonard.

Source - Collected in 1938 from Pawnee County - L.C. Newell and D.E. Atkinson.

Method of Breeding - Collections propagated through several generations. Clones of four types selected in 1948, polycrossed in isolation, and polycross tested in 1950-51. Approximately 260 clones selected from polycrossed progenies and moved to new crossing block for recombination. These clones subsequently progeny tested in 1953-54. Seed of 1959 and 1960 harvests from duplicate plants of recombination crossing block used as breeder seed (Synthetic 1) for foundation seed field establishment in 1961.

Intended Use - Range and pasture seedings.

Description - Typical of big bluestem of central prairies. Moderately long, dark-green leaves and tall flowering stalks, with forked, green to purplish inflorescence. Florets of spikelets long awned. Considerable variation in amount of pubescence in seed heads. Produces good forage yields in Nebraska; superior to native strains originating farther north and west. Seed yields and seed quality produced in cultivated rows and under irrigation superior to those of common strains of bluestem.

Adapted to - LRR H, M; PHZ 5, 6.

Released - 1963, cooperatively by Nebraska AES and USDA-ARS.

Breeder Seed/Stock - Nebraska AES and USDA-ARS, Univ. of Nebraska, Lincoln.

Certified Seed/Stock - Available.

Preparer/Additional Information - K.P. Vogel, ARS-USDA, 344 Keim Hall, Univ. of Nebraska or Agricultural Research Division, Univ. of Nebraska, Lincoln, NE 68583, (402) 472-1564, (402) 472-2045.

Rountree

Plant Materials Center, SCS, in cooperation with the Missouri AES. Carried as accession no. PI 474216.

Source - Rountree was collected from a native stand near Morehead, Monona County, IA, then evaluated, selected, and increased at the Elsberry Plant Materials Center, Elsberry, MO.

Method of Breeding - Cross-pollination.

Intended Use - Pasture, hay production, and range reseedings.

Description - Increased seedling vigor, increased leaf rust resistance, superior forage and seed production, and increased resistance to lodging.

Adapted to - LRR M, N; PHZ 4b.

Released - 1983.

Breeder Seed/Stock - Plant Materials Center, SCS, Elsberry, MO.

Certified Seed/Stock - Available.

Preparer/Additional Information - Jimmy Henry, Plant Materials Center, SCS, RR 1, Box 9, Elsberry, MO 63343, (314) 898-2012.

T-1947

Selected in 1985 at the James E. "Bud" Smith Plant Materials Center, SCS, Knox City, TX. Increased as PI 408932 for field testing. Carried as accession no. PI 408932.

Source - Original seed collected from native stand in Parker County, TX by SCS employee Earl Hogan in 1968.

Method of Breeding - Selected from 116 similar accessions for forage and seed production south of area of adaptation of Kaw.

Intended Use - Range reseeding, hay production.

Description - Grows one or two months longer than Kaw from south central OK through south central TX.

Adapted to - LRR H, I, J; PHZ 7.

Released - Proposed for 1996.

Breeder Seed/Stock - Available after 1996.

Certified Seed/Stock - Available after 1996.

Preparer/Additional Information - Plant Materials Center, SCS, Rte 1, Box 155, Knox City, TX 79529, (817) 658-3922.

Andropogon hallii Hack. - sand bluestem

Warm-season grass. Resembles big bluestem, but differs in having conspicuous hairs on panicle, more vigorous rhizomes, and greater lateral spread. Plants intermediate between typical big and sand bluestem are common. Valuable range grass on deep, sandy soils from central NE into eastern CO and south into NM and OK.

Elida

Selected at Plant Materials Center, SCS, Los Lunas, NM. Collected by J.A. Downs, G.C. Niner, and J.E. Anderson. Carried as accession no. PM-NM-14.

Source - Collection in 1956 from native stand on sand dune area near Elida, NM, at elevation of 1220 meters and annual precipitation about 400 mm.

Method of Breeding - Direct increase of original collection. Field tested as PM-NM-14.

Intended Use - Range reseeding and revegetation of disturbed areas.

Description - Somewhat variable, but more uniform than many other collections. Good foliage extending well up culms. Fairly uniform in ripening and better than other strains tested in seed production. Shows superior establishment, vigor, and production in eastern NM.

Adapted to - LRR G, H; PHZ 5.

Released - 1963, cooperatively by New Mexico AES and Plant Materials Center, SCS, Los Lunas.

Breeder Seed/Stock - SCS Plant Materials Center, Los Lunas, NM.

Certified Seed/Stock - Available.

Preparer/Additional Information - Plant Materials Center, SCS, 1036 Miller St. SW, Los Lunas, NM 87031, (505) 865-4684.

Garden

Increased at Plant Materials Center, SCS, Scottsbluff, NE - Murray Cox. Carried as accession no. PM-NB-378.

Source - Composite of several individual collections from native plants. Collected by Murray Cox and R.L. Carver, SCS, October 1, 1957.

Method of Breeding - Increase of composite of individual collections. Field tested as PM-NB-378 in several parts of NB and SD.

Intended Use - Hay production, range reseeding, and revegetation of disturbed areas on sandy soils.

Description - Vigorous, tall, leafy type. Good seed yields. adapted throughout sandhills of Nebraska and in adjacent South Dakota.

Adapted to - LRR G, H; PHZ 4b.

Released - 1960, by SCS.

Breeder Seed/Stock - SCS Plant Materials Center, Manhattan, KS.

Certified Seed/Stock - Not available. Common in commercial production.

Preparer/Additional Information - SCS, 760 S. Broadway, Salina, KS 67401, (913) 823-4511.

Goldstrike

Source - Domestic collections made in 1953 from the western and northern Sandhills of Nebraska by L.C. Newell and W.L. Tolstead and a single clone from a western Oklahoma source.

Method of Breeding - Cultivar was produced by the synthesis of two selected lines of sand bluestem tracing to the 1953 collections. One of the lines, Western Sandhills Yellow, traces to collections in the north central Nebraska sandhills. The lines were developed independently by successive generations of mass selection of rust, and the unique color of the inflorescence which characterizes the variety. The cultivar was synthesized by growing the two parents lines in alternate rows and bulk harvesting seed (Syn 1 breeder seed). Foundation seed (Syn 2) was initially grown from breeder seed but now is produced from a previous foundation seed field rogued for plant type. Certified seed is produced solely from Foundation seed.

Intended Use - Revegetating sandy sites in the central Great Plains.

Description - A typical sand bluestem that is characterized by the yellow villous character of the inflorescence. Plants are variable in height and spread from short to long rhizomes. Characteristic leaf color is glaucous blue-green in early growth and changes to yellow-green at maturity.

Adapted to - LRR northern G & H; PHZ 4, 5.

Released - 1973, cooperatively by Nebraska AES and USDA-ARS.

Breeder Seed/Stock - Nebraska AES and USDA-ARS, Univ. of Nebraska, Lincoln.

Certified Seed/Stock - Available.

Preparer/Additional Information - K.P. Vogel, ARS-USDA, 344 Keim Hall, Univ. of Nebraska or Jeff Pederson, Dept. of Agronomy, Univ. of Nebraska, Lincoln, NE 68583, (402) 472-1564, (402) 472-2811.

Woodward

Source - Traces to source nursery established by M.L. Peterson in 1942. Sources about equally divided between those in and near Woodward County, OK, and those in general vicinity of Clovis, NM. Carried as accession no. PI 469237.

Method of Breeding - Plants selected for high seed set and placed in six isolation blocks by J.R. Harlan: short-early, short-late, medium-early, medium-late, tall-early, and tall-late. Process repeated with separate populations established in 1946. Selected plants moved to six new isolation blocks. Seed from two medium blocks bulked and seeded for preliminary increase in 1949, refined somewhat by removal of excessively tall plants, and today serves as breeder seed block of Woodward sand bluestem.

Intended Use - Forage and stabilization plantings.

Description - Variable population, but with most plants similar in type. Superior to wild strains tested in flower production, seed set, lack of excessively tall plants, and leafiness. Forage yield comparable to that of better source strains. Woodward can be combined reasonably well; seed quality superior to common sources.

Adapted to - Southern Great Plains.

Released - 1955, cooperatively by Oklahoma and Kansas (Manhattan) Agricultural Experiment Stations and Plant Science Research Division, ARS.

Breeder Seed/Stock - Not available.

Certified Seed/Stock - Not available.

Preparer/Additional Information - C.L. Dewald, USDA ARS, 2000 18th Street, Woodward, OK 73801, (405) 256-7449.

***Arctagrostis latifolia* (R.Br.) Griseb. - polargrass**

Cool-season, rhizomatous, moderately robust perennial grass native to northern latitudes with circumpolar distribution. Adapted to moist and moderately wet sites. Valuable for revegetation of disturbed sites; shows promise as forage where growing seasons are short.

Alyeska (Reg. No. 61)

Selected at the Palmer Research Center, University of Alaska Fairbanks Agricultural and Forestry Experiment Station, Palmer, AK - W.W. Mitchell.

Source - Twenty-seven indigenous collections of plants made from the southwestern and western coastal region through central and southern interior Alaska.

Method of Breeding - Composite of 27 individual breeding lines of seed-propagated plants. Breeder seed is composite of equal amounts of pure, live seed from the 27 lines.

Intended Use - Revegetation purposes; forage uses in arctic to subarctic locations.

Description - First known cultivar of species. Darker colored and more northerly adapted than Kenai polargrass, a later cultivar release.

Adapted to - Northwestern coastal and central interior through arctic Alaska, and hardy across mainland Alaska including alpine and arctic regions.

Released - 1976, by the Alaska AES.

Breeder Seed/Stock - Palmer Research Center, Agricultural and Forestry Experiment Station, Palmer, AK.

Certified Seed/Stock - Available.

Preparer/Additional Information - William W. Mitchell, Palmer Research Center, Agricultural and Forestry Experiment Station, 533 E. Fireweed, Palmer, AK 99645, (907) 745-3257.

Kenai (Reg. No. 114)

Selected at Palmer Research Center of University of Alaska Fairbanks Agricultural and Forestry Experiment Station, Palmer, AK - W.W. Mitchell. Carried as accession no. PI 518659.

Source - Bulk seed collections from native stands made near Mile Post 134, Sterling Highway, and at the town of Kenai, both on Kenai Peninsula, AK, accessioned as IAS 63 and IAS 60, respectively.

Method of Breeding - The two lines were tested separately and combined as a synthetic. The two lines are perpetuated in isolated breeding lines as IAS 633 and IAS 634, and seed lots of the two lines are composited in equal amounts of pure, live seed to constitute the breeder generation.

Intended Use - For hay or silage production and for reclamation, particularly on strongly acidic soils.

Description - Generally lighter colored and more productive than Alyeska polargrass, the only other cultivar of the species.

Adapted to - Interior to southern coast of Alaska and adjacent portions of Canada. Hardy through boreal regions south of Arctic area, including alpine regions; PHZ 1, 2 and southern half of 3.

Released - 1987, by the Alaska Agricultural and Forestry Experiment Station.

Breeder Seed/Stock - Palmer Research Center, Agricultural and Forestry Experiment Station, Palmer, AK.

Certified Seed/Stock - Available.

Preparer/Additional Information - William W. Mitchell, Palmer Research Center, Agricultural and Forestry Experiment Station, 533 E. Fireweed, Palmer, AK 99645, (907) 745-3257.

***Arrhenatherum elatius* (L.) J. & K. Presl - tall oatgrass**

Cool-season bunchgrass from Europe. Used for pasture, hay, and in forage mixtures in north-eastern and north-central states and in parts of intermountain region and Pacific Northwest. Some tolerance to shade, rapid seedling development, short lived. Under irrigation, life cycle similar to that of red clover. Seed shatters at maturity.

9061649

Developed and selected at the Plant Materials Center, SCS, Quicksand, KY - Charles F. Gilbert. Also known as KY-2174, T-16677, and 9016677. PI 564692.

Source - Seven clone selections obtained from 120 accessions evaluated. The seven clones were from accessions received from Chile, Italy, and Belgium.

Method of Breeding - Material increased by space planting the seven clones to a polycross block. Breeder field established from seed harvested from polycross block. Selected for superior vigor, yield, summer regrowth, resistance to disease and insects, persistence under sustained clipping management (1983-1986), and late-winter/early-spring growth.

Intended Use - Cover and forage plant for surface mines and marginal low-fertility hillside pastures. Short-term cover on mined sites with low-fertility levels. Especially adapted to droughty soils in the Appalachian region.

Description - Dark green, leafy, and good resistance to rust and leaf spot diseases. egins spring regrowth several weeks earlier than tall fescue. Harvests taken as late as December 27 without hurting stands. More productive during hot summer months than other cool-season forages. Can be grazed in February or March during milder winters. Easily established and excellent reseeder. Performed better and was most uniform of all material, including unimproved material of Tualatin.

Adapted to - To be determined.

Released - Included in regional field planting/testing program beginning in 1993.

Breeder Seed/Stock - Plant Materials Center, SCS, Quicksand, KY.

Certified Seed/Stock - Not available.

Preparer/Additional Information - Laura Ray, Plant Materials Center, SCS, 175 Robinson Dr., Quicksand, KY 41363-9008, (606) 666-5069.

***Axonopus fissifolius* (Raddi) Kuhlm. - carpetgrass**

Warm-season, sod-forming grass indigenous to Central America and West Indies. Widely distributed in old permanent pastures in Southeastern U.S.. Used for lawns in Florida and lowlands of Coastal Plain. Lacks drought resistance; makes best growth on lowland soils. Not highly nutritious, but quality and yield improved by fertilizer applications. Varieties not available.

***Beckmannia syzigachne* (Steud.) Fern. - American sloughgrass**

Cool-season, annual or short-lived perennial bunchgrass found in cooler parts of Eurasia and North America, including the northern half of the U.S.. Adapted to marshes and other wet sites. Useful for restoration of marshland, waterfowl habitat, erosion control, and short rotation forage.

Egan

Source - The original seed collection was made by James R. Stroh, July 26, 1973. The seed was taken from a single plant growing on gravel road fill near the Steese Highway in Gold Stream Valley, AK, north of Fairbanks. The original collection notes referred to this accession as *Beckmannia erucaeformis* auct. non (L.) Host. The collection has since been renamed *B. syzigachne* (Steud.) Fern.

Method of Breeding - Increase of original selection.

Intended Use - Reclamation or erosion control plantings in seasonally wet areas such as ditches, streambeds, or fresh water shorelines. Because of the species' documented use by waterfowl, it is also recommended for plantings intended to benefit ducks and geese. While the potential for hay or forage production exists, a recommendation for such purpose is being withheld at this time.

Description - Light green grass with tufted culms native to cooler regions of North America. This species, which is usually associated with wet ground, can grow to 90 cm. in height. Stands tend to decline after four to five years, and stand decline can be hastened by competition when more aggressive grass species become established in a *Beckmannia* host.

Adapted to - LRR W, X; PHZ 3, 4.

Released - 1986, Alaska Department of Natural Resources and SCS.

Breeder Seed/Stock - Plant Materials Center, Palmer, AK.

Certified Seed/Stock - Available.

Preparer/Additional Information - Plant Materials Center, HC 02 Box 7440, Palmer, AK 99645, (907) 745-4469.

***Bothriochloa caucasica* (Trin.) C.E. Hubbard - Caucasian bluestem**

***Andropogon caucasicus* Trin.**

Warm-season bunchgrass from Russia. Used for pasture and hay in the central and southern Great Plains.

KG-40

Increased at Plant Materials Center, SCS, Manhattan, KS - D.R. Cornelius and M.D. Atkins. Carried as accession no. PI 78758.

Source - Introduced from Tiflis, in the former USSR, in 1919 as PI 78758. Seed obtained in 1934 from A.E. Aldous, KS AES, Manhattan, KS.

Method of Breeding - Increased without selection and distributed for testing as KG-40.

Intended Use - Pasture hay production, stabilization of earth structures and revegetation of disturbed areas.

Description - Good forage production; easily established and spreads well from seed; disease free.

Adapted to - LRR H, J, M; PHZ 6a.

Released - Not formally, but has come into use primarily through seed harvests from field tests.

Breeder Seed/Stock - Plant Materials Center, SCS, Manhattan, KS.

Certified Seed/Stock - Available commercially. Caucasian is not a cultivar name but simply a common name. This was tested as KG-40.

Preparer/Additional Information - SCS, 760 S. Broadway, Salina, KS 67401, (913) 823-4541.

***Bothriochloa ischaemum* (L.) Keng var. *ischaemum* - yellow bluestem**

***Andropogon ischaemum* L.**

Warm-season, semiprostrate bunchgrass from the former USSR, China, Turkey, and India. Used primarily for pasture in Texas and north to southwestern Oklahoma.

El Kan

Selected at Plant Materials Center, SCS, Manhattan, KS - D.R. Cornelius. Carried as accession no. KG-495.

Source - Seed collected in 1937 west of Howard, Elk County, KS. Exact origin unknown, but thought to have come in with cattle or hay from Texas.

Method of Breeding - Selected in comparison with other introductions of this Asiatic bluestem. Increased from original collection and tested as KG-495.

Intended Use - Summer pastures, stabilization of earth structures, revegetation of disturbed areas.

Description - Most winter-hardy strain of *Bothriochloa ischaemum* (L.) Keng tested at Manhattan, KS. Fully winter hardy in Kansas and eastern Colorado. Bunchgrass of medium leafiness and forage production; more nearly equal in forage production to sideoats than to native bluestems in eastern Kansas; not equal in production to Caucasian bluestem where latter can be grown. Low in palatability compared with native bluestems, blue grama, and sideoats grama. Only fair seed production, but usually produces two seed crops a year. Free from disease. Easily established and spreads well from seed. Adapted in Kansas, Oklahoma, eastern Colorado, and northeastern New Mexico where annual precipitation is 380 mm or more. In the south and southwest, it is not equal to King Ranch and other cultivars of this species, but its use extends north where King Ranch is not winter hardy. Will grow on sandy, medium-textured, and clay soils. Used alone as summer pasture and for stabilization of earth structures, diversions, and critical areas.

Adapted to - LRR H, M; PHZ 5b.

Released - Not formally. Very limited use from seed harvested from field test plantings.

Breeder Seed/Stock - Plant Materials Center, SCS, Manhattan, KS.

Certified Seed/Stock - Not available. Common seed is available.

Preparer/Additional Information - SCS, 760 S. Broadway, Salina, KS 67401, (913) 823-4541.

Ganada

Selected at the Plant Materials Center, SCS,, Los Lunas, NM, from an introduced collection from the Westover-Enlow expedition in 1934. Carried as accession no. PI 107017, A-1407.

Source - Collected by Westover-Enlow expedition at Tajikistan, Turkestan.

Method of Breeding - Increased at Los Lunas, NM, as PI 107017 and A-1407 for field plantings.

Intended Use - Range reseeding, dryland pasture and revegetation of disturbed areas.

Description - An erect plant 200-1500 mm tall which tends to form a large saucer-shaped clump with the stems curving upward from the perimeter. The stems are pale yellow with dark nodes. The light green leaves are mostly basal and slightly rough at the top and have long hairs scattered near the leaf base. The seedheads are slightly fan-shaped on the ends of long seed stalks. Is very drought tolerant, one of few grasses which will establish on fine textured soils in the Southern Desert Major Land Resource Area. Most of the seed is produced in Oklahoma due to blasting of seedheads in New Mexico. Appears to be well adapted in Oklahoma and may go much farther east. Out-yielded Plains in a three-year clipping test in south central Oklahoma.

Adapted to - LRR D, G, H, I, N, E; PHZ 5.

Released - 1979; cooperatively by Plant Materials Center, SCS, Los Lunas, NM, and the New Mexico, Arizona and Colorado Agricultural Experiment Stations.

Breeder Seed/Stock - Plant Materials Center, SCS, Los Lunas, NM.

Certified Seed/Stock - Available. Common seed is available.

Preparer/Additional Information - Plant Materials Center, SCS, 1036 Miller St. SW, Los Lunas, NM 87031, (505) 865-4684.

Plains

Selected at Oklahoma AES, Stillwater, ARS Cooperating - J.H. Harlan and C.M. Taliaferro.

Source - Introduced germplasm from Pakistan, Iran, Iraq, India, Turkey, and Afghanistan.

Method of Breeding - Selections made in 1962 were composited in experimental variety OWB-M.

Intended Use - Forage production.

Description - Erect, tufted perennial with narrow, mostly basal leaves; foliage color is light green to bluish green; inflorescence consists of several unbranched racemes; and culm nodes glabrous to minutely pubescent. In Oklahoma tests, higher yielding and more resistant to foliar disease organisms than King Ranch; somewhat less productive than Caucasian bluestem, but superior in animal acceptance.

Adapted to - LRR H, N, O, M; PHZ 6, 7 central Great Plains.

Released - 1970, cooperatively by Oklahoma AES and Plant Science Research Division, ARS.

Breeder Seed/Stock - Oklahoma AES.

Certified Seed/Stock - Available commercially.

Preparer/Additional Information - C.M. Taliaferro, Agronomy Dept. 368 Ag Hall, OSU, Stillwater, OK 74078, (405) 744-6410.

WW-Iron Master (Reg. No. 8)

Selected from performance trials - C.L. Dewald, Research Agronomist, ARS, Woodward, OK 73801. Carried as accession no. PI 301535 (WW-535).

Source - Introduced from Afghanistan through the Oklahoma AES and forwarded to Woodward through the Southern Regional Plant Introduction Station in 1976 as part of Regional Project S-9.

Method of Breeding - WW-Iron Master was selected from evaluation performance trials on the basis of persistence, spring vigor, leafiness, and iron efficiency. It produces more forage with less chlorosis than presently released ARS varieties of Old World bluestems when grown on iron-deficient oils.

Intended Use - WW-Iron Master is a valuable grass for beef production, both as grazed forage and hay, when used in improved pastures or rangeland plantings. It is also useful for soil stabilization and wildlife cover on iron-deficient and erodible marginal farmlands.

Description - WW-Iron Master Old World bluestem is a perennial tufted bunchgrass with an upright growth habit. It has dark green foliage with basal and cauline leaves, 3-6 mm wide and 20-30 cm long at maturity. Foliage height ranges from 0.5 to 0.75 m with seed stalks reaching lengths of 1-1.5 m. Stems are yellowish with brown-purple glabrous nodes. Compared to other Old World bluestems, WW-Iron Master has a later maturity, more and larger cualine leaves, and a darker green leaf blade color than WW-Spar, and is more robust and has a higher leaf-to-stem ratio than Ganada. WW-Iron Master has an indeterminate flowering habit and seed maturation is more clearly defined than that from Plains. Seedlings of WW-Iron Master are extremely uniform in all characteristics and no degree of sexual reproduction has been observed.

Adapted to - LRR G, H, I, J, M, N, O, P; PHZ 6, 7, 8, 9.

Released - June 1987, jointly by ARS and SCS. It has been approved by the National Grass Variety Review Board as eligible for certification.

Breeder Seed/Stock - Breeders seed will be maintained and a list of registered seed producers can be obtained from the Southern Plains Range Research Station, 2000 18th St., Woodward, OK 73801.

Certified Seed/Stock - Available. Crop Improvement Association, Oklahoma State Univ.

Preparer/Additional Information - C.L. Dewald, ARS, 2000 18th St, Woodward, OK 73801, (405) 256-7449.

WW-Spar

Selected from performance trials - C.L. Dewald, Research Agronomist, ARS, Woodward, OK. Carried as accession no. PI 301573, WW-573.

Source - Introduced from Pakistan through the Oklahoma AES and forwarded to Woodward, OK, through the Southern Regional Plant Introduction Station, Griffith, GA, in 1976 as part of Regional Project S-9.

Method of Breeding - WW-SPAR (WW-573) was selected from evaluation performance trials on the basis of persistence, spring vigor, water use efficiency, water assimilation efficiency and tugor maintenance which correlated with its demonstrated ability to maintain production longer into a drought cycle than other yellow bluestems tested.

Intended Use - WW-Spar bluestem is a valuable grass for beef production, both grazing and hay, when used in improved pastures and rangeland plantings. It is also useful for soil stabilization and for wildlife cover.

Description - WW-Spar bluestem is a perennial tufted bunchgrass with an upright growth habit. It has light green foliage with mostly basal leaves, 3-6 mm wide and 200-300 mm long at maturity. Foliage height will average about 0.50 to 0.75 m with seed stalks reaching 1-1.5 m lengths. Stems are yellowish with brown glabrous nodes. WW-Spar has non-glandular hairs about five mm long on the upper leaf surface near the collar and a short membranous ligule. WW-Spar reproduces apomictically and seedlings are uniform with no apparent degree of sexual reproduction occurring. It has an indeterminate flowering habit but the peak of flowering and seed maturation is more clearly defined than that from Plains bluestem. WW-Spar is one of the original 30 accessions used in the blend to make Plains bluestem.

Adapted to - LRR G, H, I, J, M, N, O, P; PHZ 6, 7, 8, 9. Released - WW-Spar was released jointly by the ARS and the Oklahoma AES in 1982.

Breeder Seed/Stock - Breeders seed will be maintained at the Southern Plains Range Research Station, Woodward, OK.

Certified Seed/Stock - Registered and certified seed is available through the Crop Improvement Association, Oklahoma State University and from commercial sources.

Preparer/Additional Information - C.L. Dewald, ARS, 2000 18th St, Woodward, OK 73801, (405) 256-7449.

Bothriochloa ischaemum (L.) Keng *var. songarica* (Rupr. ex Fisch. & C.A. Mey.) Celarier & Harlan - bluestem

King Ranch

Origin of King Ranch bluestem in America described in 1952 by J.R. Harlan in Oklahoma Forage Leaflet No. 11, as follows:

Material now generally in use was first noticed by Nico Diaz on King Ranch in 1937 and increased for distribution by SCS under T-3487. Recent inquiry into history of Texas yellow beardgrass, which is apparently indistinguishable from King Ranch bluestem in all respects, leaves little doubt as to original entry of grass into U.S.. History is briefly as follows:

January 11, 1917: Received by P.B. Kennedy, CA AES, Berkeley, CA, from Amoy, China. Presented by H. Hoyle Sink, American consul, Amoy. Given California number T.O. 144 and later PI 44096.

1924: PI 44096 introduced to Substation 3, Angleton, TX, by V.E. Hafner, former Bureau of Plant Industry, Washington, DC, and given Texas number T.S. 8413.

April 11, 1932: T.O. 144 received by former Division of Forage Crops and Diseases, Bureau of Plant Industry, Washington, DC, from Agronomy Department, University of California at Davis, and given F.C. number 21785.

April 11, 1935: F.C. 21785 sent to B.F. Kiltz, Oklahoma AES, Stillwater, from Beltsville, MD.

1937: F.C. 21785 obtained by U.S. Southern Great Plains Field Station, Woodward, OK, from Oklahoma AES.

1939: T.S. 8413 given name yellow beardgrass in Texas AES Bulletin No. 570 and its performance at Angleton described.

1949: F.C. 21785 given name Texas yellow beardgrass and released for certification in Texas by Texas AES. All material furnished by Texas AES to individuals or substations since 1941 originated from this source. Since original Chinese material had been grown at Substation 3, Angleton, TX as early as 1924, there is little reason to suppose that King Ranch is any other than a Chinese accession that found its way from Angleton to King Ranch sometime during 1924-37.

Selected at Kingsville, TX - N.R. Diaz; then grown at SCS Nursery, San Antonio, TX. Carried as accession no. T-3487.

Source - Original seed collected in weakened rhodesgrass pasture on King Ranch, TX where escaped bluestem had gained dominance. Country of origin unknown, but thought by some to be China. Increased for testing as T-3487.

Method of Breeding - Increased by SCS Nursery at San Antonio, TX, as T-3487.

Intended Use - Rangeland, pastureland, erosion control, and revegetation.

Description - Aggressive, tends to eliminate the competition. Grows well on rocky limestone hills, shallow soils and clays. Withstands heavy grazing. Leaf growth susceptible to rust.

Adapted to - LRR H, I, J, T, P; PHZ 7.

Released - Informally by SCS about 1941. Later certified and formally released by SCS and Texas AES.

Breeder Seed/Stock - Not available.

Certified Seed/Stock - Not available.

Preparer/Additional Information - Plant Materials Center, SCS, Rt 1 Box 155, Knox City, TX 79529, (817) 658-3922.

Bouteloua curtipendula (Michx.) Torr. - sideoats grama

Major warm-season, slightly spreading, native bunchgrass. Distributed over much of Eastern and Central U.S.; important rangegrass in central and southern Great Plains from central Nebraska to southern Texas. Grows in association with bluestems; less drought resistant than blue grama.

Butte (Reg. No. 2)

Selected at Nebraska AES, Lincoln, ARS and SCS cooperating - E.C. Conard and L.C. Newell. Carried as accession no. PI 477002.

Source - Native collections from Holt and Platte Counties, NE.

Method of Breeding - Repeated field plantings revealed superiority of seedling vigor and establishment by native collections from Holt and Platte Counties as compared with other sources. Collections eventually combined and increased for further testing as Nebraska 37. Seed distributed for testing in 1948.

Intended Use - Range/pasture reseeding.

Description - Winter hardy, long lived, relatively early maturing. Makes best growth response under long days; best adapted to areas with relatively short growing seasons. Has large caryopses; exhibits excellent seedling vigor for establishment. In eastern Nebraska, produces excellent seed crop, maturing in mid-August; matures seed before frost in western Nebraska. Rrecommended for upland plantings in north-central and western districts of Nebraska.

Adapted to - LRR F, G, M; PHZ 4.

Released - 1958, cooperatively by Nebraska AES; Plant Science Research Division, ARS; and Plant Sciences Division, SCS. Increased on limited generation basis; foundation, registered and certified, with no recertification of certified class.

Breeder Seed/Stock - Nebraska AES and USDA-ARS, Univ. of Nebraska, Lincoln.

Certified Seed/Stock - Yes.

Preparer/Additional Information - K.P. Vogel, ARS-USDA, 344 Keim Hall, Univ. of Nebraska or Jeff Pederson, Dept. of Agronomy, Univ. of Nebraska, Lincoln, NE 68583, (402) 472-1564, (402) 472-2811.

El Reno

Increased at Plant Materials Center, SCS, Manhattan, KS. Carried as accession no. KG-482.

Source - Field seed collection from native range near El Reno, OK, in 1934.

Method of Breeding - Bulk material compared with many other collections at Manhattan. Increased for testing as KG-482.

Intended Use - Range reseeding and revegetation of disturbed areas.

Description - Outstanding in leafiness, forage production, and vigor. Ranked well in disease resistance, seed production, and winter hardiness. Widely used in range seedings and widely adapted in KS, OK, and TX.

Adapted to - LRR D, G, H, J, M, N, O, P; PHZ 5.

Released - 1944, cooperatively by Plant Materials Center, SCS, Manhattan, KS, and Kansas AES.

Breeder Seed/Stock - Plant Materials Center, SCS, Manhattan, KS.

Certified Seed/Stock - Available.

Preparer/Additional Information - SCS, 760 S. Broadway, Salina, KS 67401, (913) 823-4541.

Haskell (Reg. No. 0100)

Selected in 1973 at James E. "Bud" Smith Plant Materials Center, SCS, Knox City, TX, and increased for field testing as T-420. Carried as accession no. PI 433946.

Source - Original seed collected at Haskell, TX, in 1960 by J.C. Yeary, Jr.

Method of Breeding - Selected for better rhizome production and extended southern adaptation over 54 similar accessions and cultivars.

Intended Use - Range reseeding, revegetation of disturbed areas, pastureland, and hay production.

Description - Stronger, longer rhizomes than El Reno. Prolific seed producer. Adapted as far south as Rio Grande Valley in Texas. Highly palatable.

Adapted to - LRR H, I, J, N, O, P, T; PHZ 7.

Released - 1983 by SCS, Texas AES and ARS.

Breeder Seed/Stock - Plant Materials Center, SCS, Knox City, TX.

Certified Seed/Stock - Not available.

Preparer/Additional Information - Plant Materials Center, SCS, Rte 1 Box 155, Knox City, TX 79529, (817) 658-3922.

Killdeer

Selected at the Plant Materials Center, SCS, Bismarck, ND - John McDermand. Carried as accession nos. PI 476981, PM-ND-143, and PM-ND-89.

Source - Composite of seed collected in 1956 from native stands near Bowman, Bowman County, and Killdeer, Dunn County, ND.

Method of Breeding - Initial selection based on comparison with many other field collections at the Bismarck Plant Materials Center. Phenology, forage yield, and adaptation to soil and climate were extensively examined in advanced evaluation studies and field plantings in ND, SD, and MN.

Intended Use - Grass mixtures for range and pasture, critical area, and reclamation seedings.

Description - Outstanding vigor, leafiness, fair seed production, freedom from disease, and persistence in a cold, semi-arid environment.

Adapted to - LRR F, G, K; PHZ 3a - 4a.

Released - Released informally in late 1960s by the SCS, ND.

Breeder Seed/Stock - Plant Materials Center, SCS, Bismarck, ND.

Certified Seed/Stock - Not available. Common seed available.

Preparer/Additional Information - Russell Haas, SCS, P.O. Box 1458, Bismarck, ND 58502, (701) 250-4425.

Niner

Selected at Plant Materials Center, SCS, Los Lunas, NM. Collected by G.C. Niner and J.A. Anderson in 1957. Carried as accession nos. PI 476991, T 4495,NM-28.

Source - About 6.4 km west of Socorro, NM. Elevation of the site is 1432 m and it receives 22.9 cm of precipitation.

Found growing in association with blue grama, *Bouteloua gracilis* (Willd. ex Kunth) lag. ex griffiths; black grama (*B. eriopoda* (Torr.) Torr.; sand dropseed, *Sporobolus cryptandrus* (Torr.) Gray; galleta, *Hilaria jamesii* (Torr.) Benth.; Apacheplume, *Fallugia paradoxa* (D. Don) Endl. ex Torr.; and littleleaf sumac, *Rhus microphylla* Engelm. ex Gray.

Method of Breeding - Bulk increase of native collection.

Intended Use - Range seedings, mine land reclamation, and roadside reseeding.

Description - The botanical variety *B. curtipendula* (Michx.) Torr. var. *cespitosa* Gould & Kapadia differs from the eastern variety in that is has no rhizomes. Niner is darker colored than Vaughn and matures more evenly.

Adapted to - LRR D, G, H; PHZ 5, 6, 7, 8.

Released - Formally released in 1984 by SCS and New Mexico and Colorado Agricultural Experiment Stations.

Breeder Seed/Stock - Plant Materials Center, SCS, Los Lunas, NM.

Certified Seed/Stock - Available.

Preparer/Additional Information - Plant Materials Center, SCS, 1036 Miller St. SW, Los Lunas, NM 87031, (505) 865-4684.

Pierre

Selected at the Plant Materials Center, SCS, Bismarck, ND - John McDermand. Carried as accession nos. PI 476980, PM-SD-251.

Source - Composite of seed collected 8 km west of Pierre, Stanley County, SD. Collection made in 1954 by SCS from several plants with outstanding vigor and leafiness growing in native range on south slope of shale range site. Average annual precipitation approximately 400 mm.

Method of Breeding - Initial selection based on comparison with many other field collections and the variety Butte at the Bismarck Plant Materials Center. Phenology, forage yield, and adaptation to soil and climate were extensively examined in advanced evaluation studies and field plantings in ND, SD, and MN.

Intended Use - Mixtures for range and pasture, critical area, and reclamation seedings.

Description - Outstanding in vigor, leafiness, freedom from disease, seedling vigor, and persistence in a semi-arid environment.

Adapted to - LRR F, G, M; PHZ 3b - 4b.

Released - Released informally in mid 1960s by the SCS and South Dakota AES.

Breeder Seed/Stock - Plant Materials Center, SCS, Bismarck, ND.

Certified Seed/Stock - Available.

Preparer/Additional Information - Russell Haas, SCS, P.O. Box 1458, Bismarck, ND 58502, (701) 250-4425.

Premier

Selected at Texas AES, College Station, ARS and SCS cooperating - Judd Morrow and W.G. McCully. Carried as accession no. PI 469250 and G-433.

Source - Seed collected from a single plant growing between Cuauhtemoc and Chihuahua, Mexico, in 1953.

Method of Breeding - Increased at Big Spring Field Station, Big Spring, TX, and evaluated at several research centers in TX in comparison with many other collections. Increased for testing as G-433.

Intended Use - Range/pasture reseeding.

Description - Upright, leafy type. Good seed yield. Individual spikes retained on plant make combine seed harvest practical. Germinates readily and seedlings develop rapidly after emergence. No major insect or disease problems noted. Considerable drought tolerance. Forage production equal to or slightly greater than that of other recognized varieties. Recommended for west-central Texas.

Adapted to - LRR D, I, J, G, H; PHZ 5.

Released - 1960, cooperatively by Texas AES and Plant Science Research Division, ARS.

Breeder Seed/Stock - Texas AES.

Certified Seed/Stock - Not available.

Preparer/Additional Information - Texas AES, College Station, TX 77843, (409) 845-3041.

Trailway (Reg. No. 3)

Selected at Nebraska AES, Lincoln, ARS and SCS cooperating - L.C. Newell and E.C. Conard. Carried as accession no. PI 477001.

Source - Hybrid population of sideoats grama found growing along abandoned roadway in northern Holt County, NE, by L.C. Newell in 1935.

Method of Breeding - Spaced plants of collection grown at Nebraska AES. Selection made in hybrid population for late maturity and freedom from rust. Selection carried through three generations, resulting in harvest and increase of seed from these groups combined for increase and testing as Nebraska 52.

Intended Use - Revegetation and pasture plantings alone or in mixtures with other warm season grasses.

Description - Winter-hardy, long-lived, late maturing, comparable in growth type to varieties of more southerly origin. Somewhat indeterminate as to heading and flowering responses, exhibiting considerable variability in maturity. Requires most of growing season to mature seed in eastern Nebraska; may fail to produce seed crops in regions with shorter seasons. In Nebraska, recommended for upland plantings in eastern and southern districts.

Adapted to - LRR northern H, M (Southern and Eastern Nebraska); PHZ 5.

Released - 1958, cooperatively by Nebraska AES; Plant Science Research Division, ARS; and Plant Science Research Division, SCS. Increased on limited generation basis; foundation, registered and certified, with no recertification of certified class.

Breeder Seed/Stock - Nebraska AES and USDA-ARS, Univ. of Nebraska, Lincoln.

Certified Seed/Stock - Available

Preparer/Additional Information - K.P. Vogel, ARS-USDA, 344 Keim Hall, Univ. of Nebraska or Jeff Pederson, Department of Agronomy, Univ. of Nebraska, Lincoln, NE 68583, (402) 472-1564, (402) 472-2811.

Vaughn

Selected at former SCS Nursery, Albuquerque, NM. Carried as accession nos. 476991 and A-3603.

Source - Collected from native stands near Vaughn, NM, in 1935. Method of Breeding - Bulk increase of native collection. Distributed for testing as A-3603.

Intended Use - Range reseeding and revegetation of disturbed areas.

Description - Population slightly variable, but all have erect leaf type. Good seedling vigor, easily established. More drought tolerant than El Reno, Uvalde, or Tucson for use in eastern Colorado and New Mexico.

Adapted to - LRR D, G, H, E; PHZ 4.

Released - 1940, cooperatively by New Mexico AES, University Park, and Plant Sciences Division, SCS.

Breeder Seed/Stock - Plant Materials Center, SCS, Los Lunas, NM.

Certified Seed/Stock - Available.

Preparer/Additional Information - Plant Materials Center, SCS, 1036 Miller St. SW, Los Lunas, NM 87031, (505) 865-4684.

***Bouteloua eriopoda* (Torr.) Torr. - black grama**

Major warm-season, native grass of arid and semiarid desert grasslands in Arizona, New Mexico, and Texas. Culms in contact with soil, with root of nodes under favorable conditions to form new plants. Excellent palatability and feeding value both summer and winter, and drought resistance.

Nogal

Selected at Plant Materials Center, SCS, Los Lunas, NM. Collected by G.C. Niner, SCS, and J.E. Anderson, New Mexico State University. Carried as accession no. PI 476992 and NM-44.

Source - Collected in 1957 along U.S. 85, about 78 km south of Socorro, NM, at elevation of 1345 m and with precipitation about 225 mm.

Method of Breeding - Increase of original collection.

Intended Use - Soil stabilization and range revegetation.

Description - Considerable variation in individual spaced plants, but population as whole exhibits desirable characteristics intermediate between upright, fine-stemmed, confined types and decumbent, sprawling types. Low yields of pure live seed.

Adapted to - LRR D, G, I; PHZ 5.

Released - 1971, cooperatively by Plant Materials Center, SCS, Los Lunas, NM, and New Mexico AES.

Breeder Seed/Stock - Plant Materials Center, SCS, Los Lunas, NM.

Certified Seed/Stock - Available.

Preparer/Additional Information - Plant Materials Center, SCS, 1036 Miller St. SW, Los Lunas, NM 87031, (505) 865-4684.

***Bouteloua gracilis* (Willd. ex Kunth) Lag. ex Griffiths - blue grama**

Major warm-season, native grass throughout Great Plains. Used for grazing and erosion control. Characterized by creeping growth habit, forms dense sod, and produces high-quality forage. Hardy and drought resistant. Found on various soil types; well adapted on heavy, rolling upland soils.

9055923

Department of Agronomy, Colorado State University and ARS - Dr. Robin L. Cuany and Dr. Alma Wilson. Carried as accession no. 9055923.

Source - Derived from plants collected at eight sites along and up to 167 km east of the Front Range from Cheyenne, WY, to Maxwell, NM, and Sterling, CO, to Rock Ford, CO.

Method of Breeding - A population starting from 50 plants of each of eight sources was put through two cycles of recurrent selection for caryopsis weight and seedling emergence characters including growth or six weeks on seminal root alone and capacity to develop adventitious roots when top-watered. Seed production and general forage vigor have been steady or improved. The cycle-2 synthetic is based on 76 clones and the foundation seed field was established by further selection of the 2,500 best progeny seedlings emerging through 4 cm of dry soil in greenhouse.

Intended Use - Rangeland improvement and go-back cropland reseeding in southern and central Great Plains from northern New Mexico to southern Wyoming. Also possible uses for dryland turf in parks, lawns, or golf courses.

Description - Warm-season bunchgrass adapted to the shortgrass plains of the western half of the Great Plains. Perennial with culms up to 70 cm and 2-3 spikes per culm. Cures well as standing hay. Forage productivity on range sites depends on moisture. Has shown useful improvements in emergence ability and may be able to be planted as deep as 2.5 cm through comparative data with Hachita and Lovington and 9044169. Data still being gathered. First full growing season produced seed at 73 kg in the foundation seed field.

Adapted to - LRR E, G, H, probably also D; PHZ 4, 5, 6.

Released - Not yet released, in advanced testing.

Breeder Seed/Stock - Department of Agronomy, Colorado State Univ., Fort Collins, CO 80523.

Certified Seed/Stock - Not available.

Preparer/Additional Information - Dept. of Agronomy, Colorado State Univ., Fort Collins, CO 80523, (303) 491-6832.

Alma

Developed by Dr. Alma Wilson, ARS, Fort Collins, CO, and Plant Materials Center, SCS, Los Lunas, NM. Carried as accession no. 9044169.

Source - The base population was composed of 270 plants from each of the following: Hachita, Lovington, and PMK-1483.

Method of Breeding - The primary objective was to select for heavier caryopsis weight, greater seedling emergence from deep planting depths, more seedling vigor, and greater seed and forage production. The intent was to combine the beneficial seed and seedling traits with the wide adaptation of Lovington, Hachita, and PMK-1483.

Three cycles of recurrent selection were conducted. In each cycle, 40,000 seeds were sown as polycross progenies in pots in a greenhouse and covered with 4 cm of dry soil. The seeds in pots were watered by subirrigation. About 1,500 of the most vigorous seedlings, including those with the longest adventitious roots, were selected and transplanted to the field. In the field, these plants were rated for forage yield and number of seeds per culm. Seeds were harvested from the best 50-60% of these plants, and the seeds were evaluated for caryopsis weight and fertility. From this population, 60-90 plants which excelled for all traits were selected and clonally propagated. Four clonal replicates of each selected plant were grown in isolation to produce seed for cycle 2. Selection in cycles 2 and 3 was conducted as in cycle 1, except that there were 24 clonal replicates of the 90 selected plants in the cycle 3 crossing block. Seed from the cycle 3 crossing block served as breeder seed.

Mean caryopsis weights for the base population and generations 1 and 2 were 47.2, 55.1, and 65.3 mg/100 caryopsis, respectively. Realized heritability for caryopsis weight was 84% averaged over the two cycles. Progenies varied significantly ($P < 0.01$) in caryopsis weight, percentage emergence, and length of adventitious roots; the broad sense heritabilities of these traits were 88, 65, 62%, respectively.

Intended Use - Rangeland improvement and go-back cropland reseeding in southern and central Western Great Plains.

Description - Perennial, robust, blue grama with upright growth, 40-80 cm tall, good seedling vigor. Comparisons of 9044169 with Hachita in Colorado dryland trials showed: 9044169 usually exhibited higher percentage emergence; 9044169 and Hachita did not differ for forage productivity and crude protein percentage.

Adapted to - LRR E, G, H; PHZ 5, 6.

Released - Not released; in advanced trials.

Breeder Seed/Stock - Plant Materials Center, SCS, Los Lunas, NM.

Certified Seed/Stock - Not released.

Preparer/Additional Information - Greg Fenchel, Plant Materials Center, SCS, 1036 Miller St. SW, Los Lunas, NM 87031, (505) 865-4684.

Hachita

Selected at Plant Materials Center, SCS, Los Lunas, NM. Collected by G.C. Niner and J.E. Anderson, New Mexico State University. Carried as accession nos. PI 439880 and NM-118.

Source - Collected in 1957, 51 km south of Hachita, NM, at elevation of 1340 m and with precipitation about 250 mm annually.

Method of Breeding - Increase of original collection.

Intended Use - Range reseeding and revegetation of disturbed areas.

Description - At site of collection, plants had vigorous and robust culms 0.92 m tall and leaves 150-200 mm long. Under irrigation, outstanding performance in rod-row comparison block: highest in seed and herbage production; culms 915 mm tall, leaves 460-500 mm long. Spikes considerably longer than average. Most drought tolerant of blue grama accessions tested. Adapted to - LRR D, G, H, E; PHZ 5. *Released* - 1980, cooperatively with New Mexico and Colorado AESs. Breeder Seed/Stock - Plant Materials Center, SCS, Los Lunas, NM.

Certified Seed/Stock - Available.

Preparer/Additional Information - Plant Materials Center, SCS, 1036 Miller St. SW, Los Lunas, NM 87031, (505) 865-4684.

Lovington

Selected at Plant Materials Center, SCS, Los Lunas, NM. Collected by J.A. Downs, G.C. Niner, and J.E. Anderson. Carried as accession nos. PI 476993 and A-12424.

Source - Field harvested in 1944 near Lovington, NM, at elevation of 1,220 m and with precipitation about 355 mm.

Method of Breeding - Bulk increase of source material. Tested as A-12424.

Intended Use - Range reseeding and revegetation of disturbed areas.

Description - Uniform, good leafiness, excellent seedling vigor, and fast establishment. Used as standard in evaluating other accessions of blue grama at Los Lunas, NM. Seed production under irrigation and with insect control exceeds 90 kg of pure live seed per acre. Extensive field testing showed it adapted over wide geographic area in Southwest.

Adapted to - LRR G, H, F; PHZ 5.

Released - 1963, cooperatively by New Mexico AES, University Park, and Plant Materials Center, SCS, Los Lunas, NM.

Breeder Seed/Stock - Plant Materials Center, SCS, Los Lunas, NM.

Certified Seed/Stock - Available.

Preparer/Additional Information - Plant Materials Center, SCS, 1036 Miller St. SW, Los Lunas, NM 87031, (505) 865-4684.

Brachiaria ramosa (L.) Stap - browntop millet

Warm-season annual from Asia. Used for wildbird feed and as minor forage plant in Southeastern U.S.. Not aggressive. Short season of growth; requires fertile soil and good moisture. No varieties available.

Bromus arvensis L. - field brome

Cool-season bunchgrass introduced from Europe in late 1920s. Used as cover crop and for green manure in parts of northeastern and north-central regions. Winter-hardy annual. Strong spring growth and late maturing. Develops extensive fibrous root system. Varieties not currently available in U.S.

Bromus carinatus Hook, & Arn. - California brome

Cool-season, short-lived bunchgrass indigenous to intermountain and Pacific coast regions. Palatable; used for grazing and erosion control. Considered by some botanists to be polymorphic species that includes *Bromus marginatus*.

Cucamonga (Reg. No. 13)

Selected at Plant Materials Centers, SCS, Pleasanton and San Fernando, CA - P.B. Dickey, P.E. Lemmon, and D.J. Vanderwal. Carried as accession no. P-11117.

Source - Collection by R.L. Forsyth from native stand near Cucamonga, CA, in 1939.

Method of Breeding - Mass phenotypic selection from small plot seeded at Pleasanton in 1941. Given accession number P-11117 and tested in comparison with other annual grasses.

Intended Use - Cover crop, revegetation of disturbed areas, and wildfire land rehabilitation.

Description - Self-perpetuating winter annual; pale green, very rapid developing, early maturing, with long flexuous panicles. Sheaths and leaves pilose to nearly glabrous; awns long, requiring processing of seed before it can be drilled. Best suited as self-seeding cover crop and as quick cover on droughty, low fertility sites. Limited value for forage because of its short green-feed period. Susceptible to head smut.

Adapted to - LRR C; PHZ 9.

Released - 1949, cooperatively by California AES, Davis, and Plant Materials Center, SCS, Pleasanton, CA.

Breeder Seed/Stock - Plant Materials Center, SCS, Lockeford, CA.

Certified Seed/Stock - Limited. Commercial production is being increased; native plant use is being revived because of potential use in natural restoration projects.

Preparer/Additional Information - R. Slayback, SCS, 2121-C, 2nd St, Davis, CA 95616, (916) 757-8257.

Deborah

Source - Developed by Dunns Seed and Grain, Great Britain.

Method of Breeding - Developed from two ecotypes, one from the Thames River Valley in the UK, the other from the Andes in South America. Individual plants were examined for agronomic value and a polycross was made of selected plants in 1970. Selection criteria were the number of leaves per culm, wider leaves, increased tillering over the wild type, uniformity, short internodes, high seed yields, and digestibility. Breeder seed was first formed in the F3 generation and is produced from parental clones every fourth year.

Intended Use - Hay and pasture production.

Description - This variety is octoploid. Semi-erect growth habit, non-spreading, lower growing than smooth bromegrass; larger, more numerous, elliptical tillers than those of wild ecotypes; shorter internode immediately below the panicles than wild ecotypes; glabrous nodes; dark grayish-green leaves with slight waxy bloom, wide, long, drooping; high leaf per stem ratio; sparsely pubescent sheath and blade; developing leaves rolled in the sheath; slight anthocyanin coloration; shorter plant height than wild ecotypes; medium-late maturity, later than tall fescue, earlier than crested and intermediate wheatgrass varieties. Relatively slow establishment when compared to annual grasses but quite good compared to perennial grasses; long growing; waxy, green spikelets, lax and drooping; large, high 1,000 kernel weight compared to yellowish-brown at harvest; winter hardiness good in Canada; very good drought tolerance, much better than perennial ryegrass and orchardgrass varieties tested under the same conditions. Moderately resistant to smut; American data show that it is resistant to powdery mildew and that it is susceptible to fruit fly. Deborah sweet bromegrass performed acceptably in Ontario and British Columbia trials. It has a high seed yield and very good drought tolerance. Its forage yield is lower than that of smooth bromegrass and meadow bromegrass varieties in most tests, although it performed well when compared to crested wheatgrass, intermediate wheatgrass, orchardgrass, and timothy varieties. Deborah has slower growth in May and early June than perennial ryegrass, but it is more productive than most forage grasses during the rest of the season.

Released - Dunns Seed and Grain, Great Britain. Registered in Canada in 1986 with Registration No. 2678.

Breeder Seed/Stock - Deahnfeldt Inc.

Certified Seed/Stock - Available from Topnotch Nutri Ltd.

Preparer/Additional Information - Agriculture Canada, Food Production and Protection Branch, Plant Products Division, Ottawa, Ontario K1A 0C6. Seed distributed by Topnotch Nutri Ltd., P.O. Box 1030, Abbotsford, British Columbia B2S 5B5.

***Bromus catharticus* Vahl - rescuegrass**
***Bromus unioloides* Kunth**
***Ceratochloa willdenowii* (Kunth) W.A. Weber**

Introduced annual or biennial; culms erect to spreading, as much as 100 cm tall; cultivated in the southern U.S. as a winter forage grass, dry to moderately moist waste places.

Cool-season bunchgrass from South America. Used primarily for winter pasture in southern States. Palatable. Subject to head smut. Annual growth habit, but reseeds under favorable conditions; some plants biennial or short-lived perennials.

Gala

Source - Originates from material collected in the Santiago region of central Chile. This region has Mediterranean rainfall distribution and temperatures similar to the San Francisco Bay region. Domesticated by Pyne Gould Guiness, Christchurch, New Zealand.

Method of Breeding - Recurrent selection from native material.

Intended Use - Closely related to Prairie grass and is recommended for dryland pastures where it will give improved winter growth and drought-tolerant summer pastures and allow flexible management, including tolerance to hard continuous sheep, horse, or other livestock grazing. It can provide safe pasture free of ryegrass staggers. It must be sown in early autumn or in spring to get reliable establishment and it should not be sown on wet soils or soils subject to livestock hoof damage during wet weather.

Description - Compared to Matua, Gala has at least twice as many tillers but each tiller is smaller, a similar size seed with a longer awn, tendency to flower earlier and with less aftermath seedheads. It has pale green tiller bases whereas Matua has red bases. The plant is more pubescent on the leaves and sheaths. It forms a more dense sward, the Matua being more similar in density to perennial ryegrass.

Adapted to - Naturalized in central California around Berkeley and Sonoma County. It has limited distribution in the western plains of New South Wales and southeast region of South Australia. It is widespread throughout east coast regions of South Island, New Zealand, and it is also probable that it is present in other countries but has not been distinguished from Prairie grass.

Released - Pyne Gould Guiness and DSIR. NZ.

Breeder Seed/Stock - Pyne Gould Guiness and DSIR. NZ.

Certified Seed/Stock - Pyne Gould Guiness and DSIR. NZ.

Preparer/Additional Information - Irvin H. Jacob, Cascade International Seed Company, 8483 West Stayton Road, Aumsville, OR 97325, (503) 749-1822.

Lamont (Reg. No. 7)

Selected at Delta Branch Experiment Station, Stoneville, MS, ARS cooperating - H.W. Johnson. Carried as accession no. PI 193144.

Source - La Estanzuela 157/49. Seed of this strain obtained by O.S. Aamodt from Uruguay and introduced as PI 193144 in 1950.

Method of Breeding - Mass selection. Seed harvested in May 1953 from two-year-old plants in plot seeded at Stoneville in 1951.

Intended Use - Hay and pasture production.

Description - Appears to consist largely of biennials or short-lived perennials. Consequently, provides longer grazing season during second and subsequent stands than during year of establishment or when grown as winter annual. In tests at Stoneville proved immune to head smut collections from Auburn, AL, College Station, TX, Raymond and Stoneville, MS, and Watkinsville, GA. Proved moderately susceptible to Baton Rogue, LA, collection, which appears to represent different race of head smut fungus.

Released - 1957, cooperatively by Mississippi AES, State College, and Plant Science Research Division, ARS.

Breeder Seed/Stock - Mississippi AES, State College.

Certified Seed/Stock - Available in limited quantities.

Preparer/Additional Information - Mississippi State University, Agronomy Dept., Box 5248, Mississippi State, MS 39762 (601) 375-2732.

Matua

Source - New Zealand via Europe.

Method of Breeding - A simple plant selection process from various ecotypes.

Intended Use - Pasture, hay, silage, and conservation.

Description - It is a cool-season perennial to annual, depending on climate and soil, of high quality, non-toxic forage of excellent production and palatability. Can be managed as a reseeding annual.

Adapted to - Total adaptation area is unknown, but includes Washington and Oregon states (under irrigation), plus the encompassing area from Pennsylvania to North Carolina to Texas to Kansas and back to Pennsylvania.

Released - New Zealand Ministry of Agriculture and Fisheries.

Breeder Seed/Stock - New Zealand Ministry of Agriculture and Fisheries.

Certified Seed/Stock - Available from New Zealand Ministry of Agriculture and Fisheries.

Preparer/Additional Information - R.L. Dalrymple, Noble Foundation, P.O. Box 2180, Ardmore, OK 73402, (405) 223-5810.

Nakuru

Increased at Plant Materials Center, SCS, Americus, GA. Carried as accession no. PI 195476.

Source - Introduced from Nakuru, South Africa, as PI 195476, and carrying accession nos. BN-7214 and AM-1359.

Method of Breeding - Increase of original seed lot.

Intended Use - Pasture, forage.

Description - Weak perennial, or strong reseeding annual; somewhat more robust and possessing slightly larger leaves, stems, and seed than common types. Good seed producer; characterized by above-average seedling vigor, and resistance to rust. Requires high level of soil fertility.

Adapted to - LRR T, P, N; PHZ 8.

Released - No. Only distributed for on-farm tests.

Breeder Seed/Stock - Plant Materials Center, SCS, Americus, GA.

Certified Seed/Stock - Not available.

Preparer/Additional Information - Charles M. Owsley, Plant Materials Center, SCS, 295 Morris Dr., Americus, GA 31709, (912) 924-2286.

Bromus riparius Rehmann - meadow brome

Perennial grass introduced from southeast Asia. Less strongly creeping than smooth bromegrass, leaves and stems are generally pubescent, ploriferation of basal leaves, and remains green later in the fall than smooth bromegrass. Adapted to cool, moist areas of the U.S. and Canada. Develops extensive fibrous root system. There is confusion about the correct scientific name of the released varieties discussed below. For the purpose of this publication all are identified as *Bromus riparius.*

Fleet

Source - Developed at the Agriculture Canada Research Station, Saskatoon, Saskatchewan.

Method of Breeding - The variety was formed from 67 plants selected from eight introductions of *B. riparius* of Eurasian origin. Twenty-two plants were from Krasnodar VIR K27534, 13 from Krasnodarski 8, seven from Regar, and the remainder from introductions of French and Hungarian origin. Mass selection was conducted in established space plant nurseries for seed production, reduced awn development, and reduced shattering. Further selection was carried out between 1981 and 1984 for ploidy, growth habit and fertility.

Intended Use - Hay and pasture production.

Description - Growth habit has abundance of basal tillers, similar to Regar; leaves are slightly wider than those of Regar; leaves and stems show varying degrees of pubescence; roots are restricted creeping as in Regar; seed is similar to those of Regar in size and pubescence; and regrowth is rapid. Limited data indicate that Fleet is slightly more resistant to silver top than Regar. Fleet has hay and pasture yields similar to those of Regar, but it has higher seed yields and higher seed quality ratings. Fleet is also similar to Regar in winter hardiness. It showed a slightly higher digestibility than Regar.

Adapted to - Recommended in Canada for use in British Columbia and Saskatchewan.

Released - 1987, Saskatoon Research Station. Registration no. 2778 (Canada).

Breeder Seed/Stock - Saskatoon Research Station, Saskatchewan.

Certified Seed/Stock - Available.

Preparer/Additional Information - Agriculture Canada, Food Production and Protection Branch, Plant Products Division, Ottawa, Ontario K1A 0C6. Seed distributed by SeCan Association, 200-57, Promenade Auriga, Nepean, Ontario K2E 8B2.

Paddock

Source - Developed at the Agriculture Canada Research Station, Saskatoon, Saskatchewan, from an introduction of B. riparius, VIR K27534, from Krasnodar in the former Soviet Union.

Method of Breeding - After two generations of open pollination among other meadow bromegrass strains, S-7414 was isolated in space-planted nurseries; two generations were subsequently subjected to mass selection in 1975 and 1979. Selection criteria included good vigor, good basal leaf growth, fertility, reduced awn development and some self-shedding of awns at seed maturity. Breeder seed resulted from bulking 120 selected plants. Tested as S-7414.

Intended Use - Hay and pasture production.

Description - Growth habit has an abundance of basal tillers, similar to Fleet and Regar; leaves are slightly wider than Regar, similar to Fleet; leaves, stems, and seeds have varying degrees of pubescence; roots are restricted creeper, as Fleet and Regar; seed is similar to Fleet and Regar but with slightly longer awns; and regrowth is rapid. Limited data indicate that Paddock is slightly more resistant to silver top than Regar. Paddock has similar forage yields to Fleet and Regar under simulated grazing and slightly lower seed yields than Fleet. Seed quality and digestibility are similar to Fleet.

Adapted to - Recommended in Canada for use in British Columbia and Saskatchewan.

Released - 1987, Agriculture Canada Research Station, Saskatoon, Saskatchewan.

Breeder Seed/Stock - Agriculture Canada Research Station, Saskatoon, Saskatchewan.

Certified Seed/Stock - Available.

Preparer/Additional Information - Agriculture Canada, Food Production and Protection Branch, Plant Products Division, Ottawa, Ontario K1A 0C6. Seed distributed by SeCan Association, 200-57, Promenade Auriga, Nepean, Ontario K2E 8B2.

Regar

Selected at Plant Materials Center, SCS, Aberdeen, ID. Carried as accession no. PI 172390. Previously identified in numerous publications as *Bromus biebersteinii* Roemer & J.A. Schultes.

Source - Accession PI 172390, collected near Zek, Kars Province, Turkey, in 1949, and received from Plant Introduction Station, Ames, IA, in 1957.

Method of Breeding - Fifteen clones selected from irrigated nursery in 1958. Seed multiplied for testing as P-14941.

Intended Use - Hay and pasture receiving more than 400 mm of precipitation or irrigation.

Description - Earlier heading than Manchar smooth brome, some vegetative spreading, good drought tolerance, excellent winter hardiness, and good regrowth, particularly in mid-summer. Rapid seed germination and seedling establishment. Leaves numerous, lax, dominantly basal, mildly pubescent, and light green; erect seedstalks extend above leaf mass in an open panicle. Susceptible to covered smut.

Adapted to - LRR B, D, E; PHZ 4, 5.

Released - 1966, Plant Materials Center, SCS, Aberdeen, ID.

Breeder Seed/Stock - Plant Materials Center, SCS, Aberdeen, ID.

Certified Seed/Stock - Available.

Preparer/Additional Information - Gary Young, Plant Materials Center, SCS, P.O. Box 296, Aberdeen, ID 83210-0296, (208) 397-4133.

***Bromus hordeaceus* L. ssp. *hordeaceus* - soft chess**

(*Bromus mollis* auct. non L.)

Cool-season, annual bunchgrass from Europe. Widely distributed, weedy grass. Important forage species in annual ranges of California.

Blando (Reg. No. 11)

Selected at Plant Materials Center, SCS, Pleasanton, CA - H.W. Miller and O.K. Hoglund. Now carried as accession no. PI 469232.

Source - Collected May 21, 1940, from winter annual rangeland near San Ramon, CA, by D.J. Vanderwal.

Method of Breeding - Tested in comparison with 27 other collections of *Bromus mollis* by Plant Materials Center, SCS, Pleasanton, CA, and San Fernando Nursery, SCS, CA, since fall of 1940 as P-11657.

Intended Use - Range reseeding, cover crop, revegetation of disturbed areas, and wildfire burn rehabilitation.

Description - Primary advantage over other strains is its consistent forage and seed production from year to year. During low rainfall years, its superiority was demonstrated by outperforming all other strains. In relation to the other strains tested, it is intermediate in maturity and sub-erect in growth habit.

Adapted to - LRR A, C; PHZ 9.

Released - Cooperatively by California AES, Davis, and Plant Materials Center, SCS, Pleasanton, CA.

Breeder Seed/Stock - Plant Materials Center, SCS, Lockeford, CA.

Certified Seed/Stock - Available.

Preparer/Additional Information - R. Slayback, SCS, 2121-C 2nd St., Davis, CA 95616, (916) 757-8257.

***Bromus inermis* Leyss. - smooth brome**

Major cool-season, sod-forming grass introduced from Hungary in 1884 and the Penza region of Russia in 1898 by N.E. Hanson, South Dakota Horticulture Department. Used for pasture, hay, silage, and erosion control in humid northern states to eastern ND, south to eastern KS, and extensively in northern part of intermountain region and Pacific Northwest. Grows well on fertile soils. Rated high in palatability and nutritive value. Two distinct types identified: Northern, which is adapted to western Canada and northern Great Plains, and Southern, which is adapted to corn-belt states and central Great Plains.

Achenbach

Major cool-season, sod-forming grass introduced from Hungary in 1884.

Source - Old fields tracing to original planting made in 1895 by Achenbach brothers of Washington County, KS.

Method of Breeding - Some mass selection in early generations by Achenbach brothers.

Intended Use - Pasture, hay, silage, and erosion control in humid northern states to eastern ND, south to eastern KS, and extensively in northern part of intermountain region and Pacific Northwest.

Description - Typical southern type of smooth brome. Leafy, vigorous, spreads rapidly by rhizomes to form dense, competitive sod. Heavy rhizomes of both seed and forage. Far less susceptible to leaf diseases than northern types with which it has been compared in Kansas. Most smooth brome grown in Kansas is of this strain.

Adapted to - PHZ 5 - Central U.S.

Released - Named Achenbach in 1944 by Kansas AES, Manhattan. Old fields that could be traced to Achenbach brothers' plantings were then declared eligible for certification. Those fields are the source of all Achenbach grown for certification in Kansas.

Breeder Seed/Stock - Available.

Certified Seed/Stock - Not available.

Preparer/Additional Information - Lowell Burchett, Kansas Crop Improvement Association, 2000 Kimball Ave., Manhattan, KS 66502, (913) 532-6118.

Badger

Source - Various cultivars and plant introductions.

Method of Breeding - One cycle of phenotypic selection, each for resistance to Pythium seedling damping off, resistance to brown leafspot, and high *in vitro* digestibility.

Intended Use - Hay and silage production.

Description - Excellent resistance to seedling and foliar disease, high digestibility and forage yield, and excellent seed producer. Southern to intermediate type.

Adapted to - PHZ 1, 2, 3, 4, 5 (Northeast, Central, West).

Released - 1990, Univ. of Wisconsin, Madison.

Breeders Seed/Stock - Univ. of Wisconsin, Madison.

Certified Seed/Stock - Available.

Preparer/Additional Information - Department of Agronomy, 1575 Linden Drive, UW, Madison, WI 53706-1597, (608) 262-9557.

Barton

R.R. Kalton, Land O' Lakes, Inc.

Source - An eight-clone synthetic variety stemming from Lincoln, Fischer, Southland and Jeanerette following clonal and progeny testing for desirable agronomic traits.

Method of Breeding - Parental clones selected for seedling vigor, leaf disease resistance, recovery ability, and forage and seed yield in clonal and outcross progeny trials in Iowa. Intended Use - Hay in pure stands or in legume-grass mixtures. Also for pasture and conservation use on fertile, well-drained soils.

Description - Improved forage yield, recovery capacity, seed size, seedling vigor, and leaf disease resistance compared with Lincoln and Fischer. One or two days later in blooming.

Adapted to - LRR B, E, F, H, G, K, L, M, N; PHZ 3.

Released - 1975. Breeder Seed/Stock - Maintained as foundation seed by Land O' Lakes, Inc, in Iowa.

Certified Seed/Stock - Not available.

Preparer/Additional Information - Robert R. Kalton, Land O' Lakes Research Farm, 1025 190th St., Webster City, IA 50595, (515) 543-4852.

Baylor

Selected at Rudy-Patrick Research Center, Ames, IA - R.R. Kalton.

Source - Parental clones selected from southern varieties.

Method of Breeding - Elite clones selected on basis of outcross progeny performance for forage and seed yield, recovery, seedling vigor, leafiness, and disease resistance. Seven-clone synthetic evaluated in north-central region and Canada as R.P. 101.

Intended Use - Pasture, forage, and hay.

Description - High-yielding, disease-resistant, southern type; leafy. Good in recovery and stand establishment. Improved production of high-quality seed. Same maturity as Lincoln.

Released - 1962, distributed testing by Rudy-Patrick Co.

Breeder Seed/Stock - Canada Seed Growers Association.

Certified Seed/Stock - Available.

Preparer/Additional Information - Canada Seed Growers Association, P.O. Box 8445, Ottawa, Ontario, Canada K1G 3T1.

Beacon

Source - Developed by Land O'Lakes, Inc., Webster City, Iowa.

Method of Breeding - Beacon is an eight-clone synthetic of which three were derived from Lincoln, three from Fischer, one from a large seeded synthetic developed at Iowa State University, and one from a roadside collection in Iowa of unknown origin. The parental clones were evaluated for forage vigor, yield, leaf disease resistance, seedling vigor and seed yield.

Intended Use - Hay and pasture production.

Description - Beacon is an improved southern type of bromegrass and is essentially the same in bloom dates, growth habit, and hardiness as Baylor and Lincoln. In Ontario trials, Beacon has averaged a slightly higher forage yield over 10 station-years than Saratoga, the check variety. It has outyielded Saratoga by almost 200 kg/ha in the southern stations, but has not yielded as well as Saratoga in the northern locations. In seed yield trials, at Beaverlodge Research Station, Beacon has ripened approximately one week earlier than Saratoga and Carlton. Beacon has yielded slightly less than Carlton but is superior to Saratoga in seed yield trials.

Adapted to - Recommended in Canada for use in Ontario and Quebec.

Released - Land O'Lakes, Inc., Webster City, IA.

Breeder Seed/Stock - Land O'Lakes, Inc., Webster City, IA.

Certified Seed/Stock - Available.

Preparer/Additional Information - Agriculture Canada, Food Production and Protection Branch, Plant Products Division, Ottawa, Ontario K1A 0C6. Seed distributed by Oseco Ltd., P.O. Box 219, Brampton, Ontario L6V 2L2.

Bravo

Source - Developed by Maple Leaf Mills Forage Division, Georgetown, Ontario.

Method of Breeding - It is an 11-clone synthetic variety made up of northern bromegrass types. In 1972, 31 clones of unknown origin were obtained from the University of Guelph; these were planted and 11 clones were selected for forage yield and maturity. A polycross nursery was established and breeder seed was harvested. This seed was advanced to Syn-2 and was entered for testing.

Intended Use - Hay and pasture production.

Description - Erect and leafy growth habit; medium green leaf color; slight reddish tint in the panicle; heading date is similar to Saratoga and Baylor. Bravo provides adequate forage yields in Ontario and acceptable yields in northern Ontario. This variety also produced acceptable forage yields at Winnipeg and Saskatoon. Bravo under-yielded most check varieties in trials at Charlottetown.

Adapted to - Recommended in Canada for use in Manitoba, Ontario, and Quebec.

Released - 1983, Maple Leaf Mills Forage Division, Georgetown, Ontario.

Breeder Seed/Stock - Otto Pick and Sons Seeds Ltd.

Certified Seed/Stock - Available from Pickseed Canada Inc.

Preparer/Additional Information - Agriculture Canada, Food Production and Protection Branch, Plant Products Division, Ottawa, Ontario K1A 0C6. Seed distributed by Pickseed Canada, Inc., Box 126, Richmond Hill, Ontario L4C 4X9.

Carlton

Selected at Canada Department of Agriculture Research Station, Saskatoon, Saskatchewan - R.P. Knowles.

Source - Northern common smooth brome.

Method of Breeding - Synthetic of four clones whose polycross progenies excelled in seed yield (S-4088). Enlarged in 1966 to a nine-clone synthetic (S-6324) on basis of polycross progeny tests.

Intended Use - Hay and pasture.

Description - Typical of northern type. Hay yields 5-10% and seed yields 20-30% above northern common.

Released - 1961, by Canada Department of Agriculture.

Breeder Seed/Stock - Canada Seed Growers Association.

Certified Seed/Stock - Available in quantity.

Preparer/Additional Information - Canada Seed Growers Association, P.O. Box 8445, Ottawa, Ontario, Canada K1G 3T1.

Elsberry

Plant Materials Center, SCS, in cooperation with the Missouri AES, Columbia, MO. Carried as accession no. M1-2626.

Source - SCS, Shenandoah Project, Shenandoah, IA.

Method of Breeding - Increase of original collection.

Intended Use - Cool-season grass pasture and hay production.

Description - Southern early-maturing strain of smooth brome. Best of several accessions tested at SCS Nursery, Elsberry. High forage and seed yielding, disease-resistant, early maturity. Excellent recovery after clipping. Elsberry smooth brome is used along with or in mixtures as a cool-season grass for pasture or hay and along with or in mixtures for seeding waterways and bank stabilization.

Adapted to - LRR M; PHZ 4a.

Released - 1954.

Breeder Seed/Stock - Plant Materials Center, SCS, Elsberry, MO.

Certified Seed/Stock - Not available.

Preparer/Additional Information - Jimmy Henry Date, Plant Materials Center, SCS, RR 1, Box 9, Elsberry, MO 63343, (314) 898-2012.

Jubilee

Source - Developed in Canada by Maple Leaf Mills, Agricultural Research Center, Georgetown, Ontario.

Method of Breeding - A three-clone synthetic tested in the Ontario Forage Crops Committee Trials as MLM 13009. It was derived from a collection of 46 clones that had been evaluated for forage yield, seedling vigor, leaf disease resistance, and seed yield at the Agriculture Canada Research Station, Brandon, Manitoba. The clones originated from the seed of a plant introduction from the USDA station in Beltsville, MD.

Intended Use - Hay and pasture production.

Description - Jubilee is a northern type of bromegrass intermediate in maturity between Lincoln and Carleton. Its hardiness is similar to Lincoln and Carleton. It is leafy and tall growing and has a restricted creeping growth habit. It has excellent seedling vigor and rate of establishment and recovers rapidly after cutting. No leaf diseases have been reported from Canadian trials. Seed yield in Ontario is very good. In 17 station-years of testing in the Ontario trials, Jubilee outyielded Saratoga by 211 kg/ha. It yielded consistently better than the check in northern Ontario. In tests for *in vitro* digestibility (I.V.D.), Jubilee was slightly inferior to other licensed varieties.

Adapted to - Recommended in Canada for use in Manitoba.

Released - 1979, Maple Leaf Mills, Ltd., Georgetown, Ontario.

Breeder Seed/Stock - Maple Leaf Mills, Ltd., Georgetown, Ontario.

Certified Seed/Stock - Available.

Preparer/Additional Information - Agriculture Canada, Food Production and Protection Branch, Plant Products Division, Ottawa, Ontario K1A 0C6. Seed distributed by Pickseed Canada, Inc., Box 126, Richmond Hill, Ontario L4C 4X9.

Lancaster (Reg. No. 4)

Selected at Nebraska AES, Lincoln, ARS cooperating - L.C. Newell.

Source - Clones collected from old fields of smooth brome in Nebraska.

Method of Breeding - Produced in 1943 by field hybridization of clone from five unrelated sources. Selection of clones based on previous evaluation of their sibbed and open-pollinated progenies, studies beginning with selections from old fields in 1937. Distributed as Nebraska 44 for testing.

Intended Use - Pasture and hay production.

Description - Leading smooth brome variety in forage and seed yields in tests at Lincoln, 1947-52. Showed immediate promise among several experimental synthetic varieties in early comparative tests at Nebraska AES. On fertile soils is leafy, vigorous, with finer stems and somewhat drooping panicles.

Adapted to - LRR K, M; PHZ 4, 5, 6.

Released - 1950, cooperatively by Nebraska AES and Plant Science Research Division, ARS.

Breeder Seed/Stock - Nebraska AES and USDA-ARS, Univ. of Nebraska, Lincoln.

Certified Seed/Stock - Not available.

Preparer/Additional Information - K.P. Vogel, ARS-USDA, 344 Keim Hall, Univ. of Nebraska or Jeff Pederson, Dept. of Agronomy, Univ. of Nebraska, Lincoln, NE 68583, (402) 472-1564, (402) 472-2811.

Lincoln (Reg. No. 5)

Increased at Nebraska AES, Lincoln, ARS cooperating - L.C. Newell and A.L. Frolik.

Source - Old fields of smooth brome derived from early introductions of smooth brome prior to 1898; attributed to Hungarian origin (California introduction of 1884).

Method of Breeding - Plot tests of farmer strains of smooth brome conducted in 1939-42; showed comparative superiority of locally grown southern strains as compared with strains of northern origin. Fields that showed superiority and that were traced to common origin first approved in 1941 for seed increase and later certified as Lincoln.

Intended Use - Pasture and hay production.

Description - Cool-season grass; provides abundance of early-spring pasturage and fall regrowth under favorable conditions. Rhizomatous, sod-forming type. Well adapted for conservation purposes in central latitudes as compared with less aggressive northern types. Exhibits good seedling vigor and relative ease of establishment on critical planting sites.

Adapted to - LRR K, M, N; PHZ 4, 5, 6.

Released - 1942, cooperatively by Nebraska AES and Plant Science Research Division, ARS.

Breeder Seed/Stock - Nebraska AES and USDA-ARS, Univ. of Nebraska, Lincoln.

Certified Seed/Stock - Available.

Preparer/Additional Information - K.P. Vogel, ARS-USDA, 344 Keim Hall, Univ. of Nebraska or Jeff Pederson, Dept. of Agronomy, Univ. of Nebraska, Lincoln, NE 68583, (402) 472-1564, (402) 472-2811.

Lyon (Reg. No. 6)

Selected at Nebraska AES, Lincoln, ARS cooperating - L.C. Newell.

Source - Developed from selection made in farm strains of certified Lincoln smooth brome. Later progenies became outcrossed in selection nursery to broad source germplasm of southern type smooth brome.

Method of Breeding - Single clones of Lincoln smooth brome selected for seed quality and forage type and isolated; crossed seed from progenies of these selections later outcrossed to large number of open-pollinated lines of southern type, bulked, and retested as Nebraska 36.

Intended Use - Pasture and hay production.

Description - Maintains broad adaptation of Lincoln smooth brome parental stock combined with superior seed quality, seedling vigor, and more uniformly desirable plant type. Produced larger yields of forage and seed in Nebraska tests than Lincoln. As Nebraska 36, and later as Lyon, has been tested widely since 1947, showing promise over broad range of conditions. Named after Professor T.L. Lyon, who first worked with smooth brome at Nebraska AES in 1897.

Adapted to - LRR K, M; PHZ 4, 5, 6.

Released - 1950, cooperatively by Nebraska AES and Plant Science Research Division, ARS.

Breeder Seed/Stock - Nebraska AES.

Certified Seed/Stock - Not available.

Preparer/Additional Information - Canada Seed Growers Association, P.O. Box 8445, Ottawa Ontario, Canada K1G 3T1.

Magna

Selected at Canada Department of Agriculture Research Station, Saskatoon, Saskatchewan - R.P. Knowles.

Source - Parentage is 62% from Fischer, 7% from B63 Wisconsin, and the rest from unidentifiable outcrossed sources.

Method of Breeding - Clones were polycross progeny tested for hay and seed yields at three Saskatchewan stations. The 14 best clones were selected for synthetic variety.

Intended Use - Pasture and hay.

Description - Intermediate in type between northern and southern varieties. Yields in western Canada are similar to or above those of other southern varieties, but aftermath yields are somewhat lower. Variety is characterized by high seed quality and higher seed yields than southern varieties.

Released - 1968, by Canada Department of Agriculture.

Breeder Seed/Stock - Canada Department of Agriculture Research Station, Saskatoon.

Certified Seed/Stock - Available.

Preparer/Additional Information - Canada Seed Growers Association, P.O. Box 8445, Ottawa, Ontario, Canada K1G 3T1.

Manchar (Reg. No. 10)

Selected at Plant Materials Center, SCS, Pullman, WA - J.L. Schwendiman, A.G. Law, A.L. Hafenrichter, and D.C. Tingey.

Source - Original introduction in 1935 from Kungchuling Experiment Station of South Manchurian Railway, Manchuria, China, as PI 109812.

Method of Breeding - Grown in nurseries at SCS Plant Materials Centers since 1935; subjected to mass selection and tested in uniform nurseries and strain tests since 1937 as P-177.

Intended Use - Hay and pastureland.

Description - Intermediate between weakly spreading northern types and aggressive sod-forming southern types. Maintains good balance with associated legumes; produces vigorous seedling; good yields of seed and forage; recovers rapidly after cutting. Its dark, purple-cast seeds thresh easily; seed generally heavier than that of common smooth brome.

Adapted to - LRR B; PHZ 5.

Released - 1943, as P-177, cooperatively by Idaho and Washington Agricultural Experiment Stations at Moscow and Pullman, respectively, and Plant Materials Centers, SCS Aberdeen, ID, and Pullman, WA. Named Manchar in 1946.

Breeder Seed/Stock - Plant Materials Center, SCS, Pullman, WA.

Certified Seed/Stock - Available.

Preparer/Additional Information - Plant Materials Center, SCS, Rm 104, Hulbert Agricultural Sciences Bldg, WSU, Pullman, WA 99164-6211, (509) 335-7376.

Mandan 404

Selected at Northern Great Plains Research Laboratory, ARS, Mandan, ND - G.A. Rogler.

Source - From local field collection of northern ecotypes.

Method of Breeding - Developed by selection within progeny of individual plants after two generations of single-plant selection under open pollination. Both inbred and open-pollination progeny tests made of each of eight clones constituting the cultivar.

Intended Use - Hay and pasture.

Description - Short, fine, very high in quality, light green. Not aggressive and not high-yielding, but higher in protein at Mandan at all stages of growth than the cultivar, Lincoln. Tests at Mandan show Mandan 404 to be higher in palatability than Lincoln.

Adapted to - LRR F; PHZ 3, 4.

Released - No. Included in regional testing program.

Breeder Seed/Stock - Northern Great Plains Research Laboratory, Mandan, ND.

Certified Seed/Stock - Not available.

Preparer/Additional Information - John D. Berdahl, ARS Northern Great Plains Research Laboratory, P.O. Box 459, Mandan, ND 58554, (701) 663-6445.

Polar

Selected at the Alaska AES, Palmer, ARS cooperating - H.J. Hodgson, A.C. Wilton, R.L. Taylor, and L.J. Klebesadel.

Source - Selections that trace back to hybrids between arctic brome, *Bromus inermis* Leyss. ssp. *pumpellianus* (Scribn.) Wagnon var. *pumpellianus* (Scribn.) C.L. Hitchc., and smooth brome and smooth brome sources—Manchar, Mandan 404, Colorado 144, B. *inermis* Leyss. 12, and Canadian commercial.

Method of Breeding - Clonal selections made on basis of yield and winter hardiness of polycross progeny. The variety is a 16-clone synthetic (11 clones tracing to *B. inermis* ssp. *pumpellianus* Leyss. hybrids and five clones to the smooth brome varieties identified under Source).

Intended Use - Forage and pasture seedings.

Description - Consistently superior to other northern types of smooth brome in both winter hardiness and yield. Other desirable attributes included superior lodging resistance, early-spring growth, early seed maturity, and a less aggressive spreading habit of growth. Variety characterized by very hairy nodes on about 10% of the plants and slight nodal hairiness on 60% of the plants. Seed of about 75% of the plants either slightly or very hairy. Only brome variety to survive the severe winter of 1961-62 in the Matanuska valley without serious damage.

Adapted to - LRR W, X, Y; PHZ 2, 3, 4.

Released - 1965, cooperatively by Alaska AES and Plant Science Research Division, ARS.

Breeder Seed/Stock - Alaska AES.

Certified Seed/Stock - Available.

Preparer/Additional Information - Alaska Plant Materials Center, HC 02 Box 7440, Palmer, AK 99645, (907) 745-4469.

Radisson

Source - Developed at the Agriculture Canada Research Station, Saskatoon, Saskatchewan.

Method of Breeding - Sixty-three selections of southern-type strains, primarily tracing to crosses of University of Guelph releases with Magna, were interpollinated in 1973-74 to produce the experimental line S-8792A. Fifteen of the most desirable parents were selected in 1987: 11 plants from the crosses of Guelph strains X Magna, one plant from Saratoga X Magna, two plants from Wisconsin releases, and one plant from Lincoln. These plants were interpollinated in 1977-78 to form S-8792B. Reselection in S-8792B was conducted in 1980 on the basis of vigor and seed yield. Fifty-six plants were selected to form the experimental line S-8792C, from which Radisson was derived. This variety was previously registered as Radison.

Intended Use - Hay and pasture production.

Description - Erect growth habit; rhizomatous, slightly longer leaves than that of Beacon or Tempo, slightly narrower than that of Beacon, slightly wider than that of Baylor; flag leaf similar in width to Beacon; seed slightly longer than that of Tempo, wider than that of Tempo; plant height taller than Tempo and Saratoga; maturity is similar to that of other varieties; less than 1% of plants are slightly taller with less pronounced purpling in the panicles. Radisson provided very good forage yields in Quebec trials. Limited data from Ontario and Prince Edward Island suggest that it also produced very good forage yields. In Saskatchewan, forage yields were similar to those of Carlton and Magna. However, seed yields in Saskatchewan were lower than those of Carlton and Magna and similar to those of Baylor and Rebound.

Adapted to - Under evaluation in Canada for use in Saskatchewan.

Released - Agriculture Canada Research Station, Ste-Foy, 1989 with registration no. 3154.

Breeder Seed/Stock - Agriculture Canada Research Station, Ste-Foy.

Certified Seed/Stock - Available from SeCan Association.

Preparer/Additional Information - Agriculture Canada, Food Production and Protection Branch, Plant Products Division, Ottawa, Ontario K1A 0C6. Seed distributed by SeCan Association, 200-57, Promenade Auriga, Nepean, Ontario K2E 8B2.

Rebound

Source - Developed at the University of South Dakota.

Method of Breeding - Sixty-three plants were selected in 1971 for regrowth vigor from a spaced plant nursery containing 5,623 plants of strain South Dakota 7 and 29039 plants of Saratoga. These 63 plants were grown in the greenhouse and 32 genotypes were selected on the basis of regrowth and photosynthetic rates. These plants were vegetatively propagated and established in field trials, being evaluated for two years for forage and seed yields and for regrowth. Four genotypes, originally from Saratoga, were selected and placed in isolation to produce Syn 1 (breeder) seed of the variety.

Intended Use - Hay and pasture production.

Description - Less spreading and growth habit than Saratoga; fewer reproductive tillers; medium green leaves; less lodging resistant than Saratoga; winter survival similar to Carlton; similar to Carlton in reaction to silver top; *in vitro* digestibility 2.2% greater than Carlton. Rebound showed good forage yields and low seed yields in western Canadian trials.

Adapted to - Recommended in Canada for use in Manitoba and Saskatchewan.

Released - University of South Dakota.

Breeder Seed/Stock - University of South Dakota.

Certified Seed/Stock - Available.

Preparer/Additional Information - Agriculture Canada, Food Production and Protection Branch, Plant Products Division, Ottawa, Ontario K1A 0C6. Seed distributed by John Zuelzer and Son Canada Ltd., Box 990, Carberry, Manitoba R0K 0H0.

Sac (Reg. No. 12)

Selected at Wisconsin AES, Madison, ARS cooperating - E.L. Nielsen, D.C. Smith, and P.N. Drolsom.

Source - S1 and S2 plants from older varieties and strains.

Method of Breeding - First-cycle selections polycrossed were S1 and S2 plants from older varieties and strains. Polycross tested as spaced plants and as synthetic. Second-cycle polycross and synthetic based on 81 clones selected for foliage disease reaction, vigor, leafiness, and seed production. Distributed for testing as B-81.

Intended Use - Pasture and hay plantings.

Description - Growth characteristics similar to those of southern-adapted strains; seed quality similar to that of northern-adapted strains. Good tolerance to foliage and root rot diseases. Seed production adequate.

Adapted to - PHZ 1, 2, 3, 4, 5 (W, NC, NE, and Great Plains).

Released - 1962, cooperatively by Wisconsin AES and Plant Science Research Division, ARS.

Breeder Seed/Stock - Not available. Certified Seed/Stock - Not available.

Preparer/Additional Information - Department of Agronomy, 1575 Linden Dr., Univ. of Wisconsin, Madison, WI 53706-1597, (608) 262-9557.

Saratoga (Reg. No. 8)

Selected at the Cornell University AES. Cornell University, Ithaca, NY - R.P. Murphy and S.S. Atwood.

Source - Wide collection of seed lots from plant breeders in U.S.. Parental clones: N.Y. 46-11, N.Y. 46-19, N.Y. 46-92, N.Y. 46-157, N.Y. 46-166.

Method of Breeding - Synthetic variety developed from five selected, relatively self-incompatible clones. Breeder seed produced in isolated plot from randomly planted vegetative pieces of five clones in 100 or more replications. Equal amounts of seed from each parental clone mixed together for breeder seed. Certified seed first advanced generation from foundation seed and not eligible for use as planting stock for production of any class of certified seed.

Intended Use - Hay, haylage and pasture.

Description - Vigorous, high seedling vigor, early-spring growth; quick recovery and high aftermath production after cutting. Yielded 8% more in total-season yield and 29% more in aftermath yield than Lincoln when grown alone; yielded same as Lincoln when grown in mixture with alfalfa, but higher proportion of mixture has been grass. Similar to Lincoln in yield and quality of seed and in resistance to brown spot and leaf scald, but superior to Canadian common smooth brome and Manchar.

Adapted to - Northeast and north central U.S.; PHZ 2, 3, 4, 5, higher elevations 6.

Released - 1955, by New York AES.

Breeder Seed/Stock - Department of Plant Breeding and Biometry, Cornell University AES, Cornell University, Ithaca, NY 14853

Certified Seed/Stock - Available in commerce.

Preparer/Additional Information - Royse P. Murphy, Dept. of Plant Breeding, Cornell Univ., 252 Emerson Hall, Ithaca, NY 14853-1902, (607) 255-1672.

Signal

Source - Developed at the Agriculture Canada Research Station, Saskatoon, in cooperation with Melfort Research Station and the Indian Head Experimental Farm.

Method of Breeding - This variety results from selections from Magna for seed characteristics and resistance to seed midge. Eighty-eight plant selections were made and progeny tested. Ten superior plants were selected, polycrossed and bulked to produce breeder seed.

Intended Use - Hay and pasture production.

Description - Similar to Magna in growth habit; taller than Carlton, slightly taller than Magna; leaves are similar to Magna; roots are slightly less strongly creeping; maturity is similar to Magna; lodging resistance is better than Carlton and Magna; moderately susceptible to brown spot. Hay yields of Signal in western Canada show this variety to yield similar to or better than the check varieties but with similar or slightly lower *in vitro* digestibility ratings. Seed yield and quality of Signal are higher than those of check varieties.

Adapted to - Recommended in Canada for use in British Columbia and Sackatchewan.

Released - 1983 by Agriculture Canada Research Station Saskatoon, with registration no. 2325.

Breeder Seed/Stock - Agriculture Canada Research Station Saskatoon.

Certified Seed/Stock - Available from SeCan Association.

Preparer/Additional Information - Agriculture Canada, Food Production and Protection Branch, Plant Products Division, Ottawa, Ontario K1A 0C6. Seed distributed by SeCan Association, 200-57, Promenade Auriga, Nepean, Ontario K2E 8B2.

Southland

Developed by International Seeds Inc., Halsey, OR. Carried as experimental no. LPK-composite.

Source - Selections made from old turfs in the southern and eastern parts of the U.S.

Method of Breeding - Clones evaluated for disease resistance and other characters associated with improved turf quality.

Intended Use - Permanent turf for homes, athletic fields, and commercial areas. May be used for sod production.

Description - Medium late heading, moderately fine bladed turf-type variety; capable of producing attractive, durable turf and sod; resistant to leaf spot, net blotch and crown rust.

Adapted to - LRR A, B, C, D, E, F, G, H, J, K, L, M, N, O, P, R, S, W; PHZ 4.

Released - 1985 by Hubbard Seed & Supply Co., Hubbard, OR, and Southland Sod Inc., Camarillo, CA.

Breeder Seed/Stock - International Seeds Inc.

Certified Seed/Stock - Available. U.S. PVP no. 8500184.

Preparer/Additional Information - International Seeds Inc., P.O. Box 168, Halsey, OR 97348, (503) 369-2251.

Tempo

Source - Developed at the Ottawa Research Station - Dr. W.R. Childers and R.W. Suitor. Tested as Ottawa D-9.

Method of Breeding - In 1963, the nine highest seed-yielding clones were established in a mutual pollination randomized cross block. The polycross seed was harvested in the fall of 1964 and established a plant spaced nursery to evaluate the variation for seed production. The 20 highest yielding plants were established in a polycross randomized crossing block in 1967 and the 20 high seed-yielding clones plus the checks, Redpatch and Magna, were at seven western Canada locations.

Intended Use - Hay and pasture production.

Description - The variety is similar in maturity to Saratoga. The seed heads are more reddish in color than Saratoga, reflecting the inclusion of more northern type germplasm in this high seed-yielding variety. The plant types are leafy and spreading, making the variety valuable for both hay and pasture utilization. The dry matter yield of Tempo is similar to but slightly less than Saratoga. Tempo has outyielded Baylor by 288 kg/ha in some years. Tempo is higher in percentage of protein, and also has a higher percentage of digestibility.

Adapted to - Recommended in Canada for use in British Columbia, Ontario, Quebec, and the Atlantic region.

Released - 1975 by Ottawa Research Station with Registration No. 1581.

Breeder Seed/Stock - Ottawa Research Station.

Certified Seed/Stock - Available.

Preparer/Additional Information - Agriculture Canada, Food Production and Protection Branch, Plant Products Division, Ottawa, Ontario K1A 0C6. Seed distributed by SeCan Association, 200-57, Promenade Auriga, Nepean, Ontario K2E 8B2.

York

Selected at the Cornell University AES, Cornell University, Ithaca, NY - R.P. Murphy. Carried as accession no. NY 86-B.

Source - Selection from Saratoga (New York) and Regro (South Dakota, renamed Rebound.) Parental clones: N.Y. 83-3, 83-8, 83-10, 46-19, 83-40, 83-42, 83-46, and 83-47.

Method of Breeding - Synthetic variety developed from eight selected clones. Breeder seed produced in isolated plot from randomly planted vegetative pieces of eight clones in 75 or more replications. Equal amounts of viable seed from each parental clone mixed for breeder seed. One advanced generation of this production in isolation may be designated breeder seed. Foundation seed first advanced generation from breeder seed. Certified seed from foundation or breeder seed and not eligible for use for production of any class of certified seed.

Intended Use - Hay, haylage, and pasture.

Description - Vigorous, high seedling vigor and early spring growth, quick recovery from cutting and high aftermath production. Yields similar to Saratoga when grown alone or in mixture with alfalfa. Selection of parental clones based on improvement over Saratoga for seed production and seed quality.

Adapted to - LRR northeast and north central U.S. and nearby areas; PHZ 2, 3, 4, 5, higher elevations 6 .

Released - 1989, Cornell University AES.

Breeder Seed/Stock - Department of Plant Breeding and Biometry, Cornell University AES, Cornell Univ., Ithaca, NY 14853.

Certified Seed/Stock - Initial production in 1990.

Preparer/Additional Information - Royse P. Murphy, Dept. of Plant Breeding, Cornell Univ., 252 Emerson Hall, Ithaca, NY 14853-1902, (607) 255-1672.Bromus marginatus Nees ex Steud. - mountain brome Cool-season, native bunchgrass closely related to Bromus carinatus; similar to it in soil and climatic adaptation. Short lived; but large seed and good seedling vigor. Deep, well-branched root system important in providing protection on erodible slopes.

9005308

Selected for advance testing by the Upper Colorado Environmental Plant Center, Meeker, CO. Carried as accession no. 9005308.

Source - Collected by Frank Kirschten near Garnet, Granite County, MT, in September 1976.

Method of Breeding - Distributed from small plot increase by Plant Materials Center, SCS, Bridger, MT, tested and further increased by the Upper Colorado Environmental Plant Center.

Intended Use - Revegetation of coal, oil shale, transmission corridors, improve wildlife habitat, erosion control on cropland, and as a component in rangeland seedings.

Description - Good seedling vigor, resistance to head smut, extended seed production, and longer lived than Bromar.

Adapted to - LRR D, E; PHZ 5.

Released - Not released, but distributed for field testing.

Breeder Seed/Stock - Upper Colorado Environmental Plant Center, Meeker, CO.

Certified Seed/Stock - Not available.

Preparer/Additional Information - Randy Mandel, Upper Colorado Environmental Plant Center, P.O. Box 448, Meeker, CO 81641, (303) 878-5003.

Bromar (Reg. No. 1)

Selected at Plant Materials Center, SCS, Pullman, WA - A.L. Hafenrichter, A.G. Law, and J.L. Schwendiman.

Source - Native collection made at Pullman in 1933 and assigned accession number WN-439. Selection P-3368 from this accession used in developing Bromar.

Method of Breeding - Mass selection with screening for head smut resistance. Bromar was one of four ecotypes among 69 accessions of mountain brome tested.

Intended Use - Green manure.

Description - Rapid-developing, late-maturing, perennial bunchgrass. Tall, erect, vigorous, with medium-stems and abundant, broad, well-distributed leaves. When compared with commercial strain, Bromar is taller, leafier, two weeks later in maturity; has more seeding vigor; and is earlier in spring recovery. Heavy seed and forage producer; compatible in rate of growth with sweetclover; and seed is readily de-awned. Needs chemical seed treatment before planting to control head smut, when grown for seed. Outstanding in performance in mixtures with sweetclover or red clover for pasture or green manure in short rotations.

Adapted to - LRR B, E; PHZ 5.

Released - 1946, cooperatively by Washington, Idaho, and Oregon Agricultural Experiment Stations at Pullman, Moscow, and Corvallis, respectively; Plant Materials Center, SCS, Pullman, WA; and Plant Science Research Division, ARS.

Breeder Seed/Stock - Plant Materials Center, SCS, Pullman, WA.

Certified Seed/Stock - Limited availability.

Preparer/Additional Information - Plant Materials Center, SCS, Rm. 104, Hulbert Agricultural Sciences Bldg., WSU, Pullman, WA 99164-6211, (509) 335-7376.

Bromus rubens L. - red brome

Red brome is an annual, cool-season grass that grows to a height of 15-40 cm. Sheaths and blades are pubescent and the panicle is distinctly reddish in color after maturity. Awns are 18-22 mm long.

Panoche

Selected at Plant Materials Center, SCS, Lockeford, CA - R. Clary, K. Croeni and R. Slayback.

Source - Collected from naturalized plants in Panoche Hills, Fresno County, CA in 1971.

Method of Breeding - Original seed increased and compared to other red bromes for initial establishment and seed germination.

Intended Use - Cover crop, revegetation of disturbed areas, (mine spoils), revegetation of wildfire burns.

Description - Annual, cool season grass showing superior establishment on droughty sites receiving less than 250 mm of precipitation in the Mojave Desert. Superior annual grass for the driest sites in California.

Adapted to - LRR A, C, D; PHZ 8.

Released - 1985, cooperatively by California AES, Davis, and Plant Materials Center, SCS, Lockeford, CA.

Breeder Seed/Stock - Plant Materials Center, SCS, Lockeford, CA.

Certified Seed/Stock - Available.

Preparer/Additional Information - R. Slayback, SCS, 2121-C, 2nd St., Davis, CA 95616, (916) 757-8257.

Buchloe dactyloides **(Nutt.) Engelm. - buffalograss**

Warm-season, sod-forming, native grass. Spreads by stolons. Occurs mainly in short-grass associations in Great Plains. Drought resistant; often indicative of overgrazing; adapted to grazing and erosion control on heavy soils.

609

Developed by the University of Nebraska.

Source - Selected from University of Nebraska buffalograss germplasm nursery in 1985. Germplasm had been selected from a nursery at Texas A&M, Dallas.

Method of Breeding - Selection made on basis of color, quality, rate of spread, and uniformity.

Intended Use - Warm season turfgrass for use throughout the arid or semi-arid portions of the U.S. Used in monocultures for turfgrass use in lawns, golf course roughs, or low maintenance sites.

Description - Warm season turf-type buffalograss; produces a dense, low growing, attractive turf.

Adapted to - PHZ 4, 5, 6, 7, 8, 9 under arid or semi-arid conditions. Also can be used where short-term droughts occur, if cool season turfgrasses are controlled.

Released - 1991 by the Institute of Agriculture and Natural Resources of the University of Nebraska-Lincoln.

Breeder Seed/Stock - Crenshaw & Doguet Turfgrass, Inc., Austin, TX.

Certified Seed/Stock - Available.

Preparer/Additional Information - T.P. Riordan, 377 Plant Science, University of Nebraska, Lincoln, NE 68583-0742, (402) 472-1142.

Bison

Selected at Oklahoma State University, Stillwater, OK, Oklahoma AES and USDA-ARS cooperating - C.M. Taliaferro and R.M. Ahring. Tested under the experimental designation of A-Plus.

Source - Plant selections from the Texoka and Mesa buffalograss cultivars.

Method of Breeding - Bison is a synthetic variety, derived from two male and two female clonal plants. The male and female parents of Mesa, plus a superior plant of each sex selected from Texoka, constitute the parents. Selections from Texoka were made in 1982 from the original one-acre Foundation seed increase block planted on the S.W. Forage and Livestock Research Station (SWFLRS), El Reno, OK, in the early 1970s.

Intended Use - Pasture and conservation plantings.

Description - Similar to Texoka in morphological appearance, adaptation, and forage yield and quality. Tests in Oklahoma indicated Bison to have a 24% advantage over Texoka in pure live seed yield. Tests of esterase and peroxidase isoenzymes established genetic difference between Bison and Texoka.

Adapted to - PHZ 6, 7.

Released - 1990, Oklahoma AES and USDA-ARS cooperating. Licensed to Johnston Seed Co., Enid, OK.

Breeder Seed/Stock - Oklahoma AES.

Certified Seed/Stock - Available.

Preparer/Additional Information - C.M. Taliaferro, Agronomy Dept., 368 Ag Hall, Oklahoma State Univ., Stillwater, OK 74078, (405) 744-6410.

NTDG 1

Developed by Native Turfgrass Development and the University of Nebraska.

Source - Synthetic of turf-type male and female buffalograsses selected at the University of Nebraska in 1988.

Method of Breeding - Selection of parents made on basis of color, quality, rate of spread, inflorescence production, and uniformity.

Intended Use - Warm season turfgrass for use throughout the arid or semi-arid portions of the U.S. Used in monocultures for turfgrass use in lawns, golf course roughs, or low-maintenance sites.

Description - Warm season turf-type buffalograss; produces a dense, low growing, attractive turf.

Adapted to - PHZ 4, 5, 6, 7, 8, 9 under arid or semi-arid conditions. Also can be used where short-term droughts occur, if cool season turfgrasses are controlled.

Released - No.

Breeder Seed/Stock - Johnston Seed, Enid, OK.

Certified Seed/Stock - Available.

Preparer/Additional Information - T.P. Riordan, 377 Plant Science, University of Nebraska, Lincoln, NE 68583-0742, (402) 472-1142.

NTDG 2

Developed by Native Turfgrass Development and the University of Nebraska.

Source - Synthetic of turf-type male and female buffalograsses selected at the University of Nebraska in 1988.

Method of Breeding - Selection of parents made on basis of color, quality, rate of spread, inflorescence production, and uniformity.

Intended Use - Warm season turfgrass for use throughout the arid or semi-arid portions of the U.S. Used in monocultures for turfgrass use in lawns, golf course roughs, or low-maintenance sites.

Description - Warm season turf-type buffalograss; produces a dense, low growing, attractive turf.

Adapted to - PHZ 4, 5, 6, 7, 8, 9 under arid or semi-arid conditions. Also can be used where short-term droughts occur, if cool season turfgrasses are controlled.

Released - No.

Breeder Seed/Stock - Johnston Seed, Enid, OK.

Certified Seed/Stock - Available.

Preparer/Additional Information - T.P. Riordan, 377 Plant Science, University of Nebraska, Lincoln, NE 68583-0742, (402) 472-1142.

NTDG 3

Developed by the University of Nebraska.

Source - Synthetic of turf-type male and female buffalograsses selected at the University of Nebraska in 1988.

Method of Breeding - Selection of parents made on basis of color, quality, rate of spread, inflorescence production, and uniformity.

Intended Use - Warm season turfgrass for use throughout the arid or semi-arid portions of the U.S. Used in monocultures for turfgrass use in lawns, golf course roughs, or low-maintenance sites.

Description - Warm season turf-type buffalograss; produces a dense, low growing, attractive turf.

Adapted to - PHZ 4, 5, 6, 7, 8, 9 under arid or semi-arid conditions. Also can be used where short-term droughts occur, if cool season turfgrass are controlled.

Released - No.

Breeder Seed/Stock - Johnston Seed, Enid, OK.

Certified Seed/Stock - Available.

Preparer/Additional Information - T.P. Riordan, 377 Plant Science, University of Nebraska, Lincoln, NE 68583-0742, (402) 472-1142.

NTDG 4

Developed by Native Turfgrass Development and the University of Nebraska.

Source - Synthetic of turf-type male and female buffalograsses selected at the University of Nebraska in 1988.

Method of Breeding - Selection of parents made on basis of color, quality, rate of spread, inflorescence production, and uniformity.

Intended Use - Warm season turfgrass for use throughout the arid or semi-arid portions of the U.S. Used in monocultures for turfgrass use in lawns, golf course roughs, or low-maintenance sites.

Description - Warm season turf-type buffalograss; produces a dense, low growing, attractive turf.

Adapted to - PHZ 4, 5, 6, 7, 8, 9 under arid or semi-arid conditions. Also can be used where short-term droughts occur, if cool season turfgrasses are controlled.

Released - No.

Breeder Seed/Stock - Johnston Seed, Enid, OK.

Certified Seed/Stock - Available.

Preparer/Additional Information - T.P. Riordan, 377 Plant Science, University of Nebraska, Lincoln, NE 68583-0742, (402) 472-1142.

NTDG 5

Developed by Native Turfgrass Development and the University of Nebraska.

Source - Synthetic of turf-type male and female buffalograsses selected at the University of Nebraska in 1988.

Method of Breeding - Selection of parents made on basis of color, quality, rate of spread, inflorescence production, and uniformity.

Intended Use - Warm season turfgrass for use throughout the arid or semi-arid portions of the U.S. Used in monocultures for turfgrass use in lawns, golf course roughs, or low-maintenance sites.

Description - Warm season turf-type buffalograss; produces a dense, low growing, attractive turf.

Adapted to - PHZ 4, 5, 6, 7, 8, 9 under arid or semi-arid conditions. Also can be used where short-term droughts occur, if cool season turfgrasses are controlled.

Released - No.

Breeder Seed/Stock - Johnston Seed, Enid, OK.

Certified Seed/Stock - Available.

Preparer/Additional Information - T.P. Riordan, 377 Plant Science, University of Nebraska, Lincoln, NE 68583-0742, (402) 472-1142.

Plains

Developed by Bamert Seed Co., Muleshoe, TX.

Source - Selected from production fields of Texoka at Muleshoe, TX, as well as native sites in Baca County, CO, and Clay County, NE.

Method of Breeding - Mass selection of Texoka and native plants for upright growth habit, forage yield, and speed of establishment. Plants stripped of seed which is used as breeders seed.

Intended Use - Low-maintenance turf areas, pasture renovation, and erosion control. The erect growth habit has increased forage production over Texoka.

Description - Plains averages 12 cm in height which is taller than Texoka; it is also denser and darker green than Texoka. This variety is adapted for pastures or turf areas where 15-cm maximum height is acceptable.

Adapted to - LRR A, C, D, E, G, H, I, J, L, M, N, O, P, Q, R, S, T, U; PHZ 3, 4, 5, 6, 7, 8, 9.

Released - 1992.

Breeder Seed/Stock - Bamert Seed Co., Muleshoe, TX.

Certified Seed/Stock - Available.

Preparer/Additional Information - Nick Bamert, Bamert Seed Co., Rte 3, Box 1120, Muleshoe, TX 79347, (806) 272-5506.

Texoka (Reg. No. 35)

Selected at Southern Great Plains Field Station, ARS, Woodward, OK, in cooperation with Oklahoma AES, Stillwater. Increased and evaluated at Fort Reno Experiment Station - J.R. Harlan, W.R. Kneebone, R.M. Ahring, and P.W. Voigt.

Source - Ten clones, four female and six male. The clones were individual collections, or of hybrid origin. Some came from the buffalograss collections at the Fort Hays Experiment Station, Hays, KS. Known origins range from Natoma, KS, to Spur, TX. Tested as W2.

Method of Breeding - Clones were selected because they produce a high percentage of seed-bearing plants in their progeny. The initial seed production block was established vegetatively at the Fort Reno Experiment Station in 1954. All seed of Texoka was derived from that planting. Foundation seed released to the public was Syn. 2 seed. All commercial seed is Syn. 3 or later generations.

Intended Use - Range seedings, erosion control, and turf.

Description - Texoka is a vigorous grass that produces more growth (160-190%) with more rapid spread than most collections of buffalograss to which it has been compared. It produces a high percentage of seed-bearing plants (female and monoecious) in its offspring (about 70% in the Syn. 1 and Syn. 2 generations). Thus, it produces seed yields as much as 10 times higher than that produced from range harvests.

Adapted to - LRR H (Western KS to northwestern TX); PHZ 5, 6, 7.

Released - 1974, cooperatively by ARS; SCS; and Oklahoma, Kansas, and Texas Agricultural Experiment Stations.

Breeder Seed/Stock - Not available.

Certified Seed/Stock - Not available. Commercial seed is available.

Preparer/Additional Information - Paul Voigt, ARS, 808 E. Blackland Road, Temple, TX 76502, (817) 770-6521.

Topgun

Developed by Dr. Bill Davis and licensed to Bamert Seed Co., Muleshoe, TX. Carried as PI no. BAM101.

Source - Single plant selection from Texas native. Topgun was selected in Plainview, TX, with particular reference made to low growth pattern, dense leafiness, and green color. Texoka was introduced for male pollination.

Method of Breeding - Topgun was selected in 1981, and was propagated clonally in 1982 in a greenhouse. A 0.1-acre block was established in 1983; at this time a male was needed for pollination and Texoka variety was interseeded into female clone.

Intended Use - Low-maintenance turf areas because of the prostrate growth characteristic. Specifically for lawns, parks, and erosion control.

Description - Topgun averages 8 cm in height which is shorter than Texoka. This clone has persisted for more than 15 years and shows high resistance to common root rot and nematodes.

Adapted to - LRR A, C, D, E, G, H, I, J, L, M, N, O, P, Q, R, S, T, U; PHZ 3, 4, 5, 6, 7, 8, 9.

Released - 1993.

Breeders Seed/Stock - Bamert Seed Co., Muleshoe, TX.

Certified Seed/Stock - Available.

Preparer/Additional Information - Nick Bamert, Bamert Seed Co., Rte 3, Box 1120, Muleshoe, TX 79347, (806) 272-5506.

Calamagrostis canadensis **(Michx.) Beauv. - bluejoint**

Cool-season, rhizomatous, relatively robust perennial grass; abundant throughout mainland Alaska south of the Brooks Range and south through the northern two-thirds of the U.S.. Early seral dominant on cleared or burned-over forest lands in northern latitudes. Used for reclamation and forage.

Sourdough (Reg. No. 62)

Selected at Palmer Research Center, University of Alaska Fairbanks Agricultural and Forestry Experiment Station, Palmer, AK - W.W. Mitchell.

Source - Thirty-six indigenous collections of plants, made from the southwestern and western coastal regions through the central and southern interior regions of Alaska.

Method of Breeding - Thirty-six individual breeding lines are seed-propagated and breeder seed composited of equal amounts of pure, live seed from the 36 lines.

Intended Use - Revegetation purposes.

Description - First known cultivar of the species.

Adapted to - Southern and south central regions to arctic region of Alaska. Hardy across mainland Alaska, including arctic and alpine regions; PHZ W, X and Y.

Released - 1976, by the Alaska AES.

Breeder Seed/Stock - Palmer Research Center, Agricultural and Forestry Experiment Station, Palmer, AK.

Certified Seed/Stock - Available.

Preparer/Additional Information - William W. Mitchell, Palmer Research Center, Agricultural and Forestry Experiment Station, 533 E. Fireweed, Palmer, AK 99645, (907) 745-3257.

Calamovilfa longifolia **(Hook.) Scribn. - Prairie sandreed**

Warm-season, rhizomatous, perennial grass found on sandy hills, prairies, and open woods in midwestern North America. Useful for stabilizing sandy sites subject to wind erosion; although somewhat coarse and woody, has medium value as forage and potential for improvement.

Goshen

Selected at Plant Materials Center, SCS, Bridger, MT - A.A. Thornburg, J.R. Stroh, and J.G. Scheetz. Carried as accession nos. P-15588, WY-17, PI 433949.

Source - Commercial harvest from native stand near Torrington, WY.

Method of Breeding - Direct increase of field collection.

Intended Use - Goshen is used primarily for stabilization and range revegetation on sandy soils.

Description - A leafy ecotype with excellent seed production. Summer active, late maturing. Basal leaves. Mildly rhizomatous, drought tolerant. Well adapted to sandy sites receiving more than 300 mm of annual precipitation.

Adapted to - LRR D, G, H; PHZ 4.

Released - 1976, cooperatively by Plant Materials Center, SCS, Bridger, MT, and Montana and Wyoming Agricultural Experiment Stations.

Breeder Seed/Stock - Plant Materials Center, SCS, Bridger, MT.

Certified Seed/Stock - Available.

Preparer/Additional Information - John G. Scheetz, Plant Materials Center, SCS, RR 1 Box 1189, Bridger, MT 59014, (406) 662-3579.

ND-95

Selected at the Plant Materials Center, SCS, Bismarck, ND - John McDermand, Erling T. Jacobson and Russell J. Haas. Carried as accession no. PI 477995.

Source - Seed from a field collection made in the fall 1956 in southwestern North Dakota (Bowman County).

Method of Breeding - Selection based on performance in comparison to other field collections at the Bismarck Plant Materials Center.

Intended Use - For use in range and pasture seedings and stabilization of critical areas. It may be seeded in areas to provide upland game bird cover and nesting sites. Provides good winter standing feed for grazing animals and wildlife.

Description - A perennial, native, warm-season, leafy, northern type with good vigorous growth. It normally matures seed at Bismarck with supplemental irrigation. Seed production is average for this species. Forage production is comparable to Goshen. Its dense and wiry root mass is well adapted to binding sandy soils.

Adapted to - LRR F, G; PHZ 3a - 4b.

Released - No. (Seed only available for field testing.)

Breeder Seed/Stock - Plant Materials Center, SCS, Bismarck, ND.

Certified Seed/Stock - Not available.

Preparer/Additional Information - Russell J. Haas, SCS, P.O. Box 1458, Bismarck, ND 58502, (701) 250-4425.

Pronghorn

Selection of four accessions at the SCS Plant Materials Center, Manhattan, KS, and ARS, Lincoln, NE. Carried as accession no. 9049969.

Source - Five collections, one each from Rock, Greeley, Howard and Boone Counties in Nebraska and Republic County, Kansas.

Method of Breeding - Forty-eight field collections of prairie sandreed from NE, KS, and SD were evaluated at the SCS Plant Materials Center at Manhattan, KS. Four of the accessions had superior stands, vigor, and disease resistance and were sent to the ARS grass breeding program at Lincoln, NE. Space-planted evaluation nurseries of these accessions and an ARS accession were established at Lincoln. Plants with superior culm number, leafiness, spread by rhizomes, and rust tolerance were selected for polycrossing from the Boone (48), Greeley (16), and Howard (72) populations. Clonal pieces of each selected plant (numbers in parentheses) were transplanted at random into three different isolations or crossing blocks and were allowed to cross via wind pollination. There were six different crossing blocks: Howard x Boone, Howard x Greeley, Howard x Howard, Greeley x Boone, Greeley x Greeley, and Boone x Boone. Progeny from Boone plants in each of the isolations were established in a space-planted unreplicated, progeny row evaluation nursery. The same procedure was followed for the Greeley and Howard populations. These nurseries contained from 1,000 to 3,000 plants which were evaluated for vigor, forage production, and rust resistance. Individual plants - Boone (180), Greeley (147), and Howard (180) - were selected from these nurseries and moved to separate, isolated crossing blocks. Seed was harvested from these isolations on a bulk basis. One thousand seedlings were grown from each of these bulk seed lots and were transplanted at random into rows to establish the breeder seed field in 1980.

Intended Use - Range and pasture plantings.

Description - Pronghorn is a heterogeneous cultivar with a broad genetic base that has a high degree of rust tolerance. It is adapted to and recommended for use in revegetating sandy sites in the Nebraska sandhills and northwest Kansas.

Adapted to - LRR northern half G & H sandy soils; PHZ 4, 5.

Released - 1988 - Cooperatively by ARS, SCS, and the Nebraska AES.

Breeder Seed/Stock - Nebraska AES and USDA-ARS, Univ. of Nebraska, Lincoln.

Certified Seed/Stock - Available.

Preparer/Additional Information - K.P. Vogel, ARS-USDA, 344 Keim Hall, Univ. of Nebraska or Jeff Pederson, Dept. of Agronomy, Univ. of Nebraska, Lincoln, NE 68583, (402) 472-1564, (402) 472-2811.

Chloris gayana Kunth - Rhodes grass

Warm-season, sod-forming grass introduced from Africa in 1902. Used for pasture and hay in southern Texas, to limited extent elsewhere along Gulf Coast to Florida, and under irrigation in southern Arizona and California. Although a valuable forage species that tolerates saline or alkaline conditions, distribution and use restricted by lack of winter hardiness and susceptibility to Rhodes grass scale, *Antonina graminis* Mask.

Bell

Selected at Texas A & M University Research and Extension Center, Weslaco, TX.

Source - Plant collections from old stands in southern Texas and plant introductions.

Method of Breeding - Individual plants with a high degree of tolerance to Rhodes grass scale, based on evaluations over a period of years, were bulked and increased.

Intended Use - Pasture production.

Description - Warm-season grass; upright, vigorous plants. High degree of tolerance to Rhodes grass scale as demonstrated by persistence under scale infestation.

Adapted to - LRR I, J, T; PHZ 9.

Released - 1966, by Texas AES.

Breeder Seed/Stock - Texas AES.

Certified Seed/Stock - Not available.

Preparer/Additional Information - Mark Hussey, Texas AES, Soil and Crop Science Department, College Station, TX 77843, (409) 845-3041.

Cynodon L.C. Rich. - bermudagrasses

Cynodon dactylon (L.) Pers., bermudagrass, is a major warm-season, sod-forming grass introduced from Africa in 1751 or earlier. Used for pasture, hay, lawns, general-purpose turf, and erosion control. Best adapted to relatively fertile soil in humid southern states, but found as far north as Maryland and southern part of central corn-belt states. Giant bermudagrass found in irrigated areas in southwestern U.S. appears to be diploid form of *C. dactylon.* Distinguished from common bermudagrass by greater vigor and lack of pubescence. *C. plectostachyus* (K. Schum.) Pilger, stargrass, is warm-season, stoloniferous grass from Africa. Robust, pubescent, and non-hardy. Used to a limited extent in southern Texas and southwestern U.S.. Several other species introduced for turf purposes, including *C. transvaalensis* Burtt-Davy, transvaalensis or floridagrass, *C. magennisii* Hurcombe, and *C. bradleyi* Stent. Interspecific hybridization, as noted in following descriptions, has been an important factor in the development of improved varieties.

Brazos

Carried as accession no. PI 464656.

Source - Brazos bermudagrass is an F_1 hybrid (Guymon X 9958) x (X-820). Method of Breeding - Increase of the F_1 hybrid.

Intended Use - Pasture and hay production.

Descriptions - Brazos is slower to become established than Coastal and produces less forage than Coastal until well established after which forage production is equal on heavy soils and 0-20% less on sandy soils. A one-year comparison with Tifton 44, an improved winter-hardy cultivar, showed Brazos to be superior in animal performance. The advantages over Coastal, the bermudagrass standard, are improved forage quality (dry matter digestability) and increased gain per animal. Brazos is equal or superior to coastal in stand density persistence under grazing and winter hardiness. The cultivar has larger leaves, stems and rhizomes than Coastal. The primary use of Brazos is expected to be grazing - the larger stems would not favor hay drying, but they do resist lodging in contrast to Coastal.

Adapted to - South from the Brazos River. It has survived winters satisfactorily in northwest Louisiana, north Texas and southern Oklahoma.

Released - 1982 by Texas AES in cooperation with SCS, ARS, and the Louisiana AES.

Breeder Seed/Stock - Available.

Certified Seed/Stock - Available.

Preparer/Additional Information - Soil and Crop Sciences Department, Texas AES, Texas A & M Univ., College Station, TX 77843, (409) 845-3041.

C2

Source - Parent "A" selected from a wild population in Arizona. Parent "B" selected from Arizona certified.

Method of Breeding - Topcrossing Parent "A" x Parent "B". Parent "A" (moderately self-incompatible) produces the varietal seed of C2.

Intended Use - Turf in heavy alkaline soils. Outstanding for heavy traffic areas. Planted from seed.

Description - Well adapted to highly alkaline soils and heavy traffic areas. Wide leaf blade and short internode make dense turf. C2 is very aggressive and takes over void areas rapidly. C2 is planted from seed.

Adapted to - All areas south of 36°N latitude and below 300 m elevation.

Released - 1988, D. Palmer Seed Co., Inc.

Breeder Seed/Stock - D. Palmer Seed Company, Inc.

Certified Seed/Stock - Not available, sold as common.

Preparer/Additional Information - D. Palmer Seed Company, Inc., 12466 E. Via Feliz, Yuma, AZ 85365, (602) 342-2838.

Cheyenne

Source - The cross of two infertile, self-incompatible clones, one from an old turf site in the Pacific Northwest, and the other from an open-pollinated progeny of U.S. PI 253302 (from former Yugoslavia).

Method of Breeding - Progeny from this cross were propagated in spaced plant nursery in Oregon in 1987. Five superior clones were selected. Polycross progeny of these clones were interplanted in alternating rows with the original clones in a breeders nursery in Arizona. Plants exhibiting morphology unlike the variety (6% of population) were removed prior to anthesis.

Intended Use - Turf and reclamation.

Description - Cold tolerant, leafy, dark green, seed-propagated bermudagrass, with extended green growth in early spring and fall.

Adapted to - PHZ 7, 8, 9, 10, 11 - SE and SW U.S.

Released - 1989, Jacklin Seed Co. and Pennington Seed.

Breeder Seed/Stock - Jacklin Seed Co.

Certified Seed/Stock - Available.

Preparer/Additional Information - Kim Peterson, Jacklin Seed Co., W. 5300 Riverbend Ave., Post Falls, ID 83854, (208) 773-7581 and David Lundell, Fine Lawn Research, Inc., P.O. Box 1051, Lake Oswego, OR 97034, (503) 636-2600.

Coastal (Reg. No. 1)

Selected at Georgia Coastal Plain Experiment Station, Tifton, ARS cooperating - G.W. Burton.

Source - F_1 hybrid between Tift bermudagrass (discovered by J.L. Stephens in old cotton patch near Tifton in 1929) and tall-growing introduction from Union of South Africa.

Method of Breeding - Parents interplanted to allow for maximum natural crossing. Over 5,000 seedling plants carefully screened for many traits. Few of best clones subjected to numerous replicated tests giving measures of their palatability, efficiency, yield potential, management requirements, production under grazing, etc. Tested as selection 35.

Intended Use - Grazing and hay.

Description - When compared with common bermudagrass, Coastal has larger and longer stems, stolons, and rhizomes; grows much taller; is lighter green; has deeper and more efficient root system; is more resistant to foliage diseases, root knot nematode, frost, and drought; is much more efficient in nutrient and water use; is more palatable and produces nearly twice as much forage and animal products. This superiority holds throughout most of bermudagrass belt, demonstrating wide adaptation. Produces few seed heads that rarely contain viable seed; must be propagated vegetatively.

Adapted to - LRR I, J, O, P, T, U; PHZ 8, 9.

Released - 1943, cooperatively by Georgia Coastal Plain Experiment Station and Plant Science Research Division, ARS.

Breeder Seed/Stock - Georgia Coastal Plain Experiment Station.

Certified Seed/Stock - Available.

Preparer/Additional Information - ARS Coastal Plain Experiment Station, P.O. Box 748, Tifton, GA 31793, (912) 386-3353.

Coastcross-1

Selected at Georgia Coastal Plain Experiment Station, Tifton, ARS cooperating - G.W. Burton.

Source - Coastal bermudagrass and PI 255445, obtained from A.V. Bogdan, Grassland Research Station, Kital, Kenya, in 1958.

Method of Breeding - Thousands of controlled hybrids attempted in the greenhouse between the self-incompatible parents; 381 seedlings produced. These were space-planted in droughty, deep sand and were carefully screened for many traits. Ten of the best selections included in replicated tests, giving measures of yield, digestibility (NBDMB), winter hardiness, management requirements, and quality when fed to cattle.

Intended Use - Grazing and hay.

Description - A completely sterile F_1 hybrid between Coastal bermudagrass and PI 255445, previously described as Coastal X Kenya #14; grows taller and has broader, softer leaves than Coastal; highly resistant to foliage diseases and the sting nematode; above ground stolons spread rapidly, with few, if any, rhizomes. Its forage is about 11-12% more digestible than Coastal bermuda in comparisons using nylon-bag technique. Cattle fed chopped hay or grazing Coastcross-1 have made up to 30% better daily gains than on Coastal bermuda. Coastcross-1 is less winter-hardy than Coastal bermuda and is not recommended north of an isotherm going through Macon, GA.

Adapted to - LRR T, U; PHZ 9, 10.

Released - 1967, cooperatively by Georgia Coastal Plain Experiment Station and Plant Science Research Division, ARS.

Breeder Seed/Stock - Georgia Coastal Plain Experiment Station.

Certified Seed/Stock - Available.

Preparer/Additional Information - ARS, Coastal Plain Experiment Station, P.O. Box 748, Tifton, GA 31793, (912) 386-3353.

Florico

University of Florida Institute of Food and Agricultural Science, Agricultural Research and Education Center, Ona, FL - Drs. Mislevy, Brown, Caro-Costas, Vicente-Chandler, Dunavin, Kalmbacher, Hall, Overman, Ruelke, Sonoda, Sotomayor-Rios, Stanley, Jr., and Williams. Carried as accession no. Puerto Rico 2341.

Source - Collected in 1957 by Puerto Rico Plant Breeding Dept. in Kenya, Central Africa.

Method of Breeding - Selected from material evaluated in Puerto Rico and subsequently brought to Ona, FL, in 1972 by A. Sotomayor-Rios where it was extensively evaluated by P. Mislevy and others. Previously denoted as Puerto Rico stargrass (PR 2341).

Intended Use - A perennial pasture grass adapted to south-Florida flatwoods soils which are saturated but not flooded for long periods of time.

Description - Compared to Ona stargrass, it has a higher *in-vitro* organic matter digestibility, better persistence, and good average daily gain and carry capacity. Nutritious when harvested or grazed every four to five weeks. Forage quality drops rapidly after six weeks of regrowth, and following a heavy frost. Top growth easily killed by frost. Matures rapidly. Vegetatively propagated from stem cuttings. Carries a high hydrocyanic acid potential especially when high levels of nitrogen are applied. It is a dark green pubescent perennial grass with long robust stolons. Susceptible to armyworms and grass loopers. Hay cures rapidly during favorable weather conditions. Requires a higher fertility program than bahiagrass, hemarthria, or digitgrass.

Adapted to - LRR U; PHZ 10.

Released - 1988 by University of Florida Institute of Food and Agricultural Science, Agricultural Research and Education Center, Ona, FL, USDA-ARS, TARS (Puerto Rico).

Breeder Seed/Stock - Florida Foundation Seed Producers, Inc., Greenwood, FL, and Agricultural Research and Education Center, Ona, FL.

Certified Seed/Stock - Not available.

Preparer/Additional Information - Florida AES, Agricultural Research and Education Center, Ona, FL 33865-9706, (813) 735-1314. (Circular S-361.)

Florona

University of Florida nstitute of Food and Agricultural Science, Agricultural Research and Education Center, Ona, FL - Drs. Mislevy, Brown, Overman, Sonoda, Dunavin, Hall, Kalmbacher, Ruelke, Stanley, Jr., and Williams.

Source - Origin uncertain, found in Pensacola bahiagrass planting in a former sugar cane patch.

Method of Breeding - Selected single accession. Informally denoted as Canepatch stargrass.

Intended Use - A perennial pasture grass adapted to south Florida flatwoods soils which are saturated but not flooded for long periods of time.

Description - Florona is a perennial grass which spreads by stolons forming a moderately open sod with clumps of erect stems giving a bunch effect. Higher yielding than Ona stargrass. Dry matter production is slightly higher than Ona or Florico stargrasses but forage digestibility is slightly lower. Matures rapidly and requires an intensive grazing frequency with intervals not exceeding four or five weeks. Forage quality drops rapidly after six weeks of growth and following a heavy frost. Top growth easily killed by frost. Vegetatively propagated from stem cuttings. Rapid establishment from vegetative cuttings. Carries a high hydrocyanic acid potential, especially when high levels of nitrogen are applied. Hay cures rapidly during favorable weather conditions.

Adapted to - LRR U; PHZ 10.

Released - 1988 by University of Florida Institute of Food and Agricultural Science, Agricultural Research and Education Center, Ona, FL.

Breeder Seed/Stock - Florida Foundation Seed Producers, Inc., Greenwood, FL, and Agricultural Research and Education Center, Ona, FL.

Certified Seed/Stock - Not available.

Preparer/Additional Information - Florida AES, Agricultural Research and Education Center, Ona, FL 33865-9706, (813) 735-1314. (Circular S-362)

Greenfield

Selected at Oklahoma AES, Stillwater - W.C. Elder.

Source - Selected from among a large number of common strains collected from various Oklahoma locations and planted on the Stillwater Agronomy Farm in 1948.

Method of Breeding - Single plant selection.

Intended Use - Pasture.

Description - Intermediate between coarse and very fine types of common bermudagrass. Exposed stolons purple; rhizomes short, crooked, numerous, forming dense mat. Winterhardy. Propagated vegetatively. Has good establishment capabilities and produces a dense sod. Has less forage production potential than Midland and similar cultivars.

Adapted to - LRR H, J, N; PHZ 6, 7.

Released - 1954, by Oklahoma AES.

Breeder Seed/Stock - Oklahoma AES.

Certified Seed/Stock - Commercially available, but little or none is certified.

Preparer/Additional Information - C.M. Taliaferro, Agronomy Department, 368 Ag Hall, OSU, Stillwater, OK 74078-0507, (405) 774-6410.

Guymon (Reg. No. 13)

Selected at Oklahoma State University, Stillwater, OK - C.M. Taliaferro. Carried as accession nos. 180167 and/or 183630.

Source - Guymon is a two-parental clone synthetic cultivar derived from interpollination of winter-hardy, self-incompatible accessions 9959 (PI 253302) and 12156. Accession 9959 was introduced from Yugoslavia. Accession 12156 was collected near Guymon, OK.

Method of Breeding - The two parental plants were selected on the basis of their ability to produce good seed yields when grown in mixture and cold-tolerant progeny populations.

Intended Use - A general purpose, seed-propagated, cold-tolerant variety for soil stabilization and erosion control on lawns, playgrounds, roadsides, and similar areas. Also suitable for pasture use, but produces less forage than pasture type cultivars such as Midland or Tifton 44.

Description - The outstanding feature of Guymon relative to seeded common bermudagrass originating in the Yuma and Imperial Valleys of Arizona and California, respectively, is substantially greater cold tolerance. The Syn-1 and Syn-2 progeny populations of Guymon are heterogeneous due to the heterozygosity of the non-inbred parental clones. The progeny populations retain the cold tolerance of their parents and produce a winter-hardy sod of acceptable quality for use as general purpose turf and soil stabilization.

Adapted to - LRR H, N, P, J, Q; PHZ 6, 7.

Released - 1982, Oklahoma AES.

Breeder Seed/Stock - Oklahoma AES.

Certified Seed/Stock - Available.

Preparer/Additional Information - C. M. Taliaferro, Agronomy Department, 368 Ag Hall, Oklahoma State Univ., Stillwater, OK 74078, (405) 744-6410.

Hardie (Reg. No. 11)

Selected at Oklahoma State University, Stillwater, OK - C.M. Taliaferro.

Source - Selected in 1968 as one of the best of several F_1 hybrids from crosses between Oklahoma *Cynodon* L.C. Rich. accessions 9945A x (8153 X 9953). Accession 9945A (PI 206427, *C. dactylon* (L.) Pers. var. *dactylon,* was introduced from Elazig, Turkey. Accessions 8153 and 9953, *C. dactylon* var. *afghanicus,* are variant plants of PI 223248 from Khanadad, Afghanistan.

Method of Breeding - See source.

Intended Use - Pasture and hay production.

Description - Hardie has higher forage quality than Midland. The yield potential of the two is about the same. Hardie has given better individual animal performance and produced more gain/acre/year than Midland in Oklahoma. Hardie's greatest advantage is usually realized during the first four to six weeks of the growing season both for individual animal and total weight gains. Hardie is susceptible to leaf spotting disease. Establishment difficulties may be somewhat more prevalent with Hardie compared to Midland. Hardie is an infertile, vegetatively propagated, clonal plant that grows taller and has larger rhizomes and stems and longer and broader leaves than Midland. The leaves of Hardie tend to accumulate anthocyanin pigmentation during periods of cool spring and fall weather giving it a distinct appearance from other bermudagrass cultivars during those periods. Hardie tends to produce a more open sod than Midland.

Adapted to - LRR H, N, J, P, T, U; PHZ 6, 7.

Released - 1974, Oklahoma AES.

Breeder Seed/Stock - Oklahoma AES.

Certified Seed/Stock - Available.

Preparer/Additional Information - C.M. Taliaferro, Agronomy Dept, Oklahoma State Univ., 368 Ag Hall N., Stillwater, OK 74078-0507, (405) 744-6410.

McCaleb

University of Florida, Institute of Food and Agricultural Sciences, Agricultural Research Center, Ona, FL - Hodges, Boyd, Dunavin, Kretschmer, Mislevy, and Stanley. Carried as PI 224152.

Source - Collected as vegetative material in 1955 at Frankwald, Johannesburg, South Africa by J.L. Stevens, USDA, Tifton, GA.

Method of Breeding - Single clone selected from a collection of 39 accessions established at Ona, FL, Agricultural Research Center.

Intended Use - A perennial forage grass adapted to south Florida flatwoods soils.

Description - Establishes rapidly with high forage production under high fertilization. Resists invasion by other grasses when grazed rotatively. One of first cultivars Released -d and subsequent Released -s have mainly been improvements using McCaleb as a standard.

Adapted to - LRR U; PHZ 10.

Released - 1975, University of Florida Institute of Food and Agricultural Science, Agricultural Research and Education Center, Ona, FL.

Breeder Seed/Stock - Florida Foundation Seed Producers, Inc., Greenwood, FL.

Certified Seed/Stock - Not available.

Preparer/Additional Information - Univ. of Florida, Agronomy Department, Gainesville, FL 32611, (904) 392-1814.

Midfield

Selected at Ft. Hays Branch Experiment Station, Kansas State University - Ray A. Keen. Tested under the experimental designation E-29.

Source - Presumed F_1 hybrid between cold tolerant C. *dactylon* and C. *transvaalensis* parental accessions.

Method of Breeding - Selection within large polycross progeny population. The seed parent of Midfield is a cold-hardy common bermudagrass collected in the vicinity of Hays, KS, and grown in a polycross nursery with other bermudagrass accessions. Midfield is presumed to be an F_1 hybrid of the Hays accession and one of several African bermudagrasses *(C. transvaalensis)* growing in the polycross nursery.

Intended Use - Transition zone turf on sites such as sports fields and parks.

Description - A vegetatively-propagated sterile triploid with 2n=3x=27 chromosomes. Cold-hardy, moderate to fine texture, aggressive growth habit, and wear-resistant, making it especially suitable for sports turf. Leaf color is lighter green than Midiron and Midlawn under equivalent management. Pubescence is generally lacking, although hairs occur very sparsely on the adaxial surface of leaves, appearing with somewhat greater frequency near the base. Sod strength is far superior to Midiron. Good resistance to spring dead spot disease.

Adapted to - PHZ 6, 7.

Released - 1991, Kansas and Oklahoma Agricultural Experiment Stations. Plant Patent No. 08168. Licensed to selected growers through the Kansas State University Research Foundation.

Breeder Seed/Stock - Kansas State University.

Certified Seed/Stock - Available - Kansas and Oklahoma.

Preparer/Additional Information - J.C. Pair, Horticulture Research Center, 1901 East 95th South, Wichita, KS 67233-8351, (316) 788-0492.

Midiron

Selected at Kansas State University, Manhattan, KS - Ray A. Keen.

Source - Harold L. Haekrott, Ft Hays Br. Kansas AES.

Method of Breeding - Polycross Nursery. HK-12 bermudagrass collected in N. Platte, NE. P-16 is reported to be a triploid by Dr. F. Juska, ARS, Beltsville.

Intended Use - Turf.

Description - Bright to dark green, loses color early in fall, medium coarse, associated with early dormancy, fast spreading. Most wear-resistant of winter-hardy bermudagrasses at Beltsville, MD, exceeding Tufcote and U-3 in cold hardiness and traffic tolerance.

Adapted to - LRR H; PHZ 6.

Released - 1971, Kansas AES.

Breeder Seed/Stock - Kansas AES.

Certified Seed/Stock - Available - Kansas, Missouri, and Virginia Nurseries.

Preparer/Additional Information - John C. Pair, Horticulture Research Station, Kansas State Univ., 1901 East 95th St. South, Wichita, KS 67233-8351, (316) 788-0492; or C.M. Taliaferro, Agronomy Dept, 368 Ag Hall N., Oklahoma State Univ., Stillwater, OK 74078-0507, (405) 744-6410.

Midland

Selected at the Georgia Coastal Plain Experiment Station, ARS cooperating - G.W. Burton, tested in Oklahoma by W.C. Elder and J.R. Harlan.

Source - F_1 hybrid between cold-resistant common bermudagrass from Indiana, supplied by G.O. Mott, and Coastal bermudagrass.

Method of Breeding - In 1942 enough seed of cross - Indiana bermudagrass X Coastal bermudagrass - made to give 66 F_1 hybrids evaluated for many characteristics beginning in 1943. Selection 13 was most productive and more cold-resistant than Coastal, surviving two winters at Lafayette, IN, where Coastal bermudagrass winterkilled.

Intended Use - Pasture.

Description - Taller, larger, leafier, more disease-resistant, producing more open sod than common bermudagrass. Superior to common bermudagrass in most of good traits that characterize Coastal. Midland (selection 13) less productive than Coastal where latter does not suffer stand loss because of winter-injury. Darker green, tends to produce more heads, starts growth earlier in spring than Coastal. Superiority over

Coastal in tests at Stillwater, OK, led to its release in that state. Recommended for northern part of Bermudagrass belt.

Adapted to - LRR H, J, N, O, P; PHZ 6, 7.

Released - 1953, cooperatively by Oklahoma AES, Stillwater Georgia Coastal Plain Experiment Station and Plant Science Research Division, ARS.

Breeder Seed/Stock - Georgia Coastal Plain Experiment Station. Foundation stock maintained at Oklahoma AES.

Certified Seed/Stock - Commercially available in quantity, but little or none is certified.

Preparer/Additional Information - C. M. Taliaferro, Agronomy Dept, 368 Ag Hall N., Oklahoma State Univ., Stillwater, OK 74028-0507, (405) 744-6410.

Midlawn

Selected at Ft. Hays Branch Experiment Station, Kansas State University - Ray A. Keen. Tested under the experimental designation A-22.

Source - Presumed F_1 hybrid between cold-tolerant *Cynodon dactylon* and *C. transvaalensis* parental accessions.

Method of Breeding - Selection within large polycross progeny population. The seed parent of Midlawn is a common bermudagrass collected from the Michigan State University campus and grown in a polycross nursery with other bermudagrass accessions. Midlawn is presumed to be an F_1 hybrid of the Michigan accession and one of several African bermudagrasses *(C. transvaalensis)* growing in the polycross nursery.

Intended Use - Transition zone turf on sites such as lawns and golf courses.

Description - Midlawn is a vegetatively-propagated sterile triploid (2n-3x=27 chromosomes). A cold-hardy, fine-textured, dark green cultivar with excellent turf qualities. Growth habit is less aggressive than most transition zone adapted turf bermudagrass varieties. Leaf color is usually darker green than Midfield or Midiron under equivalent management conditions. Pubescence is generally lacking, although hairs may occur very sparsely on the adaxial surface, appearing with somewhat greater frequency near the base. Good resistance to spring dead spot. Superior sod strength compared to Midiron.

Adapted to - PHZ 6, 7.

Released - 1991, Kansas and Oklahoma Agricultural Experiment Stations. Plant Patent No. 08162. Licensed to selected growers through the Kansas State University Research Foundation.

Breeder Seed/Stock - Kansas State University.

Certified Seed/Stock - Available - Kansas and Oklahoma.

Preparer/Additional Information - J.C. Pair, Horticulture Research Center, 1901 East 95th South, Wichita, KS 67233-8351, (316) 788-0492.

Midway

Selected at Kansas AES, Manhattan - R.A. Keen.

Source - Hybrid between *C. transvaalensis* and *C. dactylon.*

Method of Breeding - Single plant selected for winter-hardiness and desirable turf characteristics from over 20,000 polycross progenies.

Intended Use - Turf.

Description - A sterile triploid. Medium-textured lawn grass that produces relatively few seedheads; ascending habit contributes to slower development of thatch; more winter-hardy in Kansas than U-3; and rated as tolerant to mites and leaf spot in Arizona tests. Adapted for lawn use in southwestern Kansas.

Adapted to - LRR H; PHZ 6.

Released - 1965, by Kansas AES.

Breeder Seed/Stock - Kansas AES.

Certified Seed/Stock - Not available.

Preparer/Additional Information - J.C. Pair, Horticulture Research Center, 1901 East 95th South, Wichita, KS 67233-8351, (316) 788-0492; or C.M. Taliaferro, Agronomy Dept., 368 Ag Hall N., Oklahoma State Univ., Stillwater, OK 74078-0507, (409) 744-6410.

NuMex Sahara

Department of Agronomy & Horticulture, New Mexico State University - Arden A. Baltensperger; partial financial support from US Golf Association. Exclusively released to Farmers Marketing Corp.

Source - Selected tetraploid clones from a broad geographic area.

Method of Breeding - Genotypic and phenotypic selection toward development of a synthetic variety.

Intended Use - General purpose turfgrass for golf course fairways, parks, athletic fields, and home lawns where seeding bermudagrass is the preferred method of establishment.

Description - NuMex Sahara is a seed-propagated cultivar that has shorter internodes than Common and develops a turf that generally has greater density and increased green summer color than Common.

Adapted to - LRR I, J, P, T, O, U, N; PHZ 7, 8, 9, 10 but will often winterkill in zone 6. Released - February 1987, by the New Mexico AES.

Breeder Seed/Stock - Farmers Marketing Corp.

Certified Seed/Stock - Available.

Preparer/Additional Information - Dr. Arden A. Baltensperger, Dept. of Agronomy and Horticulture, New Mexico State Univ., Box 3Q, Las Cruces, NM 88003, (505) 646-3138.

Ona

University of Florida Institute of Food and Agricultural Science, Agricultural Research and Education Center, Ona, FL - Hodges, Mislevy, Ruelke, Stanley. Carried as PI 224566.

Source - Obtained from the Tobacco Institute, Salisbury, Rhodesia, in 1955 by J. Stevens, USDA.

Method of Breeding - Selected from nursery planting of 39 grasses due to evidence of ability to compete vigorously in ungrazed area.

Intended Use - A perennial pasture grass adapted to south Florida flatwoods soils which are saturated but not flooded for long periods of time.

Description - Higher yielding during the cool season than Pensacola bahiagrass or Pangola digit grass. It has good disease resistance and insect tolerance and an acceptable nutrient content. Vegetatively propagated. Carries a high hydrocyanic acid potential, especially when high levels of nitrogen are applied.

Adapted to - LRR U; PHZ 10.

Released - 1979; University of Florida Institute of Food and Agricultural Science, Agricultural Research and Education Center, Ona, FL.

Breeder Seed/Stock - Florida Foundation Seed Producers, Inc., Greenwood, FL.

Certified Seed/Stock - Not available.

Preparer/Additional Information - Florida AES, Ona Agricultural Research and Education Center, Ona, FL 33865-9706, (813) 735-1314.

Pee Dee 102

Increased at the Pee Dee Experiment Station, Florence, SC.

Source - Mutant from early planting of Tifgreen.

Method of Breeding - Increase of mutant clone.

Intended Use - Turf.

Description - Darker green; basic plant color turns purple with near-freezing temperature; smaller leaves and stems, and shorter internodes than Tifgreen. Tolerates closer mowing and makes a better putting surface than Tifgreen but otherwise similar to Tifgreen.

Adapted to - LRR J, O, P, T, U; PHZ 8.

Released - 1968, by South Carolina AES, Clemson.

Breeder Seed/Stock - South Carolina Foundation Seed Association, Clemson.

Certified Seed/Stock - Available.

Preparer/Additional Information - ARS Coastal Plain Experiment Station, P.O. Box 748, Tifton, GA 31793, (912) 386-3353.

Primavera

Arden A. Baltensperger. Released by Farmers Marketing Corp, Phoenix, AZ.

Source - Selected tetraploid clones from a broad geographic area.

Method of Breeding - Genotypic selection and recurrent phenotypic selection. An seven-clone synthetic cultivar.

Intended Use - General-purpose turfgrass for golf course fairways, parks, athletic fields, and home lawns where seeding bermudagrass is the preferred method of establishment.

Description - Primavera is a seed-propagated cultivar that has shorter stem internodes and narrower stem diameters than NuMex Sahara.

Adapted to - LRR I, J, P, T, O, U, N; PHZ 7, 8, 9, 10, but may winterkill in zone 6.

Released - 1989, by Farmers Marketing Corp.

Breeder Seed/Stock - Farmers Marketing Corp.

Certified Seed/Stock - Available.

Preparer/Additional Information - Dr. Arden A. Baltensperger, Dept. of Agronomy & Horticulture, New Mexico State Univ., Box 3Q, Las Cruces, NM 88003, (505) 646-3138.

Quicksand

Developed and selected at the Quicksand Plant Materials Center, SCS and University of Kentucky, Robinson Substation, Quicksand, KY; UK AES, Lexington, KY; and ARS Appalachian Soil and Water Conservation Research Laboratory, Beckley, WV - Charles F. Gilbert, Harold B. Rice, A.J. Powell, and H. Douglas Perry. Also known informally as Quicksand common bermudagrass, RS-1 bermudagrass, T34348, and 9034348. Now PI 557553.

Source - Field No. 2 of Plant Materials Center, Quicksand, KY. Thought to have been brought into Quicksand during logging operations of the old sawmill days at Quicksand in the 1920s, or with some hay during the Robinson Substation Harvest Festivals in the 1920s. Occurs in other areas on the Substation and Center as well.

Method of Breeding - Vegetative increase of original material obtained from Field No. 2.

Intended Use - Turf and heavy recreational area use and for pasture forage. Cover and forage plant on low-fertility hillside pastures.

Description - Vigorous, productive (at low fertility and nitrogen levels), extremely winter-hardy (-31°C without snow cover at Cool Ridge, WV), drought tolerant and heat resistant, dense-growing, medium to fine textured, leafy bermudagrass. Starts growth at cooler temperatures (19°C days and 12°C nights) than other strains (27°C days and 21°C nights). Initiates spring growth several weeks earlier than Tifton-44. More winter-hardy than Tifton-44, Midland, and Hardy at Quicksand, KY, Beckley, WV, and Morgantown, WV. Not as tall as true hay-types nor is it as short as the sod-types.

Adapted to - LRR C, D, H, I, J, M, N, O, P, R, S, T, U; PHZ 6, 7, 8.

Released - 1993.

Breeder Seed/Stock - Plant Materials Center, SCS, Quicksand, KY, and University of KY, AES, Lexington, KY.

Certified Seed/Stock - Not available.

Preparer/Additional Information - Laura Ray, Plant Materials Center, SCS, 175 Robinson Dr., Quicksand, KY 41363-9008, (606) 666-5069.

Royal Cape

Source - Original planting stock obtained by J.L. Stephens at Frankenwald Experiment Station, Johannesburg, Union of South Africa, on March 3, 1955. Represents direct increase of Royal Cape selected by C.M. Murray on Royal Cape Golf Course near Mowbray, Cape Province, Union of South Africa, in 1930. Introduced as PI 224147. Previous collection of Royal Cape, identified as PI 213387, received from Union of South Africa in February 1954.

Method of Breeding - Compared with named varieties and other introductions in salt basins and under mowing.

Intended Use - Turf.

Description - Dark-green, fine-leafed variety. Forms dense, wear-resistant sod and tolerates high salt concentrations. In Imperial Valley, CA, it remains green well into winter and starts growth very early in spring. Produces few, if any, seed heads in spring and none during remainder of year. Disease has not been a problem in Imperial Valley, but variety is susceptible to some leaf diseases in humid southeastern U.S..

Adapted to - High salt areas of southern California.

Released - 1960, cooperatively by University of California, Los Angeles, and Plant Science Research Division, ARS.

Breeder Seed/Stock - Southwestern Irrigation Field Station.

Certified Seed/Stock - Not available.

Preparer/Additional Information - University of CA, Los Angeles, CA 90024, (213) 825-4321.

Santa Ana

Selected by University of California, Riverside, CA - V.B. Younger.

Source - PI 213387 from Union of South Africa.

Method of Breeding - Seedling selected at UCLA in 1956 from PI 213387. Testing at locations throughout California and in several other states as RC-145.

Intended Use - Fine turf. Description - A triploid (2n=Ca 27) that does not produce viable seed as a rule. Characterized by deep blue-green color, medium-fine texture, good color retention in cool weather, and early spring growth; with good maintenance has maintained good color throughout winter in mild frost-free areas of California. Exhibits excellent tolerance to smog and to *Eriophyid* mite, and rates above average in tolerance to soil salinity. Rapid establishment;, produces smooth, even surface; and resistant to heavy foot traffic.

Released - 1966, by California AES.

Breeder Seed/Stock - California AES, Riverside.

Certified Seed/Stock - Available.

Preparer/Additional Information - University of CA, Los Angeles, CA 90024, (213) 825-4321.

Sonesta

Exclusively released as part of a germplasm release to Farmers Marketing Corp., Phoenix, AZ. Dr. Arden A. Baltensperger Subsequently sold to O.M. Scott & Sons Co. for sole ownership, naming, and production. Carried as experimental number NM-S-3.

Source - Selected tetraploid clones from a broad geographic area.

Method of Breeding - Recurrent phenotypic selection. A six-clone synthetic cultivar.

Intended Use - General purpose turfgrass for golf course fairways, parks, athletic fields, and home lawns where seeding bermudagrass is the preferred method of establishment.

Description - Sonesta is a seed-propagated cultivar that has shorter stem internodes, narrower leaf width and shorter leaf length than common. Sonesta generally develops a denser turf than NuMex Sahara or common.

Adapted to - LRR I, J, P, T, O, U, N; PHZ 7, 8, 9, 10 and where well adapted. It may sometimes winter-kill in zone 6.

Released - 1992 by O.M. Scott & Sons Co.

Breeder Seed/Stock - Farmers Marketing Corp., Phoenix, AZ.

Certified Seed/Stock - Available.

Preparer/Additional Information - Virgil Meier, O.M. Scott & Sons Co., Marysville, OH 43041, (513) 644-0011.

Sundevil

Source - Sundevil originated from an open-pollinated selection from plant intro. 20667 from Turkey.

Method of Breeding - Evaluated in a spaced plant nursery in Albany, OR, in 1987. Was evaluated for turf suitability and cold tolerance.

Intended Use - Turf and reclamation.

Description - Sundevil is a uniform, moderately low growing, seed-propagated variety that has moderate texture, medium dark green color, and medium high density under turf conditions.

Adapted to - Southern U.S.

Released - 1992, Jacklin Seed Co.

Breeder Seed/Stock - Jacklin Seed Co.

Certified Seed/Stock - Available.

Preparer/Additional Information - Kim Peterson, Jacklin Seed Co., W. 5300 Riverbend Avenue, Post Falls, ID 83854, (208) 773-7581.

Suwannee (Reg. No. 6)

Selected at Georgia Coastal Plain Experiment Station, Tifton, ARS cooperating - G.W. Burton.

Source - F_1 hybrid between Tift bermudagrass (discovered by J.L. Stephens in old cotton patch near Tifton in 1929) and tall-growing introduction from Union of South Africa.

Method of Breeding - Parents interplanted to allow for maximum natural crossing. Over 5,000 seedling plants carefully screened for many traits. Few of best clones subjected to numerous replicated tests giving measures of their palatability, efficiency, yield potential, management requirements, production under grazing, etc. Tested as selection 99.

Intended Use - Grazing and hay.

Description - Similar to Coastal, except more erect, makes more open sod, less weed resistant, less tolerant of close grazing, but more drought-resistant and definitely superior in productivity and efficiency of nutrient and water use on deep sands. eleased for use on several million hectares of these soils in the South.

Adapted to - LRR J, O, P, T; PHZ 7.

Released - 1953, cooperatively by Georgia Coastal Plain Experiment Station and Plant Science Research Division, ARS.

Breeder Seed/Stock - Georgia Coastal Plain Experiment Station.

Certified Seed/Stock - Available.

Preparer/Additional Information - ARS Coastal Plain Experiment Station, P.O. Box 748, Tifton, GA 31793, (912) 386-3353.

Tifdwarf (Reg. No. 8)

Selected at Georgia Coastal Plain Experiment Station, Tifton, ARS cooperating - G.W. Burton, J.B. Moncrief, and J.E. Elsner.

Source - Observed in 1962 growing in small patch about 460 mm in diameter on golf green at Country Club, Florence, SC, that had been planted with experimental Tifgreen received some eight years earlier from the Georgia Coastal Plain Experiment Station. Also found on greens of Country Clubs at Sea Island, GA, and Thomasville, GA, planted with same lot of Tifgreen some eight years earlier.

Method of Breeding - Believed to be a natural dwarf mutant that occurred in Tifgreen while it was being tested at the Georgia Coastal Plain Experiment Station. Apparently, each of the three golf courses mentioned above obtained a few dwarf sprigs, along with the Tifgreen (Tifton 328), sent out in 1954 for preliminary evaluation. Thoroughly evaluated in comparison with Tifgreen and Tifway.

Intended Use - Turf.

Description - Darker green; basic plant color turns purple with near-freezing temperature; and smaller leaves and stems and shorter internodes than Tifgreen. Tolerates closer mowing and makes a better putting surface than Tifgreen but otherwise similar to Tifgreen. Used on golf greens and for very fine, closely mowed lawns.

Adapted to - LRR C, D, I, J, O, P, T, U; PHZ 8.

Released - 1965, cooperatively by Georgia Coastal Plain Experiment Station and Plant Science Research Division, ARS.

Breeder Seed/Stock - Georgia Seed Development Commission, Athens, GA.

Certified Seed/Stock - Available in quantity.

Preparer/Additional Information - ARS Coastal Plain Experiment Station, P.O. Box 748, Tifton, GA 31793, (912) 386-3353.

Tiffine (Reg. No. 3)

Selected at Georgia Coastal Plain Experiment Station, Tifton, ARS cooperating - G.W. Burton and B.P. Robinson.

Source - F_1 hybrid between *C. dactylon* and *C. transvaalensis* from East Lakes Golf Course in Atlanta, GA.

Method of Breeding - Out of extensive crossing efforts involving Tiflawn bermudagrass (2n=36) and African bermudagrass (2n=18) came eight F_1 hybrids (triploids, 2n=27). Thoroughly screened under lawn and golf-green management and compared with superior selections of *C. dactylon* from golf courses. Distributed for testing as Tifton 127.

Intended Use - Turf.

Description - Lighter green, more disease resistance, and much finer texture than common bermudagrass. Superior for putting greens and fine lawns. Completely male sterile, sheds no pollen to annoy hay fever victims, and must be propagated vegetatively.

Adapted to - LRR I, J, O, P, T, U; PHZ 8.

Released - 1953, cooperatively by Georgia Coastal Plain Experiment Station and Plant Science Research Division, ARS.

Breeder Seed/Stock - Georgia Coastal Plain Experiment Station.

Certified Seed/Stock - Not available.

Preparer/Additional Information - Wayne Hanna, ARS Coastal Plain Experiment Station, P.O. Box 748, Tifton, GA 31793, (912) 386-3354.

Tifgreen (Reg. No. 5)

Selected at Georgia Coastal Plain Experiment Station, Tifton, ARS cooperating - G.W. Burton, Jim Latham, and B.P. Robinson.

Source - F_1 hybrid between superior clone from golf green at Charlotte Country Club, Charlotte, NC, and *C. transvaalensis* from East Lakes Golf Course in Atlanta, GA.

Method of Breeding - Best of several F_1 hybrids (triploids, 2n=27), involving Charlotte bermudagrass *(C. dactylon)* (2n=36) and African bermudagrass (2n=18). Thoroughly evaluated in comparison with number of bermudagrasses under golf-green management. Tested as Tifton 328.

Intended Use - Turf.

Description - Darker green and produces better putting surface than Tiffine; similar in other respects. Also used for fine lawns.

Adapted to - LRR I, J, O, P, T, U; PHZ 8.

Released - 1956, cooperatively by Georgia Coastal Plain Experiment Station and Plant Science Research Division, ARS.

Breeder Seed/Stock - Georgia Seed Development Commission, Athens, GA.

Certified Seed/Stock - Available in quantity. Preparer/ Additional Information - ARS Coastal Plain Experiment Station, P.O. Box 748, Tifton, GA 31793, (912) 386-3353.

Tiflawn (Reg. No. 4)

Selected at Georgia Coastal Plain Experiment Station, Tifton, ARS cooperating - G.W. Burton.

Source - F_1 hybrid between two selections of bermudagrass from pasture-breeding research at Georgia Coastal Plain Experiment Station.

Method of Breeding - Several hundred F_1 hybrids between short, dense, dwarf selection and larger disease-resistant type subjected to thorough screening, which involved finally evaluating best under lawn and golf-green management. Tested as Tifton 57.

Intended Use - Turf.

Description - When compared with common bermudagrass, Tiflawn spreads faster, makes denser weed-free turf, more disease- and frost-resistant, requires less fertilization, and tolerates more wear. Particularly well suited for heavy-duty turf and used on many university football fields in the South.

Adapted to - LRR I, J, O, P, T, U; PHZ 8.

Released - 1956, cooperatively by Georgia Coastal Plain Experiment Station and Plant Science Research Division, ARS.

Breeder Seed/Stock - Georgia Coastal Plain Experiment Station.

Certified Seed/Stock - Available in quantity.

Preparer/Additional Information - ARS Coastal Plain Experiment Station, P.O. Box 748, Tifton, GA 31793, (912) 386-3353.

Tifton 10

Developed by ARS and Georgia Coastal Experiment Station, Tifton, GA - Wayne Hanna, G.W. Burton, and A.W. Johnson.

Source - Vegetative introduction collected by G.W. Burton in 1974 in Shanghai, China. Selected for ease at establishment and turf quality.

Method of Breeding - Vegetatively propagated from original collection.

Intended Use - Turf.

Description - Coarse-textured, dark bluish-green foliage. Establishes rapidly by stolons.

Adapted to - LRR C, D, I, J, O, P, T, U; PHZ 7.

Released - 1988 by Georgia Coastal Plain AES and ARS.

Breeder Seed/Stock - Vegetative. Georgia Coastal Plain AES.

Certified Seed/Stock - Not available.

Preparer/Additional Information - Wayne Hanna, ARS Coastal Plain Experiment Station, P.O. Box 748, Tifton, GA 31793, (912) 386-3353.

Tifton 44 (Reg. No. 10)

Selected at the Georgia Mountain Experiment Station, Blairsville, GA.

Source - Selected as the best of several thousand F_1 hybrids screened for winter-hardiness.

Method of Breeding - F_1 hybrid from crossing Coastal and bermudagrass that had survived in Berlin, Germany, for 15 years before collection in 1966 and found to be more winter-hardy at Michigan State University than others from Europe tested at East Lansing and Lake City, MI.

Intended Use - Grazing and hay.

Description - Early grower. A little shorter, more productive and more digestible than Coastal with finer stems, more rhizomes, and denser sod. More resistant to foliage diseases than Midland.

Adapted to - LRR J, N, O, P; PHZ 7.

Released - 1978, by Georgia AES and FR-SEA-USDA.

Breeder Seed/Stock - Georgia Coastal Plain Station, Tifton, GA.

Certified Seed/Stock - Available.

Preparer/Additional Information - ARS Coastal Plain Experiment Station, P.O. Box 748, Tifton, GA 31793, (912) 386-3353.

Tifton 68 (Reg. No. 14)

Developed by ARS in cooperation with the Georgia Coastal Plain Experiment Station, Tifton, GA.

Source - Selected for rapid spread and high production.

Method of Breeding - F. hybrid from crossing PI 255450 and PI 293606, the two most digestible bermudagrasses in a collection of 500 introductions from various parts of the world.

Intended Use - Grazing and hay.

Description - Giant type with large stems, long stolons, and no rhizomes. Spreads rapidly when planted vegetatively. Higher production, digestibility, and average daily gain than Coastal.

Adapted to - LRR T, U; PHZ 9.

Released - 1984 by ARS and Georgia AES.

Breeder Seed/Stock - Georgia Coastal Plain Experiment Station, Tifton, GA.

Certified Seed/Stock - Not available.

Preparer/Additional Information - ARS Coastal Plain Experiment Station, P.O. Box 748, Tifton, GA 31793, (912) 386-3353.

Tifton 78 (Reg. No. 17)

Developed by ARS and Georgia Coastal Plain Experiment Station, Tifton, GA. Carried as accession no. PI 511312.

Source - Selected for ease of establishment, rapid spread, and early growth from a number of F_1 hybrids between Callie and Tifton 44.

Method of Breeding - F_1 hybrids from cross of Callie and Tifton 44 made in 1975.

Intended Use - Grazing and hay.

Description - Taller than Coastal, spreads much faster, establishes easier, and starts growth earlier. Similar in spread, establishment, and growth habit to Callie but produces more rhizomes, is more winter-hardy, and immune to rust.

Adapted to - LRR P, T, U; PHZ 8.

Released - May 1984 - University of GA and ARS.

Breeder Seed/Stock - Georgia Coastal Plain Experiment Station, Tifton, GA.

Certified Seed/Stock - Available.

Preparer/Additional Information - ARS Coastal Plain Experiment Station, P.O. Box 748, Tifton, GA 31793, (912) 386-3353.

Tifton 85

Source - Selected as the best of many F_1 hybrids for increased dry-matter yield and improved forage digestibility.

Method of Breeding - F_1 hybrid selected from cross between PI 290884 from South Africa and Tifton 68.

Intended Use - Grazing and hay.

Description - In a three-year study, it produced 26% more dry matter and was 11% more digestible and 10% more succulent than Coastal.

Adapted to - PHZ 8, 9.

Released - 1991, USDA-ARS and Coastal Plain Experiment Station.

Breeder Seed/Stock - Georgia Coastal Plain Experiment Station, Tifton, GA.

Certified Seed/Stock - Available.

Preparer/Additional Information - ARS Coastal Plain Experiment Station, Tifton, GA 31793, (912) 386-3353.

Tifway (Reg. No. 7)

Selected at Georgia Coastal Plain Experiment Station, Tifton, ARS cooperating - G.W. Burton, Jim Latham, and B.P. Robinson.

Source - Chance F_1 hybrid between *C. transvaalensis* and *C. dactylon* in lot of *C. transvaalensis* seed received from D. Meredith, Johannesburg, Union of South Africa, in 1954.

Method of Breeding - Thoroughly screened under lawn and golf-green management and compared with superior selections of *C. dactylon* currently available. Tested as Tifton 419.

Intended Use - Turf.

Description - Darker green, greater frost resistance, earlier spring growth, greater resistance to sod webworm and mole cricket, better herbicide tolerance, and stiffer leaf blades than Tifline or Tifgreen. Sheds no pollen to annoy hay fever victims. Well suited for fine lawns, fairways, and tees. Must be propagated vegetatively.

Adapted to - LRR I, J, O, P, T, U; PHZ 8.

Released - 1960, cooperatively by Georgia Coastal Plain Experiment Station and Plant Science Research Division, ARS.

Breeder Seed/Stock - Georgia Seed Development Commission, Athens, GA.

Certified Seed/Stock - Available in quantity.

Preparer/Additional Information - ARS Coastal Plain Experiment Station, P.O. Box 748, Tifton, GA 31793, (912) 386-3353.

Tifway II (Reg. No. 15)

Developed by ARS and Georgia Coastal Plain Experimental Station. Carried as accession no. Tifway Mutant 71-126.

Source - Selected from several promising mutants of Tifway for density and nematode resistance.

Method of Breeding - Mutant of Tifway originating from exposure of dormant Tifway sprigs to 9,000 rads of gamma irradiation, growing plants from the treated sprigs and selecting plants or sectors of plants that appeared to be different.

Intended Use - Turf.

Description - Looks like Tifway and has same desirable traits with denser sod, more root knot, resistance to ring and sting nematodes, more frost tolerance, and faster establishment from sprigs. Often greens up slightly earlier than Tifway.

Adapted to - LRR C, D, I, J, O, P, T, U; PHZ 8.

Released - April 1981, cooperatively by ARS, the Georgia Coastal Plain Experimental Station, the U.S. Golf Association Greens Section, and the U.S. Dept. of Energy.

Breeder Seed/Stock - Georgia Seed Development Commission, Athens, GA.

Certified Seed/Stock - Available.

Preparer/Additional Information - ARS Coastal Plain Experiment Station, P.O. Box 748, Tifton, GA 31793, (912) 386-3353.

Tufcote

Increased at National Plant Materials Center, SCS, Beltsville, MD - R.B. Thornton.

Source - Introductions received from Union of South Africa in 1942. Method of Breeding - Single-plant survivor selected from original source nursery; only plant that survived -28°C.

Intended Use - Heavy traffic areas, lawns, golf courses.

Description - Medium texture, medium dark green, low growth habit, and fair to good tolerance to heavy traffic. Good winter-hardiness, some tolerance to leaf diseases, and remains green longer in all than common.

Adapted to - LRR N, P, S, T; PHZ 5, 6, 7.

Released - 1962, cooperatively by SCS, National Plant Materials Center, Beltsville, MD, and Maryland AES, College Park.

Breeder Seed/Stock - Available, SCS, National Plant Materials Center, Beltsville, MD.

Certified Seed/Stock - Available.

Preparer/Additional Information - National Plant Materials Center, SCS, Bldg. 509, BARC-East, Beltsville, MD 20705, (301) 504-8175.

Dactylis glomerata L. - orchardgrass

Cool-season bunchgrass from central and western Europe. Used for hay, pasture, and silage in northeastern states south to northern part of Gulf Coast states and west to eastern edge of Great Plains. Valuable species in irrigated and high rainfall areas of intermountain region and western states. Less winter-hardy than timothy and smooth brome. Major forage grass from Pennsylvania to North Carolina and west to Iowa and Missouri.

07G23-334

ARS and University of Kentucky cooperating - P.B. Burrus, Jr. and J.F. Pedersen. Carried as accession no. 07G23-334.

Source - Cultivar Boone (Reg. No.. 4) 1986 lot of Boone Breeder Seed.

Method of Breeding - Population I consists of seven breeding lines selected from approximately 1,400 spaced-plants of two-year-old Boone breeder nursery. One cycle of phenotypic selection was made for one or more of the following characteristics: plant vigor, growth habit, maturity, panicle production, leafiness, and disease resistance.

Intended Use - Hay (more compatible with alfalfa first cut) and pasture production.

Description - Population I was selected for plant vigor, leafiness, upright growth, habit, late maturity, good panicle production, and resistance to scald and rust.

Adapted to - LRR K, L, M, N, P, R, S, T; PHZ 3, 5, 6, 7.

Released - Experimental, not released.

Breeder Seed/Stock - P.B. Burrus, Jr., Kentucky AES, ARS cooperating.

Certified Seed/Stock - Seed produced for agronomic testing consists of synthetics of polycross progeny by population.

Preparer/Additional Information - P.B. Burrus, Kentucky AES, Univ. of Kentucky, Lexington, KY 40546-0091, (606) 257-2715.

07G23-335

ARS and University of Kentucky cooperating - P.B. Burrus, Jr. and J.F. Pedersen. Carried as accession no. 07G23-335.

Source - Cultivar Boone (Reg. No. 4) 1986 lot of Boone Breeder Seed.

Method of Breeding - Population II consists of five breeding lines selected from approximately 1,400 spaced-plants of two-year-old Boone breeder nursery. One cycle of phenotypic selection was made for one or more of the following characteristics: plant vigor, growth habit, maturity, panicle production, leafiness, and disease resistance.

Intended Use - Hay and pasture production.

Description - Population II was selected for plant vigor, upright growth habit, leafiness, mid-range plant maturity, good panicle production, and resistance to scald and rust.

Adapted to - LRR K, L, M, N, P, R, S, T; PHZ 3, 5, 6, 7.

Released - Experimental, not released.

Breeder Seed/Stock - P.B. Burrus, Jr., Kentucky AES, ARS cooperating.

Certified Seed/Stock - NA. Seed produced for agronomic testing consists of synthetics of polycross progeny by population.

Preparer/Additional Information - P.B. Burrus, Jr., Kentucky AES, Univ. of Kentucky, Lexington, KY 40546-0091, (606) 257-2715.

07G23-336

ARS and University of Kentucky cooperating - P.B. Burrus, Jr. and J.F. Pedersen. Carried as accession no. 07G23-336.

Source - Cultivar Boone (Reg. No. 4) 1986 lot of Boone Breeder Seed.

Method of Breeding - Population III consists of seven breeding lines selected from approximately 1,400 spaced-plants of two-year-old Boone breeder nursery. One cycle of phenotypic selection was made for one or more of the following characteristics: plant vigor, growth habit, maturity, panicle production, leafiness, and disease resistance.

Intended Use - Hay and pasture production.

Description - Population III was selected for plant vigor, dark green color, large basal leaf production, intermediate plant maturity, fewer panicles, and resistance to scald and rust.
Adapted to - LRR K, L, M, N, P, R, S, T; PHZ 3, 5, 6, 7.

Released - Experimental, not released.

Breeder Seed/Stock - P.B. Burrus, Jr., Kentucky AES, ARS cooperating.

Certified Seed/Stock - Seed produced for agronomic testing consists of synthetics of polycross progeny by population. Preparer/Additional Information - P. B. Burrus, Kentucky AES, Univ. of Kentucky, Lexington, KY 40546-0091, (606) 257-2715.

9007238

Developed and selected at the Plant Materials Center, SCS, Quicksand, KY - Charles F. Gilbert. Also known as KY-2068 and T-7238. Now 9007238.

Source - Collected in 1979 from a naturalized stand occurring in Augusta County, VA, on a Sequoia-Berks soil.

Method of Breeding - Selected in 1987 from a collection of 24 naturalized stands in VA, WV, and MD, six commercial varieties, and 37 foreign germplasm introductions from foreign sources from well-drained shallow, low-fertility acid soils. Selected by culling undesirable off-type collections that had (1) prostrate growth form, (2) little top growth, (3) leaf diseases, or (4) little tiller production. Increase of seed made from selected clone in 1989 and 1990.

Intended Use - Drought-tolerant grass for marginal low-fertility hillside pastures. Selection based on superior vigor and growth throughout the growing season, relative resistance to disease, long-term persistence under clipping, yield, and drought tolerance.

Description - Dark green, leafy, erect, drought-tolerant; good persistence on shallow, reasonably infertile soils. Productive, somewhat better resistance to rusts and leaf spots than commercial strains; better summer growth. The next best performing entry was Jackson.

Adapted to - LRR L, M, N, P, R, S; PHZ 4, 5, 6, 7.

Released - Included in regional field planting/testing program beginning in 1993/1994.

Breeder Seed/Stock - Plant Materials Center, SCS, Quicksand, KY.

Certified Seed/Stock - Not available.

Preparer/Additional Information - Laura L. Ray, Plant Materials Center, SCS, 175 Robinson Dr., Quicksand, KY 41363-9008, (606) 666-5069.

Able

Selected at FFR Cooperative, West Lafayette, IN - J.R. Thomas and R.J. Buker.

Source - Diverse sources. Three parental clones trace to University of Illinois and one to Masshardy.

Method of Breeding - Clones selected from spaced-plant nurseries. Four parental clones identified on basis of performance in forage-yield tests in Indiana and seed-yield tests in Oregon. Tested as FFR Synthetic C.

Intended Use - Released as a hay and pasture use cultivar.

Description - Late maturity; good persistence in forage-yield tests; and acceptable forage yield, seed yield, and resistance to leaf diseases.

Adapted to - LRR A, B, C, F, G, H, J, K, L, M, N, O, P, R, S, T; PHZ 2, 3, 4, 5, 6, 7, 8.

Released - 1970, by Farmers Forage Research Cooperative.

Breeder Seed/Stock - Farmers Forage Research Cooperative.

Certified Seed/Stock - Available.

Preparer/Additional Information - Bret Winsett, FFR Cooperative, 4112 E. State Rd. 225, West Lafayette, IN 47906, (317) 567-2115.

Akaroa

Increased at Plant Materials Center, SCS, Pleasanton, CA - A.L. Hafenrichter, W.E. Chapin, and R.L. Brown. Carried as accession no. 469234.

Source - Introduced from New Zealand.

Method of Breeding - Comparative tests in Washington, Oregon, and California.

Intended Use - Pasture.

Description - In contrast to commercial, it is later maturing, finer stemmed, leafier, and shorter-growing. Seed fields appear rather variable until heading. Winter-hardy under irrigation in northern California and adjacent Nevada, but occasionally suffers from winter injury in other intermountain states.

Adapted to - LRR A, C, D; PHZ 8.

Released - 1953 - cooperatively by California AES, Davis, and Plant Materials Center, SCS, Pleasanton, CA.

Breeder Seed/Stock - Plant Materials Center, SCS, Lockeford, CA.

Certified Seed/Stock - Available.

Preparer/Additional Information - R. Slayback, SCS, 2121-C, 2nd St., Davis, CA 95616, (916) 757-8257.

Amba

Source - Danish Plant Breeding Ltd., Boelshoj, Denmark.

Method of Breeding - The first crosses were made in 1962, involving, as parental varieties: Aries, Russian No. 27863, 2

Icelandic Numbers, Szehass - File, soft leafed CB and Roskilde lines. Forty-five families were selected from the crosses on the basis of yield and persistence.

Intended Use - Hay and pasture production.

Description - Growth habit is medium to semi-erect; spring growth is early to medium. Leaves are medium green; mature plant height is tall; heading date is earlier than Prairial or Sumas; maturity is medium early. European data indicate Amba is resistant to purple eye-spot nd susceptible to leaf rust and yellow stripe rust. Amba performed well for forage as compared to the check varieties in western Canada. In Europe, Amba exhibits good winter-hardiness, seed yield and good digestibility.

Adapted to - Recommended in Canada for use in British Columbia.

Released - Danish Plant Breeding Ltd., Boilshoj, Denmark. Registered in Canada in 1986 as registration no. 2623.

Breeder Seed/Stock - Danish Plant Breeding Ltd., Boilshoj, Denmark.

Certified Seed/Stock - Available.

Preparer/Additional Information - Agriculture Canada, Food Production and Protection Branch, Plant Products Division, Ottawa, Ontario K1A 0C6. Seed is distributed by Topnotch Nutri Ltd., P.O. Box 1030, Abbotsford, British Columbia V2S 5B5.

Ambassador

Developed by International Seeds, Inc., Halsey, OR. Carried as accession no. 80A.

Source - Selections out of public varieties and various experimental lines developed by ISI.

Method of Breeding - High performing clones selected on the basis of leafiness and seed yield potential were crossed in isolation. The resulting seed was planted in an isolated field. This field was rogued for off-type plants. The seed from the remaining plants was bulked and constitutes the breeder seed.

Intended Use - Sown for pasture, either alone or in combination with legume. Also used for early season hay and silage production.

Description - Early maturing variety with superior early spring and late season forage productions. Winter-hardy and moderately resistant to leaf diseases. Ambassador has been trialed extensively in both the U.S. and Asia.

Adapted to - LRR A, B, C, L, M, N, P, R, S; PHZ 4.

Released - 1989, by International Seeds, Inc.

Breeder Seed/Stock - Maintained by International Seeds, Inc., Halsey, OR.

Certified Seed/Stock - Commercial quantities available from International Seeds, Inc.

Preparer/Additional Information - Stephen W. Johnson, International Seeds Inc., P.O. Box 168, 820 W. First St., Halsey, OR 97348, (503) 369-2251.

Arctic

Source - Developed by Dr. P.D. Jones, Courtenay, B.C., from a seed introduction from the former Soviet Union obtained through the Welsh Plant Breeding Station in 1974.

Method of Breeding - The original material was increased using seed harvested in bulk from single row plots. This seed was used to establish an evaluation block to which mass selection was applied beginning in 1976. Plant selections were made on the basis of desirable pasture growth, with plants exhibiting undesirable agronomic characteristics and leaf diseases being removed. Two hundred fifty plants were selected to interpollinate in isolation to form breeder seed of Arctic orchardgrass in 1982.

Intended Use - Hay and pasture production.

Description - Ploidy is tetraploid; growth habit is semi-erect; leaves are long and wide, are carried well up the stem, margins inclined to be rough, and they are medium green; panicle is erect, has two almost equal glumes, distinctly keeled and pointed, two outer palea with curved points and keeled from top to bottom. Susceptible to stripe rust. Arctic is taller than Kay and is similar to Kay in late maturity, forage yields, and areas of adaptation. Limited data suggest that the winter-hardiness of Arctic is slightly better on average than that of Kay.

Adapted to - Recommended in Canada for use in British Columbia.

Released - Dr. P.D. Jones, Courtenay, B.C. Registered in Canada in 1989 as registration no. 3045.

Breeder Seed/Stock - Dr. P.D. Jones, Courtenay, B.C.

Certified Seed/Stock - Available.

Preparer/Additional Information - Agriculture Canada, Food Production and Protection Branch, Plant Products Division, Ottawa, Ontario K1A 0C6. Seed is distributed by Dawson Seed Company Ltd., P.O. Box 91204, North Vancouver, British Columbia V7J 2E7.

Benchmark

Selected at FFR Cooperative - S.J. Baluch and S.D. Stratton. Carried as accession no. Syn 8501.

Source - Selections from high-yielding varieties in Indiana forage yield plots and material out of selection nurseries in Virginia.

Method of Breeding - Twelve clone synthetic selections out of spaced planted nurseries which had high polycross progeny seed and forage yield.

Intended Use - Released as a hay and pasture use variety.

Description - Dark green, high yielding orchardgrass with maturity similar to Hallmark and Potomac. It has demonstrated excellent vigor and good recovery after mowing. Rust resistance tends to be equal to or better than Hallmark and Potomac.

Adapted to - LRR A, B, C, F, G, H, J, K, L, M, N, O, P, R, S, T; PHZ 2, 3, 4, 5, 6, 7, 8.

Released - 1989 by FFR Cooperative.

Breeder Seed/Stock - FFR Cooperative.

Certified Seed/Stock - Available.

Preparer/Additional Information - Bret L. Winsett, FFR Cooperative, 4112 E. State Road 225, West Lafayette, IN 47906, (317) 567-2115.

Berber

Selected at Plant Materials Center, SCS, Lockeford, CA - B. Kay, K. Croeni and R. Slayback. Carried as accession no. PL-2-72.

Source - Received January 1972 from Waite Institute, South Australia, via Burgess L. Kay, U.C.-Davis.

Method of Breeding - Original seed increased in one-half acre block at the Plant Materials Center, SCS, Lockeford, CA. Tested in comparison with released and introduced orchardgrass strains.

Intended Use - Range reseeding, cover crop, and revegetation of disturbed areas.

Description - A dryland, cool season, perennial bunchgrass. Has good seedling vigor and shows superior seedling establishment to the Palestine, Kasbah, and Currie varieties on droughty sites.

Adapted to - LRR A, C; PHZ 9.

Released - 1981, cooperatively by California AES, Davis and Plant Materials Center, SCS, Lockeford, CA.

Breeder Seed/Stock - Plant Materials Center, SCS, Lockeford, CA.

Certified Seed/Stock - Available.

Preparer/Additional Information - R. Slayback, SCS, 2121-C, 2nd St., Davis, CA 95616, (916) 757-8257.

Cambria

Source - Selected from ecotypes in northwestern Spain.

Method of Breeding - Mass selection.

Intended Use - Pastures, in competition with clovers.

Description - Tetraploid, with late ear emergence. Prostrate growth habit, very short height at ear emergence, and very short flagleaf. Winter growth, with very good rust and virus resistance.

Adapted to - Mild winter states.

Released - Welsh Plant Breeding Station, United Kingdom.

Breeder Seed/Stock - Plant Breeding International, Cambridge, U.K.

Certified Seed/Stock - Barenbrug USA.

Preparer/Additional Information - Barenbrug USA, 32080 Old Highway 34, P.O. Box 239, Tangent, OR 97389, (800) 547-4101.

Chinook

Selected at Canada Department of Agriculture Research Station, Lethbridge, Alberta - R.W. Peake.

Source - Seed collected from 30-year-old planting near Lethbridge and from field on Hatfield Ranch, Twin Butte, Alberta, Canada.

Method of Breeding - Four clone synthetic variety selected through clonal and polycross progeny performance.

Intended Use - Primarily for irrigated pastures in southern Alberta. It has also yielded and persisted well in hay as well as pasture stands outside the irrigated districts of southern Alberta and in Montana.

Description - Early-maturing variety possessing a high level of winter-hardiness. Characterized by vigorous early spring growth and early fall dormancy. Chinook orchardgrass provided more days of grazing in Lethbridge area than other grasses. This variety is not recommended for areas where less winter-hardy orchardgrass cultivars can be grown.

Adapted to - LRR G, F; PHZ 3.

Released - 1959, by Canada Department of Agriculture.

Breeder Seed/Stock - Agriculture Canada Research Station, Lethbridge, Alberta.

Certified Seed/Stock - Distributed through the seed trade.

Preparer/Additional Information - Surya N. Acharya, Agriculture Canada, Research Station, Lethbridge, Alberta T1J 4B1, (403) 327-4561.

Comet

Developed by Northrup King Co., Minneapolis, MN - Glenn A. Page.

Source - Three clones from Boone, one from Aries, two from Sterling, and six from various experimental populations were selected to make the variety.

Method of Breeding - Twelve clones from large space planted nurseries were selected for winter-hardiness, vigor, disease reaction, and maturity. Polycross progeny from the 12 clones was tested for forage yield. Polycrossed seed retained as pre-breeders seed.

Intended Use - Hay, mixed hay, pasture.

Description - Comet is a medium-early variety, a few days later than Potomac, having good winter-hardiness and seedling vigor. Disease reaction is similar to Sterling.

Adapted to - LRR A, F, G, K, L, M, N, P, R, S; PHZ 4.

Released - 1977, by Northrup King Co.

Breeder Seed/Stock - Northrup King Co., produced in Oregon.

Certified Seed/Stock - Northrup King Co., Stanton, MN.

Preparer/Additional Information - Fred Stanley, Northrup King Co., Stanton, MN 55081, (507) 663-7639.

Dactus

Source - Developed by Weibullsholm B.V., Landskrona, Sweden.

Method of Breeding - Originated from crossings made in the 1960s between the varieties Frode and Kall. The progeny were maintained in a bulk population using selection pressure for homogeneity and forage production for several generations. Breeder seed was produced in 1976.

Intended Use - Hay and pasture production.

Description - Growth habit is semi-erect to intermediate; leaves are light to medium green; semi-erect to drooping; stem is medium length with medium length upper internode. There is a strong tendency to flower in seeding year; seed is medium sized; maturity is medium early. Dactus is a good yielding variety which outyielded the check variety Sumas in 11 station-years of trials in British Columbia.

Adapted to - Recommended in Canada for use in British Columbia.

Released - Weibullsholm, Landskrona, Sweden. Registered in Canada in 1986 as registration no. 2729.

Breeder Seed/Stock - Weibullsholm, Landskrona, Sweden.

Certified Seed/Stock - Available.

Preparer/Additional Information - Agriculture Canada, Food Production and Protection Branch, Plant Products Division, Ottawa, Ontario K1A 0C6. Seed is distributed by Topnotch Nutri Ltd., P.O. Box 1030, Abbotsford, British Columbia V2S 5B5.

Dart

Land O'Lakes, Inc - R.R. Kalton.

Source - A seven-clone synthetic derived from Sterling (3), Napier (2), Potomac (1), and a male sterile clone.

Method of Breeding - Parental clones selected on the basis of clonal and polycross progeny trials in Iowa, Minnesota, and Idaho. Breeders seed produced on parental clones in Iowa.

Intended Use - Primarily for pasture use in pure stands or legume mixtures. Also suitable for hay production.

Description - Similar in hardiness to Sterling and Potomac. Early maturing, about one day later than Sterling and somewhat more resistant to rust and other leaf diseases. Yield in pasture clipping trials averaged above Sterling and Potomac in upper Midwest trials. Generations limited to one each of breeders, foundation and certified.

Adapted to - LRR B, F, G, K, L, M; PHZ 3.

Released - 1977, Land O' Lakes, Inc.

Breeder Seed/Stock - Parental clones maintained by Land O' Lakes, Inc., in Iowa and breeders seed produced on parental clones in isolation in Iowa.

Certified Seed/Stock - Available.

Preparer/Additional Information - Robert R. Kalton, Land O' Lakes Research Farm, 1025 190th St., Webster City, IA 50595, (515) 543-4852.

Dawn

R.R. Kalton, P.A. Richardson, J. Shields. Experimental No. D57.

Source - An eight-clone synthetic variety. Parent clones derived from PI 315425 (3), PI 325302 (1), PI 315417-9 (1) and Jackson (3).

Method of Breeding - Parent clones selected on basis of clonal and polycross progeny trials for forage and seed yield, leaf disease resistance, and later maturity.

Intended Use - Forage.

Description - Dawn is a hardy, medium maturity, eight-clone synthetic variety of orchardgrass. Unlike most currently available U.S. varieties, five of the eight parental clones are selections from plant introductions of recent origin from

Russia. Dawn exhibits substantially improved rust resistance compared with Sterling and Potomac, appears tolerant to other leaf diseases, and has shown higher digestibility than Sterling in Iowa State University trials. It also yields as good or better in forage clipping trials than such commonly available varieties as Potomoc, Sterling, Hallmark, Orion, Napier, Dart, and Justus. A very winter-hardy, medium-maturity variety which may be more compatible with alfalfa maturities in grass-legume stands.

Adapted to - LRR A, B, E, F, G, H, K, L, M, N, R, S; PHZ 3, 4, 5.

Released - Fall 1989.

Breeder Seed/Stock - Parent clones maintained and breeders seed produced in Iowa at Land O' Lakes Research Farm, Webster City, IA. Foundation seed produced in Iowa or the Willamette Valley of Oregon.

Certified Seed/Stock - Available.

Preparer/Additional Information - Robert Kalton, Land O' Lakes Research Farm, 1025 190th St., Webster City, IA 50595, (515) 543-4852.

DS8

R.R. Kalton, P. A. Richardson, Vista Research.

Source - A seven-clone synthetic derived rom PI 315425 (3), PI 325302 (2), and Jackson (2).

Method of Breeding - Parental clones selected on the basis of clonal and polycross progeny performance for forage and seed traits in Iowa and Idaho. Breeders seed produced under isolation in Iowa.

Intended Use - Primarily for pasture and hay use in pure stands or legume mixtures.

Description - Superior in hardiness to Sterling, Potomac, Benchmark, and Justus. Medium maturity. Up to a week or more later than above varieties in heading depending on where grown. Superior to above in resistance to rust and other leaf disease but comparable in forage yield in pasture clipping trials. Medium maturity and strong hardiness could make it well suited for growing in mixtures with alfalfa for forage use.

Adapted to - LRR A, B, E, F, G, H, K, L, M; PHZ 3.

Released - To be released in 1992 or 1993.

Breeder Seed/Stock - Breeders I on parent clones under isolation in Iowa. Breeders II planted with Breeders I seed and produced under isolation in Iowa.

Certified Seed/Stock - Foundation (Syn 3) and certified (Syn 4) to be grown in Oregon primarily.

Preparer/Additional Information - Robert R. Kalton, Land O' Lakes Research Farm, 1025 190th St., Webster City, IA 50595, (515) 543-4852.

Frode

Developed by Swedish Seed Association, Svalof, Sweden.

Source - Material collected in central Sweden.

Method of Breeding - Four generations of mass selection. Final selection was from the best genotypes.

Intended Use - Pasture, silage and hay. Planted alone or with other grasses and/or legumes.

Description - Medium-late maturing variety heading 10-15 days later than Potomac. Very leafy and fine stemmed with good regrowth after cutting or grazing. Frode has performed very well at many locations in Canada.

Adapted to - LRR A, C, M, P, R, S; PHZ 4.

Released - 1954, by Swedish Seed Association.

Breeder Seed/Stock - Pickseed Canada, Inc., Richmond Hill, Ontario, Canada.

Certified Seed/Stock - Not available. Uncertified seed available from International Seeds, Inc., Halsey, OR.

Preparer/Additional Information - Stephen W. Johnson, International Seeds Inc., P.O. Box 168, 820 W. First St., Halsey, OR 97348, (503) 369-2251.

Hallmark

Selected at FFR Cooperative, West Lafayette, IN - J.R. Thomas and R.J. Buker.

Source - Diverse sources. Five clones trace to the following: Boone (1), Potomac (1), Univ. of Illinois (2), and Eastern States Farmers Exchange (1).

Method of Breeding - Superior clones selected from spaced-plant nurseries. Five parental clones identified on basis of forage-yield tests in Indiana and seed-yield tests in Oregon. Tested as FFR Synthetic E.

Intended Use - Released as a hay and pasture use cultivar.

Description - Vigorous, high yielding, good recovery after mowing, good seeding vigor, and superior to Boone in resistance to leaf disease; two to four days later in maturity than Potomac and three to five days later than Boone. Good seed yields.

Adapted to - LRR A, B, C, F, G, H, J, K, L, M, N, O, P, R, S, T; PHZ 2, 3, 4, 5, 6, 7, 8.

Released - 1969, by Farmers Forage Research Cooperative.

Breeder Seed/Stock - Farmers Forage Research Cooperative.

Certified Seed/Stock - Available.

Preparer/Additional Information - Bret Winsett, FFR Cooperative, 4112 E. State Road 225, West Lafayette, IN 47906, (317) 567-2115.

Haymate

Selected at FFR Cooperative, West Lafayette, IN - B.L. Winsett, S.D. Stratton and S.J. Baluch.

Source - Diverse sources. Four clones were selected out of FFR yield trials in IN, three clones were plants demonstrating resistance to rust from greenhouse rust screens, and one clone was selected from FFR yield trials in NC.

Method of Breeding - Eight clone synthetic. Superior clones selected from spaced-plant nurseries. Eight parental clones were identified on basis of maturity, progeny forage tests in Indiana, and progeny seed yield tests in Oregon. Tested as FFR synthetic OG8705.

Intended Use - A late-maturing variety for use as a hay and pasture use cultivar.

Description - OG8705 is slightly later in maturity than Rancho but earlier than Pennlate. OG8705 has routinely demonstrated excellent forage yields for a late-maturing orchardgrass.

Adapted to - LRR A, B, C, F, G, H, J, K, L, M, N, O, P, R, S, T; PHZ 2, 3, 4, 5, 6, 7, 8.

Released - 1993. Breeder Seed/Stock - FFR Cooperative.

Certified Seed/Stock - Available.

Preparer/Additional Information - Bret Winsett, FFR Cooperative, 4112 E. State Road 225, West Lafayette, IN 47906, (317) 567-2115.

Juno

Source - Developed at the Forage Section, Ottawa Research Station by Dr. W.R. Childers and R.W. Suitor. Designated Ottawa P-1 during the testing period.

Method of Breeding - The original selections were made from a night pasture which has been intensively grazed for a 15-year period without cultivation on the Borden Farm, three miles south of the Ottawa Research Station. The original seed was not identified. Twenty clones were dug up in August. These plants were divided into single shoot pieces and rooted in water. The 10 best lines which had an adequate number of shoots, were planted in a replicated test in the greenhouse, with a cutting program of four cuts at one-month intervals. Based on the number of new tillers and the total dry matter production, the seven best clones were established in the field. Further studies on leafiness and rust resistance were carried out in the field. The seven best clones were increased to produce Synthetic-1 seed. This Syn 1 seed was sown in a one and one-half acre block for the production of breeder seed. The Syn 1 seed was sent out for testing in the Ontario and Quebec Provincial Tests under the designation Ottawa P-1.

Intended Use - Hay and pasture production.

Description - The cultivar is in the same maturity range as Frode and Tardus II and, because our main program was directed toward late-maturing strains, this Ottawa P-1 strain was not a high priority. However, the results from the provincial test showed that it was promising in comparison with Frode and Tardus II, the two Swedish varieties recommended in Ontario and eastern Canada generally. This cultivar gives early vigorous growth, has excellent winter-hardiness, and has plump good-quality seed. The heads emerge from the boot from June 6-8 which is similar to Frode. The early growth is very leafy and P-1 was originally selected as a pasture-type cultivar. This cultivar has more rust resistance than most cultivars and a continuous selection program has been carried out to further increase this characteristic.

The overall uniformity of ripening has been a feature of this cultivar, an important factor in seed harvest. It was also noted that plumpness of seed was greater than for most cultivars. This character was determined by the weight-per-measured-bushel technique used in cereals.

Adapted to - Recommended in Canada for use in Ontario and Manitoba.

Released - The Forage Section, Ottawa Research Station in 1972. Registered in Canada in 1973 as registration no. 1432.

Breeder Seed/Stock - The Forage Section, Ottawa Research Station.

Certified Seed/Stock - Available.

Preparer/Additional Information - Agriculture Canada, Food Production and Protection Branch, Plant Products Division, Ottawa, Ontario K1A 0C6. Seed is distributed by SeCan Association, 200-57, Promenade Auriga, Nepean, Ontario K2E 8B2.

Justus

Developed by D.A. Sleeper, University of Missouri, Columbia, MO.

Source - Collections made from old pastures in Missouri.

Method of Breeding - Several cycles of recurrent selection for persistence and tolerance to stem rust.

Intended Use - Pastures, hay, greenchop, and silage, either alone or in combination with legumes.

Description - Medium maturity variety with tolerance to stem rust. Gives significantly higher animal gains when rust is present. Has produced high forage yields across multiple years and locations. Has given superior average daily gains in cattle feeding trials.

Adapted to - LRR A, L, M, N, R; PHZ 4.

Released - 1988 by the University of Missouri.

Breeder Seed/Stock - University of Missouri.

Certified Seed/Stock - Available. PVP #9100245.

Preparer/Additional Information - Stephen W. Johnson, International Seeds, Inc, P.O. Box 168, Halsey, OR 97348, (503) 369-2251.

K2-8

Developed by Northrup King Co., Minneapolis, MN - Jack Mings.

Source - A Russian PI.

Method of Breeding - Unknown.

Intended Use - Hay, mixed hay, and pasture.

Description - K2-8 is late-maturing winter-hardy cultivar having good resistance to leaf disease.

Adapted to - LRR A, B, E, K, L, M, R, S; PHZ 3.

Released - Not released.

Breeder Seed/Stock - Northrup King Co.

Certified Seed/Stock - Not available.

Preparer/Additional Information - Fred Stanley, Northrup King Co., Stanton, MN 55081, (507) 663-7639.

Kay

Selected at the Canada Department of Agriculture Research Station, Ottawa, Ontario - W.R. Childers.

Source - Plant introductions obtained from Academy of Science, Moscow, Russia, in 1958.

Method of Breeding - Selection in large single-plant nurseries for vigor, yield, rust resistance, and late maturity. Ninety-three plants selected and progeny tested. No significant differences were obtained for plant yield, so all plants used to produce breeder seed.

Intended Use - Hay and pasture.

Description - Tall growing with large panicles, large stems, broad and long leaves; some resistance to rust, and extremely winter-hardy. Exhibits excellent spring vigor and good growth in cool, wet springs; does less well under dry, hot conditions.

Adapted to - LRR L; PHZ 2, 3.

Released - 1970, by Canada Department of Agriculture.

Breeder Seed/Stock - Canada Department of Agriculture Research Station, Ottawa, Ontario, Canada.

Certified Seed/Stock - Available.

Preparer/Additional Information - Dr. A. R. McElroy, Forge Bldg #12, Plant Research Center, Ottawa, Ontario, (613) 995-3700.

Latar (Reg. No. 2)

Selected at Plant Materials Center, SCS, Pullman, WA - J.L. Schwendiman, R.J. Olson, and A.G. Law.

Source - Original introduction from Institute of Plant Industry, Leningrad, Russia, as PI 111536, by Westover-Enlow expedition in 1934.

Method of Breeding - Grown for three generations in nurseries at Plant Materials Center; mass selection jointly by SCS, ARS, and staff of Washington AES, Pullman, from spaced plantings in fourth generation. Tested in uniform nurseries since 1951 as P-2453.

Intended Use - Hayland and pastureland.

Description - Late-maturing, hay-type orchardgrass. Blooms and matures seed on average of 10-14 days later than commercial varieties. Leaves abundant, broad, well distributed, and noticeably light green. Vigorous and high in vegetative production. Seed production good. Lowest among seven orchardgrass varieties in lignin content and significantly higher in digestibility.

Adapted to - LRR B; PHZ 5.

Released - 1957, cooperatively by Washington and Idaho Agricultural Experiment Stations at Pullman and Moscow, respectively, and Plant Materials Centers, SCS, Aberdeen, Idaho, and Pullman.

Breeder Seed/Stock - Plant Materials Center, SCS, Pullman, WA.

Certified Seed/Stock - Available.

Preparer/Additional Information - Plant Materials Center, SCS, Rm 104, Hulbert Agricultural Sciences Bldg, WSU, Pullman, WA 99164-6211, (509) 335-7376.

Mobite

Source - Mommersteeg International B.V., Vlijmen, The Netherlands.

Method of Breeding - In 1959, seeds of the varieties Dactimo and Modac were irradiated using X-rays. Following inbreeding, a smooth-leafed mutant was found which was then crossed and backcrossed with selected plants of Dactimo and Modac. The progeny of these crosses were then inbred and smooth-leafed plants were selected. In 1966, a number of these selected plants were crossed with selected plants of the variety Holstenkamp. Following inbreeding, the smooth-leafed plants were selected and multiplied. Plant selection was made in the F3 generation, followed by plant and clone selection in the F4 generation. This was followed by progeny testing in forage trials. Seven clones were selected to form the parental material of Mobite.

Intended Use - Hay and pasture production.

Description - Ploidy is diploid; growth habit is semi-erect; leaves are yellow-green, no silicious teeth, and wide; maturity is late; winter-hardiness is similar to that of Kay; has good palatability. European data indicate that Mobite is moderately resistant to powdery mildew. Mobite is lower yielding than the check varieties, but it is later in maturity, has leaves without silicious teeth, and is, therefore, more palatable.

Adapted to - Recommended in Canada for use in Ontario and British Columbia.

Released - Mommersteeg International B.V., Vlijmen, The Netherlands. Registered in Canada in 1987 with registration no. 2749.

Breeder Seed/Stock - Mommersteeg International B.V., Vlijmen, The Netherlands.

Certified Seed/Stock - Available.

Preparer/Additional Information - Agriculture Canada, Food Production and Protection Branch, Plant Products Division, Ottawa, Ontario K1A 0C6. Seed is distributed by Dawson Seed Co. Ltd., P.O. Box 91204, North Vancouver, British Columbia V7J 2E7.

Napier

Source - North American Plant Breeders, Ames, Iowa.

Method of Breeding - Eight parental clones were selected from adapted varieties and long-term stands in Iowa. Selections were evaluated in clonal and progeny tests for desirable forage and seed attributes with emphasis on rust and pythium blight resistance, winter-hardiness, and yield. Napier was first distributed for testing in 1963 in the U.S. as experimental RP300.

Intended Use - Hay and pasture production.

Description - Napier is a medium maturing variety in the U.S. and in Ontario trials; it has been one day later than the check Frode in first flowering. Winter-hardiness has been acceptable. On the average Napier has yielded better than Frode in Ontario based on 15 station-years of data, with its best performance in southern Ontario.

Adapted to - Recommended in Canada for use in British Columbia and Ontario.

Released - North American Plant Breeders, Ames, Iowa. Registered in Canada in 1976 with Registration No. 1616.

Breeder Seed/Stock - North American Plant Breeders, Ames, Iowa.

Certified Seed/Stock - Available.

Preparer/Additional Information - Agriculture Canada, Food Production and Protection Branch, Plant Products Division, Ottawa, Ontario K1A 0C6. Seed is distributed by Oseco Ltd., P.O. Box 219, Brampton, Ontario L6V 2L2.

Orbit

Developed by Northrup King Co., Minneapolis, MN - Howard Kacwer and Glenn Page.

Source - Seven half-sibs from an Iranian PI 230116.

Method of Breeding - Clones were phenotypically selected on the basis of winter-hardiness, vigor, plant type, maturity, and disease reaction. Clones were progeny tested and the seven best clones polycrossed to produce pre-breeders seed.

Intended Use - Hay, mixed hay, pasture.

Description - Orbit is a relatively short cultivar having narrow upright leaves. It is similar to Pennlate in maturity and has good winter hardiness and resistance to leaf diseases.

Adapted to - LRR K, L, M, R, S; PHZ 3.

Released - 1975, Northrup King Co.

Breeder Seed/Stock - Northrup King Co.

Certified Seed/Stock - Not available.

Preparer/Additional Information - Fred Stanley, Northrup King Co., Stanton, MN 55081, (507) 663-7639.

Orion

Developed by Northrup King Co., Minneapolis, MN - Glenn A. Page.

Source - Orion traces back to three high-combining clones in a space planted nursery: one clone each from Nordstern, Masshardy, and PI 262459.

Method of Breeding - Self-seed of the PI 262459 clone and single cross-seed between the Nordstern and Masshardy clone were seeded into a nursery rogued, and allowed to interpollinate to produce pre-breeders seed.

Intended Use - Hay, mixed hay, pasture.

Description - Orion is a late-maturing variety. It is tall and very leafy. It has excellent winter-hardiness and resistance to leaf diseases.

Adapted to - LRR A, B, K, L, M, N, P, R, S; PHZ 3, 4.

Released - 1981, Northrup King Co.

Breeder Seed/Stock - Northrup King Co.

Certified Seed/Stock - Not available.

Preparer/Additional Information - Fred Stanley, Northrup King Co., Stanton, MN 55081, (507) 663-7639.

Paiute

USDA-FS Shrub Research Laboratory, Provo, Utah. Carried as accession no. PI 109072.

Source - Paiute was introduced from Ankara, Turkey, in 1934.

Method of Breeding - Contact the breeder.

Intended Use - Forage crop for arid rangelands.

Description - Paiute is a cool-season, shade-tolerant, long-lived bunchgrass that has an abundance of basal leaves and leafy upright stems. Under range conditions, Paiute usually develops distinct clumps and flower culms 380-460 mm tall, with leaves 250-300 mm long.

Adapted to - LRR B, D, E.

Released - 1983, Plant Materials Center, SCS, Aberdeen, ID, and USDA-FS, Provo, UT.

Breeder Seed/Stock - Plant Materials Center, SCS, Aberdeen, ID.

Certified Seed/Stock - Available.

Preparer/Additional Information - Gary Young, Plant Materials Center, SCS, Box 296, Aberdeen, ID 83210, (208) 397-4133.

Pomar

Selected at Plant Materials Center, SCS, Pullman, WA - J.L. Schwendiman. Carried as accession no. PI 111537.

Source - Introduced from Russia as PI 111537 through Westover-Enlow collection of 1934.

Method of Breeding - First selected at the Plant Materials Center, SCS, Pullman, WA, and tested as P-2454. It has since been tested as an orchard cover crop by the Plant Materials Center, SCS, Aberdeen, ID.

Intended Use - Orchard cover crop.

Description - Late-maturing, rapid-developing, long-lived, dwarf-type orchardgrass. Leaves are numerous, predominantly basal, narrow, short, and light-green. Stems are abundant, medium-fine, erect, and normally less than 610 mm tall in solid stands. Few seed stalks are formed after the first crop is cut each year. Flowers about the same time as alfalfa. Tolerance to shade, drought, insects, and diseases equal to other orchardgrass varieties. Adapted for orchard cover crop and erosion control. Clippings decompose rapidly in orchards and vineyards. Seed difficult to thresh; does not shatter.

Adapted to - LRR A, B, D; PHZ 4, 5.

Released - 1966, cooperatively by Idaho AES and Plant Materials Center, SCS, Aberdeen, Idaho.

Breeder Seed/Stock - Plant Materials Center, SCS, Corvallis, OR.

Certified Seed/Stock - Available in limited quantity. Preparer/Additional Information - Gary Young, Plant Materials Center, SCS, P.O. Box 296, Aberdeen, ID 83210, (208) 397-4133.

Potomac (Reg. No. 1)

Selected at Plant Industry Station, Beltsville, MD - R.E. Wagner, M.A. Hein, and P.R. Henson.

Source - Plants collected in 1935 from old pastures in Maryland, Virginia, West Virginia, and Pennsylvania and from strain tests conducted at Plant Industry Station.

Method of Breeding - Collections screened on basis of type, rust resistance, leafiness, persistence, and vigor; in 1940, eight plants of predominantly pasture type placed in one isolation block (Maryland pasture strain) and six plants representing erect hay types placed in another (Maryland hay strain). In 1945, plants were selected from three-year-old broadcast plots of these two strains and established in space-planted nursery together with equal number of seedlings from each of two strains. Nursery rogued and bulk seed collected for testing as Beltsville orchardgrass. Potomac represents third cycle of mass selection from 1945 nursery.

Intended Use - Pasture and hay production.

Description - Dark green, leafy, erect, similar to commercial lots in height. Productive, superior persistence, and rust-resistant.

Adapted to - LRR S, P, N; PHZ 6, 7.

Released - 1954, by Plant Science Research Division, ARS, and cooperating experiment station.

Breeder Seed/Stock - Oregon Foundation Seed and Plant Materials Project and Washington Crop Improvement Association.

Certified Seed/Stock - Available.

Preparer/Additional Information - James H. Elgin, Jr., ARS, Bldg. 005, Room 328, Beltsville, MD 20705, (301) 504-5618.

Prairial

Source - Institute of National Agronomy Research, France.

Method of Breeding - This synthetic variety was bred from two ecotypes from northwest France and is derived from 27 clones chosen amongst 12 lines after two to four generations of selfing.

Intended Use - Hay and pasture production.

Description - Growth habit is semi-erect; tillers are numerous and small; leaves are dark green, medium size, and drooping; plant height is medium; maturity is late; and good winter survival. European information indicates that Prairial has resistance to stem rust, leaf streak; susceptible to leaf spots. In trials in British Columbia, Prairial outyielded all check varieties.

Adapted to - Recommended in Canada for use in British Columbia.

Released - Institute of National Agronomy Research, France. Registered in Canada in 1983 with registration no. 2330.

Breeder Seed/Stock - Institute of National Agronomy Research, France.

Certified Seed/Stock - Available.

Preparer/Additional Information - Agriculture Canada, Food Production and Protection Branch, Plant Products Division, Ottawa, Ontario K1A 0C6. Seed is distributed by Richardson Seed Company Ltd., 4055 McConnell Dr., Burnaby, British Columbia V6A 3A7.

Prime

Developed by Northrup King Co., Minneapolis, MN - Jack Mings.

Source - Four clones from a Russian PI No. selected for late maturity, resistance to leaf diseases, yield of progeny, and phenotypic appearance.

Method of Breeding - Selected clones from the PI were progeny tested and the four best clones recombined to produce pre-breeders seed.

Intended Use - Hay, mixed hay, and pasture.

Description - Prime is a late maturing, winter-hardy variety. It is a leafy dark green cultivar having good leaf disease resistance.

Adapted to - LRR A, B, K, L, M, R, S; PHZ 3.

Released - 1975, by Northrup King Co.

Breeder Seed/Stock - Northrup King Co.

Certified Seed/Stock - Not available.

Preparer/Additional Information - Fred Stanley, Northrup King Co., Stanton, MN 55081, (507) 663-7639.

Rancho

Selected at FFR Cooperative - S.J. Baluch and S.D. Stratton. Carried as accession no. Syn O.

Source - Selections deriving from Masshardy, MII-36, miv-14, Nika, Trivolium, PI 234688, S-143

Method of Breeding - Seven-clone synthetic. Clones out of spaced planted nurseries which were selected based on clonal data and polycross progeny forage data.

Intended Use - Hay and pasture use cultivar.

Description - Late maturity, good persistence in forage yield tests, and excellent forage yield for a late-maturing orchardgrass variety. Rancho has also demonstrated very good rust resistance.

Adapted to - LRR A, B, C, F, G, H, J, K, L, M, N, O, P, R, S, T; PHZ 2, 3, 4, 5, 6, 7, 8.

Released - By FFR Cooperative.

Breeder Seed/Stock - FFR Cooperative.

Certified Seed/Stock - Available.

Preparer/Additional Information - Bret L. Winsett, FFR Cooperative, 4112 E. State Road 225, West Lafayette, IN 47906, (317) 567-2115.

Rapido

Source - Maple Leaf Mills, Georgetown, Ontario.

Method of Breeding - A six-clone synthetic derived from OSG7. Selection for spring and late fall growth, vigor, and freedom from foliar diseases began in 1975.

Intended Use - Hay and pasture production.

Description - Tetraploid; good early spring and fall growth; leaves are medium green; flag leaf is shorter and slightly wider than Hallmark; stems are taller than Hallmark, with upper stem internode slightly longer than Hallmark; panicles are larger than Hallmark; and maturity is early, similar to Juno. Rapido is a good yielding variety with better winter survival than Ina or Napier but less than Kay or Hallmark. The aftermath production of Rapido is most similar to Kay.

Adapted to - Recommended in Canada for use in British Columbia and Ontario.

Released - Otto Pick and Sons Seeds Ltd. Registered in Canada in 1988 with registration no. 2966.

Breeder Seed/Stock - Otto Pick and Sons Seeds Ltd.

Certified Seed/Stock - Available.

Preparer/Additional Information - Agriculture Canada, Food Production and Protection Branch, Plant Products Division, Ottawa, Ontario K1A 0C6. Seed is distributed by Otto Pick and Sons Seeds Ltd.

Sterling (Reg. No. 5)

Selected at Iowa AES, Ames - R.R. Kalton and M.G. Weiss.

Source - Clonal collections made in Iowa and southern Minnesota in 1941 and 1943.

Method of Breeding - Synthetic variety developed from five selected clones (Iowa 64, 120, 121, 123, and 160, all from central Iowa). Parental clones selected for superior hardiness, forage and seed yields, recovery, leafiness, and disease-resistance based on performance of clones and inbred and outcross progenies. Seed increased on limited generation basis. Breeder seed produced by natural pollination of five parent clones in planting isolated from other orchardgrass. Equal quantities of seed from each parent clone mixed for breeder seed. Certified seed is second advanced generation from breeder seed and not eligible for producing any class of certified seed.

Intended Use - Pasture, hay, green chop, and silage; can be grown alone or in combination with legumes or other grasses.

Description - Superior in stand establishment, winter-hardiness, and forage and seed yields under Iowa conditions. Particularly outstanding in seed yield. Mid-early in maturity. Moderately susceptible to rust and leaf streak.

Adapted to - LRR A, B, D, E, L, M, N, R, S, P; PHZ 4.

Released - 1960, by Iowa AES.

Breeder Seed/Stock - Iowa AES.

Certified Seed/Stock - Not available.

Preparer/Additional Information - Irving T. Carlson, Agronomy Department, Iowa State University, Ames, IA 50011, (515) 294-9653.

Sumas

Source - Sumas, known as PX 36 during the testing period, was developed through an orchardgrass breeding program at the Research Station, Agriculture Canada, Agassiz, B.C., by D.K. Taylor. The breeding program was originated and supervised from 1944 to 1955 by Dr. M.F. Clarke.

Method of Breeding - Clonal selections were made from promising introductions and some of these were crossed by the mutual bagging method. The resulting progeny were evaluated for winter-hardiness, spring vigor, and disease and lodging resistance. Selections of these populations were in turn selfed and combined in open pollinated and polycross nurseries and the resulting progeny were evaluated for uniformity and yield of forage and seed. In 1970, the Synthetic-1 and Synthetic-2 of 12 polycross lines were evaluated in advanced hay and pasture trials at the Agassiz Research Station. On the basis of three years' results, Sumas was the most promising strain, having a combination of high forage and seed yield with medium late maturity.

Sumas is a four-clone synthetic originating as follows: one clone selected from Pennlate, two from a Pennlate x Gullaker cross, and one from a Swiss Introduction x Gullaker cross.

Intended Use - Hay and pasture production.

Description - Sumas is a tall variety of orchardgrass with an attractive blue-green color. It appears to be equal to Juno in spring vigor and resistance to winter injury. In date of anthesis, Sumas is three to four days later than Juno or Sterling and equal to Hercules. Over a three-year period at Agassiz, Sumas was equal in hay and pasture yield to Sterling, equal or superior to Juno, and superior to Hercules, Kay, and Rideau when all varieties were harvested on the same date.

Although seed yield comparisons are not available with commercial varieties, a comparison with other polycrosses within the Agassiz breeding program indicates a good seed yield potential. Sumas averaged 643 lb/ac for an 11 polycross mean.

Sumas appears to produce a relatively high total yield of digestible dry matter and in trials was equal in digestibility to varieties such as Kay and Rideau and better than Sterling, Juno, and Hercules when cut on the same date.

Adapted to - Recommended in Canada for use in British Columbia, Ontario, and Quebec.

Released - Agriculture Canada, Research Station, Agassiz, B.C. Registered in Canada in 1974 with Registration No. 1535.

Breeder Seed/Stock - Agriculture Canada, Research Station, Agassiz, B.C.

Certified Seed/Stock - Available.

Preparer/Additional Information - Agriculture Canada, Food Production and Protection Branch, Plant Products Division, Ottawa, Ontario K1A 0C6.

Summer Green

Snow Brand Seed Company, Daini-Hokkai Bldg., 3-8 Higashi-Nihonbashi 3 Chome, Chuo-Ku, Tokyo 103, Japan.

Source - Thirty plants having good growth vigor and persistence were collected from old grasslands in the southern part of Chiba Prefecture, Japan, in 1967 and then another 65 were collected from the same area in 1971.

Method of Breeding - Two plants from the 1967 collection and 10 plants from the 1971 collection were crossed in an isolated field and used to start a progeny test. Seven superior clones were identified in 1976. Seed production of Syn-2 and productivity tests were begun in 1981.

Intended Use - Moderately early orchardgrass for forage.

Description - Summer Green most closely resembles Potomac and can be differentiated from it by Summer Green's heading date (which is two days later than Potomac), by its being more stem rust resistant, havings wider stalk diameter, spike length 17 mm longer than Potomac, and flag leaf 0.9 mm wider.

Adapted to - LRR A, B, C, D, E, H, M, N, O, P, R, S, T; PHZ 4, 5, 6, 7, 8.

Released - March 1985, Japan.

Breeder Seed/Stock - Maintained by Snow Brand Seed Company.

Certified Seed/Stock - Available.

Preparer/Additional Information - Susan H. Samudio, Jacklin Seed Company, W. 5300 Riverbend Ave., Post Falls, ID 83854, (208) 773-7581.

Deschampsia cespitosa **(L.) Beauv. - tufted hairgrass**

Dichanthium caricosum **A. Camus**
D. caricosum **Complex**

Cool-season, coarse perennial bunchgrass. Widespread, highly complex, and variable species adapted to many habitats in the northern hemisphere. Found in coastal marshes, along streams, and in wet mountain meadows. Key species for wetland restoration and development, riparian or shoreline plantings, and reclamation of disturbed sites.

EPC-1460

USDA-FS, Logan, UT, and the Environmental Plant Center, Meeker, CO - Dr. Ray Brown. Carried as accession no. 9024403.

Source - Collected from a native stand at Peru Creek near Dillion, CO, on alpine-subalpine meadow at 3120 m elevation.

Method of Breeding - Native collection, mass selected and tested at several sites in the Rocky Mountain Region from Montana to Colorado and Nevada. This accession appeared to have good acid tolerance and good establishment characteristics.

Intended Use - Reclamation and revegetation of mined lands and disturbed sites at high elevations and acidic soils. Also for roadside and wetland meadow use.

Description - Perennial grass, no rhizomes, inflorescence, an open panicle, culms erect over 50 cm tall; in dense tufts, leafy at base. Plants are large open crowns. Seed production is moderate to heavy at Logan, Utah.

Adapted to - LRR D, E; PHZ 2, 3, 4.

Released - Not released, for advanced trials only.

Breeder Seed/Stock - Not available.

Certified Seed/Stock - Not available.

Preparer/Additional Information - Upper Colorado Environmental Plant Center, P.O. Box 448, Meeker, CO 81641, (303) 878-5003.

Nortran (Reg. No. 113)

Selected at Palmer Research Center of the University of Alaska Fairbanks Agricultural and Forestry Experiment Station, Palmer, AK - William W. Mitchell. Carried as accession no. PI 518658.

Source - Collections of indigenous materials in: (1) Iceland (by Thorstein Tomasson, Agricultural Research Institute, Reykjavik); (2) Galena, along Yukon River in west-central Alaska; and (3) Talkeetna Mountains north of Palmer in south-central Alaska.

Method of Breeding - Breeder lines were established based on individual plant selections IAS 284 and IAS 371 out of the Icelandic collection, IAS 458 out of the Talkeetna Mountain collection, and IAS 239 bulked material of the Galena collection. Each line is seed-propagated in isolation and breeder seed comprised of pure, live seed equalling 30% IAS 371, 20% IAS 284, 25% IAS 239, and 25% IAS 458.

Intended Use - For revegetation use and low maintenance ground cover; also in turf plantings; possibly for pasture.

Description - First known cultivar of species.

Adapted to - LRR W, X, southern half of Y. (Southern, southwestern, and western coastal to north-central interior regions of Alaska and adjacent areas of Canada. Hardy through boreal regions south of Arctic, including alpine regions); PHZ 2, 3, 4.

Released - 1986, by the Alaska Agricultural and Forestry Experiment Station.

Breeder Seed/Stock - Palmer Research Center, Agricultural and Forestry Experiment Station, Palmer, AK.

Certified Seed/Stock - Available.

Preparer/Additional Information - William W. Mitchell, Palmer Research Center, Agricultural and Forestry Experiment Station, 533 E. Fireweed, Palmer, AK 99645, (907) 745-3257.

Deschampsia cespitosa (L) Beauv. ssp. *beringensis*, (Hulten) W.E. Lawr. - tufted hairgrass

Cool-season, tufted, relatively robust bunchgrass found along muddy shores in coastal habitats of northern latitudes of Canada and Alaska. Used for wetland restoration and development and for disturbed-site revegetation.

Norcoast (Reg. No. 99)

Selected at Palmer Research Center, University of Alaska Fairbanks Agricultural and Forestry Experiment Station, Palmer, AK - William W. Mitchell. Carried as accession nos. IAS 19 and IAS 242.

Source - Three bulk seed collections from native stands in the Turnagain Arm region of the Cook Inlet, southeast of Anchorage, and one bulk seed collection from the upper Knik Arm of the Cook Inlet south of Palmer, AK.

Method of Breeding - Four breeder lines are seed-propagated in isolation, accessioned as IAS 636, IAS 637, IAS 638, and IAS 639. Breeder seed is a composite of equal amounts of pure, live seed of the four lines.

Intended Use - Revegetation purposes; forage production.

Description - First known cultivar of species.

Adapted to - Maritime and southern to northern coastal regions of Alaska; also inland to southcentral Alaska. Hardy to California, 65°N. lat. along coast, marginally hardy north of there.

Released - 1981, by the Alaska AES.

Breeder Seed/Stock - Agricultural and Forestry Experiment Station, Univ. of Alaska, Fairbanks, AK.

Certified Seed/Stock - Available.

Preparer/Additional Information - William W. Mitchell, Palmer Research Center, Agricultural and Forestry Experiment Station, 533 E. Fireweed, Palmer, AK 99645, (907) 745-3257.

Dichanthelium clandestrinum (L.) Gould - deertongue

Panicum clandestinum L.

Deertongue is a perennial warm-season grass. Culms normally reach a height of 50-140 cm. Individual plants develop into vase-shaped dense clumps. Although rhizomes 5-8 cm long are produced, the base of the clumps rarely exceed 15 cm. Top width may equal plant height. In late June, a terminal panicle emerges, flowers, and produces a limited amount of seed which shatters readily. Abundant seed is produced cleistogamously and matures in late August or September.

Deertongue is a pioneer plant on disturbed sites of low fertility such as mine spoil, gravel pits, and sandy roadbanks. The species has a high tolerance of low pH, on mine spoil with a pH of 3.5.

Deertongue occurs from Maine west to Michigan and south to northern Georgia and Alabama. It is well adapted throughout the Appalachian Mountain region and east to the Virginia coast.

Tioga

Selected at Big Flats Plant Materials Center, SCS, Corning, NY. Carried as accession no. NY 4950, PI 443380.

Source - Selected from a broad collection of deertongue made in Pennsylvania, New York, and New Hampshire.

Method of Breeding - Twenty accessions, from a broad collection of deertongue, were selected for seedling vigor. A composite was formed by blending equal amounts of pure live seed from each accession.

Intended Use - For erosion control and revegetation on soils with low pH.

Description - There is variation in the size of the accessions comprising Tioga. The height varies from 50 to 140 cm. Tioga was selected for seedling vigor and rapid establishment. Tioga can tolerate a ph of 3.8 and is tolerant to aluminum and manganese toxicity.

Adapted to - LRR L, N, R, S, northern P & I, eastern part of M; PHZ 4.

Released - Tioga deertongue was released cooperatively in 1975 by the SCS, Pennsylvania AES, and New York AES.

Breeder Seed/Stock - Plant Materials Center, SCS, Corning, NY.

Certified Seed/Stock - Available.

Preparer/Additional Information - Paul Salon and John Dickerson, SCS, Big Flats Plant Materials Center, RD 1 Rte 352, Box 360A, Corning, NY 14830, (607) 562-8404.

Dichanthium annulatum (Forsk.) Stapf - bluestem

Andropogon annulatus Forsk.

Warm-season, erect to semi-decumbent bunchgrass from Union of South Africa. Shows promise as pasture plant in southern Texas.

Kleberg

Selected at Kingsville, TX - N.R. Diaz.

Source - Seed collected from weakened rhodesgrass pasture on King Ranch, TX, where escaped bluestem was dominant. Growing beside King Ranch bluestem and found at same time, 1939.

Method of Breeding - Increased at SCS Nursery, San Antonio, TX.

Intended Use - Pasture, hay production, range reseeding.

Description - Good seed producer, aggressive, air salinity tolerance, relished by cattle. Medium to fine textured soils.

Adapted to - LRR H, I, J, P, T; PHZ 8.

Released - Informally by SCS about 1944.

Breeder Seed/Stock - Plant Material Center, SCS, San Antonio, TX.

Certified Seed/Stock - Not available. Common in commercial production.

Preparer/Additional Information - Plant Materials Center, SCS, Rte 1 Box 155, Knox City, TX 79529, (817) 658-3922.

Dichanthium aristatum (Poir.) C.E. Hubbard - bluestem

Andropogon nodosus (Willem.) Nash

Warm-season, semi-decumbent bunchgrass from India. Used for pasture and hay in humid parts of the Gulf Coast plains of Texas.

Gordo

Selected at SCS Nursery, San Antonio, TX - D.H. Foster. Carried as accession no. PI 190302.

Source - Introduced from Union of South Africa as PI 190302, BN-6851. Received April 1951 as *Andropogon* sp.

Method of Breeding - Selected after comparison with several similar accessions from Africa. Increased for testing as T-20062.

Intended Use - Pastureland, hay production.

Description - Seedlings prostrate. At seeding, plants grow erect. Seedlings vigorous, establish readily. Leafy, stems up to 1.8 meters. Growth starts late spring or early summer and continues well into fall. Relished by cattle. Does well on heavy, seasonally wet soils. Responds well to fertilizer and/or irrigation.

Adapted to - LRR I, J, T; PHZ 9.

Released - Informally by SCS in 1957.

Breeder Seed/Stock - Not available.

Certified Seed/Stock - Not available. Some common seed sold by variety name.

Preparer/Additional Information - Plant Materials Center, SCS, Rte 1 Box 155, Knox City, TX 79529, (817) 658-3922.

Medio

Increased at SCS Nursery, San Antonio, TX - J.E. Smith, Jr. Carried as accession no. T-20011.

Source - Medio Creek, Bee County, TX, near State Highway 202 bridge. Lines lower benches of Medio Creek from near Berclair to Copano Bay and has evidently been in place for many years. All other except woody vegetation excluded by grass where it is established. Apparently first noticed by Dick Sentor, SCS, about 1940. Failed to reproduce it. In February 1951, S.E. Wolff, SCS Nursery, San Antonio, with Roy Boethel and Alfred Taylor, Beeville SCS Work Unit, collected five sod clumps, which were taken to San Antonio by Wolff, divided, set out in two-rod rows, and given accession number T-20011. Country of origin unknown.

Method of Breeding - Increase of bulk material from natural stand on Medio Creek.

Intended Use - Pasture, hay production, erosion control.

Description - Fine stemmed. Tolerant to alkaline soils and 500-860 mm rainfall. Grown best in clay or shallow sandy soils with clay layer within 500 mm. Prostrate stems will form sod, seed stems to 750 mm. Responds well to fertilizer and/or irrigation.

Adapted to - LRR I, J, T; PHZ 8.

Released - Informally by SCS in 1954.

Breeder Seed/Stock - Not available.

Certified Seed/Stock - Not available. Some common seed sold by variety name.

Preparer/Additional Information - Plant Materials Center, SCS, Rte 1 Box 155, Knox City, TX 79529, (817) 658-3922.

T-587

Selected at SCS Nursery, San Antonio, TX. Increased as T-587 at the James E. "Bud" Smith Plant Materials Center, Knox City, TX, 1965-68. Carried as accession no. PI 421783.

Source - World assembly made by Dr. J.R. Harlan, ARS, in the early 1950s.

Method of Breeding - Seed from 50 accessions thought to have the least cold tolerance was harvested as a composite and planted as a seed increase field. Winter temperatures from 1962 to 1980 removed several individuals. A completely unidentifiable, heterozygous composite resulted.

Intended Use - Pastureland, hay production, revegetation of disturbed areas and salt scalds, erosion control, range reseeding.

Description - More cold-tolerant than Gordo, Medio, and Kleberg. Leafier than King Ranch, leaves longer and wider. Resistant to leaf rust. High forage producer. Grows from late winter/early spring through late fall.

Adapted to - LRR H, I, J, P, T; PHZ 7.

Released - 1981 - Germplasm released by SCS and Texas Agricultural Extension Service.

Breeder Seed/Stock - Plant Materials Center, SCS, Knox City, TX.

Certified Seed/Stock - Not available.

Preparer/Additional Information - Plant Materials Center, SCS, Rte 1 Box 155, Knox City, TX 79529, (817) 658-3922.

Digitaria californica (Benth.) Henr. - Arizona cottontop

Trichachne californica (Benth.) Chase

Warm-season, native bunchgrass. Distributed from west Texas to Colorado, Arizona, and South America; found in abundance during years having good spring rains when it attains height of about 0.9 m. Light, fluffy seed produced throughout growing season when moisture favorable. Seed difficult to harvest and process. Palatable to livestock.

Digitaria ciliaris (Retz.) Koel. - crabgrass

The species are in the main good forage grasses. In the southern states, where it produces an abundant growth in late summer on fields from which crops have been gathered, it is utilized for forage and is sometimes cut for hay. Plant branching and spreading, often purplish, rooting at the decumbent base, the culms sometimes as much as 1 m. long, the flowering shoots ascending; sheaths, at least the lower, papillosepilose. Occurs in states at low and medium altitudes, more common in the East and South; temperate and tropical regions of the world. Native of Europe.

Red River

Noble Foundation - R.L. Dalrymple. Carried as selection no. RR874.

Source - Developed from seed collected from a single plant near Burneyville, OK.

Method of Breeding - Red River was a natural selection and no type of artificial breeding was used in the evolution and development of the cultivar. The plant appears to be primarily self-pollinated with capability of cross-pollination.

Intended Use - Grazing and hay during the warm season, stored field forage for fall and early winter, and for soil conservation. It is also useful in various grass and grass/legume mixtures or winter-summer double cropping. It could be used as temporary cover for certain grass plantings, lawns, etc.

Description - Red River is the first known variety of *Digitaria ciliaris* and therefore it has no relative descriptions to other cultivars. It contains the general morphological characteristics of *D. ciliaris,* but is an agronomic selection of known and better forage characteristics. The major forage area is anticipated to be from Kansas south and east to the coasts and as far north as Maryland. Production has been good in the 710-750 mm rainfall belts and above. It performs very well under irrigation. Palatability is very high. Quality is excellent. It fits well in winter crop double cropping regions, i.e., wheat-crabgrass, etc.

Adapted to - LRR A, B; PHZ 7, 8, 9, 10.

Released - 1988, by Agriculture Division, Noble Foundation, Ardmore, OK.

Breeder Seed/Stock - Agriculture Division, Samuel Roberts Noble Foundation, P.O. Box 2180, Ardmore, OK 73402.

Certified Seed/Stock - Available.

Preparer/Additional Information - R.L. Dalrymple, Noble Foundation, P.O. Box 2180, Ardmore, OK 73402, (405) 223-5810.

***Digitaria eriantha* Steud. - pangolagrass**
Digitaria decumbens Stent
Digitaria pentzii Stent

Warm-season, stoloniferous grass from Republic of South Africa. Used primarily for pasture in central and southern Florida and Caribbean region. Productive, nutritious, tolerates grazing, not frost resistant. A high-yielding species well adapted for pasture use in central and southern Florida. Must be propagated vegetatively.

Mealani

Source - University of Hawaii.

Method of Breeding - Colchicine treatment of single-node Pangola digitgrass cuttings (2n=27). This resulted in the production of a hexaploid clone (2n=54).

Intended Use - Pasture and erosion control.

Description - A creeping to decumbent, long-lived grass that grows to a height of 0.6-1.2 m. The stolons have an abundant mat of hair at the nodes. The fruiting stalks produced on the decumbent stems extend beyond the top leaves. Flowers are born on finger-like branches; very few, if any, viable seeds are produced. Spikelets are about six mm long. Stems that touch the ground root readily at the nodes producing a mat-like growth habit. The runners may grow three to six meters. Leaves are 5-225 mm long and about eight mm wide, smooth on both sides.

Adapted to - From sea level to over 800 m in Hawaii. Annual rainfall must be at least 100 cm. Mealani is more productive than Pangola digitgrass during the cool season above 730 m elevation.

Released - By Dr. U. Urata of the Hawaii Agricultural Experiment Station, University of Hawaii.

Breeder Seed/Stock - Vegetative material is maintained by the University of Hawaii and the SCS Hawaii Plant Materials Center.

Certified Seed/Stock - Not available.

Preparer/Additional Information - Robert Joy, Plant Materials Center, SCS, P.O. Box 236, Hoolehua, HI 96729, (808) 567-6378.

Pangola

S.C. Schank, E.M. Hodges, G.B. Killinger, and D.E. McCloud. Carried as accession no. PI 111110.

Source - PI 111110 vegetative planting material received from Republic of Africa in 1935.

Method of Breeding - Comparative tests and pasture plots, 1941-42.*Intended Use* - Forage.

Description - A stoloniferous perennial with straight, smooth, and very abundant leaves. The firm ligule is free-standing and 2.5-5.0 mm long. The stoloniferous vegetative stems produce roots at the nodes when they are in contact with the soil. The nodes of the stolons are very pubescent yet the leaf sheaths and blades are nearly glabrous.

Adapted to - LRR T, U; PHZ 10.

Released - 1943, by Florida Agricultural Experiment Stations.

Breeder Seed/Stock - Not available.

Certified Seed/Stock - Not available, but common is planted throughout Florida.

Preparer/Additional Information - S.C. Schank, Florida AES, 2183 McCarty Hall, Univ. of Florida, Gainesville, FL 32611, (904) 392-1823.

Slenderstem

Increased at Range Cattle Experiment Station, Ona, FL - J.E. McCaleb and E.M. Hodges. Carried as accession no. PI 300935.

Source - Vegetative planting material received in 1953 as Leesburg 5 from Watermelon and Grape Investigation Laboratory, Leesburg, FL. Designated as PI 300935 from sample sent to New Crops Research Branch, Plant Science Research Division, Beltsville, MD., in 1959. Closely resembles *Digitaria seriata* Stapf (PI 106657) that was planted by G.E. Ritchey at Gainesville, FL, in 1936. Some 70 grasses were taken to Leesburg in 1950, and it is probable that PI 106657 was included in this group.

Method of Breeding - Direct increase of accession and thorough evaluation in grazing tests.

Intended Use - Forage.

Description - Stolons produced in large numbers, similar to pangolagrass in prostrate growth habit and production of roots at each node; forms denser sod than pangolagrass; topkills at about same temperature as pangolagrass but starts regrowth early and more productive than pangolagrass and the bahiagrasses from October to May. Characterized by slender stems, scarcity of seedheads, dense upright growth in established pastures, and dusty bluish tinge (glaucous) on dense undisturbed growth. Tests on seed viability have not been made.

Adapted to - LRR T, U; PHZ 9, 10. *Released* - 1969, by Florida AES.

Breeder Seed/Stock - Agricultural Research Center, University of Florida, Ona, FL.

Certified Seed/Stock - Not available.

Preparer/Additional Information - Stanley Schank, Florida AES, 2183 McCarty Hall, Univ. of Florida, Gainesville, FL 32611, (904) 392-1823.

Taiwan

Introduced to the U.S. by the ARS, Beltsville, MD, in 1962 from the Taiwan Agricultural Research Institute, Taipei. Carried as accession no. 279651. Tested at Fort Pierce and other southern Florida locations.

Source - Obtained from the Philippines by the Taiwan Agricultural Research Institute, Taipei.

Method of Breeding - It was brought to the U.S. in 1962 by ARS, Beltsville, as a single selection denoted as Taiwan A-24 strain of Pangola digitgrass. It was placed in *D. pentzii* Stent., now *D. erianta* Steud., based on chromosome number of 18 compared to 27 for Pangola.

Intended Use - Forage grass for grazing and hay production.

Description - It closely resembles Pangola but is less winter-hardy and has a more restricted growing range than Pangola. It has larger leaves and stems than Pangola and is larger than Transvala. Exhibits better resistance to certain insects than Pangola and has high resistance to Pangola stunt virus. The 2n=18.

Adapted to - LRR U; PHZ 10.

Released - 1978, Univ. of Florida Institute of Food and Agricultural Sciences, Agricultural Experiment Stations.

Breeder Seed/Stock - Florida Foundation Seed Producers, Inc., Greenwood, FL.

Certified Seed/Stock - Not available.

Preparer/Additional Information - Dr. A.E. Kretschmer, Jr., Florida AES, 2183 McCarty Hall, Univ. of Florida, Gainesville, FL 32611, (904) 392-1823.

Transvala

University of Florida Institute of Food and Agricultural Sciences, Florida Agricultural Experiment Stations, Gainesville, FL - Drs. Schank, Boyd, Smith, Hodges, West, Kretschmer, Brolmann, and Moore. Carried as accession no. 299601.

Source - Collected in Transvala, Africa, between Nelspruit and White River by A.J. Oakes, USDA, Beltsville, MD.

Method of Breeding - Selected plant introduction after testing for Pangola stunt virus in Guyana and Suriname, South America.

Intended Use - Forage grass for grazing and hay production for peninsular Florida.

Description - It is resistant to sting nematodes and Pangola stunt virus, both of which reduce the yield of Pangola. It is a triploid (3x=27 and there is nearly complete male sterility. The grass winter-kills north of about 30°N. latitude.

Adapted to - LRR T, U; PHZ southern 9, 10.

Released - 1973, University of Florida Institute of Food and Agricultural Sciences, Florida Agricultural Experiment Stations.

Breeder Seed/Stock - Florida Foundation Seed Producers, Inc., Greenwood, FL.

Certified Seed/Stock - Not available.

Preparer/Additional Information - Florida AES, 2183 McCarty Hall, Univ. of Florida, Gainesville, FL 32611, (904) 392-1823.

Digitaria X umfolozi D.W. Hall in Turrialba

Survenola

Developed at University of Florida Institute of Food and Agricultural Sciences, Gainesville, FL - Drs. Schank, Ruelke, Ocumpaugh, Moore, and Hall. Carried as accession no. 421785.

Source - Parents received as vegetative material from Germplasm Resources Laboratory, ARS.

Method of Breeding - Female parent, *Digitaria setivalva* Roth ex Roemer & J.A. Schultes (PI 299892) crossed with male parent, *Digitaria valida* Stent., now *D. eriantha* Steud. (PI 299850) using the mutual pollination technique at the University of Florida. Crosses produced 50 seed, nine of which were viable. Most aggressive plant was denoted as X46-2 and tested extensively. Later given PI 421785. Cultivar name derived from successful adaptation in Suriname and Venezuela.

Intended Use - A forage grass for grazing and hay production for the tropics and limited areas of Florida on well-fertilized upland soils. Not recommended for Florida flatwoods soils.

Description - Has much wider leaf blades than other released digitgrass cultivars (usually 10-13 mm vs. 8 mm or less). Also has glabrous sheaths while other presently released *Digitaria* cultivars have sheaths with hairs. Panicles have 7-12 branches, rachis winged, spikelets paired on pedicels of very unequal length. Higher IVOMD than Pangola or Transvala.

Adapted to - LRR T, U; PHZ 10.

Released - 1982, University of Florida Institute of Food and Agricultural Sciences, Gainesville, FL.

Breeder Seed/Stock - Florida Foundation Seed Producers, Inc., Greenwood, FL.

Certified Seed/Stock - Not available.

Preparer/Additional Information - Florida AES, 2183 McCarty Hall, Univ. of Florida, Gainesville, FL 32611, (904) 392-1823.

Echinoohloa colona (L.) Link - jungle ricegrass

Warm-season annual bunchgrass introduced from tropical regions. Grows in moist places from New Jersey to Missouri and south to Florida and Texas, also from Oregon to southern California. Seeds eaten by birds. Leaves tender; provides good grazing.

Baldwin

Increased at Plant Materials Center, SCS, Americus, GA. Carried as accession no. PI 491424.

Source - Collection from Baldwin County, Alabama.

Method of Breeding - Direct increase of original collection, identified by accession numbers AM-430, MS-180, F-1210, and F-1459.

Intended Use - Source of food for waterfowl.

Description - More robust than most accessions tested, grows to a height of 500-610 mm. Self-seeding, produces seed in about 90 days. Adapted throughout southern region where water supplies are plentiful. Used primarily as source of food for waterfowl when grown where shattered seed will be flooded. May be used as a catch crop for forage production.

Adapted to - LRR P, N, O, T, U; PHZ 8.

Released - Not formally. Distributed for testing.

Breeder Seed/Stock - Plant Materials Center, SCS, Americus, GA.

Certified Seed/Stock - Not available.

Preparer/Additional Information - Charles M. Owsley, Plant Materials Center, SCS, 295 Morris Dr., Americus, GA 31709, (912) 924-2286.

Echinochloa frumentacea Link - Japanese millet

***Echinochloa crus-galli* (L.) Beauv. var. *frumentacea* (Link) W. Wight**

Warm-season annual from India and Japan. Limited use for green feed, silage, and hay in cooler parts of northeastern U.S.. Grows better under cool conditions than sudangrass or foxtail millet.

Chiwapa

Increased at Plant Materials Center, SCS, Coffeeville, MS, by R.S. Thornton, V.E. Aldrich, W.C. Young, and M. Byrd. Carried as accession nos. PI 434028, & MS-181.

Source - Contaminant in accession of *Setaria italica* (PI 196293) received from India.

Method of Breeding - Increase of single plant found in original accession.

Intended Use - Wildlife food, primarily ducks; also good for temporary grazing and hay.

Description - Tall growing annual (up to 1.5 m on upland and bottomland sites).

Adapted to - LRR P, O, T, N, U; PHZ 7.

Released - 1965, cooperatively by MS AES and Plant Materials Center, SCS, Coffeeville, MS.

Breeder Seed/Stock - Plant Materials Center, SCS, Coffeeville, MS.

Certified Seed/Stock - Not available. Some commercial production.

Preparer/Additional Information - David M. Lane, Jamie L. Whitten Plant Materials Center, SCS, Route 3, Box 215A, Coffeeville, MS 38922, (601) 675-2588.

Ehrharta calycina Sm. - perennial veldtgrass

Cool-season bunchgrass indigenous to the Union of South Africa. Used in range reseeding in California. Increased at California AES, Davis.

Mission (Reg. No. 10)

Selected at California AES, Davis, CA, by R.M. Love. Carried as accession no. PI 187309.

Source - Received from R.C. Rossiter, Perth, Western Australia, under accession number T.O. 1883 (PI 187309). Mission accession number T.O. 3051.

Method of Breeding - Original panicle selections reduced to polycross consisting of 25 clones. Distributed for testing as a nonshedding perennial veldtgrass.

Intended Use - Range reseeding and revegetation of disturbed areas.

Description - Differs from typical *E. calycina* in having contracted panicles and retaining larger part of seed to maturity. Somewhat shorter in stature than typical perennial veldtgrass, but otherwise has same range of variation in flowering date, growth habit, and fineness of foliage. Superior with respect to both seed yield and seed quality.

Adapted to - LRR C; PHZ 9.

Released - 1962, cooperatively by California AES and Plant Materials Center, SCS, Pleasanton, CA.

Breeder Seed/Stock - California AES and Plant Materials Center, SCS, Lockeford, CA.

Certified Seed/Stock - Not available.

Preparer/Additional Information - R. Slayback, SCS, 2121-C, 2nd St., Davis, CA 95616, (916) 757-8257.

Elymus canadensis L. - Canada wildrye

Cool-season, native bunchgrass widely distributed in U.S., especially in Great Plains, intermountain region, and Pacific Northwest. Tall, coarse, short lived. Should be cut in boot stage for best quality hay. Fair palatability. Establishes rapidly, not competitive with other grass species, used in grass mixtures to increase production from young stands of grasses that are slow to establish.

Mandan

Selected at ARS, Northern Great Plains Research Laboratory, Mandan, ND. by G.A. Rogler. Carried as accession no. Mandan 419.

Source - Seed of selected plants was collected and bulked from an upland site near Mandan, ND, in 1935.

Method of Breeding - Mass selection within progenies of two single plants for leafiness, fineness of leaves, short stature, and resistance to stem rust.

Intended Use - Mandan Canada wildrye has little current commercial usage. This cultivar establishes rapidly, has high forage and seed yields, and has been used in forage mixtures to increase production from young stands of longer-lived grasses that are slower to establish. It is adapted to sandy soils and has been used to stabilize "blow-out" areas.

Description - It has shorter stature than common Canada wildrye, but lacks other visual distinguishing characteristics. Seed of Canada wildrye has awns that must be removed before it will flow satisfactorily through a drill.

Adapted to - LRR B, G, F; PHZ 2, 3, 4.

Released - 1946, by ARS, SCS, and the North Dakota AES.

Breeder Seed/Stock - ARS, Northern Great Plains Research Laboratory, Mandan, ND 58554.

Certified Seed/Stock - Not available.

Preparer/Additional Information - John D. Berdahl, ARS, Northern Great Plains Research Laboratory, P.O. Box 459, Mandan, ND 58554, (701) 663-6445.

Elymus dahuricus Turcz. ex Griseb. - Dahurain wildrye

Cool-season, short-lived perennial bunchgrass native to Siberia, Mongolia, and northern China. Adapted to the northern Great Plains region of the U.S. and the prairie provinces of Canada. Nutritional quality and palatability similar to Altai and giant wildrye. Establishes rapidly and has quick recovery after cutting or grazing. Used in grass mixtures to increase production from young stands of grasses that are slow to establish.

Arthur

Selected at Agriculture Canada Research Station, Swift Current, Saskatchewan, by Tom Lawrence. Carried as accession no. Sc D27621.

Source - Arthur traces to plant introduction Sc 1732 which was collected by J.W. Morrison Research Branch, Agriculture Canada, Ottawa, on a trip to the People's Republic of China.

Method of Breeding - Pure line breeding method was used to develop lines for testing. Lines were compared for seedling forage yield, aftermath yield, stand persistence, seed yield, N and P content of forage, and organic matter digestibility. A cafeteria-style grazing experiment with beef cattle indicated that Arthur Dahurian wildrye was as palatable as Altai or giant wildrye.

Intended Use - Arthur Dahurian wildrye is a short-lived perennial which survived three years under dryland conditions of southern Canadian prairies. It may be used as a short rotation hay or pasture species or in alternate rows with Russian wildrye on permanent pasture stand to improve early stand productivity.

Description - The main attribute of Arthur Dahurian wildrye is its high establishment year yields compared to other wildrye grasses. It is a self-pollinated, caespitose, short-lived perennial with a chromosome complement of 2N=6X=42. There are no distinguishing morphological characteristics compared to other cultivars of Dahurian wildrye. It heads two days earlier than James at Swift Current, Saskatchewan, Canada. Forage quality of Arthur was similar to other wildrye species.

Adapted to - LRR F, G; PHZ 2, 3, 4.

Released - 1989, by Agriculture Canada.

Breeder Seed/Stock - Agriculture Canada Research Station, P.O. Box 1030, Swift Current, Saskatchewan S9H 3X2 Canada.

Certified Seed/Stock - SeCan Association, Suite 512, 885 Meadowlands Dr., Ottawa, Ontario K2C 3M2 Canada.

Preparer/Additional Information - Paul G. Jefferson, Agriculture Canada, Research Station, P.O. Box 1030, Swift Current, Saskatchewan S9H 3X2 Canada, (306) 773-4621.

James

Selected at Agriculture Canada Research Station, Swift Current, Saskatchewan, by Tom Lawrence. Carried as accession no. Sc D27620.

Source - James originates from plant introduction Sc 1732 which was collected by J.W. Morrison, Research Branch, Agriculture Canada, Ottawa, on a trip to the People's Republic of China.

Method of Breeding - The pure line breeding method was used to develop lines for testing. Lines were compared for seedling forage yield, aftermath yield, stand persistence, seed yield, N and P content of forage, and organic matter digestibility. A cafeteria-style grazing experiment with beef cattle indicated that James Dahurian wildrye was as palatable as Altai or giant wildrye.

Intended Use - James Dahurian wildrye is a short-lived perennial which survived three years at Swift Current, Saskatchewan. It may be used in alternate rows with Russian or Altai wildryes to increase early stand productivity in long-term pastures or as a short rotation hay or pasture species in grain crop systems.

Description - The main attribute of James Dahurian wildrye is its high establishment year yields compared to other wildrye grasses. It is self-pollinated, caespitose, short-lived perennial with a chromosome complement of 2N=6X=42. It has no distinguishing morphological characteristics compared to other cultivars of Dahurian wildrye. It heads two days later than Arthur at Swift Current, Saskatchewan. Forage quality of James was similar to other wildrye species.

Adapted to - LRR F, G; PHZ 2, 3, 4.

Released - 1989 by Agriculture Canada.

Breeder Seed/Stock - Agriculture Canada Experimental Farm, Indian Head, Saskatchewan SOG 2KO.

Certified Seed/Stock - SeCan Association, Suite 512, 885 Meadowlands Dr., Ottawa, Ontario K2C 3M2.

Preparer/Additional Information - Paul G. Jefferson, Agriculture Canada, Research Station, P.O. Box 1030, Swift Current, Saskatchewan S9H 3X2 Canada, (306) 773-4621.

Elymus elymoides **(Raf.) Swezey - bottlebrush squirreltail**

Sitanion hystrix **(Nutt.) J.G. Sm.**

Culms erect to spreading, rather stiff 10-50 cm tall; foliage from glabrous or puberulent to softly and densely white-pubescent. Occurs on dry plains, open woods, and rocky slopes, SD to British Columbia, south MO, TX, CA, and Mexico. At high altitudes, plants are dwarf.

9040187

Developed by the Upper Colorado Environmental Plant Center, Meeker, CO. Carried as accession no. 9040187.

Source - Collected by Larry Klock, near Pueblo, Pueblo County, CO.

Method of Breeding - Selected from row tested in dryland project at the Upper Colorado Environmental Plant Center.

Intended Use - Oil shale and coal reclamation, as well as critical area treatment and for erosion control on cropland and rangeland.

Description - Good drought tolerance, leafy, tall forage type, maturing 30 days later than common ecotypes. Fair seed production. Very likely this accession will be blended with 9040189 for the release. Only somewhat winter-hardy, but extremely hardy under droughty conditions, primarily important as a winter forage.

Adapted to - LRR D, E; PHZ 5.

Released - No. Distributed for field testing.

Breeder Seed/Stock - Breeders seed will be maintained at the Upper Colorado Environmental Plant Center.

Certified Seed/Stock - Not available.

Preparer/Additional Information - Randy Mandel, Upper Colorado Environmental Plant Center, P.O. Box 448, Meeker, CO 81641, (303) 878-5003.

9040189

Selected for advance testing by the Upper Colorado Environmental Plant Center, Meeker, CO. Carried as accession no. 9040189.

Source - Collected by Tim Carner, August 1981, near Buford, Rio Blanco County, CO.

Method of Breeding - Selected from row tested in dryland project at Upper Colorado Environmental Plant Center.

Intended Use - Oil shale and coal reclamation as well as critical area treatment and erosion control on cropland and rangeland.

Description - Fairly drought tolerant, leafy, tall forage type maturing 30 days later than common ecotypes. Good seed production. Very likely this accession will be blended with 9040187 for the release. Only somewhat winter-hardy, but extremely hardy under droughty conditions, primarily important as a winter forage.

Adapted to - LRR D, E; PHZ 5.

Released - No, distributed for testing.

Breeder Seed/Stock - Breeders seed will be maintained at the Upper Colorado Environmental Plant Center.

Certified Seed/Stock - Not available.

Preparer/Additional Information - Randy Mandel, Upper Colorado Environmental Plant Center, P.O. Box 448, Meeker, CO 81641, (303) 878-5003.

Elymus glaucus Buckl. - blue wildrye

Cool-season, native bunchgrass found from north-central region to Pacific coast. Shade tolerant, short lived. Relatively common on cut-over or burned-over land in Pacific Northwest. No varieties available.

Elymus lanceolatus (Scribn. & J.G. Sm.) Gould ssp. *lanceolatus* - streambank wheatgrass

***Elytrigia dasystachyam* (Hook.) A. & D. Love**
***Agropyron dasystachyum* (Hook.) Scribn. & J.G. Sm.**
***Agropyron riparum* Scribn. & J.M. Sm.**

Cool-season, sod-forming grass. Found from Montana to Washington and south into Nevada, Utah, and Colorado. Adapted for erosion control and general-purpose dryland turf. Drought resistant and alkali tolerant.

9006633

Selected at Plant Materials Center, SCS, Pullman, WA - J.L. Schwendiman.

Source - A native collection east of The Dalles, OR, in 1934. Previously carried as P-1822.

Method of Breeding - Selected by elimination of aberrant plants during several generations. Bulked selections increased under isolation.

Intended Use - Rangeland reseeding on coarse textured, droughty soil sites.

Description - A vigorous, blue, leafy, rapid spreading productive strain. Foliage and culms glaucus with only partial pubescence on the lemmas. Seed production is good. Chemically treated seed before planting to control head smut, when grown for seed. Well adapted to light textured soils in wind erosion areas. Chromosome number is 2n=28.

Adapted to - LRR B; PHZ 5.

Released - Not released.

Breeder Seed/Stock - Plant Materials Center, SCS, Pullman, WA.

Certified Seed/Stock - Not available.

Preparer/Additional Information - Plant Materials Center, SCS, Room 104 Hulbert Agricultural Sciences Building, Washington State Univ., Pullman, WA 99164-6211, (509) 335-7376.

Critana (Reg. No. 9)

Selected at Plant Materials Center, SCS, Bridger, MT, by A.A. Thornburg. Carried as accession nos. PI 469235, M-286, P-15581.

Source - Collected from roadside cuts near Havre, Hill County, MT, by D.E. Ryerson, 1960.

Method of Breeding - Direct increase of field collection after comparisons with approximately 60 other accessions representing native collections from Montana and Wyoming. Tested as P-15581.

Intended Use - Critana was released primarily for stabilization of disturbed areas, roadsides, airports, recreation areas, and construction sites receiving little or no maintenance.

Description - Strongly rhizomatous with excellent seedling vigor, low growing, abundant fine leaves; produces tight sod under dryland conditions but little forage. Good seed producer.

Adapted to - LRR B, D, E, F, G; PHZ 3.

Released - 1971, cooperatively by Plant Materials Center, SCS, Bridger, MT, and the Montana and Wyoming Agricultural Experiment Stations.

Breeder Seed/Stock - Plant Materials Center, SCS, Bridger, MT.

Certified Seed/Stock - Available.

Preparer/Additional Information - John G. Scheetz, Plant Materials Center, SCS, RR 1, Box 1189, Bridger, MT 59014, (406) 662-3579.

Sodar

Selected at SCS Plant Materials Centers at Aberdeen, ID, and Pullman, WA, by R.H. Stark and J.L. Schwendiman. Carried as accession no. P-2415 and PI 421021.

Source - Collected near Canyon City, Grant County, OR, by R.G. Johnson in area of 305 mm annual rainfall at elevation of 915 m.

Method of Breeding - Best of 11 accessions. Improved by mass selection and elimination of aberrants during several generations at Plant Materials Center, SCS, Aberdeen, ID. Tested as P-2415.

Intended Use - Erosion control and soil stabilization on roadsides, airports, dry canal banks, recreational areas, and grassed waterways.

Description - Drought-resistant, rhizomatous grass, particularly adapted for erosion control. Excellent seedling vigor. Narrow, tough leaves. Produces open sod highly competitive to weeds and other plants under dryland conditions and offers excellent protection against soil erosion. Used primarily on roadsides, airports, and irrigation canal banks.

Adapted to - LRR B, D, E; PHZ 4, 5.

Released - 1954, cooperatively by Idaho and Washington Agricultural Experiment Stations at Moscow and Pullman, respectively, and SCS Plant Materials Centers, Aberdeen, ID, and Pullman, WA.

Breeder Seed/Stock - Plant Materials Center, SCS, Aberdeen, ID.

Certified Seed/Stock - Available.

Preparer/Additional Information - Gary Young, Plant Materials Center, SCS, P.O. Box 296, Aberdeen, ID 83210-0296, (208) 397-4133.

Thickspike

Source - Original collection made near the Dalles, OR, on Columbia River elevation less than 150 m in 1948, called P-1822.

Method of Breeding - Eight-clone selection from P-1822, called P-14943. SCS 9021076 selected from P-14943 spaced plantings.

Intended Use - Range forage, erosion control, and weed competition.

Description - It has numerous leaves and good forage production. It is a vigorous sod former.

Adapted to - LRR B, D, E; PHZ 4, 5, 6.

Released - Expected in 1994.

Breeder Seed/Stock - SCS, Plant Materials Center, Aberdeen, ID.

Certified Seed/Stock - Not available.

Preparer/Additional Information - Gary Young, Plant Materials Center, SCS, Box 296, Aberdeen, ID 83210-0296, (208) 397-4133.

Elymus trachycaulus (Link) Gould ex Shinners - slender wheatgrass

***Agropyron trachycaulum* (Link) Malte ex H.F. Lewis var. *glaucum* (Pease & Moore) Malte**

Important cool-season bunchgrass in northern Great Plains, west to eastern Washington and Oregon, and south to Nevada, Utah, and Colorado. First native grass widely used for reseeding in western Canada and U.S.. Good seedling development, alkali tolerant, relatively short-lived, and less drought-resistant than western and crested wheatgrass. Seldom found in pure stands.

Adanac

Source - Agriculture Canada Research Station, Saskatoon, Saskatchewan.

Method of Breeding - Adanac originated from the seed of a single plant collected near Climax, Saskatchewan, in 1961. This plant was multiplied and then kept in cold storage. Breeder seed was first bulked in 1983. Tested as S-1755.

Intended Use - Hay and pasture production.

Description - Plants and seeds of Adanac are similar to those of Revenue except that Adanac is taller and has a lower leaf to stem ratio. Adanac averaged 7% higher yield than Revenue in hay yields during seven cooperative trials based on three- and four-year averages (23 station-years). Establishment, persistence, and productivity under saline conditions was superior to Revenue as measured in replicated trials conducted in the greenhouse and on a soil salinity gradient near Langham, Saskatchewan. Seed yields of adanac averaged 4% higher than Revenue based on four trial locations (19 station-years). Digestibility and protein content of hay are somewhat below those of Revenue. This may be attributed in part to a lower leaf to stem ratio and somewhat taller growth habit. Compared to Revenue, Adanac offers superior yield performance for the lighter textured, drier soil areas, superior salinity tolerance, and indicated somewhat more ease of establishment.

Adapted to - Recommended in Canada for use in Saskatchewan.

Released - Agriculture Canada, Saskatoon Research Station and the Agriculture Canada Experiment Farm, Indian Head. Registered in Canada in 1990 as registration no. 3217.

Breeder Seed/Stock - Agriculture Canada, Saskatoon Research Station and the Agriculture Canada Experiment Farm, Indian Head.

Certified Seed/Stock - Available.

Preparer/Additional Information - Agriculture Canada, Food Production and Protection Branch, Plant Products Division, Ottawa, Ontario K1A 0C6. Seed is distributed by SeCan Association, 200-57, Promenade Auriga, Nepean, Ontario K2E 8B2.

Elbee

Source - Germplasm used for development of this variety came from indigenous stands in southern Alberta and southwestern Saskatchewan.

Method of Breeding - Eight-clone synthetic variety selected through clonal and polycross progeny performance. Selection was based on aggressiveness of creeping roots, and forage and seed yields.

Intended Use - This variety is primarily for range and pasture seedings in the Brown, Dark Brown and Black soil zones in the prairie areas and for ecological repair and revegetation of industrial disturbances, roadsides, and areas that will receive no maintenance.

Description - A perennial with moderately aggressive creeping root system, excellent germinability, high seedling vigor, early spring growth, and drought tolerance. Elbee plants are tufted with erect greyish green leaves.

Released - 1980 by S. Smoliak and A. Johnston of Agriculture Canada Research Station, Lethbridge, Alberta.

Breeders Seed/Stock - Agriculture Canada Research Station, Lethbridge, Alberta.

Certified Seed/Stock - Available through the seed trade.

Preparer/Additional Information - Surya N. Acharya, Agriculture Canada Research Station, Lethbridge, AB T1J 4B1, (403) 327-4561.

Primar (Reg. No. 1)

Selected at the Plant Materials Center, SCS, Pullman, WA - A.L. Hafenrichter, J.L. Schwendiman, and A.G. Law.

Source - Collected near Beebe, MT, in 1933 by the USDA-FS.

Method of Breeding - Selected from original collection, assigned accession number P-2535, and tested with 104 other accessions.

Intended Use - Green manure.

Description - Vigorous, early-growing wheatgrass; semi-erect, long-lived, slender. Usually 10 days earlier in seed maturity and 125-250 mm taller than late commercial strains. Leaves and stems moderately coarse and glaucous gray-green. Plants high in vegetative production. Resistant to leaf rust, stem rust, and stripe rust; superior to common slender wheatgrass in resistance to head smut. Seed production moderately heavy; seeds relatively large when compared with those of ordinary strains. Adapted for use in sweet clover-grass conservation mixtures for pasture, hay, and green manure. Alkali tolerant. Adapted to short-lived dryland seedings in areas with minimum of 360 mm of rainfall.

Adapted to - LRR B; PHZ 5.

Released - 1946, cooperatively by Washington, Idaho, and Oregon Agricultural Experiment Stations at Pullman, Moscow, and Corvallis, respectively; Plant Materials Center, SCS, Pullman, WA; and ARS, Plant Science Research Division.

Breeder Seed/Stock - Plant Materials Center, SCS, Pullman, WA.

Certified Seed/Stock - Limited availability.

Preparer/Additional Information - Plant Materials Center, SCS, Room 104, Hulbert Agricultural Sciences Bldg, WSU, Pullman, WA 99164-6211, (509) 335-7376.

Pryor

Selected at the Plant Materials Center, SCS, Bridger, MT - M.E. Majerus and J.G. Scheetz. Carried as accession nos. PI 432403, M-1030.

Source - Collected by M.E. Majerus from a dry drainageway within a saline-upland range site. The collection site is in the Cottonwood Creek drainage within the Clarks Fork of the Yellowstone River Valley between the Pryor and Beartooth Mountains of south-central Montana.

Method of Breeding - Direct increase of field collection.

Intended Use - Pryor is used primarily in mixtures for conservation and reclamation plantings. This includes range revegetation, coal and bentonite reclamation, and saline-alkaline affected areas.

Description - Superior to other slender wheatgrasses in drought tolerance, saline tolerance and seedling vigor. Is earlier to mature, and has larger seed size (213,000 seeds/kg compared to 320,000 seeds/kg).

Adapted to - LRR B, D, E, F, G; PHZ 3.

Released - 1988, cooperatively by Plant Materials Center, SCS, Bridger, MT, and the Montana and Wyoming Agricultural Experiment Stations.

Breeder Seed/Stock - Plant Materials Center, SCS, Bridger, MT.

Certified Seed/Stock - Available.

Preparer/Additional Information - John G. Scheetz, Plant Materials Center, SCS, RR 1, Box 1189, Bridger, MT 59014, (406) 662-3579.

Revenue

Selected at Canada Department of Agriculture, Research Station, Saskatoon, Saskatchewan, by W.L. Crowle.

Source - Seed of single plant collected in native prairie near Revenue, Saskatchewan, in 1961. Assigned accession number E.S.S. 1558.

Method of Breeding - Selected from over 750 native and introduced collections evaluated from 1959 to 1969.

Intended Use - Pasture.

Description - Superior to Primar in establishment, salinity tolerance, forage and seed yield; and similar in maintaining good stands for three to five years of production. Characterized also by a higher leaf-to-stem ratio, dry-matter digestibility, and freedom from smut. Well suited for use on saline soils and in short rotations.

Released - 1970, by Canada Department of Agriculture.

Breeder Seed/Stock - Canada Department of Agriculture, Research Station, Saskatoon, Canada.

Certified Seed/Stock - Available commercially. Saskatchewan Wheet Pool, 2625 Victoria Ave., Regina, Saskatchewan S4L 7Y9.

Preparer/Additional Information - Canada Seed Growers Association, P.O. Box 8445, Ottawa, Ontario K1G 3T1.

San Luis

Selected at the Upper Colorado Environmental Plant Center (UCEPC), Meeker, CO. Carried as accession no. PI 483079.

Source - Collected by Glenn Niner, August 1975, near Center, San Luis Valley, Rio Grande County, CO.

Method of Breeding - Increased from original collection and rogued by the UCEPC.

Intended Use - Soil stabilization on slopes and disturbed sites above 1,800 m elevation receiving 35 cm or more of annual precipitation. Recommended for ski slopes, transmission corridors, roadsides, and mined land reclamation.

Description - Outstanding, rapid establishment and longevity.

Adapted to - LRR D, E; PHZ 5.

Released - 1984, cooperatively by Colorado, Utah, and New Mexico Agricultural Experiment Stations, SCS, and the Upper Colorado Environmental Plant Center.

Breeder Seed/Stock - Upper Colorado Environmental Plant Center, Meeker, CO.

Certified Seed/Stock - Available.

Preparer/Additional Information - Randy Mandel, Upper Colorado Environmental Plant Center, P.O. Box 448, Meeker, CO 81641, (303) 878-5003.

Elytrigia elongata (Host) Nevski - tall wheatgrass

***Agropyron elongatum* (Host) Beauv.**
***Elymus elongatus* (Host) Runemark**

Cool-season bunchgrass from Turkey and the former USSR. Used for hay and pasture in northern Great Plains and intermountain region. Tall, coarse, late-maturing species. Can be grown successfully on wet, alkaline soils. Less drought-resistant than crested wheatgrass. Good seed producer. Produces high yields, but less palatable than most wheatgrasses.

Alkar (Reg. No. 7)

Selected at Plant Materials Center, SCS, Pullman, WA - J.L. Schwendiman.

Source - Selection from PI 98526; introduced from the former USSR in 1932.

Method of Breeding - Mass selection from spaced plants of above introduction. Tested as P-2326.

Intended Use - Pasture for saline and alkaline land.

Description - Tall, very late maturing wheatgrass; stemmy bunch type. Coarse blue-green leaves. Large seeded, good seedling vigor. Very tolerant to wet, alkaline conditions and semi-arid regions of the west at 90-1830 m elevation. Fairly palatable and highly productive on sub-irrigated and irrigated saline and alkaline land.

Adapted to - LRR B; PHZ 5.

Released - 1951, under accession P-2326 for certified seed production in Idaho, Washington, and Oregon. Named Alkar in 1958 and accepted for certification in those states and California.

Breeder Seed/Stock - Plant Materials Center, SCS, Pullman, WA.

Certified Seed/Stock - Limited availability.

Preparer/Additional Information - Plant Materials Center, SCS, Room 104, Hulbert Agricultural Sciences Bldg, WSU, Pullman, WA 99164-6211, (509) 335-7376.

Jose

Selected at former SCS nursery, Albuquerque, NM, by J.A. Downs, G.C. Niner, and J.E. Anderson. Carried as accession nos. 150123 & BN-3654 & A 12465.

Source - Received from Beltsville, MD, as BN-3654 and PI 150123. Native to Eurasia, but introduced from Australia.

Method of Breeding - Mass increase of seed collected from isolated block of spaced plants after roguing atypical plants. *Intended Use* - Irrigated pasture and hay production.

Description - Uniform, leafy, medium-tall bunchgrass; not as coarse as other tall wheatgrasses tested; true-green type. Seed production not as high as some other strains tested, but forage production equal to most; limited observations indicate higher palatability. Earlier maturing than other strains tested. Is more drought tolerant than was originally realized. Has done well for five years on a FWRS dam at 1700 m elevation in a 240 mm rainfall area. Is also very salt-tolerant and is being used in areas where only very salty water is available for use.

Adapted to - LRR D, G, H, E; PHZ 4.

Released - 1965, cooperatively by New Mexico AES and Plant Materials Center, SCS, Los Lunas, NM.

Breeder Seed/Stock - Plant Materials Center, SCS, Los Lunas, NM.

Certified Seed/Stock - Available.

Preparer/Additional Information - Plant Materials Center, SCS, 1036 Miller St. SW, Los Lunas, NM 87031, (505) 865-4684.

Largo

Increased at former SCS Nursery, Albuquerque, NM, as A-1876, and at Utah AES, Logan, ARS cooperating, as PI 109452. Also carried as accession no. A-1876.

Source - Collected by Westover-Enlow expedition near Bandirma, Turkey. Introduced as *Agropyron intermedium* (Host) Beauv., now *Elymus intermedia* (Host) Nevski, and later identified as *A. elongatum* (Host) Beauv.

Method of Breeding - Bulk increase under isolation.

Intended Use - Forage production on saline and alkaline soils.

Description - Large, coarse, deep-rooted bunchgrass. Population only slightly variable, high percentage of bluish-green plants, maturing late, height 1.2-1.8 m. Widely used in Colorado, Utah, Arizona, and New Mexico for soil improvement and pasture on saline and alkaline soils. Highly productive in herbage and seed yields.

Adapted to - LRR D, G, H, E; PHZ 4.

Released - 1961, cooperatively by SCS-Los Lunas and New Mexico and Utah Agricultural Experiment Stations.

Breeder Seed/Stock - Plant Materials Center, SCS, Los Lunas, NM.

Certified Seed/Stock - Not available. Jose has almost totally replaced Largo.

Preparer/Additional Information - Plant Materials Center, SCS, 1036 Miller St. SW, Los Lunas, NM 87031, (505) 865-4684.

Nebraska 98526

Increased at SCS nurseries in cooperation with Nebraska AES, Lincoln. Tested as Nebraska 1978.

Source - PI 98526 originally grown at Colorado AES, Fort Collins, and later (1936) distributed to nurseries in Dakotas and Nebraska by SCS, PI 98526. Originally introduced into U.S. from the former USSR in 1932. Seed presented by N.I. Vavilov.

Method of Breeding - Direct increase of PI 98526 without selection.

Intended Use - Alkaline, saline soils in central and northern Great Plains and intermountain west.

Description - Erect, tall, somewhat coarse bunchgrass. Deeply veined, bluish-green leaves and coarse stems. Particularly well adapted to low, wet, or alkaline soil conditions, where it produces good yields. Also does very well in dry, upland plantings, where it appears moderately somewhat less aggressive and coarse than Turkish introductions with which it has been compared. Replaced by variety Platte which is based largely on Nebraska 98526.

Adapted to - LRR M, K, F, G, H; PHZ 3, 4, 5.

Released - First grown on Nebraska farms in 1950 for seed production under field certification by Nebraska Crop Improvement Association on recommendation of Nebraska AES, SCS, and ARS.

Breeder Seed/Stock - Nebraska AES.

Certified Seed/Stock - Not available. Replaced by variety Platte.

Preparer/Additional Information - ARS, 344 Keim Hall, Univ. of Nebraska, Lincoln, NE 68583-0910, (402) 472-1564; and Agriculture Research Division, Univ. of Nebraska, Lincoln, NE 68683.

Orbit

Selected at Canada Department of Agriculture, Research Station, Swift Current, Saskatchewan - Tom Lawrence.

Source - Nebraska 98526 (PI 98526) and seed from locally selected strains.

Method of Breeding - Twenty-six selections made within open-pollinated progeny lines that had 80% or better winter survival. Selections evaluated for winter hardiness, seed yield, and forage yield in progeny test. Variety is composite of nine open-pollinated lines and one three-clone synthetic.

Intended Use - Hay, pasture, and conservation uses in the northern Great Plains.

Description - Not distinguishable from other varieties and sources on basis of visual characteristics. Superior to Alkar and other varieties tested in winter hardiness, similar in seed and forage yield; withstands flooding for three to four weeks in Spring.

Adapted to - LRR F, G; PHZ 3.

Released - 1966, by Canada Department of Agriculture.

Breeder Seed/Stock - Canada Department of Agriculture, Research Station, Swift Current.

Certified Seed/Stock - Available.

Preparer/Additional Information - Agriculture Canada, Research Station, P.O. Box 1030, Swift Current, Saskatchewan S9H 3M2, (306) 773-4621.

Platte

Selected at Nebraska AES, Lincoln, ARS cooperating - L.C. Newell.

Source - Nebraska 98526 and a selection of unknown origin from early introductions grown at the Cheyenne Horticultural Field Station. Nebraska 98526 was an increase of PI 98526 that was presented to the USDA by N.I. Vavilov in 1932.

Method of Breeding - A seed collection was made in 1949 by E.C. Conard and L.S. Donovan from volunteer plants in grass nurseries at the Cheyenne Horticultural Field Station. The collection was characterized as a stiffly erect bunch type with good seed set. It was improved in the 1950s by periodic mass selection within spaced-planted nurseries of 800-1,000 plants and by short-row seed increases. A nursery of clones from Nebraska 98526 was developed from survivors of a planting on a saline-alkali soil at the North Platte, NE, Experiment Station and from selections in the variety from farmers' strains grown at Lincoln. This nursery of clones was screened by polycross testing to 77 clones in 1954. The resulting strain, Nebraska 98, was developed by two additional generations of mass selection and several generations of row planting on lowland saline soil on Stepens Creek bottom land near Lincoln, NE. The cultivar was synthesized by growing the two strains in alternate rows and bulk harvesting the field. Breeder seed was Syn 2 and Foundation Seed in Syn 3 generation seed. Foundation seed is only seed source for establishing certified seed fields.

Description - A winter-hardy bunchgrass spreading from short rhizomes. Inflorescence is an elongate spike with awnless lemmas. Leaves are long, narrow, non-flaccid, and upper leaf surface is deeply veined while lower leaf surface is smooth. Platte flowers in July and seed is ripe in mid to late August in the Central Plains. Plant height at seed maturity is 750-1,000 mm. It has a large seed (5-6 grams per thousand) that is easily confused with intermediate wheatgrass seed.

Intended Use - Revegetating saline-alkali soils in the Great Plains.

Adapted to - LRR G, H, F, on saline-alkali soils; PHZ 3, 4, 5.

Released - 1972, cooperatively by Nebraska AES and USDA-ARS, Univ. of Nebraska, Lincoln.

Breeder Seed/Stock - Nebraska AES and USDA-ARS, Univ. of Nebraska, Lincoln.

Certified Seed/Stock - Available.

Preparer/Additional Information - K.P. Vogel, ARS-USDA, 344 Keim Hall, Univ. of Nebraska, or Jeff Pederson, Dept. of Agronomy, Univ. of Nebraska, Lincoln, NE 68583, (402) 472-1564, (402) 472-2811.

Elytrigia intermedia **(Host) Nevski - intermediate pubescent wheatgrass**

Agropyron intermedium **(Host) Beauv.**
Agropyron trichophorum **(Link) Richt.**
Thinopyrum intermedium **(Host) Barkworth & Dewey**
Thinopyrum intermedium **ssp.** ***barbulatum*** **(Schur) Barkworth & Dewey**

Important cool-season, sod-forming grass from Eurasia. Contains pubescent wheatgrass which has short, stiff hairs on outer glumes, lemma and palea, and leaves, as well as plants without hairs. Plants integrate from one type to another. Pubescent forms tend to become more prevalent over glabrous forms in Central Asia as vegetation zones grade toward increased drought. Used for pasture and hay in northern Great Plains, west to eastern Washington, and south into Colorado and northern Kansas. Adapted in areas with annual rainfall of 380 mm or more; at some locations has grown well at elevations up to 3,050 m. Produces good hay yields, grows well with alfalfa, and suitable for erosion control. Not as drought-tolerant or winter-hardy as *Agropyron cristatum X desertorum* or *A. cristatum* and will not persist under heavy grazing pressure.

Amur

Selected at former SCS Nursery, Albuquerque, NM - J.A. Downs and T.F. Spaller. Carried as accession nos. PI 131532 & P-9838 & A-13046.

Source - PI 131532, Manchuria, China. Received as *Agropyron amurense* Drob. from SCS Nursery, Pullman, WA, P-9838. Identified by J. R. Swallen as *A. intermedium* (Host) Beauv.

Method of Breeding - Increase, under isolation, of seed from spaced plants after roguing awned and other offtype plants. In 1962 at Plant Materials Center, SCS, and New Mexico AES, Los Lunas Branch, third-generation seed was space planted and rogued to maintain uniformity. Tested as A-13046.

Intended Use - Revegetation of disturbed areas and range reseeding.

Description - Leafy, vigorous-growing, but slow sod-forming type; uniform gray green. Strong seedling vigor. High seed yield; maintains production in successive years better than other strains tested. Some introgression appears to have occurred with *A. trichophorum,* (Link) Richt., now *E. intermedia* (Host) Nevski, giving rise to a high percentage of pubescent plants. This brought into question its existing classification as *A. intermedium.*

Adapted to - LRR D, G, H, E; PHZ 4.

Released - 1952, cooperatively by New Mexico AES, University Park, and former SCS Nursery, Albuquerque.

Breeder Seed/Stock - Plant Materials Center, SCS, Los Lunas, NM.

Certified Seed/Stock - Not available. Has been replaced by other species which are either more widely adapted or better producers.

Preparer/Additional Information - Plant Materials Center, SCS, 1036 Miller St. SW, Los Lunas, NM 87031, (505) 865-4684.

Chief

Selected at Agriculture Canada Research Station, Saskatoon, Saskatchewan - R.P. Knowles.

Source - Introduction from the former USSR and Ree variety.

Method of Breeding - Mass selection for plants with high seed yields and good seed quality. Open-pollinated seed of selected plants bulked for each of five years to form five lines. After performance tests, five lines combined in isolation to produce breeder seed. Breeder seed increased through foundation to produce certified seed. Provision made in licensing to replace five basic lines with improved lines selected from same basic material.

Intended Use - Grass component of grass-alfalfa mixtures used for hay in the Parkland area of the Canadian prairie provinces. Short-term pastures that remain productive for about five years under heavy grazing pressure. Pastures under moderate to light grazing pressure remain productive for more than five years.

Description - Grayish-green type. Forage yields 5-10% and seed yields 20-30% above Ree Variety.

Adapted to - LRR F, G; PHZ 2, 3, 4.

Released - 1961, by Agriculture Canada.

Breeder Seed/Stock - Agriculture Canada, Research Station, Saskatoon, Saskatchewan S7N 0X2.

Certified Seed/Stock - Available.

Preparer/Additional Information - Scott B. M. Wright, Agriculture Canada, Research Station, 107 Science Crescent, Saskatoon, Saskatchewan S7N 0X2, (306) 343-8214.

Clarke (Reg. No. 81)

Selected at Agriculture Canada Research Station, Swift Current, Saskatchewan - T. Lawrence. Carried as accession no. ScI3733.

Source - Of the 20 clones included in Clarke, one traces to Sc 1451, a 1945 introduction received from Fort Collins, CO; 13 clones trace to Sc 1581, a 1958 introduction from the Moscow Botanical Garden, of the former USSR; five trace to Sc 1621, received as S-2275 in 1962 from R.P. Knowles, Saskatoon; and one clone traces to Sc 1623, received as S-5570 in 1962 from R.P. Knowles.

Method of Breeding - Clarke, a 20-clone synthetic, was developed through four to eight cycles of recurrent selection designed to combine winter-hardiness, drought-tolerance, establishment vigor, resistance to aphis-virus infection, good seed quality, and high forage and seed yield. The population from which Clarke was derived was subjected to extreme drought stress for two selection cycles.

Intended Use - Clarke is well adapted for hay and pasture use on either dryland or under irrigation in the Canadian prairie region and the northern Great Plains region of the U.S. Clarke maintains a favorable grass-alfalfa balance in mixtures used for hay.

Description - Clarke, a cultivar with a relatively wide genetic base, has no visual characters that distinguish it from other cultivars of intermediate wheatgrass. Clarke's main attributes are drought-tolerance, winter-hardiness, and high seed yield.

Adapted to - LRR F, G; PHZ 2, 3, 4.

Released - 1980, by Agriculture Canada.

Breeder Seed/Stock - Agriculture Canada, Research Station, Swift Current, Saskatchewan S9H 3X2.

Certified Seed/Stock - Available.

Preparer/Additional Information - Paul G. Jefferson, Agriculture Canada, Research Station, P.O. Box 1030, Swift Current, Saskatchewan S9H 3X2, (306) 773-4621.

Greenar (Reg. No. 3)

Selected at Plant Materials Center, SCS, Pullman, WA - J.L. Schwendiman.

Source - Developed from selection made in 1937 from PI 98568, which was introduced by Westover-Enlow expedition from the former USSR in 1932.

Method of Breeding - Open-pollinated selections made from planting one generation after introduction. Aberrant plants removed in following generation. Selections bulked and increased for field testing as P-2327.

Intended Use - Hayland and pastureland.

Description - Vigorous, moderate sod-forming wheatgrass; late maturing, leafy, dark green, broad-leafed, high-producing. Plants variable, but over 90% green. Less than 5% of plants show trace of pubescence. Spring recovery early and abundant; fall recovery good. Plants disease resistant, very productive, and aggressive. Widely adapted for conservation plantings on well-drained soils in dryland and irrigated areas.

Adapted to - LRR B, PHZ 5.

Released - 1945, as P-2327, cooperatively by Washington, Idaho, and Oregon Agricultural Experiment Stations at Pullman, Moscow, and Corvallis, respectively, and SCS, Plant Materials Centers, Aberdeen, ID, and Pullman, WA. Named Greenar in 1956.

Breeder Seed/Stock - Plant Materials Center, SCS, Pullman, WA.

Certified Seed/Stock - Available.

Preparer/Additional Information - Plant Materials Center, SCS, Room 104, Hulbert Agricultural Sciences Bldg., WSU, Pullman, WA 99164-6211, (509) 335-7376.

Greenleaf (Reg. No. 12)

Selected at Agriculture Canada Research Station, Lethbridge, Alberta - R.W. Peake.

Source - Commercial seed lots from Davenport, WA, and Bismarck, ND. Developed at the Agriculture Canada Research Station, Lethbridge, AB, by R.W. Peake.

Method of Breeding - Fifty-seven plants selected for forage type from space-planted nursery. Open-pollinated progenies evaluated in greenhouse for seedling vigor and tendency to creep. Progenies reduced to 14 in greenhouse tests, and to 12 plants in subsequent field tests. Twelve plants combined to form synthetic L-1747.

Intended Use - A winter-hardy cultivar for pasture and hay production. Grass component of grass-alfalfa mixtures used for hay in the Canadian prairie region and the northern Great Plains region of the U.S.. Short-term pastures that remain productive for about five years under heavy grazing pressure. Pastures under moderate to light grazing pressure remain productive for more than five years.

Description - A perennial, creeping-rooted, sod-forming grass. Foliage green to bright green; pubescence more evident than in Topar; good establishment capability and fair tolerance to soil salinity. Superior to Topar in yield of forage and seed, seedling vigor, and winter-hardiness. Similar to Mandan 759 except for improved seedling vigor. Some tolerance to saline soils and drought conditions. Greenleaf produced higher weight gain than many other grass species in a grazing study in Lethbridge.

Adapted to - LRR F, G; PHZ 2, 3, 4.

Released - 1966, by Canada Department of Agriculture.

Breeder Seed/Stock - Agriculture Canada, Research Station, Lethbridge, Alberta T1J 4B1.

Certified Seed/Stock - Available.

Preparer/Additional Information - Surya N. Acharya, Agriculture Canada, Research Station, P.O. Box 3000, Lethbridge, Alberta T1J 4B1, (403) 327-4561.

Luna (Reg. No. 6)

Selected at former SCS Nursery, Albuquerque, NM, and increased at Plant Materials Center, SCS, Los Lunas, NM - J.A. Downs, G.C. Niner, and J.E. Anderson. Carried as accession nos. 106831 & A-1115.

Source - Collected by Westover-Enlow expedition to the former USSR and Turkey in 1934. Introduced as *Agropyron popovii* Drobov, PI 106831. Identified by J.R. Swallen as *A. trichophorum* (Link) Richter, now *E. intermedia* (Host) Nevski.

Method of Breeding - Original accession space planted and rogued heavily through two generations for *E. intermedia* type. Tested as A-1115.

Intended Use - Irrigated pasture, range reseeding and revegetation of disturbed areas.

Description - Similar to intermediate wheatgrass in appearance except having varying degrees of pubescence throughout the plant. Some seedheads appear glabrous but all basal leaf blades are hairy. Less pubescent than other strains tested. Excellent seedling vigor, fast establishment, and high production of herbage. Leaves wide, lax, and dark green.

Adapted to - LRR D, G, H, E; PHZ 4.

Released - 1963, cooperatively by New Mexico AES and Plant Materials Center, SCS, Los Lunas, NM.

Breeder Seed/Stock - Plant Materials Center, SCS, Los Lunas, NM.

Certified Seed/Stock - Available.

Preparer/Additional Information - Plant Materials Center, SCS, 1036 Miller St. SW, Los Lunas, NM 87031, (505) 865-4684.

Mandan 759

Selected at ARS Northern Great Plains Research Laboratory, Mandan, ND - G.A. Rogler. Carried as accession no. PI 116252.

Source - Accession presented in 1936 by N.I. Vavilov of the former USSR, and designated PI 116252.

Method of Breeding - Seed of the original introduction, PI 116252, was increased in isolation and tested as Mandan 759.

Intended Use - Grass component of grass-alfalfa mixtures used for hay in the northern Great Plains region. Short-term pastures that remain productive for about five years under heavy grazing pressure. Pastures under moderate to light grazing pressure remain productive for more than five years. Mandan 759 has been used extensively in the northern Great Plains as a component of forage mixtures for the Conservation Reserve Program.

Description - Certified seed fields of Mandan 759 have approximately 75% plants with some degree of pubescence and approximately 25% glabrous plants. Similar in appearance to Greenleaf, except for a lower percentage of plants with bright-green foliage color.

Adapted to - LRR F, G; PHZ 2, 3, 4.

Released - Included in regional testing programs but never officially released. The North Dakota State Seed Department has allowed production of breeder, foundation, and certified seed of Mandan 759 due to recent high commercial demand for seed of this experimental strain.

Breeder Seed/Stock - ARS, Northern Great Plains Research Laboratory, Mandan, ND 58554.

Certified Seed/Stock - Available. It is anticipated that recently developed cultivars within the intermediate wheatgrass complex will replace Mandan 759.

Preparer/Additional Information - John D. Berdahl, ARS, Northern Great Plains Research Laboratory, P.O. Box 459, Mandan, ND 58554, (701) 663-6445.

Manska

Selected at ARS Northern Great Plains Research Laboratory, Mandan, ND - J.D. Berdahl. Carried as accession no. I2781.

Source - Manska traces to 11 separate commercial seed lots of Mandan 759 pubescent wheatgrass. Pubescent plants among the 11 seed lots were 20-75%, indicating that several of the seed lots had been subjected to outcrossing or seed admixture with glabrous forms of intermediate wheatgrass.

Method of Breeding - One hundred-twenty plants representing 11 commercial seed lots of Mandan 759 were selected from a space plant nursery based on vigor, resistance to leaf-spotting diseases, and pubescence. These 120 plants were tested further in replicated clonal plots for forage and seed yield, winter survival, and resistance to leaf-spotting diseases. Equal quantities of seed from 116 plants in a polycross nursery were composited to produce Mandan I2781.

Intended Use - Manska is adapted for hay and pasture use in areas of the northern and central Great Plains region of the U.S. and the Canadian prairie region that average over 350 mm annual precipitation.

Description - Foliage color of Manska is a darker green than the cultivars Greenleaf and Mandan 759. Manska has no other visual characteristics that distinguish it from other cultivars of pubescent wheatgrass. Forage quality of Manska has been high compared with other cultivars in the intermediate wheatgrass complex.

Adapted to - LRR F, G; PHZ 2, 3, 4.

Released - 1992 by ARS; SCS; the Agricultural Research Division, Institute of Agriculture and Natural Resources, University of Nebraska; and the North Dakota AES.

Breeder Seed/Stock - ARS Northern Great Plains Research Laboratory, Mandan, ND 58554.

Certified Seed/Stock - Available.

Preparer/Additional Information - John D. Berdahl, ARS, Northern Great Plains Research Laboratory, P.O. Box 459, Mandan, ND 58554, (701) 663-6445.

Oahe (Reg. No. 5)

Selected at South Dakota AES, Brookings - J.G. Ross. Carried as accession no. PI SD 20.

Source - Oahe (Oh-wah-hee) selected from Russian introduction PI 98568 obtained from Fort Collins, CO, in

1937 and released by South Dakota AES in 1945 as Ree. Introduction identified as derivative of cross between *Agropyron intermedium* and *A. trichophorum,* both now *E. intermedia.* Named after Oahe Dam on Missouri River in central South Dakota. Oahe is abbreviation of Sioux Word for "Big House," a meeting place once located near the dam.

Method of Breeding - Oahe, a four-clone synthetic, was developed from two cycles of selection for high seed set, forage, yield, and rust resistance within a population of high-seed-producing plants.

Intended Use - Oahe is well adapted for hay, pasture, and conservation purposes in the northern Great Plains of the U.S.

Description - Uniformly blue green, vigorous, rhizomatous. Oahe is drought tolerant, winter-hardy, and produces high yields of large seed.

Adapted to - LRR F, G, M; PHZ 3. *Released* - 1961, by South Dakota AES.

Breeder Seed/Stock - Foundation Seed Stocks Division, Box 2207-A, South Dakota State University, Brookings, SD 57007-1096.

Certified Seed/Stock - Available.

Preparer/Additional Information - Arvid Boe, 226 Ag Hall, Plant Science Dept., South Dakota State Univ., Brookings, SD 57007, (605) 688-4759.

Reliant (Reg. No. CV-20)

Selected at ARS Northern Great Plains Research Laboratory, Mandan, ND - J.D. Berdahl. Carried as accession no. I1813.

Source - Reliant, a six-clone synthetic, traces to a base population that was derived by intermating 24 cultivars and experimental strains within the intermediate wheatgrass complex that are adapted to the northern Great Plains region.

Method of Breeding - Eighty-one plants were selected for vigor and resistance to leaf-spotting diseases from a space-plant base population. Replicated clonal plots and open-pollinated progenies of these selections were evaluated for forage and seed yield, winter survival, resistance to leaf-spotting diseases, and forage quality. Six parent clones were selected and intermated to form a synthetic, Mandan I1813.

Intended Use - Reliant is adapted for hay and pasture use in areas of the northern Great Plains region of the U.S. and the Canadian prairie region that average over 350 mm annual precipitation. It is anticipated that Reliant will be used extensively as a component of forage mixtures for the Conservation Reserve Program and as a component of mixtures for wildlife habitat plantings.

Description - Reliant, of medium height and relatively late maturity, has no visual characteristics that distinguish it from other intermediate wheatgrass cultivars. Relative to current intermediate wheatgrass cultivars, Mandan 1813 has had high sustained forage and seed yields in long-term tests.

Adapted to - LRR F, G; PHZ 2, 3, 4.

Released - 1991 by ARS, SCS, and the North Dakota AES.

Breeder Seed/Stock - ARS, Northern Great Plains Research Laboratory, Mandan, ND 58554.

Certified Seed/Stock - Available.

Preparer/Additional Information - John D. Berdahl, ARS, Northern Great Plains Research Laboratory, P.O. Box 459, Mandan, ND 58554, (701) 663-6445.

Rush

Plant Materials Center, SCS, Aberdeen, ID. Carried as accession no. PI 281863.

Source - Received from German Botanical Garden, Berlin, in 1962, as *Agropyron junceum* (L.) Beauv..

Method of Breeding - None; accession included in several germplasm screening projects and has been outstanding in several performance traits.

Intended Use - Soil erosion control and stabilization, roadsides, mine spoils, urban sites; forage for livestock and wildlife; hayland and pastureland.

Description - An outstanding accession in seedling vigor, establishment, production and quality, and overall performance.

Adapted to - LRR B, D, E; PHZ 1, 6.

Released - No, material is available for field testing.

Breeder Seed/Stock - Plant Materials Center, SCS, Aberdeen, ID.

Certified Seed/Stock - No. Germplasm may be valuable for breeding purposes.

Preparer/Additional Information - Gary Young, Plant Materials Center, SCS, P.O. Box 296, Aberdeen, ID 83210-0296, (208) 397-4133.

Slate

Selected at Nebraska AES, Lincoln, ARS cooperating - L.C. Newell.

Source - Nebraska 50, an experimental variety tracing to PI 98568 and Amur from PI 131532.

Method of Breeding - Breeder seed of Slate was constituted by blending equal quantities of seed from two complementary, unrelated parent strains, Nebraska 50 and a derivative of Amur. Nebraska 50 consisted of 57 clones. Parent clones of both strains were selected for erect growth habit and slate-green color.

Intended Use - Slate has been grown in monoculture or as a principle component in mixtures with smooth bromegrass and alfalfa for cool-season pasture or hay in the central and portions of the northern Great Plains. Slate has been productive either with or without irrigation, but the cultivar has not been tolerant of saline-alkaline soils or extended period of flooding.

Description - Plants strongly spreading, erect at maturity; leaves broad and flat; foliage slate-green (intermediate between bright green and glaucous blue-green of other varieties). Inflorescence is well-developed spike with awnless lemmas that are either glabrous or slightly pubescent. Occasional spikelets have seeds with short awns or awn-points. Suggested for use in Central Plains area.

Adapted to - LRR E, F, G, northern H, western M; PHZ 3, 4, 5.

Released - 1969, cooperatively by Nebraska AES and USDA-ARS.

Breeder Seed/Stock - Nebraska AES and USDA-ARS, Univ. of Nebraska, Lincoln.

Certified Seed/Stock - Available.

Preparer/Additional Information - K.P. Vogel, ARS-USDA, 344 Keim Hall, Univ. of Nebraska or Nebraska Agricultural Research Division, Univ. of Nebraska, Lincoln, NE 68583, (402) 472-1564, (402) 472-2045.

Tegmar

Selected at Plant Materials Center, SCS, Pullman, WA - J.L. Schwendiman. Carried as accession no. PI 109219.

Source - Developed from accession PI 109219 collected in 1934 near Bolu, Turkey, by the Westover-Enlow expedition.

Method of Breeding - Open-pollinated selections made from planting one generation after introduction. Selections bulked and increased for field testing as P-14.

Intended Use - Erosion control and soil stability along roadsides, ditch banks, dam sites, and grassed waterways.

Description - Long-lived, late-maturing dwarf strain of intermediate wheatgrass; vigorous seedlings, good rapidly developing sod. Slightly more drought-tolerant than other intermediate wheatgrasses. Leaves are light green to blue-green and sparsely hirsute along the leaf margins and veins. Stems fine, smooth, and erect. More resistant to banksgrass mite than other intermediate wheatgrasses tested. General height about one-half that of other intermediate wheatgrass varieties. Short growth and vigorous sodding habit is suited for use as an erosion-control plant.

Adapted to - LRR B, D, E; PHZ 4, 5.

Released - 1968, cooperatively by Idaho and Washington Agricultural Experiment Stations and the SCS, Plant Materials Centers, Aberdeen, ID, and Pullman, WA.

Breeder Seed/Stock - Plant Materials Center, SCS, Aberdeen, ID.

Certified Seed/Stock - Available.

Preparer/Additional Information - Gary Young, Plant Materials Center, SCS, P.O. Box 296, Aberdeen, ID 83210-0296, (208) 397-4133.

Topar

Selected at Plant Materials Center, SCS, Pullman, WA - J.L. Schwendiman and D.S. Douglas. Carried as accession no. PI 107330.

Source - PI 107330 introduced from Tashkent, Turkestan, by Westover-Enlow expedition in 1934.

Method of Breeding - Developed by selection from spaced planting. Selections bulked and increased under isolation. All testing prior to 1953 was an accession P-41.

Intended Use - Erosion control and soil stabilization on roadsides, dry ditchbanks, dams, and grassed waterways.

Description - Vigorous-growing, late-maturing wheatgrass; sod forming and drought-resistant. Resembles and closely related to intermediate wheatgrass, but pubescent on leaves, stems, and lemmas. Forms sod more rapidly and adapted to lower fertility, higher elevations, and more alkaline sites than intermediate wheatgrass. Seedling vigor very good. Seed production moderate, and seed does not shatter easily. Adapted to shallow soils and low-fertility sites in 250-360 mm rainfall areas of West.

Adapted to - LRR B, D, E; PHZ 4, 5.

Released - 1953, cooperatively by Washington, Idaho, Oregon, and California Agricultural Experiment Stations at Pullman, Moscow, Corvallis, and Davis, respectively, and SCS, Plant Materials Centers, Aberdeen, ID, Pleasanton, CA, and Pullman, WA.

Breeder Seed/Stock - Plant Materials Center, SCS, Aberdeen, ID.

Certified Seed/Stock - Available.

Preparer/Additional Information - Gary Young, Plant Materials Center, SCS, P.O. Box 296, Aberdeen, ID 83210-0296, (208) 397-4133.

Elytrigia repens var. *repens* (L.) Desv. ex B.D. Jackson X *Pseudoroegneria spicata* (Pursh)
A. Love - RS wheatgrass

Cool-season, sod-forming, introduced grass, probably from Europe. Common, persistent weed in cultivated and abandoned fields in northern U.S.. Existing stands used for hay, pasture, and silage.

Newhy

ARS, Logan, UT - K.H. Asay and D.R. Dewey. Carried as SCS accession no. 9052852. PI no. is pending.

Source - A hybrid between quackgrass and bluebunch wheatgrass.

Method of Breeding - The initial F_1 hybrid made by D.R. Dewey in 1962 was pentaploid (2n=35), meiotically irregular, beset with chlorophyll deficiencies, and, in general, had poor vegetative vigor. Even though the hybrid was largely sterile, seedset was adequate for generations to be advanced without chromosome doubling. From the F_1 to F_5 generation, selection was based largely on fertility (seeds/spike) and was restricted to plants with characteristics of both parental species. Plants with excessive rhizome development were excluded. More intense selection was initiated in the F_5 generation to combine the vigor, productivity, salinity tolerance, and persistence of quackgrass with the drought-resistance, caespitose growth habit, seed quality, and forage quality of bluebunch wheatgrass. Two F_7-F_8 germplasm pools, designated RS-1 and RS-2, were released to breeders in 1980. Two additional cycles of selection were completed with the combined RS-1 and RS-2 breeding populations to develop the parental lines of the Newhy cultivar.

Intended Use - Recommended for moderately saline semiarid rangeland areas receiving at least 330 mm of annual precipitation. Has potential for irrigated pastures in combination with legumes such as alfalfa.

Description - Newhy is more upright and significantly less rhizomatous than quackgrass. It is more resistant to drought than quackgrass and tolerates heavier grazing pressure than bluebunch wheatgrass. Salinity tolerance of the cultivar approaches that of tall wheatgrass. It is most productive on slightly saline or alkaline range sites receiving at least 330 mm of precipitation annually. The cultivar has 2n=42 chromosomes and is meiotically regular. Rate of phenological development is intermediate to the parental species and anthesis occurs from mid to late June in nurseries near Logan, UT. Newhy produced 560 kg/ha on an irrigated site near Miles City, MT. Newhy has demonstrated excellent palatability to grazing animals and its nutritional qualities, in terms of protein content and digestibility, is equivalent or superior to other wheatgrasses. The hybrid has shown moderate susceptibility to injury by the grass billbug under soil and moisture conditions where this insect is a potential problem.

Adapted to - LRR B, D, E, F, G; PHZ 3, 4, 5, 6.

Released - ARS in cooperation with the Utah AES and SCS.

Breeder Seed/Stock - Maintained by the ARS, Forage and Range Research Laboratory, Utah State University, Logan, UT 84322-6300. Foundation seed is produced by ARS and SCS and distributed by the Utah Crop Improvement Association and the SCS.

Certified Seed/Stock - Available. Because of the morphological similarity of Newhy seed to that of quackgrass, protection has been applied for under the PVPA.

Preparer/Additional Information - K.H. Asay, ARS Forage and Range Research Laboratory, Utah State Univ., Logan, UT 84322-6300, (801) 750-3069.

Eragrostis curvula (Schrad.) Nees - weeping lovegrass

Warm-season bunchgrass introduced from Africa in 1927. Used for erosion control and pasture throughout much of the southern U.S.. Well adapted to parts of southern Great Plains. Productive, drought-resistant, relatively easy to establish, grows on low-fertility soils. Palatability low except when young. Subject to winterkilling in northern part of range in adaptation.

A variable species that includes many different forms. These forms are linked by intermediates, and this extends to E. chloromelas that represents one extreme in the range in variation.

A-67

Increased at SCS Nursery, Tucson, AZ, by C.G. Marshall. Carried as accession no. PI 477012.

Source - Seed sent to F.J. Crider by L.W. Kephart and R.L. Piemeisel from north central Tanganyika in east Africa.

Method of Breeding - Selected in initial observation plots from original seed collection and seed bulked for initial increase.

Intended Use - Range reseeding.

Description - Vigorous growing, leafy, densely tufted, long-lived bunchgrass with extensive root system and long, lax leaves. Forage and seed production greater than Lehmann or Boer lovegrass. Also, more cold-tolerant than those two species. Utilized for range, pasture, and soil protection on a wide variety of soil textures where annual precipitation exceeds 35 cm.

Adapted to - LRR D; PHZ 7, 8.

Released - Informally by SCS.

Breeder Seed/Stock - Plant Materials Center, SCS, Tucson, AZ.

Certified Seed/Stock - Not available. Sold as common.

Preparer/Additional Information - Plant Materials Center, SCS, 3241 N. Romero Rd., Tucson, AZ 85705, (602) 241-2966.

Ermelo

Seed collected in 1944 from native stand in Ermelo district of South Africa.

Source - Accession from South Africa including PI 232986 supplied by G.A. Van Dom Heever, Nigel, Transvaal, Republic of South Africa. Ermelo has been catalogued under several PIs, including 164917, 190312, 196355, 199060, 298985. The accession is maintained by the ARS PI Station, Pullman, WA.

Method of Breeding - Increase of original accessions.

Intended Use - Pasture, critical area treatment.

Description - Comparatively leafy; has performed well in plot tests and under grazing in southern Oklahoma and adjacent parts of Texas.

Adapted to - LRR H, J, P; PHZ 7.

Released - Informally from several different sources.

Breeder Seed/Stock - Not available.

Certified Seed/Stock - Not available (some common commercial supplies).

Preparer/Additional Information - Wayne Everett, SCS, P.O. Box 6567, Ft. Worth, TX 76115, (817) 334-5282.

Morpa

Selected at U.S. Southern Great Plains Field Station, ARS, Woodward, OK, in cooperation with Oklahoma AES, Stillwater, OK - W.R. Kneebone and P.W. Voigt.

Source - PI 208994, received in 1953 from Rietvlei Research Station, Transvaal, Union of South Africa.

Method of Breeding - A bulk increase of surviving winter-hardy plants, after winter of 1955-56. Identified as possessing superior acceptability in replicated animal screening tests.

Intended Use - Pasture and hay.

Description - About 8 mm taller, panicles darker, leaves slightly wider, and 7-10 days later than common weeping lovegrass. Less lignin relative to cellulose as compared with a less palatable experimental variety; produced 12-13% higher live weight gains than common and PI 206673 weeping lovegrass; and much superior to common lovegrass in average daily gains for both winter and summer grazing.

Adapted to - PHZ southern 6, Great Plains Region of 7.

Released - 1970, cooperatively by Oklahoma AES and Plant Science Research Division, ARS.

Breeder Seed/Stock - Oklahoma AES.

Certified Seed/Stock - Available.

Preparer/Additional Information - C.M. Taliaferro, Agronomy Dept., 368 Ag Hall, Oklahoma State Univ., Stillwater, OK 74078, (405) 744-6410.

Eragrostis curvula (Schrad.) Nees var. *conferta* Stapf - Boer lovegrass
Eragrostis chloromelas Steud.

Warm-season bunchgrass introduced from Union of South Africa in 1932. Adapted to warm semidesert areas in southwestern U.S.. Palatable, drought- resistant, but lacks cold tolerance. Longer lived than Lehmann lovegrass. *E. curvula* var. *conferta* grades completely into *E. curvula.* Typical specimens of *E. curvula* var. *conferta* represent one extreme in the range in variation and can be distinguished with some degree of accuracy from the great majority of "forms" of *E. curvula.* Indicative of this problem is the fact that Catalina weeping lovegrass was placed first in *E. curvula, E. chloromelas, E. curvula* var. *conferta,* then *E. curvula.*

A-84

Increased at SCS Plant Materials Centers, Tucson, AZ, by C.G. Marshall and L.P. Hamilton. Carried as accession no. PI 469222.

Source - Union of South Africa.

Intended Use - Range reseeding.

Description - Warm-season bunchgrass introduced from Union of South Africa in 1932. Adapted to warm semi-desert areas in southwestern U.S. at elevations below 1,500 m where average annual precipitation is 30 cm or more.

Adapted to - LRR D; PHZ 7, 8, 9.

Released - 1950, cooperatively by Arizona AES, Tucson, and Nursery Division, SCS.

Breeder Seed/Stock - Plant Materials Center, SCS, Tucson, AZ.

Certified Seed/Stock - Not available.

Preparer/Additional Information - Plant Materials Center, SCS, 3241 North Romero Rd., Tucson, AZ 85705, (602) 670-6491.

Catalina

Selected at Arizona AES, ARS, and SCS, Tucson, AZ - L.N. Wright. Carried as accession no. PI 203347.

Source - PI 203347, received in 1952 from Pretoria, Union of South Africa, as *E. curvula.* Classified in evaluation program as *E. curvula* var. *conferta.*

Method of Breeding - A single line, 3-17, selected from 16 accessions of Boer lovegrass and tested for seedling drought tolerance under controlled environmental conditions in a growth chamber. Final evaluation on Arizona rangelands in comparison with A-84 Boer lovegrass and Lehmann lovegrass.

Intended Use - Range reseeding.

Description - An apomictic line that is superior to A-84 and equal to Lehmann lovegrass in stand establishment. Under range conditions, forage yield has been 30% higher and of better quality than Lehmann lovegrass. Should be adapted in semiarid and arid areas of the Southwest, particularly at elevations below 1,400 m with a minimum annual rainfall of 305 mm.

Adapted to - LRR D; PHZ 7, 8, 9.

Released - 1969, cooperatively by Arizona AES; Plant Science Research Division, ARS; and Plant Sciences Division, SCS.

Breeder Seed/Stock - Plant Materials Center, SCS, Tucson, AZ.

Certified Seed/Stock - Not available.

Preparer/Additional Information - Plant Materials Center, SCS, 3241 N. Romero Rd., Tucson, AZ 85705, (602) 670-6491.

OTA-S

Source - Four clones of Boer lovegrass, three from PI 299929 and one from the cross PI 299928 x PI 299929.

Method of Breeding - The three clones of PI 299929 were from open-pollinated seed and were observed to produce variable progenies. Other similar plants produced uniform progenies (were apomictic in mode of reproduction). PI 299928 is a sexual diploid. The clone used in OTA-S is tetraploid and was apparently derived from fertilization of an unreduced egg by pollen from the tetraploid PI 299929.

Description - All clones were examined cytologically because they produced variable progenies. All four were obligate sexuals and had 40 (2n=4x=40) chromosomes. This is the only known source of sexuality at the tetraploid chromosome level in the *E. curvula* complex.

Adapted to - Not widely tested.

Released - USDA-ARS, Oklahoma and Texas Agricultural Experiment Stations.

Breeder Seed/Stock - Grassland, Soil and Water Research Lab., Temple, TX 76502.

Certified Seed/Stock - No.

Preparer/Additional Information - P.W. Voigt, USDA-ARS, 808 E. Blackland Rd., Temple, TX 76502, (817) 770-6521.

Eragrostis lehmanniana **Nees - Lehmann lovegrass**

Warm-season, slightly spreading grass introduced from Union of South Africa in 1932. Used for range reseeding in warm semideserts of southwestern U.S.. Easily established. Prostrate stems root at nodes. Smaller and less cold tolerant than Boer and weeping lovegrass.

A-68

Increased at SCS Nursery, Tucson, AZ - C.G. Marshall. Carried as accession no. PI 469223.

Source - Seed sent to F.J. Crider, National Plant Materials Center, SCS, Beltsville, MD, by M. William, Kimberley, Union of South Africa in 1932.

Method of Breeding - Increase of original line.

Intended Use - Range reseeding.

Description - Seedlings volunteer and tolerate adverse conditions better than other lovegrasses. More drought-tolerant but less cold-tolerant than Boer or weeping lovegrass.

Adapted to - LRR D; PHZ 7, 8, 9.

Released - 1950, cooperatively by Arizona AES, Tucson, AZ, and Nursery Division, SCS.

Breeder Seed/Stock - Plant Materials Center, SCS, Tucson, AZ.

Certified Seed/Stock - Not available.

Preparer/Additional Information - Plant Materials Center, SCS, 3241 N. Romero Rd., Tucson, AZ 85705, (602) 670-6491.

TEM-SD

Carried as accession no. PI 559907.

Source - Collections 459, 467, and 469 of T.B. Vorster, from the Northern Cape Province, Republic of South Africa.

Method of Breeding - Bulk harvest of seed produced in isolation from 30 plants derived through open-pollination from the three South African accessions.

Intended Use - A source of sexual reproduction in Lehmann lovegrass. Sexual germplasm is required for hybridization and development of recombinant types.

Description - TEM-SD is unique because it is diploid in chromosome number (2n=2x=20) and reproduces sexually. All other known Lehmann lovegrass germplasm has 40 or more chromosomes and reproduces by apomixis.

Adapted to - Not tested, but believed similar to A-68.

Released - USDA-ARS and Texas AES in 1991.

Breeder Seed/Stock - Germplasm maintained and distributed in limited quantity by Grassland, Soil and Water Research Lab., 808 E. Blackland Rd., Temple, TX 76502.

Certified Seed/Stock - No, a germplasm release.

Preparer/Additional Information - Paul W. Voigt, USDA-ARS, 808 E. Blackland Rd., Temple, TX 76502, (817) 770-6521.

Eragrostis superba Peyr. - Wilman lovegrass

Warm-season bunchgrass introduced from South Africa. Occurs throughout much of South Africa and also in tropical Africa. Comparatively palatable lovegrass, adapted to much the same area as Lehmann lovegrass. Starts growth later, competes less well with weeds, and seed does not ripen as uniformly as weeping lovegrass.

Palar

Increased at SCS Nursery, Tucson, AZ by L.P. Hamilton and T.F. Spaller. Carried as accession no. PI 276055.

Source - PI 276055 introduced from South Africa.

Method of Breeding - Bulk increase of original accession that was planted in 1962.

Intended Use - Range reseeding.

Description - A vigorous, leafy type with outstanding seedling vigor and good seed production. Better stand establishment obtained in range seeding trials in southern Arizona than with other lovegrass. Adapted to elevations of less than 1,200 m with a minimum annual rainfall of 25-30 cm.

Adapted to - LRR D; PHZ 7, 8, 9.

Released - 1972 by Arizona AES, ARS, and SCS.

Breeder Seed/Stock - Plant Materials Center, SCS, Tucson, AZ.

Certified Seed/Stock - Available.

Preparer/Additional Information - Plant Materials Center, SCS, 3241 N. Romero Rd., Tucson, AZ 85705, (602) 241-2966.

Eragrostis trichodes (Nutt.) Wood - sand lovegrass

Important warm-season bunchgrass. Occurs on sandy soils of central and southern Great Plains. Palatable and nutritious rangegrass, with deep root system and good drought resistance. Lacks persistence under heavy grazing. Starts growth in early spring and remains green until late fall.

Bend

Selected at Kansas AES, Manhattan, Kansas; ARS and SCS cooperating - F.L. Barnett.

Source - Collections from the Arkansas and Cimarron River basin in south-central Kansas and adjacent areas in Oklahoma.

Method of Breeding - Two cycles of selection for vigor and persistence on unirrigated sandy soils. Approximately 200 selections, tracing to 11 accessions, composited to produce variety. Tested as Kansas Experimental 3.

Intended Use - Range seedings.

Description - Uniform in maturity, good seed producer, good establishment, and relatively free of disease. Similar in appearance to local strains found in south-central Kansas. Acceptable dry matter yield and good vigor in comparative tests.

Adapted to - LRR H, M; PHZ 6, 7.

Released - 1971, cooperatively by Kansas AES; Plant Science Research Division, ARS; and Nursery Division, SCS.

Breeder Seed/Stock - Plant Materials Center, SCS, Manhattan, KS.

Certified Seed/Stock - Available in limited quantity.

Preparer/Additional Information - Kansas AES, Agronomy Dept., Manhattan, KS 66506, (913) 532-6101.

Mason

Selected from early SCS studies at Spur, TX, by J.E. Smith, Jr. Carried as accession no. PI 436990.

Source - Collected in 1957 by SCS personnel from native stand near Mason, TX.

Method of Breeding - Increase of original collection. Seed was produced first at Spur, TX and later at J.E. "Bud" Smith Plant Materials Center, Knox City, TX. Evaluated at this location during 1965-69.

Intended Use - Range reseeding.

Description - Leafier, better forage and seed producer, and starts spring growth two to three weeks earlier than common commercial sand lovegrass.

Adapted to - LRR H, I, J, T; PHZ 7.

Released - 1971 by the SCS and Texas AES.

Breeder Seed/Stock - Plant Materials Center, SCS, Knox City, TX.

Certified Seed/Stock - Not available.

Preparer/Additional Information - Plant Materials Center, SCS, Rte 1, Box 155, Knox City, TX 79529, (817) 658-3922.

Nebraska 27

Increased at Nebraska AES, Lincoln, NE; ARS and SCS cooperating - L.C. Newell and E.C. Conard.

Source - Collection from native meadow in northern Holt County, NE.

Method of Breeding - Selections moved to Lincoln in 1935 by L.C. Newell and Elver Hodges. Seed from these plants was later increased at North Platte and Waterloo, NE, by E.C. Conard.

Intended Use - Range seedings.

Description - Winter-hardy, relatively long-lived strain of sand lovegrass. In Nebraska, plantings survived and maintained stands superior to plantings made with seed from more southern sources. Well adapted to range of soil types. Application of phosphorus fertilizers to sandy soils of low fertility usually increases yields of forage and seed. Produces highly palatable, nutritious forage. Best used in mixtures with other warm-season, native grasses, such as gramas or bluestems.

Adapted to - LRR G, H; PHZ 5, 6.

Released - 1949, cooperatively by Nebraska AES; Plant Science Research Division, ARS; and Nursery Division, SCS.

Breeder Seed/Stock - Nebraska AES and USDA-ARS, Univ. of Nebraska, Lincoln.

Certified Seed/Stock - Available.

Preparer/Additional Information - K.P. Vogel, ARS-USDA, 344 Keim Hall, Univ. of Nebraska or Jeff Pederson, Dept. of Agronomy, Univ. of Nebraska, Lincoln, NE 68583, (402) 472-1564, (402) 472-2811.

Eragrostis trichophora **Coss. & Dur. - atherstone lovegrass**

Eragrostis atherstonii **Stapf**

Warm-season perennial that includes both bunchgrass and stoloniferous forms. Widely distributed from Republic of South Africa to south tropical Africa.

Cochise

Increased at Plant Materials Center, SCS, Tucson, AZ - L.P. Hamilton and T.F. Spaller. Carried as accession no. PI 276033.

Source - PI 276033 introduced from Pretoria, South Africa, via Australia.

Method of Breeding - Bulk increase of seed from initial observation block. First increase block planted in 1964. Tested as A-16753.

Intended Use - Range reseeding.

Description - Large, vigorous bunchgrass with characteristics of both Lehmann and weeping lovegrasses. Larger, more productive than Lehmann, with longer green period; finer stems and leaves than weeping. Very good seedling vigor and cold-hardiness. Adapted to elevations of 450-1,850 m with at least 25-30 cm of annual precipitation.

Adapted to - LRR D; PHZ 7, 8, 9.

Released - 1979, SCS and Arizona AES.

Breeder Seed/Stock - Plant Materials Center, SCS, Tucson, AZ.

Certified Seed/Stock - Available.

Preparer/Additional Information - Plant Materials Center, SCS, 3241 N. Romero Rd., Tucson, AZ 85705, (602) 241-2966.

Eremochloa ophiuroides **(Munro) Hack. - centipedegrass**

Warm-season, sod-forming grass from China. (Seed found in baggage of Frank N. Meyer, USDA plant explorer who disappeared on his fourth trip to China in 1916.) Well adapted to soils and climatic conditions of southern U.S.;

survives as far north as northern AL and Raleigh, NC; grows well on poor soils. Adapted for use as low-maintenance, general-purpose turf and lawn grass. Propagated vegetatively or by seed. Not desirable as pasture plant.

Au-Centennial

Source - Irradiation of common seed.

Method of Breeding - Selection from irradiated seedlings.

Intended Use - Lawns and fine turf.

Description - Dwarf cultivar with improved color and density. Vegetatively propagated.

Adapted to - PHZ 7, 8, 9, 10.

Released - Ray Dickens and J. Pedersen

Breeder Seed/Stock - Ray Dickens.

Certified Seed/Stock - Available.

Preparer/Additional Information - Dr. Ray Dickens, AL AES, Auburn Univ., Auburn, AL 36849, (205) 844-4100.

Oklawn

Oklahoma AES, Oklahoma State University, Stillwater, OK - W.W. Huffine and W.C. Elder.

Source - Germplasm accessions thought to have been included in test of lawn grasses planted at Stillwater, OK, before the drought of the mid-1930s.

Method of Breeding - About 1935, vegetative material from lawn grass test previously planted on the OSU campus was used for erosion control on the Agronomy Research Station at Stillwater, OK. In 1949, a collection from this planting was included in the turfgrass nursery and subsequently identified as a well adapted centipedegrass.

Intended Use - Turf.

Description - Persistent under adverse environmental conditions. Exhibits tolerance to heat, drought, insects, and diseases. Grows in partial shade and full sun; adapted to soils that are moderately acid and of medium fertility. Produces satisfactory turf without high-level maintenance. Bluish green, medium textured, slow growing; forms relatively dense sod resistant to bermudagrass encroachment and weed invasion. Maximum height of growth about 75-100 mm. Normally established with sprigs or sod, but can be established from seed.

Adapted to - LRR P, N, J, T, U; PHZ 7, 8, 9, 10.

Released - 1965, by Oklahoma AES.

Breeder Seed/Stock - Oklahoma AES.

Certified Seed/Stock - No, but still a few plantings subject to certification.

Preparer/Additional Information - C. M. Taliferro, Agronomy Dept., 368 Ag Hall N., Oklahoma State Univ., Stillwater, OK 74078-0507, (405) 774-6410.

Festuca arizonica Vasey - Arizona fescue

Arizona fescue is a native, long-lived bunchgrass. Culms are densely tufted in large bunches 30-100 cm tall. Leaves are numerous, usually elongate, scabrous, filiform, and involute. Propagation is by seed. Natural range of adaptation of the species is from Colorado south to Texas and west to Nevada at elevations of 1,700-2,800 m. It is found on shallow clay loam to loam and sandy to gravelly soils mainly in association with ponderosa pine. It has an extensive, tough, fibrous root system which makes it a valuable plant for soil stabilization. It can tolerate extended dry periods, especially in the spring months.

Redondo

Selected at Plant Materials Center, SCS, Los Lunas, NM. Collected by J.A. Downs and G.C. Niner. Carried as accession no. PI 469218, NM-5.

Source - Original seed was collected in 1956 by J.A. Downs from the ponderosa pine zone on the Baca Land Grant west of Los Alamos, NM, at an elevation of 2,140 m and 355 mm precipitation.

Method of Breeding - Bulk increase of native collection.

Intended Use - Roadside stabilization, range seeding and mine revegetation.

Description - Redondo was selected for superior seed production and establishment success but otherwise has no morphological differences with other Arizona fescue strains.

Adapted to - LRR D, E, G, H; PHZ 4, 5, 6.

Released - 1973, New Mexico State Univ., Colorado State Univ. AES, SCS, NM State Highway Department.

Breeder Seed/Stock - Plant Materials Center, SCS, Los Lunas, NM.

Certified Seed/Stock - Available.

Preparer/Additional Information - Plant Materials Center, SCS, 1036 Miller St. SW, Los Lunas, NM 87031, (505) 865-4684.

Festuca arundinacea Schreb. - tall fescue

Major cool-season bunchgrass from Europe. Used for pasture, hay, general-purpose turf, and erosion control throughout humid parts of northern U.S. and under irrigation in arid regions. Adapted in southern states, but of limited value in southern coastal plain. Vigorous; grows well on both wet and dry sites; does best on heavy soils. Palatability often problem in mixtures; should be grazed close for best animal acceptance and feeding value.

Adventure

Pure Seed Testing, Inc. Hubbard, OR - Dr. William Meyer.

Source - An advanced generation synthetic derived from the progenies of 17 clones. All of these clones were selected from old turf areas in the northeastern U.S.

Method of Breeding - One hundred forty-five parental clones were allowed to interpollinate in an isolated nursery. Seed production is limited to two generations.

Intended Use - Turf.

Description - An attractive, leafy, turf-type fall fescue with good density, a medium dark green color, and a moderately low growth habit. It has good shade adaptation and good heat and cold tolerance.

Adapted to - LRR A, B, C, D, E, G, N, P, R, S; PHZ 6.

Released - 1982 - Pure Seed Testing, Inc., Hubbard, OR.

Breeder Seed/Stock - Dr. William A. Meyer, Pure Seed Testing, Inc., Hubbard, OR.

Certified Seed/Stock - Available.

Preparer/Additional Information - Frank Berns, Warren's Turf Nursery, Box 459, Suisun City, CA 94585, (707) 422-5100.

Alta (Reg. No. 1)

Selected at Oregon AES, Corvallis, ARS cooperating - H.A. Schoth.

Source - In 1918, some of the more promising lines of tall fescue from nursery at Pullman, WA, were established at Corvallis. Seed obtained from M. Heinricks, Pullman. Three lines used PI 19728, PI 24838, and PI 25206. PI 19728 received January 24, 1907, from A. LeCoq & Co., Darmstadt, Germany. PI 24838 from commercial lot of about 225 kg of seed purchased from Peppard Seeds, Inc., Kansas City, MO, March 9, 1909. PI 25206 from a seed lot presented by G. Bitter, Director, Botanic Garden, Bremen, Germany, and received by Plant Introduction Section, USDA, March 26, 1909.

Method of Breeding - Evolved as ecotype selection. Material mentioned above planted in Spring 1918. Noted to have made exceptionally fine growth during first season. Received special mention in annual reports from 1919 through 1922. In Winter of 1922-23, it suffered severe winterkilling. Surviving plants put together and became source seed of Selection 7 (changed to FC 29366). Remained under this selection number until given name Alta in 1940. Selected for ability to remain green during dry summers in western Oregon and for high yields of forage.

Intended Use - Irrigated pastures.

Adapted to - Western parts of Oregon and Washington.

Released - 1940, cooperatively by Oregon AES and Plant Science Research Division, ARS.

Breeder Seed/Stock - Not available.

Certified Seed/Stock - Available.

Preparer/Additional Information - Crop and Soil Science Dept., Crop Science Building 107, Oregon State Univ., Corvallis, OR 97331-3002, (503) 737-4513.

Amigo

Developed by Pure-Seed Testing, Hubbard, OR.

Source - Parental clone from Rutgers University and selections from Apache, Bonanza, and a pythium blight-resistant clone from NJ.

Method of Breeding - Phenotypic recurrent selection and progeny turf trials. Selections for leaf spot and brown patch resistance. Final selections were made for dark green plants with resistance to crown and stem rust and high seed yielding ability.

Intended Use - Amigo is intended for use in turf where irrigation potential is limited. Amigo performs well in sod production, golf course roughs, home lawns, and parks. Amigo is tolerant to sites with full sun to partial shade condition. Amigo has also shown persistence at cutting heights as low as 110 mm.

Description - Amigo tall fescue is a dense, dwarf, bright blue-green cultivar. It has very stiff upright foliage with a low vertical growth rate. Amigo has exhibited improved shade tolerance compared to other tall fescues. Amigo has excellent summer density. Resistance to brown patch and pythium has been good. Leafspot resistance is very good. Amigo has a higher percentage of living ground cover than other tall fescues. Color retention is good in the fall.

Adapted to - LRR A, B, C, D, E, F, G, H, I, J, K, L, M, N, O, P, Q, R, S, T, U, V; PHZ 4, 5, 6, 7, 8, and coastal California.

Released - 1989 by Pure Seed Testing. Licensed to NK Lawn and Garden.

Breeder Seed/Stock - Pure Seed Testing, Hubbard, OR.

Certified Seed/Stock - NK Lawn and Garden, Minneapolis, MN, and Tangent, OR.

Preparer/Additional Information - Eric K. Nelson, P.O. Box 300, 33731 Hwy 99E, Tangent, OR 97389, (503) 928-2393.

Anthem

Source - From vegetative clones originating out of Mustang, Bonanza, from Athens, GA, and from collections in western Oregon.

Method of Breeding - Extensive evaluations for type disease resistance, and seed yield in spaced plantings; phenotypic matched matings over three generations towards the desired type.

Intended Use - Especially well adapted to the upper southeastern U.S. for lawns, etc., either as a single component or blended with other dark tall fescues.

Description - Excellent dark green color, strongly upright growth habit, and excellent turf diversity and quality.

Adapted to - PHZ 5, 6, 7, 8.

Released - 1991 by Turf Merchants, Inc.

Breeder Seed/Stock - Turf Merchants, Inc. and Pure Seed.

Certified Seed/Stock - Available.

Preparer/Additional Information - Fred B. Ledeboer, Ph.D., TMI, 33390 Tangent Loop, Tangent, OR 97389, (800) 421-1735.

Apache

Source - New Jersey, Illinois, and Georgia.

Method of Breeding - Advanced generation synthetic of six clones from three cycles of phenotypic recurrent selection.

Intended Use - Cool season turf areas, low maintenance turf, athletic fields, and golf courses.

Description - Apache has an upright leaf orientation, widely adapted, and good resistance to leaf spot and crown rust.

Adapted to - PHZ 5, 6, 7, 10, 11.

Released - Turf-Seed, Inc.

Breeder Seed/Stock - Pure Seed Testing, Inc.

Certified Seed/Stock - Available. PVP No. 8400143.

Preparer/Additional Information - Turf-Seed, Inc., P.O. Box 250, Hubbard, OR 97032, (503) 651-2130.

Aquara

Source - Original material from old turf in Georgia. Developed by Rutgers University.

Method of Breeding - Synthetic variety based on 27 clones originating from 12 plants in turf from Georgia.

Intended Use - Permanent turf.

Description - High heat and drought stress tolerance, high endophyte level.

Adapted to - PHZ 4, 5, 6, 7.

Released - 1990 by O.M. Scott & Sons Co.

Breeder Seed/Stock - O.M. Scott & Sons Co.

Certified Seed/Stock - Available.

Preparer/Additional Information - Virgil Meier, O.M. Scott & Sons Co., Marysville, OH 43041, (513) 644-0011.

Arid (Reg. No. 31)

Developed at the New Jersey AES - Cyril R. Funk, Jr. Carried as accession no. PI 509069.

Source - An advanced generation synthetic cultivar selected from the progenies of 88 clones. The plants were collected from old, low-maintenance turfs in Alabama, Georgia, New Jersey, North Carolina, Pennsylvania, and Virginia.

Method of Breeding - Recurrent selection of 88 selected clones. Selection was based on high seed-yield potential, uniform maturity, and desirable plant-type.

Intended Use - Turf-type tall fescue intended for use on home lawns, parks, athletic fields, roadsides, golf courses, and school playgrounds.

Description - Arid is three days earlier than Rebel in Oregon and Idaho and is also significantly taller than Rebel at maturity, and has a greater flag leaf height. It also has a significantly narrower flag leaf and leaf blades than Rebel. Arid has a significantly shorter lemma length and smaller seed than Rebel. Arid is a persistent, leafy, moderately low-growing, turf-type tall fescue with moderately dark green color, medium fine leaf texture, and medium high density. Arid has good heat and drought tolerance, good winter-hardiness, and the ability to perform well under conditions of limited fertility. It has good cool weather color retention into late fall and early winter, and it has improved shade tolerance.

Adapted to - LRR A, B, D, E, G, N, P, R, S; PHZ 6.

Released - 1986 by Jacklin Seed Company, Post Falls, ID.

Breeder Seed/Stock - Jacklin Seed Company, Post Falls, ID.

Certified Seed/Stock - Available.

Preparer/Additional Information - Mark Sellmann, Jacklin Seed Co., W. 5300 Riverbend Ave., Post Falls, ID 83854, (208) 773-7581.

Au Triumph (Reg. No. 21)

Developed by Alabama AES, Auburn University, AL. Carried as accession no. AF-5.

Source - Tall fescue introductions established at the Alabama University Plant Breeding Unit, Tallahassee, AL.

Method of Breeding - Tall fescue introductions were evaluated for vigor, winter growth, regrowth potential, and disease-resistance. Twelve genotypes were selected to form synthetic variety.

Intended Use - Pasture and hay production, with or without legumes.

Description - Early maturing cultivars; leafy with very upright growth. Superb winter forage production; yields more than winter dormant varieties. Au Triumph is free of the fungal endophyte which is associated with fescue toxicosis.

Adapted to - LRR A, C, N, O, P; PHZ 7.

Released - August 1981, by Alabama AES, Auburn University, AL.

Breeder Seed/Stock - Controlled by Auburn University.

Certified Seed/Stock - Available from International Seeds Inc., Halsey, OR. Widely used in South America, Australia, and New Zealand, PVP No. 8400017.

Preparer/Additional Information - Stephen W. Johnson, International Seeds Inc., P.O. Box 168, 820 W. First St., Halsey, OR 97348, (503) 369-2251.

Austin

Source - One hundred seventy-five clones collected from old turfs of the eastern U.S. by Rutgers University.

Method of Breeding - Advanced generation synthetic cultivar developed at the New Jersey AES with Dr. C. Reed Funk.

Intended Use - Turf, either alone or in blends.

Description - Early maturing, vigorous, dark green, medium texture, excellent heat and drought tolerance, and good disease resistance.

Adapted to - Southeast and southwest U.S., lower transition states.

Released - 1990, by Barenbrug/Normarc.

Breeder Seed/Stock - Willamette Valley Plant Breeders and Barenbrug USA.

Certified Seed/Stock - Available from Barenbrug USA.

Preparer/Additional Information - Barenbrug USA, P.O. Box 239, Tangent, OR 97389, (503) 926-5801.

Aztec

KWS-AG, Einbeck, West Germany.

Source - Collections in old turf areas in South Carolina, Georgia, and Tennessee, plus progeny from selected clones of Bonanza and Mustang. Carried as experimental no. KWS-VDA.

Method of Breeding - Clonal progeny evaluation in space plant nursery and turf; evaluation of vegetative propagules from best turf plots for color, texture, and disease-resistance; inter-crossing of most desirable plants, recurrent selection, synthetic polycross.

Intended Use - Ornamental, athletic, and recreational turf wherever turf-type tall fescues are adapted; lacks cold-hardiness in northern areas without adequate snow cover.

Description - Dark-colored, upright, intermediate foliare longation and overall plant height; forms medium-fine textured, dense sward under close mowing and moderately high fertility; very attractive for a tall fescue turf. Excellent turf performance, dark color with very good sward density.

Adapted to - The moderately temperate zone; PHZ 4, 5, 6, 7 (some restrictions in high humidity areas).

Released - 1990, KWS-AG.

Breeder Seed/Stock - KWS-AG.

Certified Seed/Stock - Distributed by TMI, Tangent, OR.

Preparer/Additional Information - Fred B. Ledeboer, PhD, TMI, 22068 Case Rd. NE, Aurora, OR 97002, (503) 678-2597.

Barcel

Source - Material collected from old pastures in The Netherlands in 1959.

Method of Breeding - Selection of plants, clones, and families. The synthetic variety is based on 13 clones.

Intended Use - As a smooth-leafed variety, it is suitable for hay, silage, and pastures.

Description - Very late date of ear emergence. Very smooth, palatable leaves. High digestibility and good disease resistance.

Adapted to - Cental and northeastern U.S. PHZ 1, 2, 3, 4, 5.

Released - Barenbrug.

Breeder Seed/Stock - Barenbrug Holland, Oosterhout, The Netherlands.

Certified Seed/Stock - Barenbrug USA.

Preparer/Additional Information - Barenbrug USA, P.O. Box 239, Tangent, OR 97389, (503) 926-5801.

Barnone

Source - Ecotypes.

Method of Breeding - Mass selection on fine leaf and good sod density in the ecotypes from the east coast of the U.S.

Intended Use - Cool season turfgrass. Very good in sportfields and parks.

Description - A medium dark green color and a fine leaf texture. It has a good Net Blotch resistance and wear tolerance. It has a good turf performance with a good density.

Adapted to - Southern U.S.; PHZ 5, 6, 7, 8, 9.

Released - Barenbrug.

Breeder Seed/Stock - Barenbrug Holland, Oosterhout, The Netherlands.

Certified Seed/Stock - Available from Barenbrug USA.

Preparer/Additional Information - Barenbrug USA, P.O. Box 239, Tangent, OR 97389, (503) 926-5801.

Barvetia

Source - Material collected in old patures in The Netherlands in 1975.

Method of Breeding - A three-clone synthetic in which the clones were chosen on the basis of clonal performance.

Intended Use - Summer forage production in the northern states because of its very high digestibility.

Description - Hexaploid, middle-late variety, with a long flag leaf and a very long culm. Leaf color in autumn of sowing year is mid-dark green. Very erect leaves.

Adapted to - Central and northeastern U.S., but not to southern states. PHZ 1, 2, 3, 4, 5.

Released - 1986, Barenbrug.

Breeder Seed/Stock - Barenbrug Holland, Oosterhout, The Netherlands.

Certified Seed/Stock - Barenbrug USA.

Preparer/Additional Information - Barenbrug USA, P.O. Box 239, Tangent, OR 97389, (503) 926-5801.

Beltsville 16-1 (Reg. No. GP24)

ARS Beltsville Agricultural Research Center (BARC), Beltsville, MD - J.J. Murray and N.R. O'Neill.

Source - Forty-six Plant Introductions (PI) from 21 countries and 53 clones selected from an old BARC nursery and from turf areas in Maryland, Virginia, and North Carolina.

Method of Breeding - Open-pollinated progenies of clones from PIs and old turf areas. Forty-six clones selected to produce synthetic variety.

Intended Use - Parental germplasm for further varietal development for use on lawns, parks, athletic fields, etc.

Description - Beltsville 16-1 has a semiprostrate growth habit, medium leaf texture, a moderate vertical growth rate, and medium-dark green color. It has good resistance to crown rust and moderate resistance to leafspot and brown patch. Beltsville 16-1 has good drought and heat tolerance and very good color retention under low temperatures in the fall and rapid greenup in the spring. In widely scattered turfgrass evaluation trials, it has provided high quality turf with improved density, texture, and color under moderate to low maintenance levels.

Adapted to - LRR A, B, C, D, E, F, G, H, J, M, N, O, P, Q, R, S; PHZ 3, 4, 5, 6, 7, 8.

Released - May 1982, ARS, Beltsville, MD.

Breeder Seed/Stock - ARS, Beltsville Agricultural Research Center, Beltsville, MD.

Certified Seed/Stock - Not available.

Preparer/Additional Information - Kevin N. Morris, National Turfgrass Evaluation Program, BARC-West, Bldg. 001, Room 333, Beltsville, MD 20705, (301) 504-2125.

Bonsai E

KWS-AG, Einbeck, West Germany.

Source - Selected high tiller-count progeny of three clones that were derived from progenies of Mustang, a clone collected in a farm driveway in Aurora, OR, and a clone collected on the University of Georgia campus in Athens. Carried as experimental no. KWS DW-3.

Method of Breeding - Progeny evaluation in space plant nursery for color, tillering ability, reduced plant height, and texture; desirable plants were isolated for inter-crossing; two cycles of recurrent selection for desirable traits.

Intended Use - Permanent turf in the transition zone for ornamental, recreational, and athletic purposes; sod production in California.

Description - Very fine textured, dark colored, low growing with very high sward density; very slow shoot elongation during short days/excellent drought tolerance; early tillering primarily extra vaginal, and very compact. Early development after seeding is very compact, seeding rates of no more than 4 lbs/1,000 sq.ft. are recommended to obtain tillering benefit. Now with high endophyte.

Adapted to - The moderately temperate zone; PHZ 4, 5, 6, 7 (some limitations in high humid areas).

Released - 1989, KWS-AG.

Breeder Seed/Stock - KWS-AG.

Certified Seed/Stock - Distributed by TMI, Tangent, OR.

Preparer/Additional Information - Fred B. Ledeboer, Ph.D., TMI, 22068 Case Rd. NE, Aurora, OR 97002, (503) 678-2597.

Bonsai Plus

Source - Selected out of extensive spaced plantings of Bonsai.

Method of Breeding - Testing and evaluating for coarse plant type in spaced planting for increased resistance to crown rust, net blotch, and stem rust, phenotypic matched matings over two generations.

Intended Use - Turf such as industrial and park areas, golf course roughs.

Description - Very dark green, medium coarse texture, holds color well even with medium to low fertility.

Adapted to - Wherever turf-type tall fescues are adapted.

Released - 1992, KWS-AG, Einbeck, Germany.

Breeder Seed/Stock - KWS Seeds, Inc. and TMI.

Certified Seed/Stock - Will be available in 1993.

Preparer/Additional Information - Fred B. Ledeboer, Ph.D., TMI, 22068 Case Road NE, Aurora, OR 97002, (503) 678-2597.

Cajun (Reg. No. 36)

Developed and released jointly by the Alabama AES, Auburn University and International Seeds Inc., Halsey, OR. Carried as accession no. PI 520749 (trialed as TTFL).

Source - Au Triumph tall fescue.

Method of Breeding - Two cycles of phenotypic selection for late anthesis.

Intended Use - Pasture and hay production, seeded alone or in combination with legumes.

Description - Early maturing cultivar with a leafy, upright growth habit. Provides high early season and late fall forage yields. Anthesis is five to seven days later than Au Triumph, making it more adapted for seed production in Oregon. Cajun is free of the fungal endophyte associated with fescue toxicity. Adapted to a slightly broader area than Au Triumph.

Adapted to - LRR A, C, N, O, P; PHZ 6.

Released - July 1987, by Auburn University and International Seeds Inc., Halsey, OR.

Breeder Seed/Stock - Produced and maintained by International Seeds Inc., Halsey, OR.

Certified Seed/Stock - Available from International Seeds Inc., Halsey, OR. PVP. No. 8700178.

Preparer/Additional Information - Stephen W. Johnson, International Seeds Inc., P.O. Box 168, 820 W. First St., Halsey, OR 97348, (503) 369-2251.

Carefree

Developed by International Seeds Inc., Halsey, OR.

Source - Material collected in the southern and eastern U.S. and derivatives of commercial cultivars.

Method of Breeding - Multiple cycles of mass selection with emphasis on fine blade width and freedom from stem rust incited by *Puccinia graminis* Pers.

Intended Use - Permanent turf for home and commercial lawns, golf course roughs, and athletic fields.

Description - Very fine bladed, late maturing turf-type variety capable of forming an attractive, dense turf; excellent resistance to stem rust and superior seed yield potential.

Adapted to - LRR A, B, C, D, E, G, H, J, K, L, M, N, O, P, R, S, W; PHZ 4.

Released - 1988, by Mid Valley Agricultural Products, Corvallis, OR.

Breeder Seed/Stock - Maintained by International Seeds Inc.

Certified Seed/Stock - Available through Mid Valley Agricultural Products. U.S. Plant Variety Protection has been applied for #8900259.

Preparer/Additional Information - Stephen W. Johnson, International Seeds Inc., P.O. Box 168, 820 W. First St., Halsey, OR 97348, (503) 369-2251.

Chesapeake

Source - Three clones from Georgia and one clone from Ohio. Developed in O.M. Scott breeding program.

Method of Breeding - Four-clone synthetic variety.

Intended Use - Permanent turf.

Description - Excellent heat and drought stress tolerance, high endophyte level.

Adapted to - PHZ 4, 5, 6, 7.

Released - 1985, O.M. Scott & Sons Co.

Breeder Seed/Stock - O.M. Scott & Sons Co.

Certified Seed/Stock - Available.

Preparer/Additional Information - Virgil Meier, O.M. Scott & Sons Co., Marysville, OH 43041, (513) 644-0011.

Chieftain

Chieftain was developed by Pickseed West, Inc., Tangent, OR - Gerard W. Pepin. It is licensed to Roberts Seed Co., Tangent, OR.

Source - An advanced-generation synthetic cultivar selected from the polycross progenies of eight clones. The material originated from breeding programs at Pickseed West, Inc. or the New Jersey AES. Selection was made based on attractive appearance, a rich dark green color, medium texture and density, low growth, and freedom from disease.

Method of Breeding - An eight-clone synthetic variety.

Intended Use - Turf on golf courses, grounds, sod farms, athletic fields, parks, and home lawns. The high level of durability, low growth habits, and relatively low water and nitrogen requirements make it ideal for low-maintenance, high-traffic areas.

Description - A semi-dwarf variety with excellent persistence and durability. The variety produces a medium low-growing, medium density turf of a bright, dark green color, good color retention into cold weather, good winter-hardiness, and early-medium spring green-up. Multiple pest resistance, including good resistance to net blotch and moderate resistance to brown patch.

Adapted to - LRR A, B, C, D, E, F, G, H, I, K, L, M, N, O, R, S; PHZ 2, 3, 4, 5, 6, 7.

Released - Commercial seed was first available in 1988.

Breeder Seed/Stock - Breeder seed is maintained at Pickseed West, Inc., Tangent, OR.

Certified Seed/Stock - A registered and protected variety of Roberts Seed Co., Tangent, OR.

Preparer/Additional Information - Robert Simerly, Roberts Seed Company, 33095 Highway 99E, P.O. Box 206, Tangent, OR 97389, (503) 926-8891.

Cimarron

Source - The original parent plants were six clones of low growing tall fescue selected from old turf areas in New Jersey, Illinois, and Georgia. The seedlings from these crosses were moved to space plant nurseries to initiate cycles of recurrent phenotypic selection for early maturity, low growth habit, dark color, and improved seed production. After the fifth cycle of selection, 210 clones were selected as the parents of Cimarron.

Method of Breeding - Cimarron turf-type tall fescue is an advanced generation synthetic cultivar resulting from five cycles of recurrent selection.

Intended Use - Athletic fields, home lawns, golf course roughs, commercial and public turf areas, playgrounds, and parks. Cimarron blends well with other premium turf-type tall fescues such as Trailblazer and Wrangler.

Description - Cimarron is a moderately low growing turf-type tall fescue showing improved turf performance, density, and a dark green color, as well as persistence. It has shown improved drought tolerance because of its deep root system. It has good turf density and a moderately fine texture compared to other varieties. This variety has shown improved resistance to leaf spot, brown patch, and crown rust. Cimarron has moderate tolerance to stem rust. It has very good heat, cold, wear, and shade tolerance.

Adapted to - PHZ 3, 4, 5, 6, 7.

Released - 1989 by Pure Seed Testing, Inc.

Breeder Seed/Stock - Pure Seed Testing, Inc.

Certified Seed/Stock - Available.

Preparer/Additional Information - Art Wick, LESCO, Inc., 20005 Lake Rd., Rocky River, OH 44116, (216) 333-9250.

Clemfine (Reg. No. 22)

Developed by Lofts Seed, Inc., from germplasm obtained from Clemson University.

Source - Parental clones from the Sand Hills of North and South Carolina in 1970.

Method of Breeding - Three open pollinated progenies selected and placed in breeder nurseries then pollinated with SYN-1 breeder seed in 1977.

Intended Use - Medium- to low-maintenance turfs, lawns, parks, and roadsides.

Description - Medium textured, turf-type tall fescue with improved heat and drought tolerance as well as darker color and greater density over K-31.

Adapted to - LRR all except I, U, T; PHZ 4, 5, 6, 7.

Released - 1982, Lofts Seed, Inc., Bound Brook, NJ.

Breeder Seed/Stock - Lofts Seed, Inc., Bound Brook, NJ.

Certified Seed/Stock - Available.

Preparer/Additional Information - Lofts Seed Inc., Chimney Rock Rd., P.O. Box 146, Bound Brook, NJ 08805, (908) 560-1590.

Coronado

Source - Rebel tall fescue and collections from Alabama, Georgia, Idaho, Kansas, Kentucky, Maryland, Missouri, Mississippi, New Jersey, North Carolina, Ohio, Pennsylvania, Tennessee, Texas, and Virginia.

Method of Breeding - Advanced generation synthetic selected from the maternal progenies of 12 clones selected for excellent floret fertility, stem rust resistance, very dark color, and low growth profile.

Intended Use - Cool-season turf areas, low-maintenance turf, athletic fields, and golf courses.

Description - Very dark, fine-leaved, dense turf with reduced vertical growth.

Adapted to - PHZ 5, 6, 7, 10, 11.

Released - Turf-Seed, Inc.

Breeder Seed/Stock - Pure Seed Testing, Inc.

Certified Seed/Stock - Available.

Preparer/Additional Information - Turf-Seed, Inc., P.O. Box 250, Hubbard, OR 97032, (503) 651-2130.

Courtenay

Source - Developed by P. Jones of Courtenay, B.C.

Method of Breeding - This variety was derived from a seed introduction from the former U.S.S.R. which was obtained through the Welsh Plant Breeding Station in 1974. This seed was collected from old land race material in the Leningrad (St. Petersburg) region. The original material was increased using seed harvested in bulk from single row plots. Mass selection was used to eliminate undesirable plants for a period of several generations. The individual plant selections were made on the basis of desirable pasture growth habit. Further mass selection was carried out to remove plants which exhibited undesirable agronomic characteristics and/or leaf disease. These plants were also screened for the presence of endophyte for three years. Endophyte-free plants and plants developed from heat-treated seed were put into a space planted nursery. This plant material was allowed to interpollinate in isolation to form the breeder seed of Courtenay.

Intended Use - Hay and pasture production.

Description - It is hexaploid; growth habit is semi-erect, some rhizomatous plants; leaves are light green, long and broad; plant height is taller than Kentucky 31; and maturity is medium.

Adapted to - Recommended in Canada for use in British Columbia.

Released - S.G. Bonin, Courtenay, B.C. Registered in Canada in 1987 with Registration No. 2889.

Breeder Seed/Stock - S.G. Bonin, Courtenay, B.C.

Certified Seed/Stock - Available from Dawson Seed Co. Ltd.

Preparer/Additional Information - Agriculture Canada, Food Production and Protection Branch, Plant Products Division, Ottawa, Ontario K1A 0C6. Seed distributed by Dawson Seed Co. Ltd., P.O. Box 91204, North Vancouver, British Columbia V7J 2E7.

Crossfire

Source - Selected out of source populations of tall fescue germplasm in western Oregon.

Method of Breeding - Synthetic cultivar development via polycross and progeny testing.

Intended Use - General purpose cool season turfgrass.

Description - Cultivar has a moderately late heading date for seed production. It has been highly rated in national turf trials for having good summer density and dark green color.

Adapted to - PHZ 6, 7, 8, and western part of northeast U.S., transition zone.

Released - July 1989 by Pickseed West Inc.

Breeder Seed/Stock - Pickseed West Inc.

Certified Seed/Stock - Available.

Preparer/Additional Information - Pickseed West Inc., P.O. Box 888, Tangent, OR 97389, (503) 926-8886.

Dovey

Source - Ecotypes from southeastern France.

Method of Breeding - Mass selection.

Intended Use - Hay and silage, extensive pastures. Good choice for low-maintenance and set-aside mixtures.

Description - Hexaploid, very early variety. Erect growth habit. Short, narrow flagleaf. Very early spring growth, very high rust resistance, and very good persistence.

Adapted to - Large area of adaptability, including good performance under southern conditions.

Released - Plant Breeding International, United Kingdom.

Breeder Seed/Stock - Plant Breeding International, Cambridge, U.K.

Certified Seed/Stock - Barenbrug USA.

Preparer/Additional Information - Barenbrug USA, 32080 Old Highway 34, P.O. Box 239, Tangent, OR 97389, (800) 547-4101.

Duke

Developed for Cascade International Seed Co. and Jonathan Green and Sons, Inc., by Dr. C. Reed Funk, New Jersey AES, New Brunswick.

Source - The parental clones of Duke trace their maternal lineage to plants selected from old turfs in New Jersey (12 clones), Alabama (8 clones), North Carolina (8 clones), Georgia (7 clones), and Kansas (2 clones) during the period of 1962-80. The 37 parental clones of Duke were selected from spaced-plant nurseries at Adelphia, NJ, immediately prior to anthesis in the late spring of 1987 by Dr. Reed Funk and Ted Proehl of Jonathan Green and Sons, Inc.

Method of Breeding - Seed harvested from 27 clones was used to establish a single-plant progeny turf trial at North Brunswick, NJ, and also an isolated spaced-plant nursery near Brownsville, OR, during the fall of 1987. A second cycle of selection pressure was conducted at Brownsville directed towards improving seed yield, phenotypic uniformity, and resistance to disease, including net blotch and stem rust.

Intended Use - Improved turfgrass areas throughout the U.S. and other adapted areas.

Description - Plants of Duke were selected on the basis of fine stems, narrow leaves, a bright rich dark green color, relative freedom from disease, medium plant height, a high percentage of reproductive tillers, medium reproductive maturity, and an upright growth profile. Duke (910 mm) is shorter than Trailblazer (1,170 mm), Finelawn GL (1,370 mm), and Mojave (970 mm). Duke has somewhat wider leaves than Montauk, but turf quality, color, and disease resistance are not significantly different.

Adapted to - It has been planted successfully in southern Canada and as far south as South Carolina and Texas. Duke expands the areas where turf types of tall fescue might be used with satisfaction.

Released - 1988 to Cascade International Seed Co. to establish a breeder field near Lebanon, OR.

Breeder Seed/Stock - Willamette Valley Plant Breeders, Inc., Brownsville, OR.

Certified Seed/Stock - Cascade International Seed Co.

Preparer/Additional Information - Irvin H. Jacob, Cascade International Seed Company, 8483 West Stayton Road, Aumsville, OR 97325, (503) 749-1822.

Earthsave

Source - Numerous accessions in eastern states and out of Bonsai and Mustang.

Method of Breeding - Extensive evaluations for type, seed yield and disease resistance in clonal spaced planting; phenotypic matched mating over three generations.

Intended Use - Fine lawn type turf.

Description - Exceptionally fine texture for tall fescue, slow foliar elongation especially in spring and fall, very dense sod formation.

Adapted to - Wherever turf-type tall fescues are adapted. Less adapted in high humidity and high temperature situations.

Released - KWS-AG, Einbeck, Germany.

Breeder Seed/Stock - KWS Seeds, Inc. and TMI.

Certified Seed/Stock - Available.

Preparer/Additional Information - Fred B. Ledeboer, Ph.D., 22068 Case Road NE, Aurora, OR 97002, (503) 678-2597.

Eldorado

Source - Alabama, North Carolina, and New Jersey and dwarf selections from Olympic and Monarch.

Method of Breeding - Advanced generation synthetic resulting in 21 clones selected for early maturity, dark color, and dwarf growth habit.

Intended Use - Cool season turf areas, low maintenance turf, athletic fields, and golf courses.

Description - High seed yielding, early maturing dwarf type with good turf performance in most areas of the U.S.

Adapted to - PHZ 5, 6, 7, 10, 11.

Released - Turf-Seed, Inc.

Breeder Seed/Stock - Pure Seed Testing, Inc.

Certified Seed/Stock - Available. PVP No. 9000133.

Preparer/Additional Information - Turf-Seed, Inc., P.O. Box 250, Hubbard, OR 97032, (503) 651-2130.

Emperor

Source - Selected out of source populations of tall fescue germplasm in western Oregon.

Method of Breeding - Synthetic cultivar development via polycross and progeny testing.

Intended Use - General purpose cool season turfgrass.

Description - Characterized by having a very late heading date, very dark green color, a below average vertical elongation growth rate, and a very attractive appearance in turf as reflected by excellent turf quality ratings in replicated trials.

Adapted to - PHZ 6, 7, 8, western Northeast, transition zone.

Released - Zajac Performance Seeds.

Breeder Seed/Stock - Pickseed West Inc.

Certified Seed/Stock - Available.

Preparer/Additional Information - Zajac Performance Seeds, 33 Sicomac Road, North Haledon, NJ 07508, (201) 423-1660.

Era

Developed by International Seeds Inc., Halsey, OR.

Source - Experimental lines and breeding composites derived from plants collected from old turfs throughout the U.S.. Carried as experimental no. PX-3.

Method of Breeding - Two cycles of phenotypic mass selection following the cross of four elite parent clones.

Intended Use - Permanent turf for home lawns, athletic fields, golf course roughs, industrial areas, and roadsides. Also may be used for sod production.

Description - Medium dark green, upright growing turf-type; very fine bladed with leaf widths approaching those of some Kentucky bluegrasses; forms a dense, fine textured, high quality turf with a somewhat reduced rate of vertical growth. Moderately resistant to stem rust.

Adapted to - LRR A, B, C, D, E, G, H, J, K, L, M, N, O, P, R, S, W; PHZ 4.

Released - 1990, by International Seeds Inc.

Breeder Seed/Stock - Maintained by International Seeds Inc.

Certified Seed/Stock - Available from International Seeds Inc. U.S. Plant Variety Protection has been applied for #9000091.

Preparer/Additional Information - Stephen W. Johnson, International Seeds Inc., P.O. Box 168, 820 W. First St., Halsey, OR 97348, (503) 369-2251.

Falcon

Rutgers University and Pure Seed Testing.

Source - Collections of old turf areas in Georgia, New Jersey, Pennsylvania, and Virginia.

Method of Breeding - Selection of superior turf type plants under low mowing in New Jersey. Sixteen-clone synthetic polycross.

Intended Use - Turf.

Description - Finer leaf blades and better turf quality than K-31.

Adapted to - Transition zone.

Released - 1979, Rutgers University and Pure Seed Testing.

Breeder Seed/Stock - Rutgers University and Pure Seed Testing.

Certified Seed/Stock - Available.

Preparer/Additional Information - Mike McCarthy, E.F. Burlingham & Sons, P.O. Box 217, Forest Grove, OR 97116, (503) 359-9368.

Falcon II

Burlingham Research Farm - M. McCarthy.

Source - Germplasm based on selected clones from Falcon tall fescue crossed with superior performing selections from turf trials from Virginia.

Method of Breeding - A synthetic polycross based on several cycles of recurrent selection. Criteria included dark green color, disease resistance, and slower growth habit.

Intended Use - Primarily turf, also low-maintenance areas and erosion control.

Description - Improved resistance to brown patch leaf spot diseases and rust. It has a very dark green color and maintains an attractive turf under limited water and maintenance.

Adapted to - PHZ 6, 7, 8, 9, 10 throughout the transition zone.

Released - 1992, McCarthy-Burlingham Research.

Breeder Seed/Stock - McCarthy-Burlingham Research.

Certified Seed/Stock - Available in 1994.

Preparer/Additional Information - Mike McCarthy, E.F. Burlingham & Sons, P.O. Box 217, Forest Grove, OR 97116, (503) 359-9368.

Fawn

Selected at Oregon AES, Corvallis - R.V. Frakes and J.R. Cowan.

Source - Named varieties and foreign introductions.

Method of Breeding - Selection conducted in space-planted introduction nursery and clonal tests. Progenies evaluated and eight parental clones selected for high chromogen content, high crude protein, high seed yield, low self-fertility, and desirable phenotypic appearance. Included in comparative tests as Oregon Synthetic E.

Intended Use - Forage production.

Description - More spring vigor, earlier maturity, and greater height in spring than Alta and Kentucky 31. In 1962 and 1963, produced 22% and 15% more forage, respectively, than Alta. Average seed yield for these two years exceeded that of Alta by 36%. Adapted for seed and forage production in Willamette Valley, OR.

Adapted to - Pacific northwest.

Released - 1964, by Oregon AES.

Breeder Seed/Stock - Oregon AES.

Certified Seed/Stock - Available.

Preparer/Additional Information - Department of Crop and Soil Sciences, Crop Science Building 107, Oregon State Univ., Corvallis, OR 97331-3002, (503) 737-4513.

Festorina

Source - Developed by D.J. Van der Have B.V., Kapelle, Netherlands.

Method of Breeding - Material was collected in 1955 in old pastures in the southwestern part of The Netherlands. After seed multiplication, these ecotypes were established in yield trials. Concurrently, based on preliminary yield evaluations, the 10 highest yielding ecotypes were interpollinated and the polycross progeny evaluated for disease resistance, uniformity, and seed yield. Breeder seed was first harvested in 1961. Tested as HRZ 1.

Intended Use - Hay and pasture production.

Description - Hexaploid; dark green leaves, medium to long flag leaf length, medium to wide flag leaf width; stem is medium to long upper internode; there are no rhizomes; ligule is short, membranous; auricles are usually large with fine hairs; panicle is erect, numerous medium to long branches and spikelets; heading date is early to medium; and plant height is tall. Festorina forage-type tall fescue has outyielded the check varieties in trials conducted in Ontario and in British Columbia.

Adapted to - Recommended in Canada for use in British Columbia.

Released - D.J. Van der Have, Kapelle, The Netherlands. Registered in Canada in 1990 with registration no. 3259.

Breeder Seed/Stock - D.J. Van der Have, Kapelle, The Netherlands.

Certified Seed/Stock - Available from Oseco Inc. and Dawson Seed Co. Ltd.

Preparer/Additional Information - Agriculture Canada, Food Production and Protection Branch, Plant Products Division, Ottawa, Ontario K1A 0C6. Seed distributed by Oseco Inc., P.O. Box 219, Brampton, Ontario L6V 2L2 and Dawson Seed Co. Ltd., P.O. Box 91204, North Vancouver, British Columbia V7J 2E7.

Finelawn 1

Source - Plants collected from old turf areas in Alabama, Georgia, Pennsylvania, and New Jersey contributed to the germplasm of Finelawn 1.

Method of Breeding - Advanced generation synthetic cultivar derived from the progenies of 16 clones.

Intended Use - Turf.

Description - Medium density and texture, bright rich green color, moderately low growing, good shade, drought adaptation, and very good resistance to leaf spot, brown patch, and crown rust.

Adapted to - Cool, humid regions, transitional zones, warm humid regions, coast to coast.

Released - 1981, Dr. Bill Meyer - Pure Seed Testing.

Breeder Seed/Stock - Pure Seed Testing.

Certified Seed/Stock - Available.

Preparer/Additional Information - Pennington Seed Inc. of Oregon, P.O. Box 386, Lebanon, OR 97355, (503) 451-5261.

Finelawn 5GL

Source - Old turf areas in Georgia.

Method of Breeding - An advanced-generation synthetic cultivar resulting from four cycles of recurrent selections. After the fourth cycle of selection, 54 clones were selected as parents of Finelawn 5GL.

Intended Use - Turf.

Description - Semi-dwarf turf type, very good wear tolerance, performs well under low maintenance, and superior resistance to leaf spot and brown patch.

Adapted to - PHZ 2, 3, 4, 5, 6, 7 - cool humid regions, transitional zones, coast to coast.

Released - 1987, Dr. Bill Meyer of Pure Seed Testing.

Breeder Seed/Stock - Pure Seed Testing.

Certified Seed/Stock - Available.

Preparer/Additional Information - David Lundell, Fine Lawn Research, Inc., P.O. Box 1051, Lake Oswego, OR 97034, (503) 636-2600; or Pennington Seed Inc. of Oregon, P.O. Box 386, Lebanon, OR 97355, (503) 451-5261.

Finelawn 88

Source - Willamette Valley Plant Breeders.

Method of Breeding - A synthetic variety derived from the maternal progenies of 435 elite clones prior to the time anthesis.

Intended Use - Turf.

Description - Dark green color, semi-dwarf, moderately fine leaf blade, good disease resistance, excellent heat and drought tolerance, quick establishment, and excellent turf.

Adapted to - PHZ 4, 5, 6, 7, 8 - the northern cool humid regions; winter overseeding.

Released - 1992 by Kevin McVeigh, Willamette Valley Plant Breeders.

Breeder Seed/Stock - Kevin McVeigh, Willamette Valley Plant Breeders.

Certified Seed/Stock - Available.

Preparer/Additional Information - David Lundell, Fine Lawn Research, Inc., P.O. Box 1051, Lake Oswego, OR 97034, (503) 636-2600.

Finelawn Petite

Source - Reed Funk Nursery, Adelphia, NJ, Rutgers University.

Method of Breeding - Petite came from 57 parent selection. Superior selections were made and seed of these selections were planted. Two hundred superior selections were made from these lines. Further roguing was done before anthesis and Petite is the result.

Intended Use - Turf.

Description - Dark forest green color, low (dwarf) growth habit, fine texture. It has above average disease resistance.

Adapted to - PHZ 2, 3, 4, 5, 6, 7 - Cool humid regions, transitional zone.

Released - Judy Brede, Innovative Turf Research, Inc.

Breeder Seed/Stock - Pennington Seed Inc.

Certified Seed/Stock - Available.

Preparer/Additional Information - David Lundell, Fine Lawn Research, Inc, P.O. Box 1051, Lake Oswego, OR 97034, (503) 636-2600.

Fuego

Source - Developed at D.J. Van der Have, Kapelle, The Netherlands.

Method of Breeding - Ecotypes were collected in the south of France in 1973. After multiplication, these ecotypes were put in a yield trial for several years. Three clones from the best forage producing ecotypes were combined to form Fuego. Breeder seed was first harvested in 1977. Tested as HRz 11.

Intended Use - Hay and pasture production.

Description - Leaves are medium to dark green, medium flexibility; panicle is erect, medium to long, with numerous branches and spikelets, awnless; there are no rhizomes; plant height is tall; upper stem internode is medium to long; head formation has medium to strong tendency (approximately 50%) to form heads in year of sowing. In British Columbia tests in 1988 and 1989, Fuego has been shown to yield between 103% and 131% of Alta. Fuego also outyielded Barcel in all trials and outyielded Courtenay on average.

Adapted to - Under evaluation in Canada for use in British Columbia.

Released - D.J. Van der Have, Kapelle, The Netherlands. Registered in Canada in 1990 with registration no. 3243.

Breeder Seed/Stock - D.J. Van der Have, Kapelle, The Netherlands.

Certified Seed/Stock - Available from Topnotch Nutri Ltd.

Preparer/Additional Information - Agriculture Canada, Food Production and Protection Branch, Plant Products Division, Ottawa, Ontario K1A 0C6. Seed distributed by Topnotch Nutri Ltd., P.O. Box 1030, Abbotsford, British Columbia B2S 5B5.

Gala

Developed by International Seeds Inc., Halsey, OR. Carried as experimental no. FPW 83/ISI-865.

Source - Five parent clones; three selected from International Seeds experimental lines and two from commercial varieties.

Method of Breeding - Mass selection for several generations following cross of parent clones. Selection criteria were narrow blade width, dense growth habit, and early maturity.

Intended Use - Provide turf for home lawns, parks, industrial sites, athletic fields, and golf course roughs.

Description - A moderately early flowering, improved turf-type tall fescue variety with an upright growth habit. Moderately dark green and fine bladed, it produces quality turf.

Adapted to - LRR A, B, C, D, E, G, H, J, K, L, M, N, O, P, R, S, W; PHZ 4.

Released - 1990 by International Seeds Inc.

Breeder Seed/Stock - Maintained by International Seeds Inc.

Certified Seed/Stock - Certified and uncertified seed available in commercial quantities from International Seeds. U.S. Plant Variety Protection has been applied for #9000063.

Preparer/Additional Information - Stephen W. Johnson, International Seeds Inc., P.O. Box 168, 820 W. First St., Halsey, OR 97348, (503) 369-2251.

Georgia No. 5

University of Georgia and Plant Materials Center, SCS, Americus, GA. Carried as accession no. GA No. 5 (UGA).

Source - Five lines collected from Sumter Co., GA, Baltimore, MD, Brooksville, FL, and Big Flats, NY.

Method of Breeding - Synthetic from polycross of five lines.

Intended Use - Pasture, hay production, and turf.

Description - Endophyte fungus-infested fescue that persists in the coastal plains of the southeast for superior forage production. Highly recommended for winter overseeding in bermudagrass and bahiagrass pastures and as a general purpose turf.

Adapted to - LRR T, P; PHZ 7 (southern coastal plain).

Released - 1992, University of Georgia, and Plant Materials Center, SCS, Americus, GA.

Breeder Seed/Stock - Dr. Joe Bouton, Univ. of Georgia, Department of Agronomy, Room 3111, Miller Plant Science Bldg., Athens, GA 30602.

Certified Seed/Stock - Available.

Preparer/Additional Information - Pennington Seed Inc. of Oregon, P.O. Box 386, Lebanon, OR 97355, (503) 451-5261.

Goar

Selected at Imperial Valley Field Station, El Centro, CA - L.G. Goar. Carried as accession nos. T.O. 899, P-13847.

Source - Original material came from D. Dagen of Budapest, Hungary, to Professor Southworth of University of Manitoba, Winnipeg, Canada. Received by P. B. Kennedy of California AES, Davis, in March 1925.

Method of Breeding - Planted at El Centro in 1941; tall fescue types selected. Seed of this type received from L. G. Goar by Plant Materials Center, SCS, Pleasanton, CA, and assigned accession number P-13847. Tested there in cooperation with California AES since 1946.

Intended Use - Irrigated pasture.

Description - Early-maturing, vigorous, rather coarse bunchgrass, with high fertility level. Strong seedling vigor. Well adapted to heavy-textured alkaline soils. Grows better during periods of high summer temperature than do other strains of tall fescue.

Adapted to - LRR A, C; PHZ 8.

Released - Certified by California Crop Improvement Association in 1946.

Breeder Seed/Stock - Plant Materials Center, SCS, Lockeford, CA.

Certified Seed/Stock - Limited. This strain just recently being revived because of demand for a heat-tolerant variety.

Preparer/Additional Information - R. Slayback, SCS, 2121-C, 2nd St., Davis, CA 95616, (916) 757-8257.

GQ

Source - Selections were made in the 1970s and 1980s from Georgia, Mississippi, Texas, and Virginia.

Method of Breeding - Twenty-three selections made on basis of uniform anthesis dates and negatively selected for stem rust.

Intended Use - Turf.

Description - GQ is low growing, with high density, medium dark green color, medium-fine leaf texture, and good seed yield.

Adapted to - LRR A, B, D, E, G, N, P, R, S; PHZ 6.

Released - 1992, Jacklin Seed Co.

Breeder Seed/Stock - Jacklin Seed Co.

Certified Seed/Stock - Available.

Preparer/Additional Information - Kim Peterson, Jacklin Seed Co., W. 5300 Riverbend Ave., Post Falls, ID 83854, (208) 773-7581.

Guardian

Guardian turf type tall fescue was developed by Pickseed West, Inc. of Tangent, OR - Gerard W. Pepin. The variety is licensed to Roberts Seed Co. of Tangent, OR. Carried as experimental no. 84SPN.

Source - Guardian turf-type tall fescue was developed beginning in 1984. Twenty clones were selected from a spaced plant nursery in 1986 for dark green color, low growth habit, and freedom from disease. Based on further evaluation, four clones were later discarded, making Guardian 84SPN a 16-clone variety. Germplasm originated from various commercial varieties and Rutgers University experimental varieties.

Method of Breeding - A 16-clone synthetic variety.

Intended Use - Turf on golf courses, grounds, sod farms, athletic fields, parks, and home lawns. The high level of durability, low growth habits, and relatively low water and nitrogen requirements make it ideal for low-maintenance, high-traffic areas.

Description - A semi-dwarf variety with excellent persistence, durability, a very rich, dark green color, moderately fine leaf texture, and moderately dense turf. Guardian has multiple pest resistance, including good resistance to leaf spot, brown patch, and pithium.

Adapted to - LRR A, B, C, D, E, F, G, H, I, K, L, M, N, O, R, S; PHZ 2, 3, 4, 5, 6, 7.

Released - 1989.

Breeder Seed/Stock - Breeder seed at Pickseed West, Inc., Tangent, OR.

Certified Seed/Stock - Certified Guardian is available from Roberts Seed Company, Tangent, OR. Guardian is a registered and protected variety of Roberts Seed Company, Tangent, OR.

Preparer/Additional Information - Robert Simerly, Roberts Seed Company, 33095 Highway 99E, P.O. Box 206, Tangent, OR 97389, (503) 926-8891.

Houndog (Reg. No. 28)

Cooperatively developed by International Seeds Inc. and the New Jersey AES.

Source - Plants collected in Kentucky and Tennessee along with other germplasms developed by the New Jersey AES.

Method of Breeding - Three hundred progeny from seven parents selected for turf quality combined to form synthetic variety.

Intended Use - Medium- to low-maintenance permanent turf either in full sun or in light to moderate shade.

Description - Leafy, persistent turf-type capable of producing an attractive, moderately dense turf with excellent tolerance to heat, drought, and moderate shade. Good resistance to net blotch and brown patch diseases. Has traditionally performed very well in the transition zone between the areas of optimum adaptation for cool season and warm season turfgrasses.

Adapted to - LRR A, B, C, D, E, G, H, J, K, L, M, N, O, P, R, S, W; PHZ 4.

Released - 1982, by International Seeds Inc.

Breeder Seed/Stock - Maintained by International Seeds Inc.

Certified Seed/Stock - Available in quantity from International Seeds Inc. PVP No. 8300011.

Preparer/Additional Information - Stephen W. Johnson, International Seeds Inc., P.O. Box 168, 820 W. First St., Halsey, OR 97348, (503) 369-2251.

Hubbard 87

Source - Germplasm obtained from the New Jersey AES was used in the development of Hubbard 87.

Method of Breeding - Selections were made for lower growth habits, a reduced rate of vertical growth, rich dark green color, improved appearance during heat and drought stress, good cold weather color retention, improved resistance to disease, stress tolerance, and seed production.

Intended Use - For turf in medium- to low-maintenance conditions and for penetration in dense soils in regions where tall fescue is well adapted for turf use.

Description - A leafy turf-type tall fescue of medium reproductive maturity and capable of producing a persistent, moderately dense, attractive, medium-to-low growing turf with medium texture and a medium-dark green color. Moderately good resistance to net blotch and large brown patch. Hubbard 87 has good seedling vigor on warm soils and is adapted to a wide range of soils. Produces less thatch than vigorous cultivars of Kentucky bluegrass and chewings fescue.

Adapted to - PHZ 3, 4, 5, 6, 7, 8.

Released - 1989 by Hubbard Seed & Supply Co.

Breeder Seed/Stock - Hubbard Seed & Supply Co., Hubbard, OR.

Certified Seed/Stock - Available. PVP registration No. 9100001.

Preparer/Additional Information - Gordon W. & Sharon K. Jones, Hubbard Seed & Supply Co., P.O. Box 310, 13346 Whiskey Hill Road NE, Hubbard, OR 97032, (503) 981-4136.

ITR90-2

Source - Rutgers University breeding program.

Method of Breeding - Phenotypic recurrent selection.

Intended Use - Turf.

Description - Very low growing (double-dwarf), dark green, high performance turf.

Adapted to - PHZ 4, 5, 6, 7, 8, 9, 10 - west Texas to California.

Released - 1990, Innovative Turf Research, Inc.

Breeder Seed/Stock - Pennington Seed Inc.

Certified Seed/Stock - Available.

Preparer/Additional Information - Pennington Seed Inc. of Oregon, P.O. Box 386, Lebanon, OR 97355, (503) 451-5261.

Jaguar II

Source - Selections from Jaguar.

Method of Breeding - Advanced generation synthetic from three cycles of recurrent selection for improved seed yield, crown rust resistance, darker color, and improved turf quality over Jaguar.

Intended Use - Cool season turf areas, low maintenance turf, athletic fields, and golf courses.

Description - Jaguar II has improved seed yield, a darker green color on turf and improved turf quality. Good fusarium patch resistance.

Adapted to - PHZ 5, 6, 7, 10, 11.

Released - Zajac Performance Seeds.

Breeder Seed/Stock - Zajac Performance Seeds.

Certified Seed/Stock - Available. PVP No. 8900242.

Preparer/Additional Information - Zajac Performance Seeds, 33 Sicomac Rd., North Haledon, NJ 07508, (201) 423-1660.

Kentucky 31

Increased at Kentucky AES, Lexington, KY - E.N. Fergus.

Source - William Suiter's farm in Menifee County, KY. Collected by E. N. Fergus in 1931 for testing at Kentucky AES. Apparently grown on Suiter's farm since 1887.

Method of Breeding - Increase of original collection.

Intended Use - Pasture, hay, and revegetation.

Description - Wide adaptation to soil types and temperature extremes. Suited to upper South, where it remains green all year with occasional exception of midsummer months. Very productive, but not too palatable. Excellent for erosion control.

Adapted to - LRR M, N, S, L, R, T, P, K; PHZ 4, 5, 6, 7.

Released - Kentucky AES.

Breeder Seed/Stock - Available.

Certified Seed/Stock - Available.

Preparer/Additional Information - Dr. A.J. Powell, AES, Univ. of Kentucky, Lexington, KY 40506, (606) 257-9000.

Lancer

Source - Selection was made in each cycle for an improved dark green color, finer leaf texture, dwarf growth habit, high reproductive tiller number, leaf spot resistance, stem rust resistance, brown patch resistance, and drought avoidance.

Method of Breeding - An advanced generation synthetic cultivar.

Intended Use - Well adapted throughout cool humid and cool arid zones and well adapted in the warm arid zones as compared to other cool season turfgrasses. Moderately well adapted in warm humid and warm arid zones.

Description - Lancer Narrow Leaf Dwarf Turf-Type Tall fescue is a new finer leaf, lower growing, darker green tall fescue with a slower leaf extension rate and a shorter mature plant height. Lancer exhibits a narrower leaf than other dwarf tall fescues which in combination with lower growth habit and slower growth rate results in a dense fine textured turf. Lancer has good heat tolerance, and its deep roots provide good drought avoidance. Once established, lancer forms a very tough, wear tolerant turf. This variety has shown good resistance to leaf spot, crown rust, stem rust, and fusarium patch and moderately good resistance to brown patch. It has shown good tolerance to heat, cold, wear, and shade. Lancer's drought tolerance ranks equal to most standard turf-type tall fescue varieties.

Adapted to - PHZ 3, 4, 5, 6, 7.

Released - 1991, Pickseed, Inc.

Breeder Seed/Stock - Pickseed, Inc.

Certified Seed/Stock - Available.

Preparer/Additional Information - Art Wick, LESCO, Inc., 20005 Lake Rd., Rocky River, OH 44116, (216) 333-2950.

Leprechaun

Leprechaun was developed by Pickseed West, Inc., Tangent, OR - Gerard W. Pepin. The variety is licensed to Roberts Seed Co., Tangent, OR.

Source - Tillers collected from old turf plots in Adelphia, NJ.

Method of Breeding - Leprechaun is an eight-clone synthetic cultivar. Tillers from old turf plots at Adelphia, NJ, were allowed to inter-pollinate and the resulting progeny were screened for low growth, dark color, and good seed production. The eight parent clones were selected after three generations of recurrent selection.

Intended Use - Turf on golf courses, grounds, sod farms, athletic fields, parks, and home lawns. The high level of durability, low growth habits, and relatively low water and nitrogen requirements make it ideal for low-maintenance, high-traffic areas.

Description - Leprechaun is a dwarf variety with excellent persistence and durability. Has a much below average mature plant height with a slower than average varietal growth rate and very dark green color. More dense and with a finer leaf texture than most cultivars. Additional evaluation is underway including multiple pest resistance.

Adapted to - LRR A, B, C, D, E, F, G, H I, K, L, M, N, O, R, S; PHZ 2, 3, 4, 5, 6, 7.

Released - 1990 in limited quantities.

Breeder Seed/Stock - Breeders seed is maintained at Pickseed West, Inc., Tangent, OR.

Certified Seed/Stock - Application for acceptance into the Oregon Seed Certification system is planned. Certified seed became available in 1991 from Roberts Seed Co., Tangent, OR. Application for U.S. Plant Variety Protection is anticipated.

Preparer/Additional Information - Robert Simerly, Roberts Seed Company, 33095 Highway 99E, P.O. Box 206, Tangent, OR 97389, (503) 926-8891.

Lexus

Source - Sixty-two clonal lines developed by New Jersey AES and Rutgers University.

Method of Breeding - Synthetic cultivar originally developed by the New Jersey AES with Dr. C. Reed Funk.

Intended Use - Turf, either alone or in blends.

Description - Dwarf growth characteristics, very dark green, dense, fine leafed, superior disease resistance, drought and heat tolerant, excellent mowing characteristics, and turf quality.

Adapted to - Southeast and southwest U.S., transition states.

Released - 1993, Barenbrug USA.

Breeder Seed/Stock - Barenbrug USA.

Certified Seed/Stock - Available in 1994.

Preparer/Additional Information - Barenbrug USA, P.O. Box 239, Tangent, OR 97389, (503) 926-5801.

Marksman

McCarthy-Burlingham Research Farms.

Source - Selections were made in individual space plant nurseries for lower-growing types from materials collected throughout the hot, humid areas of the transition zone.

Method of Breeding - Recurrent selections of top performing clones. Evaluated in Virginia for turf quality (brown patch, pythium, and leaf spot resistance) and crossed and selected in Oregon for seed production.

Intended Use - Turf, erosion control, and low-maintenance areas (e.g., roadsides).

Description - Low growing, dark color under limited fertility, tolerated limited drought, and disease resistant.

Adapted to - PHZ 6, 7, 8, 9, 10 - transition zone.

Released - 1992, McCarthy-Burlingham Research.

Breeder Seed/Stock - McCarthy-Burlingham Research.

Certified Seed/Stock - Available in 1994.

Preparer/Additional Information - Mike McCarthy, E.F. Burlingham & Sons, P.O. Box 217, Forest Grove, OR 97116, (503) 359-9368.

Martin

Developed by D.A. Sleeper, University of Missouri, Columbia, MO.

Source - Tall fescue plants collected in the south-central U.S.

Method of Breeding - A broad-based, medium- early maturity, crown-rust resistant population with high herbage and seed yield was developed. From this population two superior clones were selected as the basis for the variety Martin. These two clones were selected for high forage yield, high magnesium content, and low potassium-to-magnesium [K/OCa-to-Mg] ratio.

Intended Use - Pasture and hay production, seeded alone or in combination with legumes.

Description - Medium-early, widely adapted variety; very leafy with excellent forage production. Summer forage yields are higher than for cultivars developed from Mediterranean germplasm; improved resistance to crown rust. Free of the endophyte which is associated with fescue toxicosis. Average daily gains significantly higher than Kentucky 31 in animal feeding trial.

Adapted to - LRR A, C, D, J, K, L, M, N, O, P, Q, R, S; PHZ 3, 4, 5, 6, 7.

Released - 1986 by University of Missouri.

Breeder Seed/Stock - Available from University of Missouri.

Certified Seed/Stock - Available. PVP number 8700162.

Preparer/Additional Information - Stephen W. Johnson, International Seeds, Inc, P.O. Box 168, Halsey, OR 97348, (503) 369-2251.

Maximize

Source - Developed by Pure Seed Testing Inc., Hubbard, OR, and Van der Have B.V., Rilland, The Netherlands.

Method of Breeding - An advanced generation synthetic variety resulting from two cycles of recurrent selection. Seed from ecotypes collected in southeast France in 1976 were evaluated in a spaced plant trial for disease resistance, forage yield, and palatability. The best, most attractive plants with similar maturity were selected to form experimental synthetics by cloning. In 1979, a second cycle of selection was performed in which five clones were selected as the parental material of Maximize. Breeder seed was first produced in 1980.

Intended Use - Hay and pasture production.

Description - Growth habit is erect; leaves are medium light green with rough margins, medium course, shorter than those of Kentucky 31; stem has shorter internodes than Kentucky 31; panicle is narrow-tapering, open, nodding, shorter than that of Kentucky 31, awned; plant height is shorter than Kentucky 31; rhizomes are rare; maturity is medium early, similar to Kentucky 31; and the photoperiod is sensitive. In British Columbia trials, Maximize outyielded all the check varieties on average. In contrast, in the Ontario trials, it consistently underyielded the test mean.

Adapted to - Recommended in Canada for use in British Columbia.

Released - Pure Seed Testing Inc., Hubbard, OR.

Breeder Seed/Stock - Turf Seed Inc., Hubbard, OR.

Certified Seed/Stock - Available.

Preparer/Additional Information - Agriculture Canada, Food Production and Protection Branch, Plant Products Division, Ottawa, Ontario K1A 0C6. Seed distributed by Dawson Seed Co., P.O. Box 91204, North Vancouver, British Columbia V7J 2E7.

Mesa

This was developed by the New Jersey AES - Cyril R. Funk, Jr.

Source - An advanced generation synthetic cultivar selected from progenies of eight clones collected from old, low maintenance turfs in Georgia, Idaho, New Jersey, and Pennsylvania. Also, some germplasm originated from Rebel tall fescue.

Method of Breeding - Recurrent selection based on attractive appearance, medium texture, medium density, soft leaves, freedom from disease, and a bright, dark green color. Eight clones were then transferred to a replicated, isolated spaced-plant nursery where 25% of the plants were removed to improve uniformity.

Intended Use - Recommended for use on home lawns, parks, athletic fields, roadsides, institutional grounds, golf courses, and school play areas.

Description - Capable of producing an attractive turf of medium texture and density with a rich, dark green color. Closely resembles Rebel II tall fescue, but it has a longer, mature culm length and sheath length. It also has a narrower flagleaf and a greater leaf-to-width ratio of its flagleaf. Mesa also has a greater length-to-width ratio of its seed. It has better turfgrass quality than Adventure, Maverick, Pacer, and Irident and better resistance to net blotch disease than Maverick and Pacer. Mesa also has better resistance to pink snow mold than Apache. It has a significantly longer culm and panicle than Mojave.

Adapted to - LRR A, B, C, D, E, G, N, P, R, S; PHZ 6.

Released - 1987, Jacklin Seed Company and Jonathan Green and Sons, Inc.

Breeder Seed/Stock - Jacklin Seed Company, Post Falls, ID.

Certified Seed/Stock - Available.

Preparer/Additional Information - Kim Peterson, Jacklin Seed Co., W. 5300 Riverbend Ave., Post Falls, ID 83854, (208) 773-7581.

Mini-Mustang

Source - Parents were selected from source populations in western Oregon.

Method of Breeding - Synthetic cultivar development via polycross and subsequent progeny testing to attain uniformity.

Intended Use - Cool season turf grass.

Description - Mature seed production plants exhibit short, upright seed culms with good seed production potential.

Adapted to - PHZ 6, 7, 8, and western part of northeastern U.S., transition zone.

Released - August 1990, Pickseed West Inc.

Breeder Seed/Stock - Pickseed West Inc.

Certified Seed/Stock - Available.

Preparer/Additional Information - Pickseed West Inc., P.O. Box 888, Tangent, OR 97389, (503) 926-8886.

Mojave

Developed by International Seeds Inc., Halsey, OR.

Source - Breeding composites derived from crossing selections out of commercial varieties. Carried as experimental no. ISI-822.

Method of Breeding - Plants evaluated for characteristics associated with high turf quality; five plants selected to produce synthetic variety.

Intended Use - Turfgrass for lawns, athletic fields, golf course roughs, and roadsides.

Description - Moderately fine bladed, late heading turf-type with a dense, leafy growth habit. Produces a handsome, persistent turf. Resistant to stem rust.

Adapted to - LRR A, B, C, D, E, F, G, H, J, K, L, M, N, O, P, R, S, W; PHZ 4.

Released - 1985, by Mid Valley Agricultural Products, Corvallis, OR.

Breeder Seed/Stock - Maintained by International Seeds Inc.

Certified Seed/Stock - Available from Mid Valley Agricultural Products. U.S. PVP No. 8500183

Preparer/Additional Information - Stephen W. Johnson, International Seeds Inc., P.O. Box 168, 820 W. First St., Halsey, OR 97348, (503) 369-2251.

Monarch

Source - Georgia, New Jersey, and selections from Olympic.

Method of Breeding - Advanced generation synthetic of 58 clones resulting from three cycles of recurrent selection.

Intended Use - Cool-season turf areas, low-maintenance turf, athletic fields, and golf courses.

Description - Moderate dwarf-type with good turf performance across the U.S. Good seed yield, dark green color.

Adapted to - PHZ 5, 6, 7, 10, 11.

Released - Turf-Seed, Inc.

Breeder Seed/Stock - Pure Seed Testing, Inc.

Certified Seed/Stock - Available. PVP No. 8700121.

Preparer/Additional Information - Turf-Seed, Inc., P.O. Box 250, Hubbard, OR 97032, (503) 651-2130.

Montauk

Developed for Cascade International Seed Co. and Jonathan Green and Sons, Inc., Farmingdale, NJ - Dr. Kevin McVeigh, Willamette Valley Plant Breeders, Inc., Brownsville, OR, and Dr. C. Reed Funk, New Jersey AES, New Brunswick.

Source - The parental germplasm of Montauk traces its maternal origin to plants selected from old turfs in Georgia (18 clones), New Jersey (seven clones), North Carolina (three clones), Alabama (two clones), Maryland (two clones), and Kansas (one clone) during the period 1962-1980. Thirty-three parental clones were selected from a large spaced-plant nursery at Adelphia, NJ, just prior to anthesis in the late spring of 1987 by Dr. Kevin McVeigh.

Method of Breeding - After removal to an isolated crossing nursery at North Brunswick, NJ, seed was harvested from the 31 plants showing the best floret fertility. Progenies of these selected clones were established in closely mowed turf trials in NJ, and in isolated spaced-plant nursery near Brownsville, OR, from plants which demonstrated improved seed yield, uniformity, and resistance to diseases including stem rust.

Intended Use - Improved turfgrass areas throughout the U.S. and other adapted areas.

Description - Montauk is an advanced generation synthetic cultivar with a bright dark green color, a medium-tall upright growth profile, a high percentage of reproductive tillers, and attractive appearance. It is a semi-dwarf plant which exhibits a severe reduction in stem elongation, internode length, and leaf growth in turf applications. Leaf blades are easily mowed because they are soft fibered and oriented vertically in the lawn. This erect growth habit is somewhat unusual in tall fescues. Montauk has a significantly shorter first internode length (151 mm) than Duke (176 mm), Mojave (251 mm), and Finelawn GL (261 mm). The flag leaf of Montauk is finer (4.9 mm) than Duke (5.3 mm).

Adapted to - It has been planted successfully in southern Canada and as far south as South Carolina and Texas. Montauk expands the areas where turf types of tall fescue might be used with satisfaction.

Released - 1988 by Dr. Kevin McVeigh to Cascade International Seed Co. to establish a breeder field near Lebanon, OR.

Breeder Seed/Stock - Willamette Valley Plant Breeders Inc., Brownsville, OR.

Certified Seed/Stock - Available.

Preparer/Additional Information - Irvin H. Jacob, Cascade International Seed Company, 8483 West Stayton Road, Aumsville, OR 97325, (503) 749-1822.

Mozark

Developed by D.A. Sleeper, University of Missouri, Columbia, MO.

Source - Plant introduction from Algeria and France and the varieties Kentucky 31 and Kenmont.

Method of Breeding - Four clones for synthetic variety selected from a variety of germplasms. Selection was based on vigor after defoliation, high seed and herbage yield, and resistance to crown rust.

Intended Use - Pasture and hay production with or without legumes.

Description - Medium-early cultivar, good forage production, and good resistance to crown rust. Broadly adapted variety. Free of the endophyte associated with fescue toxicosis. Superior average daily gains across multiple years in animal feeding trials.

Adapted to - LRR A, C, D, J, K, L, M, N, O, P, R, S; PHZ 3, 4, 5, 6, 7.

Released - 1987 by University of Missouri.

Breeder Seed/Stock - University of Missouri.

Certified Seed/Stock - Available. PVP number 8700163.

Preparer/Additional Information - Stephen W. Johnson, International Seeds, Inc, P.O. Box 168, Halsey, OR 97348, (503) 369-2251.

Murietta

Source - Selections from Bonanza and New Jersey.

Method of Breeding - Advanced generation synthetic of 34 plants selected for late maturity and a dwarf growth habit.

Intended Use - Cool season turf areas, low maintenance turf, athletic fields, golf courses.

Description - Good resistance to powdery mildew, crown rust, reduced clippings with dense turf. Excellent turf performance in arid regions.

Adapted to - PHZ 5, 6, 7, 10, 11.

Released - Turf-Seed, Inc.

Breeder Seed/Stock - Pure Seed Testing, Inc.

Certified Seed/Stock - Available. PVP No. 8800135.

Preparer/Additional Information - Turf-Seed, Inc., P.O. Box 250, Hubbard, OR 97032, (503) 651-2130.

Mustang

Source - Original germplasm contributing parents to the cultivars came from two advanced selection populations in NJ.

Method of Breeding - Synthetic cultivar development via polycross and subsequent progeny testing and selection to attain uniformity.

Intended Use - Cool season turfgrass.

Description - Mustang belongs to the first group of turf-type tall fescues. It was originally selected from source plants that persisted well under close mowing regimes.

Adapted to - PHZ 6, 7, 8, western Northeast, and transition zone.

Released - Pickseed West Inc.

Breeder Seed/Stock - Pickseed West Inc.

Certified Seed/Stock - Available.

Preparer/Additional Information - Pickseed West Inc., P.O. Box 888, Tangent, OR 97389, (503) 926-8886.

Nuicro

Source - Numerous accessions from old turfs in the eastern and southeastern U.S.

Method of Breeding - Extensive evaluations for desired turf type with compact growth in space plant nurseries, phenotypic matched matings over four generations with emphasis toward dark green compact growth with moderately fine texture.

Intended Use - Throughout tall fescue adaptation zone, better performance in west than in the humid eastern and southeastern sections.

Description - Compact, dark green with excellent turf quality; good resistance to crown rust and net blotch.

Adapted to - Wherever tall fescues for turf are adapted.

Released - February 1992, KWS-AG, Einbeck, Germany.

Breeder Seed/Stock - KWS Seeds, Inc. and TMI.

Certified Seed/Stock - Available in 1993.

Preparer/Additional Information - Fred B. Ledeboer, Ph.D., TMI, 22068 Case Road NE, Aurora, OR 97002, (503) 678-2597.

Oasis (SP)

Source - Ecotype selections were made by Dr. Reed Funk in the late 1970s and the early 1980s from Georgia, Mississippi, Texas, and Virginia.

Method of Breeding - Fifteen thousand progeny from a 3,000-space plant nursery of 1987 were planted in 1988. Fifteen plants were selected for uniform anthesis dates and negatively selected for stem rust and put into a crossing block. From this crossing block, one acre of breeders field was planted in 1989 near Albany, OR.

Intended Use - Turf.

Description - Oasis is a low-growing, medium textured tall fescue with moderately good resistance to netblotch and brown patch.

Adapted to - LRR A, B, D, E, G, N, P, R, S; PHZ 6.

Released - 1992, Jacklin Seed and Outdoor Equipment Co.

Breeder Seed/Stock - Jacklin Seed Co.

Certified Seed/Stock - Available.

Preparer/Additional Information - Kim Peterson, Jacklin Seed Co., W. 5300 Riverbend Ave., Post Falls, ID 83854, (208) 773-7581.

Olympic

Source - Alabama, North Carolina, and New Jersey.

Method of Breeding - Advanced generation synthetic derived from the progenies of eight clones. Selected for drought tolerance, soft leaves, and disease resistance.

Intended Use - Cool-season turf areas, low-maintenance turf, athletic fields, and golf courses.

Description - Broadly adapted, performing well across the U.S., improved resistance to crown rust and netblotch.

Adapted to - PHZ 5, 6, 7, 10, 11.

Released - Turf-Seed, Inc.

Breeder Seed/Stock - Pure Seed Testing, Inc.

Certified Seed/Stock - Available. PVP No. 8100168.

Preparer/Additional Information - Turf-Seed, Inc., P.O. Box 250, Hubbard, OR 97032, (503) 651-2130.

Olympic II

Source - Selections from Olympic, Apache, and old turf areas in New Jersey.

Method of Breeding - Advanced generation synthetic of the progenies of 13 clones selected for improved seed yield and turf performance.

Intended Use - Cool-season turf areas, low-maintenance turf, athletic fields, and golf courses.

Description - Good tolerance of heat, cold, shade and wear; good spring greenup and fall density.

Adapted to - PHZ 5, 6, 7, 10, 11.

Released - Turf-Seed, Inc.

Breeder Seed/Stock - Pure Seed Testing, Inc.

Certified Seed/Stock - Available. PVP No. 9000134.

Preparer/Additional Information - Turf-Seed, Inc., P.O. Box 250, Hubbard, OR 97032, (503) 651-2130.

Pacer (Exp. No. ISI-CJ)

Developed by International Seeds Inc., Halsey, OR.

Source - Breeding composites derived from commercial varieties and experimental lines.

Method of Breeding - Polycross progenies evaluated under turf conditions. Six elite clones selected to produce synthetic cultivar.

Intended Use - Permanent turf for home lawns, parks, athletic fields, golf course roughs, and industrial sites. May be used in full sun or partial shade.

Description - Lower growing, moderately dark green, late heading turf-type variety; dense compact growth habit; especially well adapted to the northern transition zone; resistant to brown patch and crown rust. Good winter color retention.

Adapted to - LRR A, B, C, D, E, G, H, J, K, L, M, N, O, P, R, S, W; PHZ 4.

Released - 1987 by International Seeds Inc.

Breeder Seed/Stock - Maintained by International Seeds Inc.

Certified Seed/Stock - Available in quantity from International Seeds Inc. U.S. PVP No. 8700199

Preparer/Additional Information - Stephen W. Johnson, International Seeds Inc., P.O. Box 168, 820 W. First St., Halsey, OR 97348, (503) 369-2251.

Penngrazer

Source - Old Kentucky 31 seed production fields in Kentucky.

Method of Breeding - Polycross of 13 clones.

Intended Use - Forage.

Description - Five percent better forage yield than Kentucky 31, no endophyte. Superior crown rust resistance.

Adapted to - PHZ 4, 5, 6, 7, 8 - eastern U.S.

Released - 1985, FFR Cooperative.

Breeder Seed/Stock - Pennington Seed Inc.

Certified Seed/Stock - Available.

Preparer/Additional Information - Pennington Seed Inc. of Oregon, P.O. Box 386, Lebanon, OR 97355, (503) 451-5261.

Phoenix

Source - Twenty-six clones selected from Rebel, and old turfs in Australia, Kansas, Idaho, and the southeast U.S.

Method of Breeding - Advanced generation synthetic cultivar developed at the New Jersey AES with Dr. C. Reed Funk.

Intended Use - Turf, either alone or in blends.

Description - Dark green, medium texture, excellent heat and drought tolerance, winter-hardy, and good disease resistance.

Adapted to - Southeast and southwest U.S., transition states.

Released - 1990, Barenbrug USA.

Breeder Seed/Stock - Willamette Valley Plant Breeders and Barenbrug USA.

Certified Seed/Stock - Available from Barenbrug USA.

Preparer/Additional Information - Don Herb and Matt Herb, Barenbrug USA, P.O. Box 239, Tangent, OR 97389, (503) 926-5801.

Phyter

Selected at FFR Cooperative - S.J. Baluch and S.D. Stratton.

Source - Seven hundred tall fescue plants selected out of old KY-31 seed fields in 1983. Fields were estimated to be at least 15 years old.

Method of Breeding - Eleven-clone synthetic. Original 700 clones were selected based on color and vigor. The 11% clones were selected on polycross progeny evaluations and clonal evaluations on spring vigor, summer and fall color, and moisture content of regrowth herbage.

Intended Use - Released as a low endophyte hay and pasture use cultivar.

Description - Similar in maturity to KY-31. Rust resistance better than Kenhy and similar to KY-31. Phyter is also slightly darker in color than KY-31.

Adapted to - LRR A, B, C, F, G, H, J, K, L, M, N, O, P, R, S, T; PHZ 2, 3, 4, 5, 6, 7, 8.

Released - 1988 by FFR Cooperative.

Breeder Seed/Stock - FFR Cooperative.

Certified Seed/Stock - Available.

Preparer/Additional Information - Bret L. Winsett, FFR Cooperative, 4112 E. State Road 225, West Lafayette, IN 47906, (317) 567-2115.

Pixie

Source - An advanced generation, synthetic cultivar selected from progenies of 47 sister lines.

Method of Breeding - Selected clones were evaluated in a nursery in New Jersey.

Intended Use - Turf.

Description - Superior density resistance to summer stress and moderately fine leafed with dark green color.

Adapted to - LRR A, B, D, E, G, N, P, R, S.

Released - 1992 by Jacklin Seed Co.

Breeder Seed/Stock - Jacklin Seed Co.

Certified Seed/Stock - Available.

Preparer/Additional Information - Kim Peterson, Jacklin Seed Co., W. 5300 Riverbend Ave., Post Falls, ID 83854, (208) 773-7581.

Rebel

Developed by Lofts Seed, Inc., from germplasm obtained from the New Jersey Agricultural Experiment Station.

Source - Plants collected from old turf areas in New Jersey and other states and from trispecies hybrids of tall fescue, meadow fescue, and perennial ryegrass obtained from the U.S. Regional Pasture Research Laboratory, University Park, PA.

Method of Breeding - Clones were initially evaluated under close mowing, then singling out the most attractive plants, they underwent three cycles of recurrent selection for disease resistance, persistence, attractiveness, and performance in turf trials.

Intended Use - Medium- to low-maintenance turf in sun or moderate shade, home lawns, parks, athletic fields, and sod farms.

Description - An attractive, turf-type cultivar with persistent turf of greater density, finer texture, and slower rate of vertical growth than other K-31 tall fescue.

Adapted to - LRR all except I, U, T; PHZ 4, 5, 6, 7.

Released - 1981 - Lofts Seed, Inc., in cooperation with the New Jersey AES.

Breeder Seed/Stock - Lofts Seed, Inc.

Certified Seed/Stock - Available.

Preparer/Additional Information - Lofts Seed Inc., Chimney Rock Road, P.O. Box 146, Bound Brook, NJ 08805, (908) 560-1590.

Rebel 3D

Source - Plants selected from, or related to, Rebel tall fescue were mated with plants selected from old turfs of the eastern U.S.

Method of Breeding - Progenies from crosses of the above described germplasm sources were subjected to varying number of cycles of population improvement including phenotypic recurrent selection in spaced-plant nurseries followed by progeny trials conducted in mowed turf.

Intended Use - Turf, recommended as an improved lawn grass in temperate and transition climates.

Description - Compared with some extreme dwarf tall fescues, Rebel 3D shows improved resistance to and more rapid recovery from diseases such as brown patch. Rebel 3D also shows excellent performance under irrigation, in warm to hot arid climates typical of much of California.

Adapted to - Temperate and transition climates.

Released - 1992 by Lofts Seed Inc.

Breeder Seed/Stock - Lofts Seed Inc.

Certified Seed/Stock - Available.

Preparer/Additional Information - Marie Pompei, Lofts Seed Inc., P.O. Box 146, Bound Brook, NJ 08805, (908) 560-1590.

Rebel II (Reg. No. 30)

Developed by Lofts Seed, Inc., using germplasm obtained from the New Jersey Agricultural Experiment Station. Carried as accession no. 508284; R-2.

Source - Attractive disease-resistant plants from old turfs located in Alabama, Georgia, Idaho, New Jersey, North Carolina, Ohio, and Virginia.

Method of Breeding - A modified back-crossing program combined with varying cycles of recurrent restricted phenotypic and genotypic selection for stress tolerance, disease resistance, attractive appearance, and improved turf performance.

Intended Use - Medium- to low-maintenance turfs in full sun or light to moderate shade for lawns, parks, athletic fields, sod farms, and roadsides.

Description - An improved cultivar of Rebel. It shows improved resistance to crown rust and net blotch, a darker green color, higher seed yield potential, and turf that is finer in texture and slightly more dense. It has shown good heat tolerance, winter-hardiness, and improved performance under close mowing.

Adapted to - LRR all except I, U, T; PHZ 4, 5, 6, 7.

Released - 1986, Lofts Seed, Inc.

Breeder Seed/Stock - Lofts Seed, Inc., with the cooperation of the New Jersey Agricultural Experiment Station.

Certified Seed/Stock - Available.

Preparer/Additional Information - Lofts Seed Inc., Chimney Rock Rd, P.O. Box 146, Bound Brook, NJ 08805, (908) 560-1590.

Rebel, Jr.

Developed by Lofts Seed, Inc., Bound Brook, NJ, using germplasm obtained from the New Jersey Agricultural Experiment Station.

Source - Rebel cultivar and plants selected from old turf plots in Alabama, Georgia, Kansas, Maryland, Mississippi, New Jersey, North Carolina, Ohio, Pennsylvania, Tennessee, Texas, and Virginia.

Method of Breeding - An advanced generation of synthetic cultivars from 91 clones. Plants selected for attractiveness, seed yield potential, and disease resistance were inter-pollinated or top-crossed with plants selected from or related to Rebel. Resulting progenies were closely mown. Tillers selected were then used to initiate new cycles of recurrent selection.

Intended Use - Medium- to low-maintenance turf, lawns, parks, roadsides, sod farms, and athletic fields in full sun to moderate shade.

Description - A lower growing, persistent, and attractive turf-type cultivar as compared to Rebel. It shows good wear tolerance, good seedling vigor, and good tolerance of heat and drought. Also exhibits an improved dark green color.

Adapted to - LRR all except P, I, U, T; PHZ 4, 5, 6, 7.

Released - 1990, Lofts Seed, Inc.

Breeder Seed/Stock - Lofts Seed, Inc., in cooperation with the New Jersey AES.

Certified Seed/Stock - Available.

Preparer/Additional Information - Lofts Seed Inc., Chimney Rock Rd, P.O. Box 146, Bound Brook, NJ 08805, (908) 560-1590.

Richmond

Developed by International Seeds Inc., Halsey, OR.

Source - The turf-type tall fescue variety Houndog. Carried as experimental no. HBM.

Method of Breeding - Derived from genotypes of Houndog. Selection of plants showing increased tiller density and shorter flag leafs followed by multiple cycles of mass selection.

Intended Use - Medium- to low-maintenance turf in full sun to partial shade. Because it is free of endophyte, it is suitable for forage use and has demonstrated exceptional quality as a pasture grass for grazing animals.

Description - Early maturing variety with moderately dark green, fine textured leaves developed for use in turf. Good seed yield potential. High resistance to brown patch and other fungus related diseases.

Adapted to - LRR A, B, C, D, E, G, H, J, K, L, M, N, O, P, R, S, W; PHZ 4.

Released - 1987 by Cascade International Seeds, Aumsville, OR, and Jonathan Green & Sons, Farmingdale, NJ.

Breeder Seed/Stock - International Seeds Inc.

Certified Seed/Stock - Available from Cascade International Seeds and Jonathan Green & Sons. U.S. Plant Variety Protection No. 8900301.

Preparer/Additional Information - Stephen W. Johnson, International Seeds Inc., P.O. Box 168, 820 W. First St., Halsey, OR 97348, (503) 369-2251.

Safari

Source - Kansas, Georgia, New Jersey, Idaho, and North Carolina.

Method of Breeding - Advanced generation synthetic resulting from two cycles of phenotypic recurrent selection.

Intended Use - Cool season turf areas, low-maintenance turf, athletic fields, and golf courses.

Description - Improved brown patch resistance, good powdery mildew resistance, and good summer performance.

Adapted to - PHZ 5, 6, 7, 10, 11.

Released - Turf-Seed, Inc.

Breeder Seed/Stock - Pure Seed Testing, Inc.

Certified Seed/Stock - Available. PVP No. 9100080.

Preparer/Additional Information - Turf-Seed, Inc., P.O. Box 250, Hubbard, OR 97032, (503) 651-2130.

Sapphire

Source - U.S.

Method of Breeding - Plants selected from plant selection field based on similar morphology, with selected plants intermated and then evaluated in turf trials. Variety based on six clones.

Intended Use - Home lawns, parks, and playgrounds.

Description - Medium green cultivar, good tillering ability.

Adapted to - PHZ 5, 6, 7.

Released - D.J. Van der Have B.V., The Netherlands.

Breeder Seed/Stock - D.J. Van der Have B.V.

Certified Seed/Stock - Available from Advanta Seeds West, Inc.

Preparer/Additional Information - Kenneth Hignight, Advanta Seeds West, Inc., 33725 Columbus Street S.E., P.O. Box 1496, Albany, OR 97321-0452, (503) 967-8923.

Shenandoah

Developed by R. J. Peterson Enterprises Inc., Hillsboro, OR, using germplasm from the New Jersey AES, New Brunswick, NJ. Carried as experimental no. PE 7 E.

Source - Plant collection from old lawns, pastures, and similar turfs throughout Connecticut, Delaware, Maryland, New Jersey, New York, Pennsylvania, Virginia, West Virginia, and Washington, DC.

Method of Breeding - An advanced-generation synthetic cultivar selected from progenies of 55 clones. Selection was directed toward improved seed yield, a moderately dark green color, an upright growth profile, relative freedom from disease, medium height, and absence of leaf roll under heat and drought stress.

Intended Use - Turf use on lawns and sports fields in the transition zone.

Description - A stable and uniform variety. No off-types or variants have been observed in multiplication of Shenandoah. Turf trials and seed fields show acceptable uniformity.

Adapted to - Transition Zone.

Released - 1990 by R.J. Peterson Enterprises Inc., Forest Grove, OR.

Breeder Seed/Stock - R. J. Peterson Enterprises, Inc., with the assistance of the New Jersey AES.

Certified Seed/Stock - Available 1990.

Preparer/Additional Information - Robert J. Peterson, R.J. Peterson Enterprises, P.O. Box 312, Forest Grove, OR 97116, (503) 357-4336.

Shortstop

Source - Source material for the cultivar came from recurrent selection populations at Tangent, OR.

Method of Breeding - Synthetic cultivar development via polycross and subsequent progeny testing to attain uniformity.

Intended Use - Cool-season turf.

Description - It has relatively slower and lower growth habit than other turf-type tall fescue cultivars.

Adapted to - PHZ 6, 7, 8, and parts of 5.

Released - 1989 by Pickseed West Inc.

Breeder Seed/Stock - Pickseed West Inc.

Certified Seed/Stock - Available.

Preparer/Additional Information - Pickseed West Inc., P.O. Box 888, Tangent, OR 97389, (503) 926-8886.

Silverado

Source - Georgia, New Jersey, and selections from Olympic.

Method of Breeding - Advanced-generation synthetic from the progenies of 15 plants selected for a dwarf mature plant height.

Intended Use - Cool-season turf areas and low-maintenance turf, such as athletic fields and golf courses.

Description - Reduced clippings, denser turf quality, finer leaf texture, and a dark green color.

Adapted to - PHZ 5, 6, 7, 10, 11.

Released - Turf-Seed, Inc.

Breeder Seed/Stock - Pure Seed Testing, Inc.

Certified Seed/Stock - Available. PVP No. 8800130.

Preparer/Additional Information - Turf-Seed, Inc., P.O. Box 250, Hubbard, OR 97032, (503) 651-2130.

Southern Cross

Snow Brand Seed Company, Daini-Hokkai Bldg., 3-8 Higushi-Nihonbashi 3 Chome, Chuo-Ku, Tokyo 103, Japan.

Source - About 500 clones were collected in old grasslands in the southern part of Chiba prefecture, Japan, for good growth vigor and persistence between 1971 and 1975.

Method of Breeding - Source population was screened for good regrowth vigor, heat and disease resistance in 1976 and 1977. Nineteen clones were selected and put in a polycross block. These were progeny tested between 1977 and 1979 and seven clones were selected in 1980 to be used in producing Syn 1 seed by polycross. In 1982, Syn-2 seed production began.

Intended Use - Forage.

Description - Southern Cross is most similar to Fawn. It differs from Fawn in the following characteristic: heading date is three to seven days later than Fawn, leaf width is wider, spike length is longer, and Southern Cross is more resistant to crown rust and net blotch than Fawn.

Adapted to - LRR A, B, C, D, E, H, I, G, M, N, P, R, S; PHZ 4, 5, 6, 7, 8.

Released - 1984, Japan.

Breeder Seed/Stock - Produced and maintained by Snow Brand Seed Company.

Certified Seed/Stock - Available.

Preparer/Additional Information - Susan H. Samudio, Jacklin Seed Company, W. 5300 Riverbend Ave., Post Falls, ID 83854, (208) 773-7581.

SR 8200

Seed Research of Oregon and Crop Science Dept., Rutgers University - M.F. Robinson, Leah A. Brilman, and C.R. Funk.

Source - Germplasm selected from old turf throughout the U.S., maternal lineage of 75 clones traced to plants collected from New Jersey (30 plants), Georgia (27 plants), Maryland (seveb plants), Tennessee (four plants), North Carolina (two plants), one plant each from Alabama, Idaho, Kansas, Mississippi, and Ohio.

Method of Breeding - Population improvement using cycles of phenotypic and genotypic recurrent selection under stress from frequent close mowing and diseases. Three cycles of assortive mating toward selection of lower-growing plants. 75 plants from a 10,000- spaced plant nursery were selected for upright growth, high seed yield, dark green color, late maturity, and disease resistance produced Syn I seed. The resulting 4,622 progeny were evaluated for stem rust resistance, uniformity, and seed yield with 474 progeny from 32 clones used in the production of Syn I breeder seed.

Intended Use - Medium-maintenance turf on home lawns, parks, sports fields, roadsides, school play areas, and institutional grounds where tall fescue is adapted for turf. Since it has a reduced vertical growth rate, it can be used in areas that require less mowing. Mixtures with compatible Kentucky bluegrasses are also recommended.

Description - Lower-growing dwarf-type tall fescue with a rich dark green color. Upright growth habit. Late reproductive maturity. Improved resistance to current races of stem rust. Greater density and slower rate of vertical elongation than many turf-type tall fescues.

Adapted to - LRR A, B, C, D, E, F, G, H, K, L, M, N, R, S; PHZ 4, 5, 6, 7, 8, 10.

Released - 1990 by Seed Research of Oregon.

Breeder Seed/Stock - Seed Research of Oregon, Corvallis, OR.

Certified Seed/Stock - Available.

Preparer/Additional Information - Dr. Leah A. Brilman, Seed Research of Oregon, Inc., P.O. Box 1416, Corvallis, OR 97339, (800) 253-5766.

SR 8300

Source - Old turf sites in the mid-Atlantic region, with best performers in turf in Maryland used for additional cycles of selection. Five superior clones allowed to intercross and entered as Bel 86-2 in NTEP.

Method of Breeding - Progeny of the five clones planted in Corvallis, OR, and rogued for uniformity based on semi-dwarf growth habit, dark green color, medium to fine leaf texture, very strong tillering, seed yield potential, and relative freedom from spot and stem rust.

Intended Use - Cool-season turfgrass in transitional zone and north. May be seeded alone or with other improved tall fescues, bluegrass, or zoysiagrasses.

Description - Has demonstrated superior performance in Maryland, Virginia, and California. Semi-dwarf with rapid tillering, leaf spot resistance, tolerance to brown patch (little injury and rapid recovery), and drought and heat tolerant. Maintains tall fescue performance in transitional zone with dwarf growth habit.

Adapted to - PHZ 5, 6, 7, 8, 9, NE, CA, Midwest.

Released - 1992, Seed Research of Oregon, Inc.

Breeder Seed/Stock - Seed Research of Oregon, Inc.

Certified Seed/Stock - Available.

Preparer/Additional Information - Dr. Leah A. Brilman, Seed Research of Oregon, Inc., P.O. Box 1416, Corvallis, OR 97339, (503) 757-2663.

Stargrazer

Selected at FFR Cooperative, West Lafayette, IN - B.L. Winsett, S.D. Statton, and S.J. Baluch.

Source - Trace back to selections out of KY-31.

Method of Breeding - Eight-clone synthetic. Superior clones selected from spaced-plant nurseries. Eight parental clones were identified on basis of summer color, progeny forage tests in Indiana, and clonal seed yield tests in Oregon. Tested as FFR synthetic TF8501.

Intended Use - Low endophyte hay and pasture.

Description - Similar in maturity to Kenhy. Rust resistance better than Kenhy or KY-31. Stargrazer also demonstrates better summer regrowth potential than KY-31.

Adapted to - LRR A, B, C, F, G, H, J, K, L, M, N, O, P, R, S, T; PHZ 2, 3, 4, 5, 6, 7, 8.

Released - 1993 by FFR Cooperative.

Breeder Seed/Stock - FFR Cooperative.

Certified Seed/Stock - Available.

Preparer/Additional Information - Bret L. Winsett, FFR Cooperative, 4112 E. State Rd. 225, West Lafayette, IN 47906, (317) 567-2115.

Stef

Source - Developed by Poznan Plant Breeding, Grunwaldzka, Poland.

Method of Breeding - The breeding material was gathered from old fields in Vilnus region of Poland beginning in 1939 and subsequently screened over several generations of mass and recurrent selection for vigor and productivity. This was followed by individual plant selections for winter hardiness and drought tolerance. Final selections were made in 1955 based on progeny evaluation for seed and forage yield. Breeder seed was first bulked in 1955. This variety is sold in Poland as Brudzynska/Stef.

Intended Use - Hay and pasture production.

Description - Erect, medium dense growth habit; leaves are vivid green, 30-39 cm long, 1-2 cm wide, rough edges, light green sheath, light pubescence; medium to late flowering; stems are long, nodes have anthocyanin pigment; medium to late maturity. Limited U.S. data show good winter survival, but moderate susceptibility to stem rust.

Adapted to - Recommended in Canada for use in British Columbia and Ontario.

Released - Poznan Plant Breeding, Szelejewo, Poland. Registered in Canada in 1990 with Registration no. 3312.

Breeder Seed/Stock - Poznan Plant Breeding, Szelejewo, Poland.

Certified Seed/Stock - Available from Oseco Inc.

Preparer/Additional Information - Agriculture Canada, Food Production and Protection Branch, Plant Products Division, Ottawa, Ontario K1A 0C6. Seed distributed by Oseco Inc., P.O. Box 219, Brampton, Ontario L6V 2L2.

Taurus

KWS-AG, Einbeck, West Germany.

Source - Collections in old turf area in North Carolina, South Carolina, and Georgia plus selected clones of Mustang and Bonanza. Experimental number Poly V.

Method of Breeding - A 147-clone synthetic variety developed through restricted recurrent selection over three cycles. Plants were evaluated based on phenotypic characteristics such as growth habit, leaf texture, color, overall height, production of reproductive tillers, and disease resistance.

Intended Use - Ornamental, recreational, and athletic turf wherever turf-type tall fescues are adapted; lacks cold hardiness in northern areas without adequate snow cover.

Description - Medium dark-colored, strongly upright with intermediate plant height, forms a dense sward with upright foliage of medium texture.

Adapted to - The moderately temperate zone; PHZ 4, 5, 6, 7, 8 (some limitations in high humidity areas).

Released - 1989, Turf Merchants, Inc.

Breeder Seed/Stock - Turf Merchants.

Certified Seed/Stock - Available.

Preparer/Additional Information - Fred B. Ledeboer, Ph.D., TMI, 33390 Tangent Loop, Tangent, OR 97389, (800) 421-1735.

Tempo

Source - Clonal selection from old turfs in Maryland, Florida, and Georgia.

Method of Breeding - Advanced generation synthetic cultivar developed by John Rutkai, NPI Agriculture Service Corporation.

Intended Use - Turf, either alone or in blends.

Description - Vigorous, medium green, medium texture, excellent heat and drought tolerance, good disease resistance, and very persistent.

Adapted to - Southeast and southwest U.S., transition states.

Released - 1983, Barenbrug USA/Normarc.

Breeder Seed/Stock - Barenbrug USA.

Certified Seed/Stock - Available from Barenbrug USA.

Preparer/Additional Information - Barenbrug USA, P.O. Box 239, Tangent, OR 97389, (503) 926-5801.

Thunderbird

Pure Seed Testing.

Source - Collections originating from Spain, Turkey, Japan, UK, Tunisia, Israel, Australia, New Jersey, California, Georgia, and Oregon.

Method of Breeding - One hundred twenty-one parent plants used in a polycross.

Intended Use - Turf.

Description - Dark green, intermediate type, well adapted over a large area.

Adapted to - PHZ 6, 7, 8, 9, 10 - transition zone.

Released - 1989, Pure Seed Testing.

Breeder Seed/Stock - Pure Seed Testing.

Certified Seed/Stock - Available.

Preparer/Additional Information - Mike McCarthy, E.F. Burlingham & Sons, P.O. Box 217, Forest Grove, OR 97116, (503) 359-9368.

Titan

Seed Research of Oregon, NJ AES - M.F. Robinson, C.R. Funk, Bruce B. Clark and R.W. Duell.

Source - Tall fescue plants selected from mowed turfs and closely grazed pastures from an extensive plant exploration program during the period of 1970-76 over a wide area of the U.S. The 207 maternal plants of Titan are descended from selections made in 11 states as follows: GA-87, MS-40, MD-19, KS-18, OH-12, TX-10, VA-6, PA-6, NC-3, KY-3, NJ-3. Carried as experimental no. SR 8000.

Method of Breeding - Germplasm from the above program were subjected to various cycles of population improvement including phenotypic recurrent selection and progeny evaluation in turf trials. After a period of summer stress, 3,696 tillers were selected from turf plots and transferred to a spaced-plant nursery. From these, 233 clones were selected and placed in an isolated crossing block. Seed was harvested from the 207 plants showing the highest floret fertility. This seed was used to establish an isolated spaced-plant nursery with selection to improve uniformity, attractiveness, disease resistance, and seed yield potential.

Intended Use - Medium- to low-maintenance turf on home lawns, parks, sports fields, roadsides, school play areas, and institutional grounds in adapted areas.

Description - Leafy, turf-type tall fescue that produces a high quality, persistent turf with medium density and leaf texture and medium dark green color. Very good summer stress tolerance, good winter-hardiness and moderate tolerance of close mowing. Good wear tolerance and performance under varying light intensities. Moderate resistance and good recovery from large brown patch disease. Little or no thatch and good performance at low fertility levels. Seed lots with high levels of endophyte show enhanced resistance to insects and increased stress tolerance.

Adapted to - LRR A, B, C, D, E, F, G, H, J, K, L, M, N, O, P, R, S, T; PHZ 4, 5, 6, 7, 8, 10.

Released - 1987 by Seed Research of Oregon.

Breeder Seed/Stock - Seed Research of Oregon, Corvallis, OR.

Certified Seed/Stock - Available.

Preparer/Additional Information - Dr. Leah A. Brilman, Seed Research of Oregon, Inc., P.O. Box 1416, Corvallis, OR 97339, (800) 253-5766.

Tomahawk

Source - Alabama, New Jersey, Georgia, North Carolina, Pennsylvania, Ohio, Idaho, Kansas, Tennessee, Texas, Virginia, and Mississippi.

Method of Breeding - Advanced generation synthetic resulting from 11 cycles of recurrent selection for dwarf stemmy upright plants with fine leaves.

Intended Use - Cool-season turf areas, low-maintenance turf, athletic fields, and golf courses.

Description - Excellent turf quality, very dark green color, improved resistance to crown rust and net blotch, and reduced vertical growth.

Adapted to - PHZ 5, 6, 7, 10, 11.

Released - Turf-Seed, Inc.

Breeder Seed/Stock - Pure Seed Testing, Inc.

Certified Seed/Stock - Available. PVP No. 9100179.

Preparer/Additional Information - Turf-Seed, Inc., P.O. Box 250, Hubbard, OR 97032, (503) 651-2130.

Tradition

Source - Eight robust, persistent clones from old turf areas in Georgia.

Method of Breeding - Recurrent phenotypic selection for disease resistance and endophyte.

Intended Use - Turf.

Description - Very good performance in southeastern U.S.

Adapted to - PHZ 5, 6, 7, 8.

Released - 1987, Pure Seed Testing, Inc.

Breeder Seed/Stock - Pennington Seed Inc.

Certified Seed/Stock - Available.

Preparer/Additional Information - Pennington Seed Inc. of Oregon, P.O. Box 386, Lebanon, OR 97355, (503) 451-5261.

Trailblazer

Source - Plant collected from old turf stands in Alabama, North Carolina, and New Jersey contributed to the parental germplasm of Trailblazer. Selection was made in each cycle for a dark green color, dwarf growth habit, high reproductive tiller number, resistance to leaf spot and crown rust, and drought avoidance. After the fourth cycle of selection, 225 individual clones were chosen as the parents of Trailblazer.

Method of Breeding - It is an advanced-generation synthetic cultivar from the progenies of five clones.

Intended Use - Athletic fields, golf course roughs, commercial and public turf areas, home lawns, parks, and playgrounds. Blends well with other quality turf-type tall fescues such as Wrangler and Cimarron.

Description - A new, lower growing, darker green tall fescue with a slower leaf extension rate and a shorter mature plant height. The result is a very fine textured, dense, premium quality turf. Blends well with other turf-type tall fescues and is well-suited to use in tall fescue-bluegrass blends. Developed through years of plant breeding, the dwarf turf-type tall fescues are a real breakthrough in this widely adapted species. This variety has shown good resistance to leaf spot and crown rust, and moderately good resistance to brown patch disease. It has shown good tolerance to cold, wear and shade, and moderately good heat tolerance. Trailblazer's drought tolerance ratings in the 1988 USDA National Turfgrass Evaluation Report ranked equal to such varieties as Olympic, Wrangler, Bonanza, Cimarron, Apache, Tribute, and Rebel II.

Adapted to - PHZ 3, 4, 5, 6, 7.

Released - 1986, Pure Seed Testing, Inc.

Breeder Seed/Stock - Pure Seed Testing, Inc.

Certified Seed/Stock - Available.

Preparer/Additional Information - Art Wick, LESCO, Inc., 20005 Lake Road, Rocky River, OH 44116, (216) 333-9250.

Trailblazer II

Source - Plants collected from old turf stands in Alabama, North Carolina, and New Jersey contributed to the parental germplasm of Trailblazer II. Selection was made in each cycle for an improved dark green color, dwarf growth habit, high reproductive tiller number, leaf spot resistance, stem rust resistance, brown patch resistance, and drought avoidance.

Method of Breeding - It is an advanced generation cultivar.

Intended Use - Athletic fields, golf course roughs, commercial and public turf areas, home lawns, parks, and playgrounds. Blends well with other quality turf-type tall fescues such as Wrangler, Cimarron, and Trailblazer.

Description - A new, lower growing, darker green tall fescue with a slower leaf extension rate and a shorter mature plant height. The result is a very fine textured, dense, premium quality turf. Blends well with other turf-type tall fescues and is well-suited for use in tall fescue-bluegrass blends. Developed through years of plant breeding, the dwarf turf-type tall fescues are a real breakthrough in this widely adapted species. This variety has shown good resistance to brown patch leaf spot, crown crust, stem rust, and fusarium patch. It has shown good cold, wear, and shade tolerance and good heat tolerance. Trailblazer II's drought tolerance ranked equal to most standard turf-type tall fescue varieties.

Adapted to - PHZ 3, 4, 5, 6, 7.

Released - Pure Seed Testing, Inc., 1989.

Breeder Seed/Stock - Pure Seed Testing, Inc.

Certified Seed/Stock - Available.

Preparer/Additional Information - Art Wick, LESCO, Inc., 20005 Lake Road, Rocky River, OH 44116, (216) 333-2950.

Tribute

Named in honor of Peter S. Loft, former President of Lofts Seed, Inc. Developed by Lofts Seed, Inc., Bound Brook, NJ, using germplasm obtained from the New Jersey Agricultural Experiment Station. Carried as accession no. SB, PL-1.

Source - Old turf plots in Georgia, Kansas, Maryland, Michigan, New Jersey, North Carolina, Ohio, Pennsylvania, Tennessee, Texas, and Virginia.

Method of Breeding - An advanced generation synthetic cultivar selected from the progenies of 37 clones.

Intended Use - Medium- to low-maintenance turfs in full sun to moderate shade for home lawns, parks, athletic fields, sod farms, and roadsides.

Description - A moderately low-growing, turf-type cultivar that produces moderately dark green turf of medium leaf texture and medium plant density.

Adapted to - LRR all except I, U, T; PHZ 4, 5, 6.

Released - 1988, Lofts Seed, Inc.

Breeder Seed/Stock - Lofts Seed, Inc., with cooperation of the New Jersey AES.

Certified Seed/Stock - Available.

Preparer/Additional Information - Lofts Seed Inc., Chimney Rock Road, P.O. Box 146, Bound Brook, NJ 08805, (908) 560-1590.

Twilight

KWS-AG, Einbeck, West Germany.

Source - Collections in old turf areas in South Carolina, Georgia, Alabama, and Arkansas in 1985 and 1986, plus progeny from selected clones of Mustang.

Method of Breeding - Extensive testing of clonal progeny for color and disease resistance; evaluation of progeny from phenotypically matched mating in isolation, back-crossing to the darkest clones and polycrossing of most desirable progeny after three cycles of recurrent mass selection.

Intended Use - Ornamental, recreational, and athletic turf wherever turf-type tall fescues are adapted; lacks cold-hardiness in northern areas without adequate snow cover.

Description - The blue-green color hue of Twilight is quite distinct and to date not known to exist in any other tall fescue variety. Semi-upright of intermediate height and medium late maturity; texture is medium coarse as adult plant; turf texture is medium fine because of excellent tiller production.

Adapted to - The moderately temperate zone; PHZ 4, 5, 6, 7, 8 (some limitations in high humidity areas).

Released - 1991 by Turf Merchants.

Breeder Seed/Stock - Turf Merchants.

Certified Seed/Stock - Available.

Preparer/Additional Information - Fred B. Ledeboer, Ph.D., TMI, 33390 Tangent Loop, Tangent, OR 97389, (800) 421-1735.

Vegas

Source - Nine clones selected from progeny from turf plots at Adelphia, NJ.

Method of Breeding - Synthetic cultivar developed at Pickseed West with Dr. G.W. Pepin.

Intended Use - Turf, either alone or in blends.

Description - Dark green, medium fine texture, low upright growth, improved disease resistance, excellent heat and drought tolerance, and high endophyte for insect resistance.

Adapted to - Southeast and southwest U.S., lower transition states.

Released - 1990, Barenbrug/Normarc.

Breeder Seed/Stock - Pickseed West and Barenbrug USA.

Certified Seed/Stock - Available from Barenbrug USA.

Preparer/Additional Information - Barenbrug USA, P.O. 239, Tangent, OR 97389, (503) 926-5801.

Virtue

Source - Rutgers University breeding program. Connecticut, Delaware, Maryland, New Jersey, New York, Pennsylvania, Virginia, West Virginia, Washington, DC, and Idaho. Three selected populations.

Method of Breeding - Seven to 13 cycles of phenotypic recurrent selection.

Intended Use - High-performance turf, cool-season turf areas, low-maintenance turf, athletic fields, and golf courses.

Description - Very dark green, semi-dwarf growth habit. Improved summer performance, brown patch, leaf spot and crown rust resistance.

Adapted to - PHZ 5, 6, 7, 8 - transition zone to west coast.

Released - 1990, Pure Seed Testing, Inc.

Breeder Seed/Stock - Pennington Seed Inc. PVP No. 9200133.

Certified Seed/Stock - Available.

Preparer/Additional Information - Pennington Seed Inc. of Oregon, P.O. Box 386, Lebanon, OR 97355, (503) 451-5261.

Willamette

Developed by International Seeds Inc., Halsey, OR.

Source - Selections made from old parks, pastures, and other turf in the central and eastern U.S. Carried as experimental no. TF-805.

Method of Breeding - Individual plants screened for color, leaf blade width, and disease resistance; polycross progeny evaluated under turf management. Five elite clones were selected to produce synthetic cultivar.

Intended Use - Medium- to low-maintenance turf for homes, commercial areas, athletic fields, and roadsides.

Description - Medium-late maturing, moderately fine bladed variety. Produces an attractive, durable turf sward in areas where tall fescue is adapted.

Adapted to - LRR A, B, C, D, E, G, H, J, K, L, M, N, O, P, R, S, W; PHZ 4.

Released - 1985, by Willamette Seed and Grain Co., Albany, OR.

Breeder Seed/Stock - Maintained by International Seeds Inc.

Certified Seed/Stock - Available in quantity from Willamette Seed Co. U.S. PVP No. 8500150

Preparer/Additional Information - Stephen W. Johnson, International Seeds Inc., P.O. Box 168, 820 W. First St., Halsey, OR 97348, (503) 369-2251.

Winchester

Rutgers University and Pure Seed Testing.

Source - Breeding program at Rutgers and Pure Seed Testing.

Method of Breeding - Cycles of recurrent selection resulting in a synthetic polycross of parent lines.

Intended Use - Turf.

Description - Darker, shorter, and more disease resistant than Falcon.

Adapted to - Transition zone.

Released - 1986, Rutgers and Pure Seed Testing.

Breeder Seed/Stock - Rutgers and Pure Seed Testing.

Certified Seed/Stock - Available.

Preparer/Additional Information - Mike McCarthy, E.F. Burlingham & Sons, P.O. Box 217, Forest Grove, OR 97116, (503) 359-9368.

Winchester II

McCarthy-Burlingham Research Farm.

Source - Material was collected from old turf plots in the transition zone and used in a breeding program with selections of Winchester.

Method of Breeding - Superior performing clones were polycrossed and cycled with selections made for outstanding turf performance in New Jersey, Virginia, and Oregon.

Intended Use - Turf, roadsides, and low-maintenance turf requiring good wear tolerance.

Description - Attractive dark green color and has a short growth habit. It has very good wear tolerance, disease resistance, and drought tolerance.

Adapted to - PHZ 6, 7, 8, 9, 10 - transition zone.

Released - 1992, McCarthy-Burlingham Research.

Breeder Seed/Stock - McCarthy-Burlingham Research.

Certified Seed/Stock - Available in 1994.

Preparer/Additional Information - Mike McCarthy, E.F. Burlingham & Sons, P.O. Box 217, Forest Grove, OR 97116, (503) 359-9368.

Wrangler

Developed at the New Jersey AES - Cyril R. Funk Jr.

Source - Plant collections were made from old lawns, pastures, and similar turfs throughout Connecticut, Delaware, Maryland, New Jersey, New York, Pennsylvania, Virginia, West Virginia, and Washington, D.C., beginning in 1961. Accessions were also obtained from the U.S. Plant Introduction Program. Tri-species hybrids of tall fescue, meadow fescue, and perennial ryegrass were obtained from the U.S. Regional Pasture Research Laboratory, University Park, PA. Carried as experimental no. LB-1.

Method of Breeding - Wrangler is a 14-clone advanced-generation synthetic variety with 10 of the parental clones tracing maternally to plants selected from T-6 tall fescue. The 14 parental clones were selected because of their attractive dark green color, freedom from disease, fine leaves, high shoot density, and reduced vertical growth rate.

Intended Use - A turf-type tall fescue intended for use on home lawns, parks, athletic fields, roads, commercial and public turf areas, golf course roughs, and school play areas. Blends well with other quality turf-type tall fescues such as Trailblazer and Cimarron.

Description - Moderately low growing, semi-dwarf turf-type tall fescue with a medium dark green color, a medium high density, and medium fine leaf blade. Wrangler closely resembles Arid tall fescue but the culm length, panicle length, and flag leaf are shorter than Arid at maturity. Wrangler has more branches at the second lowest node in the panicle and longer internode length directly beneath the panicle. Wrangler also has significantly longer and wider seeds than Arid. It also has a better turf quality than Arid in trials in New Jersey and Idaho. It has good cool-temperature color retention and good early spring color; moderately tolerant of close mowing and produces little thatch. Performs well in full sun or moderate shade. Wrangler has shown improved resistance to leaf spot, net blotch, and brown patch disease, as well as good heat, cold, wear, and shade tolerance.

Adapted to - LRR A, B, D, E, G, N, P, R, S; PHZ 6.

Released - August 1987, Jacklin Seed Company and Rutgers University.

Breeder Seed/Stock - Jacklin Seed Company, ID, and Rutgers University.

Certified Seed/Stock - Available.

Preparer/Additional Information - Mark Sellmann, Jacklin Seed Co., W. 5300 Riverbend Ave., Post Falls, ID 83854, (208) 773-7581; or Art Wick, LESCO, Inc. 20005 Lake Road, Rocky River, OH 44116, (216) 333-9250.

Festuca filiformis **Pourret - slender fescue**

F. ovina **L. var.** ***tenuifolia*** **(Sibthorp) Sm.**
F. tenuifolia **Sibthorp**

Cool-season bunchgrass from Europe. Fine, hairlike leaves, dense growth, adapted to dry, poor soils. Adapted for use in shady lawns.

Barok

Developed by Barenbrug Holland N.V., Arnhem, The Netherlands.

Source - Collections from lawns, pastures, and other uncultivated areas. Dutch ecotypes.

Method of Breeding - Selection and evaluation of individual plants, clones, and families under turf conditions.

Intended Use - Cool-season turfgrass used in mixtures. High traffic areas, low-maintenance, heat and drought tolerant. Used primarily in cemeteries and the steep slopes in southern California.

Description - Very narrow leafed and drought resistant, requiring minimum mowing. Suitable for roadsides on sandy soils. It has a prostrate growth habit and a light green leaf color. It produces an attractive turf under bad growing conditions.

Adapted to - PHZ 1, 2, 3, 4, 5, 6, 7, 8 - Throughout U.S.

Released - 1964, Barenbrug.

Breeder Seed/Stock - Barenbrug Holland N.V.

Certified Seed/Stock - Available from Barenbrug USA.

Preparer/Additional Information - Barenbrug USA, P.O. Box 239, Tangent, OR 97389, (503) 926-5801.

Festuca idahoensis **Elmer - Idaho fescue**

Cool-season, native bunchgrass found from Washington and Montana south to central California and Colorado. Prevalent at higher elevations in Montana, Idaho, and Utah. Valuable rangegrass, palatable in spring, cures well on stem, and makes good fall forage.

P-6435

Selected at Plant Materials Center, SCS, Pullman, WA - R.J. Olson and J.L. Schwendiman.

Source - Collected from native Ponderosa pine grassland association near Winchester, ID, by D. Hendrick in 1938.

Method of Breeding - Accession outstanding among 61 in collection. Improved by mass selection during several generations.

Intended Use - Rangeland revegetation.

Description - Vigorous, long-lived perennial; bunch-type fescue. Excellent seedling vigor, strong root system. Dark-green, basal, abundant leaves. Seed culms spreading, abundant, up to 91 cm in height. Large, awned seeds. Seed production much better than that of any strain previously found.

Adapted to - B.

Released - Maintained for testing.

Breeder Seed/Stock - Plant Materials Center, SCS, Pullman, WA.

Certified Seed/Stock - Not available.

Preparer/Additional Information - Plant Materials Center, SCS, Rm. 104, Hulbert Agricultural Sciences Bldg., WSU, Pullman, WA 99164-6211, (509) 335-7376.

Trident

Developed by International Seeds Inc., Halsey, OR.

Source - Collections made from old turfs in the southern and eastern U.S., along with derivatives of commercial cultivars. Carried as experimental ISI-813.

Method of Breeding - Six parental clones selected to produce synthetic variety following screening in spaced plant nursery and evaluation of polycross progeny in turf plots.

Intended Use - Turfgrass for lawns, athletic fields, and golf course roughs.

Description - Dark green, upright growing, late heading turf-type. Winter color retention in turf is frequently superior to that of other tall fescue varieties.

Adapted to - LRR A, B, C, D, E, G, H, J, K, L, M, N, O, P, R, S, W; PHZ 4.

Released - 1988 by Seed Research of Oregon, Corvallis, OR.

Breeder Seed/Stock - Maintained by International Seeds Inc.

Certified Seed/Stock - Available from Seed Research of Oregon. U.S. PVP No. 8700177

Preparer/Additional Information - Stephen W. Johnson, International Seeds Inc., P.O. Box 168, 820 W. First St., Halsey, OR 97348, (503) 369-2251.

Festuca ovina **L. - sheep fescue**

Cool-season bunchgrass indigenous in Northern Hemisphere. Used as durable turfgrass on sandy soils and for erosion control in northern states. Cold and drought tolerant; succeeds better than most grasses on sandy, gravelly soils. Grazed well in early spring, but not widely used for pasture.

Bighorn

Source - Old turf areas in New Jersey.

Method of Breeding - Advanced-generation synthetic resulting from three cycles of phenotypic recurrent selection for improved turf performance, a powder blue color and less wiry types.

Intended Use - For cool-season turf and low-maintenance areas.

Description - Powder blue color, with improved uniformity and turf quality.

Adapted to - PHZ 2, 3, 4, 5, 6, 7, 8.

Released - Turf-Seed, Inc.

Breeder Seed/Stock - Pure Seed Testing, Inc.

Certified Seed/Stock - Available. PVP No. 8800064.

Preparer/Additional Information - Turf-Seed, Inc., P.O. Box 250, Hubbard, OR 97032, (503) 651-2130.

Covar (Reg. No. 16)

Selected at Plant Materials Center, SCS, Pullman, WA - J.L. Schwendiman.

Source - PI 109497; south of Konya, Turkey; introduced by Westover-Enlow expedition in 1934.

Method of Breeding - Selections from spaced plantings in which aberrant types eliminated.

Intended Use - Soil erosion control and revegetation of disturbed land.

Description - A dwarf, blue-green, densely tufted, erect growing perennial with abundant fine stems. Leaves are narrow, dense, short, stiff, basal, and abundant. Covar is shorter, more uniform, and has a deeper blue color than other sheep fescue selections. As a fine-leafed fescue, Covar is somewhat slow to establish. Once established, it is very persistent, competitive, winter-hardy and drought tolerant. Adapted to well-drained soils with a mean annual precipitation of 25 cm.

Adapted to - LRR B; PHZ 5.

Released - 1977, cooperatively with the Washington Agricultural Research Center, Idaho and Oregon Experiment Stations, and the Plant Materials Center, SCS, Pullman, WA.

Breeder Seed/Stock - Plant Materials Center, SCS, Pullman, WA.

Certified Seed/Stock - Available.

Preparer/Additional Information - Plant Materials Center, SCS, Rm 104, Hulbert Agricultural Sciences Bldg, WSU, Pullman, WA 99164-6211, (509) 335-7376.

MX-86

Source - MX-86 originated from selections made from a single lot of common sheep fescue from Germany.

Method of Breeding - Prospective seedlots were screened for rapidity of germination. Seed was planted in a field near Spokane, WA. A single uniform plant morphology was selected to produce parental materials.

Intended Use - Turf.

Description - MX-86 is a medium blue-green variety, selected for enhanced seedling vigor and seed yield.

Adapted to - Northeast, north-central, and Pacific northwest, and central U.S.

Released - 1989, Jacklin Seed Company.

Breeder Seed/Stock - Jacklin Seed Company.

Certified Seed/Stock - Available.

Preparer/Additional Information - Kim Peterson, Jacklin Seed Co., W. 5300 Riverbend Ave., Post Falls, ID 83854, (208) 773-7581.

SR 3000

Seed Research of Oregon and New Jersey AES - M.F. Robinson, B.B. Clarke, D.C. Saha, R.W. Duell, and C.R. Funk. Carried as accession no. PI 525460.

Source - Parental germplasm from collections and varieties were entered into a population improvement program initiated in 1968.

Method of Breeding - Advanced-generation synthetic selected from the polycross progeny of five clones derived from above program. The population improvement program included varying cycles of screening for attractive appearance and disease resistance in greenhouse trials and spaced-plant nurseries in NJ and OR. Polycross progeny trials under turf management and programs to maintain a high level of the endophyte which enhances resistance to many turfgrass insects were also performed.

Intended Use - Turf use in regions for which fine fescues are adapted. It is especially recommended for sites that are infertile, acid, or droughty or areas where supplemental irrigation and fertilization are not appropriate. It performs well in full sun and light or moderate shade.

Description - Moderately aggressive, persistent, turf-type hard fescue with fine leaves and a medium bright green color. It has a reduced rate of leaf elongation and a low growth profile which helps it to produce an attractive, dense low-growing turf with little or no supplemental irrigation or fertilization and reduced mowing requirements. It has good shade tolerance and can persist with tree root competition. It has adapted to a wide range of well-drained soils, including acid soils of low fertility. It has shown good winter-hardiness and improved summer performance in NJ. It has good resistance to powdery mildew, anthracnose, netblotch, and red thread. With high levels of endophyte, it has enhanced resistance to many turf insects.

Adapted to - LRR A, B, C, D, E, F, G, H, K, L, M, N, P, R, S, W; PHZ 2, 3, 4, 5, 6, 7, 8.

Released - 1987 by Seed Research of Oregon.

Breeder Seed/Stock - Seed Research of Oregon under conditions to maintain high levels of endophyte.

Certified Seed/Stock - Available.

Preparer/Additional Information - Dr. Leah A. Brilman, Seed Research of Oregon, Inc., P.O. Box 1416, Corvallis, OR 97339, (800) 253-5766.

Festuca pratensis Huds. - meadow fescue

Festuca elatior L. pro parte

Cool-season bunchgrass from Europe. Used for pasture and erosion control in humid parts of the northern U.S. Grows well on moist, fertile soils, but subject to rust damage. Neither so high-yielding nor so persistent as tall fescue. Varieties developed in Europe and elsewhere not generally adapted in U.S.

Bartura

Source - Ecotypes from Hungary.

Method of Breeding - Mass selection in Hungary.

Intended Use - In mixtures with Lolium perenne or as a pure variety for pastures and meadows.

Description - Diploid. Semi-erect growth. Good resistance to crown rust. Winder-hardy with good spring growth.

Adapted to - Central and northeastern U.S. PHZ 1, 2, 3, 4, 5.

Released - Research Institute for Irrigation, Szarvas, Hungary.

Breeder Seed/Stock - Research Institute for Irrigation, Szarvas, Hungary.

Certified Seed/Stock - Barenbrug USA.

Preparer/Additional Information - Barenbrug USA, P.O. Box 239, Tangent, OR 97389, (800) 547-4101.

Ensign

Selected at Canada Department of Agriculture Research Station, Ottawa, Ontario - R.M. MacVicar.

Source - Basic nursery established from seed lots obtained from various European sources.

Method of Breeding - Synthetic variety built up by combining several desirable clones selected in selfed-line breeding program.

Intended Use - Pasture and erosion control plantings.

Description - Tall, upright, uniform growth, leafy basal growth. Considered equal to most other strains in forage production; outstanding in seed production.

Released - 1944, by Canada Department of Agriculture.

Breeder Seed/Stock - Canada Department of Agriculture Research Station, Ottawa, Ontario.

Certified Seed/Stock - Available in quantity.

Preparer/Additional Information - Canada Seed Grower's Association, P.O. Box 8445, Ottawa, Ontario K1G 3T1.

Mimer

Developed by Plant Breeding Institute, Weibullsholm, Landskrona, Sweden.

Source - Indigenous plants.

Method of Breeding - Mass selection. Breeder seed produced generation after generation in isolation under natural conditions. Considered to be in genetic equilibrium under these conditions. Undesirable plants rogued prior to anthesis. Seed of remaining plant bulked to form breeder seed.

Intended Use - Forage and pasture.

Description - Similar to Ensign and common in growth pattern; similar to Ensign in maturity, but more resistant to leaf rust; higher yielding than Ensign or common; although not as leafy as common, almost as high in percentage of protein.

Released - Plant Breeding Institute, Weibullsholm. Introduced into Canada by Ontario Seed Cleaners and Dealers, Ltd., Toronto.

Breeder Seed/Stock - Plant Breeding Institute, Weibullsholm, Sweden. *Certified Seed/Stock* - Available.

Preparer/Additional Information - Canada Seed Growers Association, P.O. Box 8445, Ottawa, Ontario K1G 3T1.

Trader

Selected at Canada Department of Agriculture Research Station, Ottawa, Ontario - R.M. MacVicar and D.R. Gibson.

Source - Basic nursery established from European varieties and strains.

Method of Breeding - Synthetic variety developed from 15 progeny-tested clones.

Intended Use - Forage and pasture.

Description - Leafy, somewhat later in maturity than common types. Good basal growth and recovery characteristics. Considerable resistance to leaf rust in comparison with common.

Released - 1963, by Canada Department of Agriculture.

Breeder Seed/Stock - Canada Department of Agriculture Research Station, Ottawa, Ontario.

Certified Seed/Stock - Available in limited quantity.

Preparer/Additional Information - Canada Seed Trade Association, 2948 Baseline Rd., Ottawa, Ontario K2H 8T5.

Festuca rubra L. - red fescue

Important cool-season, sod-forming grass introduced from Europe. Used for lawns, general purpose turf, and erosion control. Occurs in pastures in northern states and Pacific northwest in relatively moist, cool areas. Grows over wide range of soil types; valued for its shade tolerance. Not highly palatable.

Arctared

Selected at Alaska AES, ARS cooperating.

Source - Single plant collected in 1957 from Matanuska Valley near Palmer.

Method of Breeding - Open-pollinated seed harvested from superior plants in space-plant progeny test of collections; used to establish turf evaluation test. Variety traces to open-pollinated seed from a number of plants within single progeny.

Intended Use - Lawns, disturbed site revegetation.

Description - Outstanding winter-hardiness surviving without injury when all introduced varieties were damaged very seriously or completely eliminated. Produces dense, medium-textured turf; medium green. Good establishment, excellent seedling vigor, and early spring growth; no disease problems observed at Palmer.

Adapted to - LRR W, X, Y; PHZ 2, 3, 4.

Released - 1965, cooperatively by University of Alaska and Plant Science Research Division, ARS.

Breeder Seed/Stock - Alaska AES.

Certified Seed/Stock - Available.

Preparer/Additional Information - Alaska Plant Materials Center, HC 02 Box 7440, Palmer, AK 99645, (907) 745-4469.

Bargena

Developed by Barenbrug Holland N.V., Arnhem, The Netherlands.

Source - Collections from lawns, pastures, and other cultivated areas. Ecotypes and commercial seed.

Method of Breeding - Selection and evaluation of individual plants, clones, and families.

Intended Use - Cool-season turfgrass, home lawns, cemeteries, and ballparks; any place where a low-maintenance and drought-tolerant turf is needed. May be used as a component in mixtures and blends.

Description - Rather late maturing; soft, green-colored leaves; suitable for lawns. It has a long flag leaf and produces a good turf with good mowing qualities.

Adapted to - PHZ 1, 2, 3, 4, 5, 6, 7.

Released - 1965, Barenbrug.

Breeder Seed/Stock - Barenbrug Holland N.V.

Certified Seed/Stock - Available from Barenbrug USA.

Preparer/Additional Information - Barenbrug USA, P.O. Box 239, Tangent, OR 97389, (503) 926-5801.

Bargena II

Source - Ecotypes.

Method of Breeding - Mass selection of ecotypes.

Intended Use - Cool-season turfgrass. Suitable for lawns and parks. May be seeded in mixtures.

Description - Performs very well in low-maintenance mixtures.

Adapted to - Southwest, northeast, northwest, central, and mountain states; PHZ 1,2, 3, 4, 5, 6, 7.

Released - 1992, Barenbrug.

Breeder Seed/Stock - Barenbrug Holland, Oosterhout, The Netherlands.

Certified Seed/Stock - Available from Barenbrug USA.

Preparer/Additional Information - Barenbrug USA, P.O. Box 239, Tangent, OR 97389, (503) 926-5801.

Boreal (Reg. No. 6)

Selected at Canada Department of Agriculture Research Station, Beaverlodge, Alberta - C.R. Elliott.

Source - Commercial seed fields in Peace River region of Alberta, that were seeded 10-15 years earlier to Olds variety.

Method of Breeding - Open-pollinated seed of 300 lines used to establish progeny test. Ratings taken on seedling vigor, winter injury, uniformity, disease resistance (primarily leaf rusts), and seed and herbage yields. The 36 top-yielding clones with similar maturity included in polycross nursery. Equal quantities of seed from each clone used to establish breeder seed plot.

Intended Use - Dual purpose lawn and grazing forage.

Description - More uniform and stronger creeper than Olds and Duraturf varieties. In western Canada produces 12% more seed and 11% more herbage than Olds. Good seedling vigor and early spring growth; recommended for turf and pasture.

Adapted to - Southern Alberta, Saskatchewan, Manitoba; PHZ 2, 3.

Released - 1966, by Canada Department of Agriculture.

Breeder Seed/Stock - Canada Department of Agriculture Research Station, Beaverlodge, Alberta.

Certified Seed/Stock - Available. United Grain Growers Ltd., Seed Division, Box 6030, Station C, Edmonton, Alberta T5B-4K5.

Preparer/Additional Information - Dr. Nigel Fairey, Agriculture Canada Research Station, P.O. Box 29, Beaverlodge, Alberta T0H-0C0, (403) 354-2212.

Cindy

Developed by Cebeco Zaden B.V., Vlijmen, The Netherlands.

Source - Selections made from turfs in northern Europe.

Method of Breeding - Plants with good turf quality components crossed with those showing good seed yield potential; families were evaluated and four lines were selected to make up the variety Cindy.

Intended Use - Seeded primarily as a component in turfgrass mixtures for lawns, golf courses, and industrial/commercial sites. Also used in combination with a low-growing perennial ryegrass as a low-maintenance ground cover for orchards, vineyards, and roadsides.

Description - Medium maturing, moderately dark green variety. Produces turf superior in quality to that of older creeping red fescue varieties, especially at lower mowing heights. Resistant to powdery mildew, pythium blight, fusarium blight, and crown rust. Combines well with other fine fescue species, Kentucky bluegrass, and perennial ryegrass.

Adapted to - LRR A, B, C, E, F, G, L, M, N, R, S, W; PHZ 3.

Released - European release by Cebeco Zaden B.V., Vlijmen, The Netherlands. In the U.S., 1990, by International Seeds, Inc., Halsey, OR.

Breeder Seed/Stock - Maintained by Cebeco Zaden B. V.

Certified Seed/Stock - International Seeds, Inc. U.S. PVP #8900061.

Preparer/Additional Information - Stephen W. Johnson, International Seeds, Inc., P.O. Box 168, 820 W. First St., Halsey, OR 97348, (503) 369-2251.

Dawson

Developed by the Sports Turf Research Institute, Bingly, England.

Source - Old English lawn.

Method of Breeding - Selection of plants from old turf. Progeny from selected plants tested for turf quality.

Intended Use - For mixes with other fescues in high salinity soils and lower maintenance areas.

Description - Short-rhizome type, with open-bladed, dark-green leaves. In lawns, forms fine-leafed, dense turf.

Adapted to - Most of the U.S.

Released - 1968, by D. J. van der Have.

Breeder Seed/Stock - D. J. van der Have, The Netherlands.

Certified Seed/Stock - OECD certified (Northrup, King & Co.)

Preparer/Additional Information - Rick Meirers, NK Lawn and Garden, 1490 Industrial Way SW, Albany, OR 97321, (800) 568-8873.

Ensylva

Developed by Cebeco Zaden B.V., Vlijmen, The Netherlands.

Source - Selections out of old turfs in The Netherlands.

Method of Breeding - Selections interpollinated and progenies tested for turf quality, disease resistance and seed yield. High performing plants combined to produce synthetic cultivar.

Intended Use - Primarily used as a component in turfgrass mixtures for home lawns and commercial and industrial sites. Also used alone or in combination with a low-growing perennial ryegrass as a ground cover for orchards, roadsides and other low-maintenance sites.

Description - Extremely fine leafed, strong creeping red fescue, capable of producing an attractive, dense and fine textured turf under a wide range of conditions. Exhibits greater density than common strong creeping red fescues. Exhibits excellent winter appearance and hardiness along with improved persistence.

Adapted to - LRR A, B, C, E, F, G, L, M, N, R, S, W; PHZ 3.

Released - 1978 by International Seeds Inc., Halsey, OR.

Breeder Seed/Stock - Maintained by Cebeco Zaden B.V., Vligmen, The Netherlands.

Certified Seed/Stock - Certified and uncertified seed available from International Seeds Inc.

Preparer/Additional Information - Stephen W. Johnson, International Seeds Inc., P.O. Box 168, 820 W. First St., Halsey, OR 97348, (503) 369-2251.

Flyer

Source - Ten clones from old turf areas in northwest U.S. and seven clones from old turf areas in southern U.S.

Method of Breeding - Selections from space planted nurseries based on attractive appearance, resistance to leaf spot and rust, seed head production, and floret fertility to create a synthetic cultivar.

Intended Use - Cool-season turfgrass to provide shade and drought tolerance.

Description - Consistently one of the highest ranked creeping red fescues in national turf trials. Improved dark green color, leaf spot and red thread resistance.

Adapted to - PHZ 2, 3, 4, 5, 6, 7 - east to west coast.

Released - 1986, Pure Seed Testing, Inc.

Breeder Seed/Stock - Pennington Seed Inc.

Certified Seed/Stock - Available.

Preparer/Additional Information - Pennington Seed Inc. of Oregon, P.O. Box 386, Lebanon, OR 97355, (503) 451-5261 and David Lundell, Fine Lawn Research, Inc., P.O. Box 1051, Lake Oswego, OR 97034, (503) 636-2600.

Franklin

Source - Mommersteeg International B.V., Vlijmen, The Netherlands.

Method of Breeding - Franklin strongly creeping red fescue, and was tested as MOMFrr 234. This variety was developed from five parental clones originating from an ecotype collected in the U.S. in 1972. These clones were observed in a nursery from 1973 to 1974 before synthetics were formed. Single plant selection and progeny testing were used to develop Franklin. Plants were chosen for their turf quality and seed yield. The final cross was made in 1975 and the breeder seed was first bulked in 1979.

Intended Use - Playing fields, lawns, and other turf uses.

Description - Semi-prostrate growth habit; octoploid; wide leaf, medium to dark green; medium to long head; and medium to long stems. Franklin is a medium to late heading, medium to dark green variety adapted for use in home lawns, fairways, and roadsides. Seed yield of Franklin is good.

Adapted to - Under evaluation in Canada for use in Manitoba.

Released - Mommersteeg International B.V., Vlijmen, The Netherlands. Registered in Canada in 1990, Registration No. 3196.

Breeder Seed/Stock - Mommersteeg International B.V., Vlijmen, The Netherlands.

Certified Seed/Stock - Available from Brett-Young Seeds Ltd.

Preparer/Additional Information - Agriculture Canada, Food Production and Protection Branch, Plant Products Division, Ottawa, Ontario K1A 0C6. Seed distributed by Brett-Young Seeds Ltd., Box 99, St-Norbert Station, Winnipeg, Manitoba R3V 1L5.

Hector

Source - U.S. germplasm.

Method of Breeding - Ecotypes collected in U.S. Material observed in Plant Nursery. Plants were selected and synthetics formed. Tested in turf trials and row trials. Based on five families.

Intended Use - Turf.

Description - Dark green color; quick establishing; dense; and good divot recovery.

Adapted to - PHZ 2, 3, 4, 5, 6.

Released - Mommersteeg International, B.V., The Netherlands.

Breeder Seed/Stock - Mommersteeg International, B.V.

Certified Seed/Stock - Available from Advanta Seeds West, Inc.

Preparer/Additional Information - Kenneth Hignight, Advanta Seeds West, Inc., 33725 Columbus Street S.E., P.O. Box 1496, Albany, OR 97321-0452, (503) 967-8923.

Herald

Developed by Cebeco Zaden, Vlijmen, The Netherlands.

Source - Ecotypes collected in Europe.

Method of Breeding - Pedigree selection synthetic based on 24 high seed-yielding clones.

Intended Use - Primary usage is as a component in turfgrass mixtures for home and institutional lawns and golf courses. Also used in mixtures for orchards, roadsides, and other low-maintenance areas.

Description - Medium green, medium-late maturity, capable of making a dense high-quality turf; has good seeding vigor and good resistance to dollar spot. Lower growing than some strong creeping fescue varieties.

Adapted to - LRR A, B, C, E, F, G, L, M, N, R, S, W; PHZ 3.

Breeder Seed/Stock - Van Engelen Zaden, The Netherlands.

Certified Seed/Stock - Available.

Preparer/Additional Information - Stephen W. Johnson, International Seeds Inc., P.O. Box 168, Halsey, OR 97348, (503) 369-2251.

Illahee (Reg. No. 2)

Selected at Oregon AES, Corvallis, ARS cooperating - H.A. Schoth.

Source - Seed lot imported in 1937 from England by Oscar Loe, Silverton, OR.

Method of Breeding - Comparative tests at Corvallis, OR, and Beltsville, MD.

Intended Use - Turf.

Description - Turf variety, with fine stems and leaves, slow rate of spread. Produces dense, uniform, fine-textured turf. At Beltsville established more rapidly from fall seedling and more cold tolerant than five varieties of red fescue, including Oregon-grown commercial, with which it was compared.

Adapted to - LRR A, B, E, R; PHZ 3, 4, 5.

Released - 1950, cooperatively by Oregon AES and Plant Science Research Division, ARS.

Breeder Seed/Stock - Not available.

Certified Seed/Stock - Available.

Preparer/Additional Information - Department of Crop and Soil Sciences, Crop Science Building 107, Oregon State Univ., Corvallis, OR 97331-3002, (503) 737-4513.

Jasper

Source - From naturally segregating turf plot population in NJ.

Method of Breeding - Synthetic cultivar development via polycross and progeny test to attain uniformity.

Intended Use - Cool-season turf grass.

Description - One of a small number of commercially available cultivars in this species, exhibits good color and quality ratings in turf trials since 1986.

Adapted to - PHZ 8, and eastern areas of 5, 6 and 7.

Released - 1989 by Pickseed West Inc.

Breeder Seed/Stock - Pickseed West Inc.

Certified Seed/Stock - Available.

Preparer/Additional Information - Pickseed West Inc., P.O. Box 888, Tangent, OR 97389, (503) 926-8886.

Marker

Developed by Cebeco Zaden, Vlijmen, The Netherlands.

Source - Collection of plants from sheep grazed dikes in the southwestern part of The Netherlands and northwest Belgium.

Method of Breeding - Several generations of mass selection for tillering, tolerance to close mowing, persistency, and dark green color. Synthetic based on seed from 29 plants.

Intended Use - Component in turfgrass mixtures for commercial sites, home lawns, and golf courses. Very well suited for golf course fairways and tees.

Description - Very dark green variety with a dense, compact growth habit. Combines well with other fine fescues, Kentucky bluegrass, and perennial ryegrass. Tolerates low mowing and is somewhat tolerant of saline conditions. Darkest genetic color in 1989 National Turf Evaluation Program, national fineleaf fescue test.

Adapted to - LRR A, B, C, E, G, L, M, N, R, S, W; PHZ 3.

Released - Europe by Cebeco Zaden; 1992 in U.S. by International Seeds, Inc.

Breeder Seed/Stock - Cebeco Zaden, Vlijmen, The Netherlands.

Certified Seed/Stock - Available.

Preparer/Additional Information - Stephen W. Johnson, International Seeds Inc., P.O. Box 168, Halsey, OR 97348, (503) 369-2251.

Medallion

Source - Developed at Zwaan and de Wiljes B.V., Scheemda, The Netherlands.

Method of Breeding - A five-clone synthetic originating from 15 plants selected in Denmark. The breeding techniques used were mass selection and clonal selection. Selection was based on turf quality and seed yield and was selected for five cycles. The five clones selected were combined to form a synthetic variety (Syn-1) in 1967. Tested as Rapid.

Intended Use - Playing fields, lawns, and other fine turf areas.

Description - Semi-prostrate to prostrate growth habit; octaploid; dark green leaves; narrow, long, medium to broad flag leaf; needle-like folds; slight waxy bloom, medium anthocyanin; medium to tall height; strongly creeping rhizomes; and early to medium heading. German data indicate good resistance to red thread and to pink snow mold.

Released - Zwaan and de Wiljes B.V., The Netherlands. Registered in Canada in 1990 with registration no. 3258.

Breeder Seed/Stock - Zwaan and de Wiljes B.V., Scheemda, The Netherlands.

Certified Seed/Stock - Available from Brett-Young Seeds Ltd.

Preparer/Additional Information - Agriculture Canada, Food Production and Protection Branch, Plant Products Division, Ottawa, Ontario K1A 0C6. Seed distributed by Brett-Young Seeds Ltd., Box 99, St-Norbert Station, Winnipeg, Manitoba R3V 1L5.

Pennlawn (Reg. No. 3)

Selected at Pennsylvania AES, University Park - H.B. Musser.

Source - Individual plants selected from university golf-course fairways. This turf seeded approximately 30 years previously with seed of European origin.

Method of Breeding - Screening tests of source material received from England, Hungary, Canada, and U.S.. Sod plugs collected from established turf and included in tests. Turf-quality tests of approximately 50 strains established at University Park and Beltsville, MD. Three superior strains (on basis of data from two locations) isolated for increase. Strains identified as F-55 (38), F-74 (38), and F-78 (38).

Intended Use - General turf use.

Description - Synthetic variety; produces better turf than any of original parents. Not immune to leaf spot diseases, but decidedly tolerant of them; not attacked severely and recovers rapidly. Good foliage density, rapid spread, ability to withstand close clipping.

Adapted to - Cool temperate regions in the eastern U.S.

Released - 1954, by Pennsylvania AES.

Breeder Seed/Stock - Not available.

Certified Seed/Stock - Certified and common available.

Preparer/Additional Information - Pennsylvania State Univ., Agronomy Dept., University Park, PA 16802, (814) 865-6541.

Rainier

Selected at Oregon AES, Corvallis, ARS cooperating - H.A. Schoth.

Source - Accession received in 1938.

Method of Breeding - Increase of original accession.

Intended Use - Turf.

Description - High seed yielder, stiff stems, good turf developer, long lived, uniform dark green, rapid grower. Resistant to common leaf, stem, and head diseases in Pacific Northwest.

Adapted to - LRR A, B, E; PHZ 3, 4.

Released - 1944, cooperatively by Oregon AES and Plant Science Research Division, ARS.

Breeder Seed/Stock - Not available.

Certified Seed/Stock - Available.

Preparer/Additional Information - Department of Crop and Soil Sciences, Crop Science Building 107, Oregon State Univ., Corvallis, OR 97331-3002, (503) 737-4513.

Recent

Source - Developed by Cebeco Zaden B.V., The Netherlands.

Method of Breeding - This turf-type variety originated from native ecotype selections made in Zeeland. In 1972, a parent plant was cloned and its F_1 was tested in a turf trial at Elst. The selection criteria were good turf quality and seed production capabilities. This variety was tested as ERU 496.

Intended Use - Hay and pasture production.

Description - Hexaploid; medium growth habit; leaves are open and rather narrow in width, with pale green color and medium anthocyanin in leaf sheath; flag leaf is medium length and rather narrow in width; late flowering; medium length heads, awns present; stems are medium length with medium internode length; and rhizomes are present. This is a strongly creeping red fescue.

Released - Cebeco Zaden B.V., The Netherlands. Registered in Canada in 1990 with registration no. 3256.

Breeder Seed/Stock - Cebeco Zaden B.V., the Netherlands.

Certified Seed/Stock - Available from Tib Szego Associates Ltd.

Preparer/Additional Information - Agriculture Canada, Food Production and Protection Branch, Plant Products Division, Ottawa, Ontario K1A 0C6.

Reptans

Developed by Plant Breeding Institute, Weibullsholm, Landskrona, Sweden.

Source - Collections from southern Sweden.

Method of Breeding - Mass selection.

Intended Use - Lawns, turf, and erosion control.

Description - Dark green, good spread, and reported as resistant to helminthosporium and fusarium in Sweden. In Canada, equal to Olds variety in hardiness, recovery after cutting, color, and disease resistance and superior in seedling vigor. Apparently well adapted in northern Alberta and British Columbia.

Released - Plant Breeding Institute, Weibullsholm. Distributed in Canada by Ontario Seed Cleaners and Dealers, Ltd.

Breeder Seed/Stock - Plant Breeding Institute, Weibullsholm.

Certified Seed/Stock - Available in Canada. Canada Seed Growers Association, P.O. Box 8445, Ottawa, Ontario K1G 3T1.

Preparer/Additional Information - Canada Seed Growers Association, P.O. Box 8445, Ottawa, Ontario K1G 3T1.

Ruby

Developed in The Netherlands - D.J. van der Have.

Source - Collection of western European varieties.

Method of Breeding - Plants evaluated for creeping character, plant type, and polycross progeny evaluated under turf conditions. Plants screened for rust resistance, summer color, and turf quality. Six clones selected to form synthetic variety.

Intended Use - For mixed plantings with bluegrasses and ryegrasses in lower maintenance and shady areas.

Description - Vigorous growth, rather broad leaves, long rhizomes. Open, creeping habit produces a more open sod than Pennlawn and comparable varieties. Characterized by rapid seedling establishment and excellent shade tolerance.

Adapted to - Most of the U.S., especially in the north.

Released - 1967, by D.J. van der Have.

Breeder Seed/Stock - D.J. van der Have.

Certified Seed/Stock - Available.

Preparer/Additional Information - Rick Meirers, Medalist America, 1490 Industrial Way SW, Albany, OR 97321, (800) 568-8873.

Salem

Developed for Cascade International Seed Co. and Jonathan Green and Sons Inc. - Dr. C. Reed Funk, New Jersey AES, New Brunswick.

Source - New Jersey AES, Rutgers University. Selected from spaced-plant nurseries from plants improved by recurrent selection.

Method of Breeding - A synthetic variety selected clones were planted at Brownsville, OR, in the Fall of 1987. Syn II seed was collected and used to establish a breeder field near Lebanon, OR.

Intended Use - Wherever cool-season grasses are adapted for turfgrass areas.

Description - A dark green, prolific variety that is medium-late maturity. In the National Turfgrass Evaluation Program, it demonstrated outstanding turf quality, including improved mowing quality, rapid leaf elongation, improved leaf spot resistance, improve drought tolerance (dormancy), an extremely dense turf, earliest spring green-up, and darkest green genetic color.

Adapted to - It has been planted successfully in southern Canada and as far south as North Carolina. It is useful along with other cool-season turfgrasses.

Released - 1988 by Dr. Kevin McVeigh to Cascade International Seed Co. to establish a breeder field near Lebanon, OR.

Breeder Seed/Stock - Cascade International Seed Co.

Certified Seed/Stock - Available.

Preparer/Additional Information - Irvin H. Jacob, Cascade International Seed Company, 8483 West Stayton Road, Aumsville, OR 97325, (503) 749-1822.

Shademaster

Source - A synthetic of the nine clones was made and called 433 and planted in a mowed turf trial in Oregon. Plants showing improved turf performance and leaf spot resistance were sprigged out of the 433 turf plots. Seed from these plants was used to establish an isolated nursery for recurrent selection for seed yield, leaf spot resistance, and an attractive dark green color.

Method of Breeding - Shademaster Creeping Red Fescue is an advanced generation synthetic variety derived from the progenies of nine clones. These clones trace back to collections from old turf areas in Tennessee and other southern parts of the U.S.

Intended Use - Recommended for all turfgrass areas where environmental factors limit optimum performance of Kentucky bluegrass and turf type perennial ryegrass; well adapted for use in difficult to maintain areas such as roadside banks and shaded park settings where frequent mowing is not necessary; excellent complement in a quality general use turf mixture as well as shady turf mixtures.

Description - It has a bright medium green color with improved winter color and spring green-up. It has shown better heat tolerance than most other spreading varieties and has shown very good cold tolerance. Shademaster has the ability to recover from drought by regrowth from extensive underground rhizomes. This variety has good shade tolerance and will perform well in mixtures with improved Kentucky bluegrasses and perennial ryegrasses. Shademaster has shown improved resistance to leaf spot, lead rust, and powdery mildew. Shademaster tolerates drought, low fertility, and shaded conditions very well and maintains its green color throughout most summers even under low management.

Adapted to - PHZ 3, 4, 5, 6, 7.

Released - 1987, Seed Testing, Inc.

Breeder Seed/Stock - Pure Seed Testing, Inc.

Certified Seed/Stock - Available.

Preparer/Additional Information - Art Wick, LESCO, Inc., 20005 Lake Road, Rocky River, OH 44116, (216) 333-9250.

Sylvester

Source - U.S. ecotypes.

Method of Breeding - Synthetic variety formed by grouping similar clones. Tested in turf trials and row trials.

Intended Use - Turf.

Description - High seed-yielding cultivar.

Adapted to - PHZ 2, 3, 4, 5, 6.

Released - Mommersteeg International, The Netherlands.

Breeder Seed/Stock - Mommersteeg International, The Netherlands.

Certified Seed/Stock - Available from Advanta Seeds West, Inc.

Preparer/Additional Information - Kenneth Hignight, Advanta Seeds West, Inc., 33725 Columbus Street S.E., P.O. Box 1496, Albany, OR 97321-0452, (503) 967-8923.

Valda

Developed by Cebeco Zaden B. V., Vlijmen, The Netherlands.

Source - Collections of wild materials.

Method of Breeding - Clones evaluated for turf quality and seed yield potential. High performing plants were selected to create synthetic variety.

Intended Use - Turf and ground cover especially for infertile soils which receive little water; also used in dry, shady areas. Good choice for roadsides.

Description - Exceptionally dark green, very fine bladed, late maturing variety. Has shown good resistance to red thread across multiple years and locations.

Adapted to - LRR A, B, C, D, E, G, L, M, N, P, R, S, W; PHZ 4.

Released - 1979, in The Netherlands, by Cebeco Zaden B.V.; 1989, in the U.S., by International Seeds Inc., Halsey, OR.

Breeder Seed/Stock - Maintained by Cebeco Zaden B.V.

Certified Seed/Stock - Available from International Seeds Inc.

Preparer/Additional Information - Stephen W. Johnson, International Seeds Inc., P.O. Box 168, 820 W. First St., Halsey, OR 97348, (503) 369-2251.

Vista

Source - Old turf areas in Tennessee and other southern parts of the U.S. Selections from Flyer which is from old turf areas in the northern U.S.

Method of Breeding - Advanced-generation synthetic derived from the progenies of 20 clones selected after three cycles of phenotypic recurrent selection for medium green color, resistance to leaf spot, and good floret fertility.

Intended Use - Cool-season turfgrass for shady areas and low maintenance areas.

Description - Improved resistance to leaf spot, and red thread, reduced thatch accumulation, and good spring green up.

Adapted to - PHZ 2, 3, 4, 5, 6, 7, 8.

Released - Zajac Performance Seeds.

Breeder Seed/Stock - Zajac Performance Seeds.

Certified Seed/Stock - Available. PVP No. 8800136.

Preparer/Additional Information - Zajac Performance Seeds, 33 Sicomac Rd., North Haledon, NJ 07508, (201) 423-1660.

Festuca rubra **L. ssp. *falax* Thuill. - chewings fescue**
Festuca rubra **L. *var. commutata* Gaudin**

Cool-season bunchgrass from Europe. Used for lawns and general-purpose turf in humid northern states. Growth habit more erect than red fescue.

Atlanta

Source - The Netherlands.

Method of Breeding - Unknown.

Intended Use - Golf course roughs, home lawns, parks, and cemeteries.

Description - Good tolerance to shade, low fertility, low water, good density.

Adapted to - PHZ 2, 3, 4, 5, 6.

Released - D.J. Van der Have, B.V., The Netherlands.

Breeder Seed/Stock - D.J. Van der Have B.V.

Certified Seed/Stock - Available from Advanta Seeds West, Inc.

Preparer/Additional Information - Kenneth Hignight, Advanta Seeds West, Inc., 33725 Columbus Street S.E., P.O. Box 1496, Albany, OR 97321-0452, (503) 967-8923.

Banner

Source - Germplasm developed by Rutgers University.

Intended Use - Turf.

Description - Good turf performance; good shade tolerance.

Adapted to - PHZ 2, 3, 4, 5, 6.

Released - E.F. Burlingham & Sons.

Breeder Seed/Stock - E.F. Burlingham & Sons.

Certified Seed/Stock - Available.

Preparer/Additional Information - Virgil Meier, O.M. Scott & Sons Co., Marysville, OH 43041, (513) 644-0011; or George Burlingham, E.F. Burlingham & Sons, P.O. Box 217, Forest Grove, OR 97116, (503) 357-2141.

Banner II

Developed by McCarthy-Burlingham Research and Rutgers University.

Source - Banner crossed with endophyte containing parent.

Method of Breeding - Parents crossed to add endophyte, then back crossed four cycles then selections made during three cycles for turf quality and disease resistance.

Intended Use - Turf, low maintenance, erosion control in vineyards, orchards, etc.

Description - Attractive color; mildew, rust and leaf spot resistant. Endophyte enhanced for insect resistance.

Adapted to - Areas where fine fescues and bluegrass are used.

Released - 1991 by McCarthy-Burlingham Research.

Breeder Seed/Stock - McCarthy-Burlingham Research.

Certified Seed/Stock - Available from O.M. Scott and Sons.

Preparer/Additional Information - Mike McCarthy, E.F. Burlingham & Sons, P.O. Box 217, Forest Grove, OR 97116, (503) 359-9368.

Bargreen

Source - Ecotypes.

Method of Breeding - In 1974 selection in ecotypes under turf conditions on uniformity and seed yield.

Intended Use - Cool-season turfgrass. The variety is very suitable for lawns.

Description - Performs well in lawns and Scottish greens.

Adapted to - Southwest, northeast, northwest, central and mountain states; PHZ 1, 2, 3, 4, 5, 6, 7.

Released - Barenbrug.

Breeder Seed/Stock - Barenbrug Holland, Oosterhout, The Netherlands.

Certified Seed/Stock - Available from Barenbrug USA.

Preparer/Additional Information - Barenbrug USA, P.O. Box 239, Tangent, OR 97389, (503) 926-5801.

Barnica

Source - Commercial seed.

Method of Breeding - Selection of space plants.

Intended Use - Cool season turfgrass. Suitable for lawn mixtures.

Description - A moderate, early flowering turf type with good mowing qualities.

Adapted to - Southwest, northwest, northeast, central and mountain states; PHZ 1, 2, 3, 4, 5, 6, 7.

Released - Barenbrug.

Breeder Seed/Stock - Barenbrug Holland, Oosterhout, The Netherlands.

Certified Seed/Stock - Available from Barenbrug USA.

Preparer/Additional Information - Barenbrug USA, P.O. Box 239, Tangent, OR 97389, (503) 926-5801.

Bridgeport

Source - Collected from old turfs in northeast U.S.

Method of Breeding - Twenty-five clone advanced-generation synthetic cultivar originally developed by the New Jersey AES with C. Reed Funk.

Intended Use - Turf, either alone or in a blend.

Description - Dark green, upright growth, good disease resistance, high endophyte for insect resistance, winter hardy, shade tolerant.

Adapted to - Northeast U.S., upper transition states.

Released - 1990, Barenbrug/Normarc.

Breeder Seed/Stock - Available from Willamette Valley Plant Breeders and Barenbrug USA.

Certified Seed/Stock - Available from Barenbrug USA.

Preparer/Additional Information - Don Herb/Matt Herb, Barenbrug USA, P.O. Box 239, Tangent, OR 97389, (503) 926-5801.

Camaro

Source - Mommersteeg International B.V., P.O. Box 1, 5250 AA Vlijmen, The Netherlands.

Method of Breeding - Originated from a Dutch ecotype. A synthetic variety based on eight clones.

Intended Use - Turf.

Description - Very fine bladed leaf texture, excellent tolerance to leafspot, red thread, summer patch, pink patch, and anthracnose. Withstands drought conditions and does well in low-maintenance areas.

Adapted to - PHZ 2, 3, 4, 5, 6, 7.

Released - 1982, Mommersteeg International, The Netherlands.

Breeder Seed/Stock - Mommersteeg International.

Certified Seed/Stock - Available.

Preparer/Additional Information - David Lundell, Fine Lawn Research, Inc., P.O. Box 1051, Lake Oswego, OR 97034, (503) 636-2600.

Cascade

Increased at Oregon AES, Corvallis - R.V. Frakes.

Source - Oregon-grown common chewings fescue.

Method of Breeding - Seed selected from 16 fields of common chewings, tracing to the New Zealand source, and carefully examined for chewings seed characteristics. Equal amounts of seed from 12 sources used to establish basic seed for authentic source of chewings fescue.

Intended Use - Turf.

Description - Does not creep, dark green, fine leafed; responds to 37 mm clipping height and high soil fertility. Produces desirable, durable, high-quality turf that is comparable with authentic sources of chewings fescue.

Adapted to - LRR A, B, E; PHZ 3, 4.

Released - 1966, by Oregon AES.

Breeder Seed/Stock - Oregon AES.

Certified Seed/Stock - Available.

Preparer/Additional Information - Department of Crop and Soil Sciences, Crop Science Building 107, Oregon State Univ., Corvallis, OR 97331-3002, (503) 737-4513.

Center

Developed by Cebeco Zaden B.V., Vlijmen, The Netherlands.

Source - Selections made from turfs in northern Europe.

Method of Breeding - Mass selection within elite family material. Major emphasis in selection included increased density and improved mowing qualities.

Intended Use - Cool-season turfgrass; especially useful in shady and infertile areas. Also used in mixtures for overseeding dormant bermudagrass turfs.

Description - Medium dark green, fine textured chewings fescue which produces an attractive turfgrass sward. This variety has good year-round density and improved winter color.

Adapted to - LRR C, D, I, J, O, P, T, U, V, overseeding dormant warm-season turfs. A, B, C, E, G, L, M, N, R, S, permanent turf; PHZ 4.

Released - 1984, first release in United Kingdom by Cebeco Zaden B.V.; 1987, U.S. release by International Seeds Inc., Halsey, OR.

Breeder Seed/Stock - Maintained by Cebeco Zaden B. V.

Certified Seed/Stock - Available through Turf Merchants Inc., Tangent, OR. U. S. PVP No. 8700077.

Preparer/Additional Information - Stephen W. Johnson, International Seeds Inc., P.O. Box 168, 820 W. First St, Halsey, OR 97348, (503) 369-2251.

Dover

Northrup King, now known as NK Lawn and Garden - Howard E. Kaerwer and Dr. Eric K. Nelson.

Source - Single clone siblings from Long Island, NY.

Method of Breeding - Mass selection cycles for resistance to *Crambus trisecta* and *Bipolaris sorokiniana.* Final cycle of selection for attractive dark green color, seedling vigor, and seed yield characteristics.

Intended Use - A low-maintenance turf for home lawns, parks, roadsides, and golf course roughs. Excellent seedling vigor makes it suitable in mixture with Kentucky bluegrasses and perennial ryegrasses for turf in both sun and heavy shade situations. Excellent mowing quality and tolerance to close mowing make Dover suitable for use in dormant bermudagrass overseeding in mixture with perennial ryegrasses.

Description - Excellent germination energy. Good seedling vigor. Dark green color. Excellent shade tolerance. Resistance to leaf spot and sod webworm damage. Excellent seed yield characteristics.

Adapted to - LRR A, B, C, D, E, F, G, H, K, L, M, N, O, P, R, S, T, W; PHZ 3, 4, 5, 6, 7, 8.

Released - 1990 by NK Lawn and Garden.

Breeder Seed/Stock - Maintained by NK Lawn and Garden, Tangent, OR.

Certified Seed/Stock - NK Lawn and Garden, Minneapolis, MN, Tangent, OR, or Chicago, IL.

Preparer/Additional Information - Eric K. Nelson, NK Lawn and Garden, P.O. Box 300, 33731 Highway 99E, Tangent, OR 97389, (503) 928-2393.

ECO

McCarthy-Burlingham Research Farm.

Source - Germplasm is from a diverse source collected in Europe and North America. Selection of parental clones was from large nurseries in Oregon.

Method of Breeding - Cycles of recurrent selection for turf quality and seed production.

Intended Use - Turf, erosion control, cover in orchards, vineyards, etc.

Description - Endophyte enhanced, good recovery after drought conditions, disease resistant, good turf quality and mowability, shade tolerant, and a good seed yielder.

Adapted to - Northern climates suited to fine fescue and bluegrass.

Released - 1993 by McCarthy-Burlingham Research.

Breeder Seed/Stock - McCarthy-Burlingham Research.

Certified Seed/Stock - Available in 1995.

Preparer/Additional Information - Mike McCarthy, E.F. Burlingham & Sons, P.O. Box 217, Forest Grove, OR 97116, (503) 359-9368.

Enjoy

Developed by Van Engelen Zaden B.V. - Vlijmen, The Netherlands.

Source - Ecotypes collected in the central part of The Netherlands.

Method of Breeding - Selections were intercrossed and the resulting progeny were subjected to mass selection pressure for disease resistance, color retention, seed yield, and turf and mowing quality.

Intended Use - Turf for lawns, golf courses, and roadsides. Well suited for use on infertile soils and in shady areas. Also used in mixtures for overseeding dormant warm-season turf.

Description - Very fine bladed, medium maturing variety. Forms a dense, medium dark green turf. Tolerates low mowing heights well; may be cut as low as 5 mm when used in an overseeding situation. Resistant to fusarium blight, red thread, and leaf spot.

Adapted to - LRR A, B, C, E, L, M, N, R, S, permanent turf. C, D, G, I, J, O, P, T, U, overseeding dormant warm-season turfs; PHZ 4.

Released - 1982, The Netherlands by Van Engelen Zaden B.V.; 1987 in the U.S. by International Seeds Inc.

Breeder Seed/Stock - Van Engelen Zaden B.V.

Certified Seed/Stock - Certified and uncertified seed available from International Seeds Halsey, OR.

Preparer/Additional Information - Stephen W. Johnson, International Seeds Inc., P.O. 168, 820 W. First St., Halsey, OR 97348, (503) 369-2251.

Jamestown

Developed at the University of Rhode Island - Dr. Skogley. Carried as accession no. RI-6.

Source - Parental clones were found on an abandoned golf course at the Beavertail Golf Course and Country Club, on the island of Jamestown located in the Narraganset Bay off the coast of Rhode Island.

Method of Breeding - Comparative tests from first-generation seed. Parent material maintained vegetatively. Distributed for testing as Rhode Island 6.

Intended Use - Winter overseeding of dormant bermudagrass on golf course greens and tees; permanent home lawns, golf courses, and roadsides.

Description - Improved, moderately dark-green, profusely tillering, fine leafed and low growing variety with excellent persistence under low cutting.

Adapted to - LRR all except O, I, J, P, T, U; PHZ 3, 4, 5, 6, 7.

Released - 1974, by Rhode Island AES.

Breeder Seed/Stock - Lofts Seed, Inc., with the cooperation of the RI AES.

Certified Seed/Stock - Available.

Preparer/Additional Information - Lofts Seed Inc., Chimney Rock Road, P.O. Box 146, Bound Brook, NJ 08805, (908) 560-1590.

Jamestown II

Source - Jamestown and LF-1, which was developed from plants selected from Longfellow Park in Cambridge, MA. LF-1 is closely related to the cultivar Longfellow.

Method of Breeding - Jamestown II was developed using a modified backcross procedure with plants of Jamestown as the recurrent parents and plants selected from LF-1 as the donor parents.

Intended Use - For medium- to low-maintenance turf under varying light intensities. Can also be used to overseed dormant bermudagrass for winter color in the southern U.S.

Description - Excellent for lower maintenance turfs; contains an endophyte for insect resistance and improved stress tolerance and persistence.

Adapted to - Wherever cool-season grasses are adapted and for winter overseeding in warm-season grass areas.

Breeder Seed/Stock - Rhode Island AES.

Certified Seed/Stock - Available.

Preparer/Additional Information - Marie Pompei, Lofts Seed, Inc., P.O. Box 146, Bound Brook, NJ 08805, (908) 560-1590.

Koket

Developed by Momersteeg.

Source - European selections.

Intended Use - Turf.

Description - Moderately good turf quality, good seed yield, and shade tolerant.

Adapted to - Northern climates adapted to fine fescue use areas.

Released - Momersteeg.

Breeder Seed/Stock - Momersteeg.

Certified Seed/Stock - Available.

Preparer/Additional Information - Mike McCarthy, E.F. Burlingham & Sons, P.O. Box 217, Forest Grove, OR 97116, (503) 359-9368.

Longfellow

Developed by the New Jersey AES, New Brunswick, NJ, and International Seeds, Inc., Halsey, OR.

Source - Old lawn in Massachusetts.

Method of Breeding - Multiple cycles of mass selections.

Intended Use - Permanent cool-season turf; especially useful in shaded and low fertility areas; also may be used as a component in mixtures for overseeding dormant bermudagrass.

Description - Persistent, dark green, moderately late flowering cultivar; shows good drought tolerance and improved summer performance; moderately resistant to red thread and typhula blight, susceptible to powdery mildew. Longfellow has been highly rated in trials at multiple locations. U. S. plant variety protection has been applied for.

Adapted to - LRR A, B, C, E, G, L, M, N, R, S, permanent turf. D, I, J, O, P, T, U, overseeding dormant warm-season turf; PHZ 4.

Released - 1990 by International Seeds Inc.

Breeder Seed/Stock - International Seeds Inc.

Certified Seed/Stock - Available. U.S. PVP No. 9000258.

Preparer/Additional Information - Stephen W. Johnson, International Seeds Inc., P.O. Box 168, 820 W. First St., Halsey, OR 97348, (503) 369-2251.

Mary

Source - The Netherlands.

Method of Breeding - Selected plants from an ecotype collection were interpollinated. Following a three- year turf trial, top performers were selected. Based on four clones.

Intended Use - Golf course roughs, home lawns, and parks.

Description - Dark green color, quick establishment, good resistance to melting out and red thread.

Adapted to - PHZ 2, 3, 4, 5, 6.

Released - 1982 by Mommersteeg International B.V., The Netherlands.

Breeder Seed/Stock - Mommersteeg International B.V.

Certified Seed/Stock - Available from Advanta Seeds West, Inc.

Preparer/Additional Information - Kenneth Hignight, Advanta Seeds West, Inc., 33725 Columbus Street S.E., P.O. Box 1496, Albany, OR 97321-0452, (503) 967-8923.

Molinda

Source - Germplasm originated from The Netherlands.

Method of Breeding - Synthetic variety based upon seven families.

Intended Use - Turf.

Description - Maintains good density under low levels of fertility, water stress, and low pH.

Adapted to - PHZ 2, 3, 4, 5, 6.

Released - 1992 by O.M. Scott & Sons Co.

Breeder Seed/Stock - Advanta Seeds West, Inc.

Certified Seed/Stock - Available.

Preparer/Additional Information - Virgil Meier, O.M. Scott & Sons Co., Marysville, OH 43041, (513) 644-0011.

Nimrod

Developed by Cebeco Zaden, Vlijmen, The Netherlands.

Source - Ecotype from an old golf course on poor, dry, and sandy soil in the center of The Netherlands.

Method of Breeding - Mass selection for tillering, persistence, and tolerance to close mowing for several generations. Eleven clones selected for synthetic.

Intended Use - Turf for golf courses, lawns, and roadsides. Suited for use in both close-cut high-quality turfs and low-maintenance areas.

Description - Moderately dark green, medium maturity variety. Capable of forming a dense attractive turf. Has a lower growth habit than many chewings fescues.

Adapted to - LRR A, B, C, E, L, M, N, R, S; PHZ 4.

Released - 1991 in Europe by Van Engelen Zaden; expected U.S. release in 1994 by International Seeds, Inc.

Breeder Seed/Stock - Cebeco Zaden, Vlijmen, The Netherlands.

Certified Seed/Stock - Not yet available.

Preparer/Additional Information - Stephen W. Johnson, International Seeds, Inc., P.O. Box 168, Halsey, OR 97348, (503) 369-2251.

Scarlet

Source - Netherlands germplasm.

Method of Breeding - Synthetic variety.

Intended Use - Home lawns, golf course roughs, and parks.

Description - Tolerates close mowing, good summer color.

Adapted to - PHZ 2, 3, 4, 5, 6.

Released - D.J. Van der Have, B.V., The Netherlands.

Breeder Seed/Stock - D.J. Van der Have.

Certified Seed/Stock - Available from Advanta Seeds West, Inc.

Preparer/Additional Information - Kenneth Hignight, Advanta Seeds West, Inc., 33725 Columbus Street S.E., P.O. Box 1496, Albany, OR 97321-0452, (503) 967-8923.

Shadow

Source - Old turf areas throughout the northeastern U.S. and two European sources.

Method of Breeding - Seedlings were screened for resistance to powdery mildew in the greenhouse for three cycles of recurrent selection and two cycles for seed yield and attractive appearance as space plants.

Intended Use - Cool-season turf areas, shady areas, and low-maintenance areas.

Description - Improved resistance to powdery mildew, good shade tolerance, good resistance to candii stripe rust.

Adapted to - PHZ 2, 3, 4, 5, 6, 7, 8.

Released - Turf-Seed, Inc.

Breeder Seed/Stock - Pure Seed Testing, Inc.

Certified Seed/Stock - Available. PVP No. 8100155.

Preparer/Additional Information - Turf-Seed, Inc., P.O. Box 250, Hubbard, OR 97032, (503) 651-2130.

Southport

Developed for Cascade International Seed Co. and Jonathan Green and Sons, Inc. - Dr. C. Reed Funk, New Jersey AES, Rutgers University.

Source - New Jersey AES, Rutgers University. Selected from spaced-plant nurseries from plants improved by recurrent selection.

Method of Breeding - A synthetic variety; selected clones were planted at Brownsville, OR, in the fall of 1987. Syn II seed was collected and used to establish a breeder field near Lebanon, OR, in the fall of 1987.

Intended Use - Wherever cool-season grasses are adapted for turfgrass areas.

Description - A dark green, prolific variety that is medium-late maturity. In the National Turfgrass Evaluation Program, it demonstrated outstanding turf quality, including improved red thread resistance, and noticeably wider leaf texture.

Adapted to - It has been planted successfully in southern Canada and as far south as North Carolina. It is useful along with other cool-season turfgrasses.

Released - 1993.

Breeder Seed/Stock - Cascade International Seed Co.

Certified Seed/Stock - Available.

Preparer/Additional Information - Irvin H. Jacob, Cascade International Seed Company, 8483 West Stayton Road, Aumsville, OR 97325, (503) 749-1822.

SR 5000

Seed Research of Oregon - M.F. Robinson and Leah A. Brilman. Accession no. is not assigned at present.

Source - Germplasm NJC-1 (primary pollen parent) - old turfs in warmer regions in eastern U.S. down to Atlantic. Germplasm NJ6281 - maternal plants trace back to endophyte containing plants selected from an old lawn in Cambridge, MA, pollinated with plants from Banner, Jamestown, Shadow, Waldorf, NJC-2. Germplasm NJ7142 - same maternal germplasm as NJ6281 pollinated with Banner, Shadow and NJC-1.

Method of Breeding - Advanced-generation synthetic cultivar developed in a population improvement program using varying cycles of phenotypic and genotypic selection and the use of maternal parents that contained an endophytic fungus *Acremonium sp.* Three elite germplasms were developed from this program: NJC-1 (without endophyte), NJ6281 (high endophyte), and NJ7142 (high endophyte) These were planted in a spaced-plant nursery with every other row NJC-1 as a primary pollen source and alternate rows NJ6281 or NJ7142. Selection was performed in this population for powdery mildew resistance, seed yield, floret fertility, uniformity, and dark green color. Seed harvested from selected plants of NJ6281 and NJ7142 to ensure high endophyte levels.

Intended Use - Cool season turfgrass in mixtures with other fine fescues, perennial ryegrasses, Kentucky bluegrasses, and colonial bentgrasses. Low-maintenance turf areas, especially areas with dry, infertile, and acid soils. Shade environments. Maritime climates, areas of cool summers with SR 5000 showing improved summer stress tolerance in additional areas. Mixtures with Kentucky bluegrass and perennial ryegrass.

Description - Improved summer performance compared to other chewings fescues. Improved resistance to powdery mildew for better shade tolerance. Endophyte-enhanced for resistance to many turf insects, improved summer survival and fall recovery, and reduced weed invasion. Very good resistance to net blotch.

Adapted to - LRR A, B, C, D, E, F, G, H, K, L, M, N, R, S, W; PHZ 3, 4, 5, 6, 7, 8 - Northeast, midwest, Pacific northwest, California.

Released - 1989 by Seed Research of Oregon, Inc. and Rutgers University.

Breeder Seed/Stock - Seed Research of Oregon, Corvallis, OR.

Certified Seed/Stock - Available.

Preparer/Additional Information - Dr. Leah A. Brilman, Seed Research of Oregon, Inc., P.O. Box 1416, Corvallis, OR 97339, (800) 253-5766.

SR 5100

Source - Superior lower-growing chewings fescue selected from a large population at Rutgers University. Moved to an isolate crossing block (29 clones).

Method of Breeding - Progeny of above material placed in a large spaced-plant nursery of 100 progeny of each clone (2,900 plants). Population extensively rogued for freedom from powdery mildew, lower growing height of leaves, seed yield, choke, and dark green color. Seed from the progeny of the six clones with endophyte used for breeder seed.

Intended Use - Cool-season turfgrass in mixtures with other fine fescues, perennial ryegrasses, Kentucky bluegrass, and colonial bentgrasses.

Description - Lower growing, dark green chewings fescue with improved disease resistance, good stress tolerance and less clippings.

Adapted to - PHZ 3, 4, 5, 6, 7, 8, NE, Midwest, Pacific Northwest, CA.

Released - 1993, Seed Research of Oregon, Inc. and Rutgers University.

Breeder Seed/Stock - Seed Research of Oregon, Inc.

Certified Seed/Stock - Available.

Preparer/Additional Information - Dr. Leah A. Brilman, Seed Research of Oregon, Inc., P.O. Box 1416, Corvallis, OR 97339, (503) 757-2663.

Tiffany

Source - Endophyte sources from New Jersey and selections from Shadow (E).

Method of Breeding - Advanced generation synthetic of 11 clones selected for a dwarf-growth habit.

Intended Use - Cool-season turf areas, shady areas, and low-maintenance areas.

Description - Good seedling vigor, spring and fall density, and winter color; improved drought tolerance and resistance to candii stripe rust, red thread, leaf spot, dollar spot, summer patch, and pink patch.

Adapted to - PHZ 2, 3, 4, 5, 6, 7, 8.

Released - Turf Seed Inc.

Breeder Seed/Stock - Pure-Seed Testing, Inc.

Certified Seed/Stock - Not currently available. PVP # pending.

Preparer/Additional Information - Turf Seed Inc., P.O. Box 250, Hubbard, OR 97032, (503) 651-2130.

Trophy

Developed by Van Engelen Zaden Vlijmen, The Netherlands.

Source - Ecotype from the eastern part of The Netherlands.

Method of Breeding - Mass selection in offspring from original ecotype. Variety ultimately based on seed from one high seed-yielding clone with good turf quality top-crossed with 13 half-sibs.

Intended Use - Turf for both high- and low-maintenance areas may be planted in shaded areas and on infertile soils.

Description - Medium maturity and moderately dark green variety; produces turf tolerant of close cutting and low fertility. Has good spring green up.

Adapted to - LRR A, B, C, E, L, M, N, R, S permanent turf; C, D, I, J, O, P, T, U, V overseeding warm-season turf; PHZ 4.

Released - Europe by Van Engelen Zaden; expected U.S. release in 1994 by International Seeds, Inc.

Breeder Seed/Stock - Cebeco Zaden B.V.

Certified Seed/Stock - Available.

Preparer/Additional Information - Stephen W. Johnson, International Seeds, Inc., P.O. Box 168, Halsey, OR 97348, (503) 369-2251.

Victory

Source - Source material obtained from a population improvement program at Rutgers University, NJ.

Method of Breeding - Synthetic cultivar development via polycross and subsequent progeny testing to attain uniformity.

Intended Use - Cool season turfgrass.

Description - Exhibits good powdery mildew resistance and a late heading date in comparison with other cultivars.

Adapted to - PHZ 8, and eastern areas of 5, 6 and 7.

Released - 1985 by Pickseed West Inc.

Breeder Seed/Stock - Pickseed West Inc.

Certified Seed/Stock - Available.

Preparer/Additional Information - Pickseed West Inc., P.O. Box 888, Tangent, OR 97389, (503) 926-8886.

Waldorf

Source - Northeast Netherlands.

Method of Breeding - Progeny testing of ecotypes.

Intended Use - Golf course roughs, home lawns, and parks.

Description - Widely adapted, good leaf spot resistance, and low-medium nitrogen requirement.

Adapted to - PHZ 2, 3, 4, 5, 6.

Released - D.J. Van der Have B.V., The Netherlands.

Breeder Seed/Stock - D.J. Van der Have B.V.

Certified Seed/Stock - Available from Advanta Seeds West, Inc.

Preparer/Additional Information - Kenneth Hignight, Advanta Seeds West, Inc., 33725 Columbus Street S.E., P.O. Box 1496, Albany, OR 97321-0452, (503) 967-8923.

Festuca trachyphylla **(Hack.) Krajina - hard fescue** ***Festuca longifolia*** **auct. non Thuill.**

Festuca ovina L. var. _duriuscula_ auct. non (L.) W.D.J. Koch

Cool-season bunchgrass from Europe. Introduced into North America; originating in the Baltic states to the Carpathian Mountains through central Europe to the Atlantic. Useful in erosion control and soil improvement in parts of Pacific Northwest. Tougher leaves and less drought-tolerant than sheep fescue.

Attila

Source - Numerous collections from old turfs in Oregon and several southeastern U.S. states.

Method of Breeding - Extensive evaluation in spaced-plant nursery for desired dark green and compact growth habit, phenotypic matched mating over two generations and concurrent evaluation in turf plots.

Intended Use - Turf for low-maintenance areas, roadsides, and golf courses outside of roughs.

Description - Short growth, compact, dark green, very good seed yield, and disease resistance to leaf spot and anthracnose.

Adapted to - North of the Mason-Dixon line.

Released - 1992 by KWS-AG, Einbeck, Germany.

Breeder Seed/Stock - KWS Seeds, Inc. and TMI.

Certified Seed/Stock - Available in 1993.

Preparer/Additional Information - Fred B. Ledeboer, Ph.D., TMI, 22068 Case Road NE, Aurora, OR 97002, (503) 678-2597.

Aurora

Source - Old turf areas of the northeastern U.S. and selections from C-26 and Scaldis.

Method of Breeding - Advanced-generation synthetic resulting from progeny tests of clones selected for reduced vertical leaf growth, disease resistance, and increased seed head production.

Intended Use - Cool-season turf, moderate- and low-maintenance areas, shady areas.

Description - Performs very well in shady areas, excellent density with reduced vertical growth.

Adapted to - PHZ 2, 3, 4, 5, 6, 7, 8.

Released - Turf-Seed, Inc.

Breeder Seed/Stock - Pure Seed Testing, Inc.

Certified Seed/Stock - Available. PVP No. 8600070.

Preparer/Additional Information - Turf-Seed, Inc., P.O. Box 250, Hubbard, OR 97032, (503) 651-2130.

Barcrown

Source - Ecotypes.

Method of Breeding - In 1973, selection of ecotypes on turf quality and uniformity.

Intended Use - Cool-season turfgrass. Good performance under low-maintenance and under-green conditions.

Description - Very fine leafed, it gives a dense and fine turf. Good disease resistance, drought tolerance, good wear-resistance, and a good red thread resistance.

Adapted to - Southwest, northeast, northwest, central and mountain states; PHZ 1, 2, 3, 4, 5, 6, 7.

Released - Barenbrug.

Breeder Seed/Stock - Barenbrug Holland, Oosterhout, The Netherlands.

Certified Seed/Stock - Available from Barenbrug USA.

Preparer/Additional Information - Barenbrug USA, P.O. Box 239, Tangent, OR 97389, (503) 926-5801.

Bardur

Source - Crossing of ecotypes of the U.S. and Biljart.

Method of Breeding - Selection made on basis of color, seed yield and persistence.

Intended Use - Cool-season turfgrass. Suitable for lawns, parks, and roadsides.

Description - It is an early flowering turf-type with a good seed yield and a very good persistence.

Released - 1992, Barenbrug.

Breeder Seed/Stock - Barenbrug Holland, Oosterhout, The Netherlands.

Certified Seed/Stock - Available from Barenbrug USA.

Preparer/Additional Information - Barenbrug USA, P.O. Box 239, Tangent, OR 97389, (503) 926-5801.

Barreppo

Source - Dutch ecotypes.

Method of Breeding - Selection of ecotypes and families under turf conditions. Afterwards, two clones were selected because of their remarkable performance under 75% shade.

Intended Use - Cool-season turfgrass. May be used for mixtures in the shade.

Description - It has a very good winter color, a good drought tolerance, and a good resistance against snowmold. Excellent performer under artificial shade.

Released - Barenbrug.

Breeder Seed/Stock - Barenbrug Holland, Oosterhout, The Netherlands.

Certified Seed/Stock - Available from Barenbrug USA.

Preparer/Additional Information - Barenbrug USA, P.O. Box 239, Tangent, OR 97389, (503) 926-5801.

Biljart

Selected at NVH Mommersteegs, Vlijmen, The Netherlands - W.A. Eschauzier.

Source - Plant collections made in Europe.

Method of Breeding - Selections screened for uniformity of texture and flowering and growth habit. Fourteen clones selected on the basis of performance in polycross progeny tests.

Intended Use - Cool-season turfgrass.

Description - Synthetic with moderate to high resistance to dollar spot and leaf spot. Seedling development slower than red and chewings fescue. Forms low growth profile, has good tolerance to shade and high summer temperatures. *Released* - NVH Mommersteegs.

Breeder Seed/Stock - NVH Mommersteegs.

Certified Seed/Stock - Available. Distributed in the U.S. by O.M. Scott & Sons Co.

Preparer/Additional Information - Virgil Meiers, O.M. Scott & Sons, Marysville, OH 43041, (513) 644-0011.

Brigade

Source - Germplasm originated from ecotypes collected from central area of The Netherlands.

Method of Breeding - Synthetic variety based on four clones.

Intended Use - Turf.

Description - Maintains excellent density under low levels of fertility and water stress. Performs well in shade.

Adapted to - PHZ 2, 3, 4, 5, 6.

Released - 1992, O.M. Scott & Sons Co.

Breeder Seed/Stock - Advanta Seeds West, Inc.

Certified Seed/Stock - Available.

Preparer/Additional Information - Virgil Meier, O.M. Scott & Sons Co., Marysville, OH 43041, (513) 644-0011.

Durar (Reg. No. 4)

Selected at Plant Materials Center, SCS, Pullman, WA - V.B. Hawk and J.L. Schwendiman.

Source - Collected from 1934 V.B. Hawk planting on Eastern Oregon Branch Experiment Station, Union.

Method of Breeding - Mass selection for several generations. Increased for field testing as P-2517.

Intended Use - Soil erosion control and revegetation of disturbed lands in areas of 35-70 cm of annual precipitation.

Description - Tall, semi-erect, densely tufted, perennial bunchgrass. Differs from sheep fescue in its smoother, wider, longer, firmer leaf blades. Related to chewings fescue, but more drought resistant and more densely tufted. Heavy root producer, shade tolerant. Basal, harsh foliage. Consistently high seed production. In mixtures with alfalfa, it is very heavy root producer. Widely adapted to well drained soil.

Adapted to - LRR B, D, E; PHZ 5.

Released - 1949 as P-2517, cooperatively by Washington, Idaho, and Oregon Agricultural Experiment Stations at Pullman, Moscow, and Corvallis, respectively, and the Plant Materials Center, SCS, Pullman, WA. Named Durar in 1963.

Breeder Seed/Stock - Plant Materials Center, SCS, Pullman, WA.

Certified Seed/Stock - Available.

Preparer/Additional Information - Plant Materials Center, SCS, Room 104, Hulbert Agricultural Sciences Bldg., WSU, Pullman, WA 99164-6211, (509) 335-7376.

EcoStar (HSY)

Source - Rutgers and Jacklin breeding program.

Method of Breeding - Recurrent selection for high seed- yield and disease resistance.

Intended Use - Turf and reclamation.

Description - Exhibits excellent texture and color. EcoStar is a moderately dense variety with resistance to dollar spot.

Adapted to - Southern Canada, northern and central U.S.

Released - 1993 by Jacklin Seed Co.

Breeder Seed/Stock - Jacklin Seed Co.

Certified Seed/Stock - Available in 1994.

Preparer/Additional Information - Kim Peterson, Jacklin Seed Co., W. 5300 Riverbend Ave., Post Falls, ID 83854, (208) 773-7581.

Eureka

Developed by Van Engelen Zaden B.V., Vlijmen, The Netherlands.

Source - Ecotype out of The Netherlands.

Method of Breeding - Recurrent selection for non-blue color and low growth, synthetic variety finally based on 83 plants.

Intended Use - Ground cover and turf; especially in areas where a grass which requires few mowings is desired. Also suited for use in infertile, acidic soils and in dry shady areas.

Description - Fine bladed, dark green and very low-growing variety. Mature plant height is significantly less than many other hard fescue cultivars. Maturity is early, relative to other hard fescues.

Adapted to - LRR A, B, C, D, E, F, G, L, M, N, P, R, S, W; PHZ 4.

Released - 1991 in Europe by Van Engelen Zaden B.V.; U.S. release, 1992 by International Seeds Inc., Halsey, OR.

Breeder Seed/Stock - Maintained by Cebeco Zaden, B. V.

Certified Seed/Stock - Available. U.S. PVP applied for No. 9100120.

Preparer/Additional Information - Stephen W. Johnson, International Seeds Inc., P.O. Box 168, 820 W. First St., Halsey, OR 97348, (503) 369-2251.

Gladiator

McCarthy-Burlingham Research.

Source - Selections of superior performing plants from alternating space-plant nurseries and turf trials.

Method of Breeding - Polycross containing a diverse germplasm of parental clones selected over several cycles of recurrent selection based on phenotypical performance.

Intended Use - Turf, erosion control, cover crop in orchards, nurseries, vineyards.

Description - Shade tolerant, disease resistant, attractive color, good turf quality, low growing, adapted to low-maintenance situations as well as manicured turf.

Adapted to - Northeast, Pacific northwest, and northern areas.

Released - 1993 by McCarthy-Burlingham Research.

Breeder Seed/Stock - McCarthy-Burlingham Research.

Certified Seed/Stock - Available in 1995.

Preparer/Additional Information - Mike McCarthy, E.F. Burlingham & Sons, P.O. Box 217, Forest Grove, OR 97116, (503) 359-9368.

HF 9032

Northrup King Co., now known as NK Lawn and Garden - Howard E. Kaerwer and Dr. Eric K. Nelson.

Source - F_4 population from progeny of HDUG 6, Scaldic and Centurion x Biljart cross.

Method of Breeding - Mass selection and progeny tests.

Intended Use - HF 9032 was developed for low-maintenance turf, such as home lawns, parks, roadsides, orchards, and golf course roughs.

Description - Excellent germination energy, high-percentage green ground cover, and drought tolerance. Excellent leaf spot and red thread resistance. Pleasant dark green color. Excellent overall quality turf.

Adapted to - LRR A, B, C, D, E, F, G, H, K, L, M, N, O, P, R, S, T, W; PHZ 3, 4, 5, 6, 7, 8.

Released - 1992 by NK Lawn and Garden.

Breeder Seed/Stock - NK Lawn and Garden, Tangent, OR.

Certified Seed/Stock - NK Lawn and Garden, Minneapolis, MN, Tangent, OR, or Chicago, IL.

Preparer/Additional Information - Eric K. Nelson, P.O. Box 300, 33731 Hwy 99E, Tangent, OR 97389, (503) 928-2393.

Nordic

Source - Collections from old-turf areas in U.S. and selections from Biljart and Scaldis.

Method of Breeding - Advanced-generation synthetic from the polycross progenies of 60 clones selected from spaced-plant nurseries of 10 different germplasm composites.

Intended Use - Cool-season turfgrass for moderate- and low-maintenance areas and shady areas.

Description - Nordic has good summer and winter performance, good seedling vigor, and improved resistance to leaf spot, dollar spot, and red thread.

Adapted to - PHZ 2, 3, 4, 5, 6, 7, 8.

Released - 1993, Zajac Performance Seeds.

Breeder Seed/Stock - Zajac Performance Seeds.

Certified Seed/Stock - Available. PVP # pending.

Preparer/Additional Information - Zajac Performance Seeds, 33 Sicomac Road, North Haledon, NJ 07508, (201) 423-1660.

PST-4HD

Source - Endophyte sources topcrossed by Aurora and Waldina.

Method of Breeding - Advanced generation synthetic of 15 clones selected for low-growth habit and high seed yield potential.

Intended Use - Cool-season turf areas, shady areas, and low-maintenance areas.

Description - Performs well in shade; has good winter color, seedling vigor, and high density. Improved resistance to candii stripe rust, red thread, pink patch, summer patch, dollar spot, and anthracnose.

Adapted to - PHZ 2, 3, 4, 5, 6, 7, 8.

Released - Turf Seed Inc.

Breeder Seed/Stock - Pure-Seed Testing, Inc.

Certified Seed/Stock - Not currently available. PVP #9300182.

Preparer/Additional Information - Turf Seed Inc., P.O. Box 250, Hubbard, OR 97032, (503) 651-2130.

Reliant (Reg. No. 25)

Developed and released by Lofts Seed, Inc., using germplasm obtained from the New Jersey AES. Carried as accession no. FL-1.

Source - Cultivars of European origin and germplasm accessions collected from old turfs from the New Jersey Agricultural Experiment Station.

Method of Breeding - A 43-clone synthetic variety developed by screening 30,000 seedlings for resistance to powdery mildew. One hundred thirty-eight of the best plants were cross pollinated in a second cycle, with 43 of the best plants being transplanted to produce breeders seed.

Intended Use - A low-maintenance grass for full sun and moderate shade areas. Adapts to poor, infertile soil where irrigation, fertilization, and mowing are not practical. Works well on slopes, home lawns, parks, roadsides, golf course roughs, and out-of-play areas.

Description - A leafy, persistent, turf-type hard fescue, producing dense, low-growing, fine-textured turf with limited irrigation and fertilization; shows improved resistance to powdery mildew, anthracnose, net blotch, and red thread. Especially useful on poor and infertile soils.

Adapted to - LRR all except O, I, J, P, T, U; PHZ 3, 4, 5, 6, 7.

Released - 1981, Lofts Seed, Inc.

Breeder Seed/Stock - Lofts Pedigreed Seed in co-operation with the New Jersey Agricultural Experiment Station.

Certified Seed/Stock - Available.

Preparer/Additional Information - Lofts Seed Inc., Chimney Rock Road, Box 146, Bound Brook, NJ 08805, (908) 560-1590.

Scaldis

Source - Netherlands collection.

Method of Breeding - Synthetic variety formed by combining clones of like general characteristics. Bred by Van der Have, B.V.

Intended Use - Turf for golf course roughs, shade areas, and low-maintenance areas.

Description - Low nitrogen and water inputs required, shade tolerant, dark green.

Adapted to - PHZ 2, 3, 4, 5, 6.

Released - D.J. Van der Have, B.V., Vlijmen, The Netherlands.

Breeder Seed/Stock - D.J. Van der Have, B.V.

Certified Seed/Stock - Available from Advanta Seeds West, Inc.

Preparer/Additional Information - Kenneth Hignight, Advanta Seeds West, Inc., 33725 Columbus Street S.E., P.O. Box 1496, Albany, OR 97321-0452, (503) 967-8923.

Silvana

Source - Netherlands germplasm.

Method of Breeding - Ecotypes planted and matching types were combined. Syn 1 seed was harvested and turf trials established and evaluated. Top performers were selected and Syn 2 seed produced.

Intended Use - Golf course roughs, shade areas, and low-maintenance areas.

Description - Low nitrogen and water input. Shade tolerance is good, and slow vertical growth.

Adapted to - PHZ 2, 3, 4, 5, 6.

Released - 1973, D.J. Van der Have, The Netherlands.

Breeder Seed/Stock - D.J. Van der Have.

Certified Seed/Stock - Available from Advanta Seeds West, Inc.

Preparer/Additional Information - Kenneth Hignight, Advanta Seeds West, Inc., 33725 Columbus Street S.E., P.O. Box 1496, Albany, OR 97321-0452, (503) 967-8923.

Spartan

Source - Original germplasm utilized for the development of Spartan came from the NJ AES.

Method of Breeding - Synthetic cultivar development via polycross and subsequent progeny testing to attain uniformity.

Intended Use - Cool-season turfgrass.

Description - Evolved from a population improvement program stressing the development of a cultivar having good performance on poor soils.

Adapted to - PHZ 7, 8, and parts of 5 and 6.

Released - 1984 by Pickseed West Inc.

Breeder Seed/Stock - Pickseed West Inc.

Certified Seed/Stock - Available.

Preparer/Additional Information - Pickseed West Inc., P.O. Box 888, Tangent, OR 97389, (503) 926-8886.

SR 3100

Source - SR 3000 breeder block with original seed from Dr. C. Reed Funk contained 100 plants that were dwarfer and darker green. These plants planted in isolation and allowed to intercross.

Method of Breeding - A second generation of 720 plants was established in 1987. Undergoing selection in 1989 for diseases, maturity, dwarf, and dark green habit resulted in 400 plants. Rouged for long-term yield potential in 1990 and 1991 yielding 297 plants which were classified for endophyte level. These were intercrossed with seed from the 161 E plants used for breeder seed.

Intended Use - Cool-season turfgrass in all areas where hard fescues are used for low-maintenance sites and lawns alone or in combination with other species.

Description - Dark green, dwarf hard fescue with high endophyte levels. Reduced mowing and maintenance. Top ranked turf quality in 1989 NTEP.

Adapted to - PHZ 3, 4, 5, 6, 7, 8. NE, Midwest, Pacific Northwest, California.

Released - 1993, Seed Research of Oregon, Inc.

Breeder Seed/Stock - Seed Research of Oregon, Inc.

Certified Seed/Stock - Available.

Preparer/Additional Information - Dr. Leah A. Brilman, Seed Research of Oregon, Inc., P.O. Box 1416, Corvallis, OR 97339, (503) 757-2663.

Waldina

Source - Netherlands collections.

Method of Breeding - Synthetic variety.

Intended Use - Low-maintenance turf sites.

Description - Resistant to red thread, dollarspot, and leaf spot at low levels of water and fertility.

Adapted to - PHZ 2, 3, 4, 5, 6.

Released - D.J. Van der Have, B.V., The Netherlands.

Breeder Seed/Stock - D.J. Van der Have, B.V.

Certified Seed/Stock - Available from Advanta Seeds West, Inc.

Preparer/Additional Information - Kenneth Hignight, Advanta Seeds West, Inc., 33725 Columbus Street S.E., P.O. Box 1496, Albany, OR 97321-0452, (503) 967-8923.

Warwick

Developed for Cascade International Seed Co. and Jonathan Green and Sons Inc. - Dr. C. Reed Funk, New Jersey AES, New Brunswick.

Source - New Jersey AES, Rutgers University. Selected from spaced-plant nurseries from plants improved by recurrent selection.

Method of Breeding - A synthetic variety; selection was based on improved turf quality and disease resistance and highest possible levels of endophyte. Selected clones were planted at Brownsville, OR, in the fall of 1987. Syn II seed was collected and used to establish a breeder field near Lebanon, OR, in the fall of 1988.

Intended Use - Wherever cool-season grasses are adapted for turfgrass areas.

Description - A dark green, prolific variety that is of medium-early maturity. In the National Turfgrass Evaluation Program, it demonstrated outstanding turf quality including improved resistance to red thread and very high percentage of living ground cover in summer and fall.

Adapted to - It has been planted successfully in southern Canada and as far south as North Carolina. It is useful along with other cool-season grasses for improved turf areas, especially under low maintenance and infrequent mowing.

Released - 1993.

Breeder Seed/Stock - Cascade International Seed Co.

Certified Seed/Stock - Available.

Preparer/Additional Information - Irvin H. Jacob, Cascade International Seed Company, 8483 West Stayton Road, Aumsville, OR 97325, (503) 749-1822.

Hemarthria altissima (Poir.) Stapf & C.E. Hubbard - limpograss

Warm-season, productive stoloniferous perennial grass, usually found along edges of swamps in tropical Asia and Africa. Good forage producer and especially palatable in the fall, retaining good quality when feed value of other grasses has declined. Plant may smother if not grazed or harvested periodically.

Bigalta (Reg. No. 54)

Joint introduction by University of Florida Institute of Food and Agricultural Sciences Agricultural Experiment Stations and Plant Materials Center, SCS, Brooksville, FL. Carried as accession no. PI 299995.

Source - Collected in the Limpopo River valley region of Eastern Transvaal in South Africa by A.J. Oakes of the USDA Plant Introduction Service.

Method of Breeding - Selection of a single collected clone.

Intended Use - Where adapted, this grass provides fresh grazing and conserved winter forage following frost. Strong growth in moist areas indicates potential use in filter strips in water quality improvement plantings.

Description - Fewer but broader leaves, larger stems, and has less cold tolerance than Redalta and Greenalta. Similar in appearance to Floralta, but has a less intense purple coloration under cool temperature stress or fertility deficiency. It is the last to initiate growth in the spring and is best adapted to wet soils; it will grow even where the soil surface is intermittently covered with water. Does not tolerate overgrazing. It is tetraploid (2n=36) vegetatively propagated.

Adapted to - LRR T, U; PHZ southeastern 9, 10.

Released - 1978, a joint release of University of Florida Institute of Food and Agricultural Sciences Agricultural Experiment Stations and the Plant Materials Center, SCS, Brooksville, FL.

Breeder Seed/Stock - Florida Foundation Seed Producers, Inc., Greenwood, FL, Agricultural Research Center, Ona, FL, and Plant Materials Center, SCS, Brooksville, FL.

Certified Seed/Stock - Not available.

Preparer/Additional Information - Plant Materials Center, SCS, 14119 Broad St., Brooksville, FL 34601, (904) 796-9600.

Floralta (Reg. No. 107)

University of Florida Institute of Food and Agricultural Sciences, Agricultural Experiment Stations, Gainesville, FL - Drs. Quesenberry, Ocumpaugh, Ruelke, Dunavin, and Mislevy. Carried as accession no. PI 508285.

Source - Collected in the Limpopo River valley region, Kruger National Park, Republic of South Africa by A.J. Oakes, USDA, Plant Introduction Service.

Method of Breeding - Selection of a single clone from a collection.

Intended Use - Grazing and hay production on the poorly drained sandy soils of peninsular Florida.

Description - Floralta is superior to other released limpograsses in total dry matter yield. It is superior to Redalta or Greenalta in *in vitro* organic matter digestion. It was selected for superior persistence under grazing. Similar in appearance to Bigalta with wider leaves and larger stems than Redalta and Greenalta. Under stress, it has more intense purple coloration than Bigalta, but not the intense red of Redalta. It is tetraploid (2n=36). High production, when well fertilized. Competitive with broadleaf and grassy weeds. Adapted to wet soil. Superior late fall and early spring production. Vegetative planting materials dry rapidly when exposed to sun. Crude protein content can be low with inadequate N fertilization. Not well adapted to dry, deep, sandy soils. Difficult to cure for hay at advanced stages of growth.

Adapted to - LRR T, U; PHZ southeastern 9, 10.

Released - 1984, by University of Florida Institute of Food and Agricultural Sciences Agricultural Experiment Stations, Gainesville, FL.

Breeder Seed/Stock - Florida Foundation Seed Producers, Inc., Greenwood, FL, Agricultural Research Center, Ona, FL, and Plant Materials Center, SCS, Brooksville, FL.

Certified Seed/Stock - Not available.

Preparer/Additional Information - Plant Materials Center, SCS, 14119 Broad St., Brooksville, FL 34601, (904) 796-9600.

Greenalta (Reg. No. 53)

Joint introduction by University of Florida Institute of Food and Agricultural Sciences Agricultural Experiment Stations and the Plant Materials Center, SCS, Brooksville, FL. Carried as accession no. PI 299994.

Source - Collected in the Limpopo River valley region of Eastern Transvaal in South Africa by A.J. Oakes of the USDA, Plant Introduction Service.

Method of Breeding - Selection of a single collected clone.

Intended Use - Where adapted, this grass provides fresh grazing and conserved winter forage following frost. Strong growth in moist areas indicates potential use in filter strips in water quality improvement plantings.

Description - It retains a medium dark green color at maturity even under stress conditions. Plant form is similar to Redalta, except leaves are slightly wider. Cold tolerance is also similar. Performs well on wet and alternating wet/dry sites. It is diploid (2n=18) vegetatively propagated.

Adapted to - LRR T, U; PHZ southeastern 9, 10.

Released - 1978, joint release of University of Florida Institute of Food and Agricultural Sciences Agricultural

Experiment Stations and the Plant Materials Center, SCS, Brooksville, FL.

Breeder Seed/Stock: Florida Foundation Seed Producers, Inc., Greenwood, FL, Agricultural Research Center, Ona, FL, and Plant Materials Center, SCS, Brooksville, FL.

Certified Seed/Stock - Not available.

Preparer/Additional Information - Plant Materials Center, SCS, 14119 Broad St., Brooksville, FL 34601, (904) 796-9600.

Redalta (Reg. No. 52)

Joint development by University of Florida Institute of Food and Agricultural Sciences Agricultural Experiment Stations and the Plant Materials Center, SCS, Brooksville, FL. Carried as accession no. PI 299993.

Source - Collected in the Limpopo River valley region of Eastern Transvaal in South Africa by A.J. Oakes of the USDA, Plant Introduction Service.

Method of Breeding - Selection from a collection.

Intended Use - Where adapted, this grass provides fresh grazing and conserved winter forage following frost. Strong growth in moist areas indicates potential use in filter strips in water quality improvement plantings.

Description - It has exhibited the most cold resistance of the limpograsses, having survived winter temperatures of -18°C at Coffeeville, MS. However, best growth is from upper central Florida south. Distinguished from other limpograsses by characteristic red color at advanced stages of growth or under stress conditions, and usually finer stems and narrower leaves. It is diploid (2n=18) vegetatively propagated and presently is the best adapted variety for drier upland sands. Also, it does well on wetter sites.

Adapted to - LRR T, U; PHZ southeastern 9, 10.

Released - 1978, joint release of University of Florida Institute of Food and Agricultural Sciences, Agricultural Experiment Stations and the Plant Materials Center, SCS, Brooksville, FL.

Breeder Seed/Stock - Florida Foundation Seed Producers, Inc., Greenwood, FL, Agricultural Research Center, Ona, FL, and Plant Materials Center, SCS, Brooksville, FL.
Certified Seed/Stock - Not available.

Preparer/Additional Information - Plant Materials Center, SCS, 14119 Broad St., Brooksville, FL 34601, (904) 796-9600.

Heteropogon contortus (L.) Beauv. ex. Roem. & J.A. Schultes - tanglehead

This tufted perennial grass is found in rocky hills and canyons from Texas to Arizona within the U.S. where it is utilized as a forage species. Called *pili* in the Hawaiian Islands, it is an important range grass in the drier areas and was used as thatch. Also present throughout the tropical-to-warmer regions of the world.

Rocker

Source - Originally collected from a native stand on the Rocker M ranch, Cochise County, Arizona, in October 1984, at T24S, R9E, SW1/4 of Sec. 8. approximately 275 m north of Geronimo Trail at an elevation of 1,280 m.

Method of Breeding - Tanglehead is an obligate apomict. Rocker was chosen out of a mass selection block in 1991.

Intended Use - Control rill and gully erosion on degraded rangeland, critical areas, abandoned cropland, and disturbed roadside areas.

Description - Large, robust plant; vigorous growth; later flowering period; tolerance of drought-like conditions; and early spring greenup.

Adapted to - PHZ 7, 8, 9 - southwest and south U.S. (Arizona to Texas) and PHZ 11 in the Hawaiian Islands.

Released - 1992, Tucson Plant Materials Center, SCS.

Breeder Seed/Stock - Tucson Plant Materials Center.

Certified Seed/Stock - Available.

Preparer/Additional Information - Bruce Munda or Mark Pater, Tucson Plant Materials Center, SCS, 3241 N. Romero Road, Tucson, AZ 85705, (602) 670-6491.

Hilaria jamesii (Torr.) Benth. - galleta grass

Slightly spreading, native grass. Found from Wyoming and Nevada south to California and western Texas. Produces good forage yields in New Mexico and Arizona. Palatable during summer growing season, but harsh and unpalatable after growth ceases. Spreads by rhizomes under favorable conditions. Drought resistant; tolerant of heavy grazing.

Viva

Selected at Plant Materials Center, SCS, Los Lunas, NM - G.C. Niner, J.E. Anderson, and W.R. Oaks. Carried as accession nos. PI 476995, A-12413.

Source - Collected in 1944 by an SCS field crew from a native stand near New Kirk, NM, at an elevation of 1,750 m and 230 mm precipitation.

Method of Breeding - Bulk increase of source material. Selected in comparison rows and field studies for improved seedling vigor, herbage production, and seed production.

Intended Use - In range seedings, mined land reclamation and erosion control, revegetation plantings such as highway and pipeline rights-of-way, and other disturbed sites.

Description - Improved seedling vigor and higher seed production differentiates Viva from other strains. Although a southern geographic line, Viva has excellent cold tolerance and drought tolerance as far north as the Canadian border.

Adapted to - LRR D, E, G, F, H; PHZ 4.

Released - Formally released in 1979, cooperatively with SCS and New Mexico and Colorado Agricultural Experiment Stations.

Breeder Seed/Stock - Plant Materials Center, SCS, Los Lunas, NM.

Certified Seed/Stock - Available.

Preparer/Additional Information - Plant Materials Center, SCS, 1036 Miller St. SW, Los Lunas, NM 87031, (505) 865-4684.

***Hordeum vulgare* L. - cereal barley**

Widely domesticated cool-season cereal grass adapted to temperate regions of the world. In addition to its primary use as a food and forage crop, it also has high conservation value, with some cultivars (listed elsewhere in this publication) developed specifically for use as winter cover crops, green manure, and erosion control on disturbed sites.

Seco

Developed by ARS - R.T. Ramage. Carried as accession no. PI 508552.

Source - World barley collection.

Method of Breeding - Male sterile facilitated recurrent selection from composite XXXIX.

Intended Use - Wildlife habitat improvement and erosion control on abandoned cropland and critical areas in low-moisture areas.

Description - Robust, six-rowed, rough-awned, spring barley. Culms are erect, 75-120 cm tall. The spike is lax and non-waxy. The root crown is 2.5-5.0 mm below the soil surface, and is capable of ratooning. Root development extends beyond 1.8 m under favorable conditions.

Adapted to - LRR B, D, F; PHZ 4, 5, 6, 7, 8, 9, 10.

Released - 1987, SCS, AZ AES and ARS.

Breeder Seed/Stock - ARS, Tucson, AZ.

Certified Seed/Stock - Available.

Preparer/Additional Information - Plant Materials Center, SCS, 3241 N. Romero Rd., Tucson, AZ 85705, (602) 241-2966.

***Koeleria macrantha* (Ledeb.) J.A. Schultes - crested hairgrass**
***Koeleria cristata* auct. pro pate non Pers.**

A tufted perennial found as a scattered component of the western North America range south to Mexico and utilized as forage. Also widely distributed in the temperate regions of Eurasia. Very plentiful in western Mongolia.

Barkoel

Source - Ecotypes.

Method of Breeding - Selection in clones after three years of testing under turf conditions. Clones were chosen on the basis of a good winter color and good seed yield.

Intended Use - Cool-season turfgrass. Suitable for low-maintenance areas, lawns, and extensive greens.

Description - The variety can withstand dry conditions and has a good, dense turf quality. Performs well with less nitrogen.

Released - Barenbrug.

Breeder Seed/Stock - Barenbrug Holland, Oosterhout, The Netherlands.

Certified Seed/Stock - Available from Barenbrug USA.

Preparer/Additional Information - Barenbrug USA, P.O. Box 239, Tangent, OR 97389, (503) 926-5801.

***Leptochloa dubia* (Kunth) Nees - green sprangletop**

Green sprangletop is a native, warm-season, essentially biennial grass species that, due to its relative ease of establishment, rapid growth, and palatability to livestock, is included in many range seedings to take the grazing pressure off of the slower-developing perennial species until they are well established. Culms wiry, erect, 50-100 cm tall; panicle of few to many spreading or ascending racemes 3-12 cm long. Rocky hills and canyons; sandy soil, as in southern Florida; Oklahoma and Texas to Arizona, south through Mexico; Argentina. Racemes of cleistogamous spikelets are often found in the sheaths.

Van Horn

Selected in 1980 at the James E. "Bud" Smith Plant Materials Center, Knox City, TX. Increased as T-3926 for field testing. Carried as accession no. PI 441106.

Source - Original seed collected from native stand near Van Horn, TX, by SCS employees Rhett Johnson and Steve Holtz in 1975. Carried as accession no. PI 441106.

Method of Breeding - Selected from among 26 similar accessions and cultivars for better forage quality and seed production.

Intended Use - Range reseeding and revegetation of disturbed sites.

Description - Cures with higher forage quality which is retained well into winter in drier regions. Often reaches 5-6 ft. height on favorable sites. Excellent "nurse" crop for slower establishing perennials.

Adapted to - LRR D, G, H, I, J, T; PHZ 7.

Released - 1989 by SCS, Texas Agricultural Extension Service and ARS.

Breeder Seed/Stock - Plant Materials Center, SCS, Knox City, TX.

Certified Seed/Stock - Not available.

Preparer/Additional Information - Plant Materials Center, SCS, Rte 1 Box 155, Knox City, TX 79529, (817) 658-3922.

Leymus angustus **(Trin.) Pilger - Altai wildrye**

Cool-season, long-lived perennial grass native to western Siberia and the Altai mountain region that borders a portion of the People's Republic of China from the Mongolian Peoples' Republic. Tolerant of cold, drought, and saline conditions. Adapted to clay and clay-loam soils of the Canadian prairie region and the northern Great Plains and intermountain-western regions of the U.S.. Mildly rhizomatous, roots penetrate 3.0-4.3 m and use moisture to that depth. Coarse, erect basal leaves retain much of their nutritive value throughout the summer and fall. Useful for fall and winter grazing; leaves project above shallow snow and stay erect in deep snow, making them accessible to grazing cattle. Slow-developing seedling can result in establishment problems.

Eejay

Selected at Agriculture Canada Research Station Swift Current, Saskatchewan - Tom Lawrence. Carried as accession no. Sc A3772.

Source - Eejay traces back to two introductions from the the former USSR: Ottawa No. 1927-117 was collected by A.C. Woronow on the Steppe of Kustanay in 1937. Ottawa No. 1114 was collected at Voronezh in the former USSR.

Method of Breeding - Eejay, a 12-clone synthetic, was developed through four cycles of recurrent selection for high seed yield, high forage yield, freedom from leaf spot diseases, and good seed quality within a green (nonglaucous) population.

Intended Use - Eejay is similar to Prairieland in end-use characteristics. It has good curing quality and is used for late fall and early winter grazing on the northern Great Plains and intermountain region. It is salt-tolerant and adapted to clay-loam and loam sites with a high water table. It is productive in wet, saline meadows.

Description - Eejay produces on average 8% more forage and 15% more seed than Prairieland. It is similar to Prairieland in morphological appearance except that a greater proportion of Eejay plants are distinctly nonglaucous.

Adapted to - LRR F, G; PHZ 2, 3, 4.

Released - 1989 by Agriculture Canada.

Breeder Seed/Stock - Agriculture Canada Experimental Farm, Indian Head, Saskatchewan SOG 2K0.

Certified Seed/Stock - SeCan Association, Suite 512, 885 Meadowlands Dr., Ottawa, Ontario K2C 3M2. Registration for sale in Canada is pending sufficient breeders seed production in 1989.

Preparer/Additional Information - Paul G. Jefferson, Agriculture Canada, Research Station, P.O. Box 1030, Swift Current, Saskatchewan S9H 3X2, (306) 773-4621.

Pearl

Developed at Agriculture Canada Research Station, Swift Current, Saskatchewan - Tom Lawrence. Carried as accession number Sc A3761.

Source - Pearl originates from two plant introductions from the former USSR: Ottawa No. 1927-117 was collected by A.C. Woronow on the Steppe of Kustanay in 1937. Ottawa No. 1114 was collected at Voronezh in the former USSR.

Method of Breeding - Pearl is a four-clone synthetic from the fourth cycle of recurrent selection for high seed yield, high emergence from deep seeding, high forage yield, freedom from leaf spot diseases, and good seed quality within a blue (glaucous) population.

Intended Use - Pearl has good curing qualities for late fall and early winter grazing on the northern Great Plains and intermountain region. It is productive on wet saline meadows, is salt tolerant, and is best adapted to clay and clay-loam sites with a high water table.

Description - Pearl yields 6% less forage but 23% more seed than Prairieland. It is morphologically similar to Prairieland except that a majority of plants exhibit a glaucous appearance.

Adapted to - LRR F, G; PHZ 2, 3, 4.

Released - 1989 by Agriculture Canada.

Breeder Seed/Stock - Agriculture Canada Research Station, Swift Current, Saskatchewan S9H 3X2.

Certified Seed/Stock - SeCan Association, Suite 512, 885 Meadowlands Dr., Ottawa, Ontario K2C 3M2. Registration for sale in Canada is pending sufficient breeders seed production in 1989.

Preparer/Additional Information - Paul G. Jefferson, Agriculture Canada, Research Station, P.O. Box 1030, Swift Current, Saskatchewan S9H 3X2, (306) 773-4621.

Prairieland (Reg. No. 55)

Selected at Agriculture Canada Research Station, Swift Current, Saskatchewan - Tom Lawrence. Carried as accession number Sc 3717.

Source - Prairieland originates from two introductions from the former USSR: Ottawa No. 1927-117 was collected by A.C. Woronow on the Steppe of Kustany in 1937; Ottawa No. 1114, from Voronezh, was brought to Canada by Dr. H. T. Gussow in 1934.

Method of Breeding - Prairieland, a 22-clone synthetic, was developed by recurrent selection among spaced plants for high forage yields, high seed yields and seed quality, and resistance to leaf spot diseases. Recurrent selection was practiced for seedling emergence from a 5 cm depth in soil in a greenhouse.

Intended Use - Prairieland has good curing qualities and generally is used for fall and winter grazing in the Canadian prairie region and the northern Great Plains and intermountain-west regions of the U.S. Prairieland is relatively salt tolerant and is best adapted to clay-loam and clay soils.

Description - Prairieland, the first known cultivar of Altai wildrye, has much higher seed yields than unselected populations of this species. Prairieland has characteristic green foliage color but no other visual characteristics that distinguish it from other experimental strains of Altai wildrye.

Adapted to - LRR B, D, E, F, G; PHZ 2, 3, 4.

Released - 1976, by Agriculture Canada.

Breeder Seed/Stock - Agriculture Canada, Research Station, Swift Current, Saskatchewan S9H 3X2.

Certified Seed/Stock - Phillips Seed Farm Ltd., Box 249, Tisdale, Saskatchewan SOE 1TO.

Preparer/Additional Information - Paul G. Jefferson, Agriculture Canada, Research Station, P.O. Box 1030, Swift Current, Saskatchewan S9H 3X2, (306) 773-4621.

Leymus arenarius **(L.) Hochst. - beach wildrye**

Coarse and very glaucous culms, extensive, creeping rootstocks, can grow to 1.5 m in height. Seed head is dense, stiff 10-30 cm long and 1-3 cm wide. Similar to E. mollis Trin. Occurs on sandy beaches in arctic North America and in Greenland to the coast of Massachusetts, the shores of Lake Michigan, and Lake Superior.

Reeve

Source - Mass selection from a 6-m row of PI 345978, originally received from the U.S. Plant Introduction Office. The original material was collected by the University of Oslo Botanical Garden in Norway.

Method of Breeding - Open-pollinated increased from original selection.

Intended Use - Revegetation of sandy soils, such as stabilization of coastal sand dunes.

Description - Erect, perennial, strongly rhizomatous, sod-forming grass growing to 100-130 cm tall. Reeve is more rigid and has a bluer color than Alaskan collections of this species. Unlike other collections, Reeve produces commercially useable quantities of seed.

Adapted to - South-central to coastal Alaska and the Aleutian Islands. LRR W, Y (southern half); PHZ 3, 4, 5, 6, 7, 8.

Released - 1991, by Alaska Department of Natural Resources.

Breeder Seed/Stock - Alaska Plant Materials Center.

Certified Seed/Stock - Available.

Preparer/Additional Information - Stoney Wright, Manager, Alaska Plant Materials Center, State of Alaska, HC 02 Box 7440, Palmer, AK 99645, (907) 745-4469.

Leymus cinereus **(Scribn. & Merr.) A. Love - basin wildrye**

Elymus cinereus **Scribn. and Merr.**

Cool-season, slightly spreading, native grass. Distributed throughout much of the western U.S.; common on alkaline soils. Tall, coarse, long lived. Relatively poor in palatability and seed set.

Magnar

Selected at Plant Materials Center, SCS, Pullman, WA - J.L. Schwendiman. Carried as accession no. P-5797 and PI 469229.

Source - University of Saskatchewan, Saskatoon, Saskatchewan, Canada, in 1938.

Method of Breeding - Selection of vigorous types during several generations.

Intended Use - Range forage for livestock and wildlife, thermal protection for livestock and wildlife, and wind barrier.

Description - Robust, vigorous, blue, tall-growing, leafy. Broad, coarse leaves. Large stems and seedheads; good seed production; seed grows rapidly. Good seedling vigor. Grows well on saline soils.

Adapted to - LRR B, D, F, G; PHZ 4, 5.

Released - 1979, Plant Materials Center, SCS, Aberdeen, ID.

Breeder Seed/Stock - Plant Materials Center, SCS, Aberdeen, ID.

Certified Seed/Stock - Available.

Preparer/Additional Information - Gary Young, Plant Materials Center, SCS, Box 296, Aberdeen, ID 83210-0296, (208) 397-4133.

Trailhead

Increased at Plant Materials Center, SCS, Bridger, MT - A.A. Thornburg. Carried as accession nos. PI 478831, P-15590, M-27.

Source - Collected near Roundup, MT, on a sub-irrigated range site by Sterle Dale in 1960.

Method of Breeding - Direct increase of field collection. Selected from among 125 accessions representing native collections in Montana and Wyoming.

Intended Use - Early spring and winter grazing in dryland or sub-irrigated conditions. Adapted for use in native mixtures for reclamation of drastically disturbed areas. Has potential use in grass row barriers.

Description - Vigorous, disease-free, dark green, excellent early-spring forage production. Abundant broad, semi-lax, leaves. Judged as palatable to livestock. Good seedling vigor. Seed production fair. Adapted to moderately saline-alkaline to acid soils. More drought tolerant than other cultivars of basin wildrye.

Adapted to - LRR B, D, F, G; PHZ 3.

Released - 1991, cooperatively by Plant Materials Center, SCS, Bridger, MT, and Montana and Wyoming Agricultural Experiment Stations.

Breeder Seed/Stock - Plant Materials Center, SCS, Bridger, MT.

Certified Seed/Stock - Available.

Preparer/Additional Information - John G. Scheetz, Plant Materials Center, SCS, RR 1 Box 1189, Bridger, MT 59014, (406) 662-3579.

Leymus mollis (Trin.) Hara - American dunegrass

***Elymus mollis* Trin.**

Strong, stout culms, 60-120 cm tall, with many overlap basal ping leaves, vigorous, widely creeping rhizomes, and firm blades 7-12 mm wide. Occurs on northern coastal dunes in North America south to central California, and on the coasts of Japan and Siberia. It is also found along the shores of the Great Lakes. Although the native stands are course and somewhat open, they produce excellent sand stabilization from wind erosion. It is very similar to *Elymus arenarius* L., European dunegrass. Its principal value is in cold climates where *Ammophila breviligulata* Fern., American beachgrass, or *A. arnaria* (L.) Link, European beachgrass, are not adapted.

Benson

Source - Parental seed collected from an isolated stand on a beach segment near the U.S. Coast Guard Loran Station near Narrow Cape, Kodiak Island, Alaska.

Method of Breeding - Vegetatively propagated by separating and transplanting rhizomatous sprigs. Benson does not produce commercially useable quantities of seed.

Intended Use - Revegetation and reclamation of sandy coastal area where natural erosion prevents traditional seedings.

Description - Erect, leafy, perennial, rhizomatous, sod-forming grass reaching 95-110 cm in height. Distinctly greener color than Reeve. Low seed production; propagated with sprigs.

Adapted to - Alaskan coastal areas southwesterly through the Aleutian Islands and north to Kotzebue. LRR W, Y (south of Brooks Range); PHZ 3, 4, 5, 6, 7, 8.

Released - 1991, Alaska Department of Natural Resources.

Breeder Seed/Stock - Alaska Plant Materials Center.

Certified Seed/Stock - Available in limited quantities.

Preparer/Additional Information - Stoney Wright, Manager, Plant Materials Center, State of Alaska, HC 02 Box 7440, Palmer, AK 99645, (907) 745-4469.

***Leymus racemosus* (Lam.) Tzvelev - mammoth wildrye**
Elymus giganteus Vahl.

Cool-season, sod-forming grass from the former USSR. Used for inland dune stabilization in Pacific northwest and occasionally for ornamental purposes. Drought resistant, coarse. Unpalatable; not adapted for forage.

Volga (Reg. No. 9)

Selected at Plant Materials Center, SCS, Pullman, WA. Carried as accession no. PI 108491.

Source - Lower Volga region of the former USSR and collected by Westover-Enlow expedition in 1934.

Method of Breeding - Selection of most vigorous plants from above introduction during several generations followed by vegetative reproduction of most desirable type. Tested as P-208.

Intended Use - Sand dune stabilization and soil stabilization.

Description - Tall, coarse, green, creeping. Non-palatable to livestock. Long lived on inland sand dunes, where it will stop sand movement and provide permanent cover. Grown from seed or propagated vegetatively. Rate of increase from culms under proper cultural conditions is 15-to-1 in the first year. When established from seed, seedlings show excellent vigor.

Adapted to - LRR B, D, E; PHZ 4, 5.

Released - Vegetative material distributed in 1949 for inland dune control by Plant Materials Center, SCS, Pullman, WA. Seed not released.

Breeder Seed/Stock - Vegetative material and seed from bulked selections, SCS, Plant Materials Center, Pullman, WA.

Certified Seed/Stock - Plant Materials Center, SCS, Meeker, CO.

Preparer/Additional Information - Gary Young, Plant Materials Center, SCS, Box 296, Aberdeen, ID 83210-0296, (208) 397-4133.

***Leymus triticoides* (Buckl.) Pilger - beardless wildrye**
Elymus triticoides Buckl.

Cool-season, sod-forming native grass. Distributed at low and medium elevations from Montana to Washington and south to west Texas and California. Important rangegrass.

Rio

Selected at Plant Materials Center, SCS, Lockeford, CA - K. Croeni, R. Slayback, and D. Dyer.

Source - Collected in 1973 from native stand in Stratford, Kings County, CA, by Clarence Finch.

Method of Breeding - Direct increase of field collection after initial evaluation.

Intended Use - Soil stabilization on channel, stream, and river slopes and restoration of roadside, riparian, and rangeland areas.

Description - Rio is a native, cool-season, perennial grass with superior seed viability and vigorous rhizomes.

Adapted to - PHZ 8, 9, 10, 11 - central and coastal California.

Released - 1991, cooperatively by California AES and Plant Materials Center, SCS, Lockeford, CA.

Breeder Seed/Stock - Plant Materials Center, SCS, Lockeford, CA.

Certified Seed/Stock - Available.

Preparer/Additional Information - Robert D. Slayback, 2121-C, 2nd Street, Davis, CA 95616, (916) 757-8257.

Shoshone

Increased at Plant Materials Center, SCS, Bridger, MT - A.A. Thornburg. Carried as accession nos. PI 434040, WY-5, P-15594.

Source - Collected in 1958 at Riverton, WY, fairgrounds by J.L. McWilliams and Aubry Stanton. Possibly a seeded stand established in the 1940s from an unknown source.

Method of Breeding - Direct increase of field collection.

Intended Use - Shoshone is used primarily for forage, stabilization, or cover on wet or wet-saline-alkaline soils. This includes pastureland; saline-affected, irrigated cropland; and dryland, saline-seep discharge areas.

Description - An exceptionally leafy, fine-stemmed, high forage producer; rhizomes especially vigorous, extending 1.8 m in one season; leaves broad, lax, and dark green. Comparatively high seed production for this species; seed plump and heavy. Seed dormancy requires fall dormant planting in a moist seedbed.

Adapted to - LRR B, D, E, G; PHZ 3.

Released - 1980, cooperatively by Plant Materials Center, SCS, Bridger, MT, and the Montana and Wyoming Agricultural Experiment Stations.

Breeder Seed/Stock - Plant Materials Center, SCS, Bridger, MT.

Certified Seed/Stock - Available.

Preparer/Additional Information - John G. Scheetz, Plant Materials Center, SCS, RR 1, Box 1189, Bridger, MT 59014, (406) 662-3579.

Lolium L. - *Festuca* L. derivatives - hybrid tall fescue-ryegrass

Hybrids derived from crosses involving *L. perenne* ssp. *multiflorum* (Lam.) Husnot x *F. arundinacea* Schreb., *L. perenne* L. x *F. arundinacea* Schreb., and *L. perenne* L. x *F. pratensis* Huds. outcrossed to *F. arundinacea* Schreb. Other combinations may be used in producing hybrids between these two genera.

Bison

Developed by International Seeds Inc., Halsey, OR, and Michigan State University - Dr. Fred Elliot.

Source - Tetrelite intermediate ryegrass.

Method of Breeding - Winter-hardy lines selected from Tetrelite intermediate ryegrass were crossed and the resulting seed was increased.

Intended Use - High-quality pasture and hay production for cattle, sheep, and horses. Used with alfalfa as a nurse crop. Also used for renovating older alfalfa stands.

Description - Tall, broad-leafed, vigorous tetraploid; moderately cold tolerant and very productive. Life of stand is one to five years depending on environment and management. Bison is a 2N=4X=28 tetraploid.

Adapted to - LRR A, K, L, M, N, P, R, S, T, W; PHZ 3.

Released - 1983 by International Seeds Inc.

Breeder Seed/Stock - Maintained by International Seeds Inc. PVP No. 8400038.

Certified Seed/Stock - Available from International Seeds, Inc.

Preparer/Additional Information - Stephen W. Johnson, International Seeds Inc., P.O. Box 168, 820 W. First St., Halsey, OR 97348, (503) 369-2251.

Johnstone (Reg. No. 23)

ARS and University of Kentucky - R.C. Buckner, J.A. Boling, P.B. Burrus, Jr., L.P. Bush, and R.A. Hemken. Carried as accession no. PI mixture of G1-316 and G1-307(318).

Source - Blend of syn-2 seed from two low perloline strains accessioned G1-316 and G1-307 (318).

Method of Breeding - Accession G1-316 consists of seven parental clones of Kenhy and accessions G1-307 and/or G1-318 consist of 29 parental clones derived by outcrossing *F. annual* and perennial ryegrass and tall fescue hybrids to their 2n=56 chromosome amphiploids and to tall fescue.

Intended Use - Pasture and hay.

Description - Johnstone has from 50-60% less perloline content than Kenhy or Kentucky 31.

Adapted to - LRR L, M, N, O; PHZ 2, 3, 4, 5, 6, 7.

Released - 1983 by Kentucky AES and the ARS.

Breeder Seed/Stock - Kentucky AES; Lexington, KY.

Certified Seed/Stock - Available commercially.

Preparer/Additional Information - Jeff Pedersen, Agronomy Department, University of Kentucky, Lexington, KY 40542-0091, (606) 257-3144.

Kenhy (Reg. No. 12)

ARS and University of Kentucky - R.C. Buckner, P.B. Burrus, Jr., and L.P. Bush.

Source - The 42-chromosome derivatives of annual ryegrass x tall fescue hybrids.

Method of Breeding - Kenhy is a synthetic of progenies of 11 42-chromosome derivatives of annual ryegrass x tall fescue hybrids which were selected for plant vigor, soft lax leaves, and high moisture content for forage during drought stress.

Intended Use - Pasture and hay production.

Description - Kenhy has seedling vigor characteristics of tall fescue. When managed as hay and pasture, it had 12% higher dry matter yields than Kentucky 31. It is higher in digestibility, and lower in crude fiber and lignin than Kentucky 31. Stomatal frequency is lower than for Kentucky 31.

Adapted to - LRR L, M, N, O; PHZ 2, 3, 4, 5, 6, 7.

Released - 1977 by Kentucky AES and the ARS.

Breeder Seed/Stock - Kentucky AES, Lexington, KY.

Certified Seed/Stock - Available commercially.

Preparer/Additional Information - Jeff Pedersen, Agronomy Department, University of Kentucky, Lexington, KY 40502-0091, (606) 257-3144.

Tetrelite

Developed by Michigan State University.

Source - Annual and perennial ryegrass germplasm assembled at East Lansing, MI.

Method of Breeding - Select annual and perennial plants crossed, F_1, seedlings treated with colchicine to produce tetraploids which were interpollinated. Winter-hardy selections were made to form the basis for the variety.

Intended Use - Pasture and hay production, planted alone or in combination with legumes. Also used for silage and haylage.

Description - Vigorous, broad-leafed tetraploid intermediate; moderately winter-hardy; acts as a short-lived perennial in most environments; highly productive and very palatable. Used extensively in South America.

Adapted to - LRR A, K, L, M, N, P, R, S, T, W; PHZ 3.

Released - 1969 by Michigan State University.

Breeder Seed/Stock - Maintained by Michigan State University.

Certified Seed/Stock - Available in quantity from International Seeds, Inc.

Preparer/Additional Information - Stephen W. Johnson, International Seeds Inc., P.O. Box 168, 820 W. First St., Halsey, OR 97348, (503) 369-2251.

Lolium perenne L. - perennial ryegrass

Important cool-season bunchgrass from Europe. Well adapted in Pacific northwest. Widely used in mixtures for pasture, hay, lawns, and erosion control. Does best in cool, moist regions with mild winters; grows well on heavy soils; tolerates heavy grazing. Nutritious, palatable. Not recommended as lawngrass.

246-Sunrye

Source - Selections from Citation II, Manhattan II, and collections from old turf areas in New Jersey.

Method of Breeding - Advanced generation synthetic variety from four cycles of selection for high seed head number and disease resistance.

Intended Use - Cool-season turfgrass; also for overseeding of dormant warm-season turfs in southern U.S.

Description - Increased seed yield provides good transition in overseeding areas.

Adapted to - PHZ 3, 4, 5, 6, 7, 8, 9 10.

Released - Turf-Seed, Inc.

Breeder Seed/Stock - Pure Seed Testing, Inc.

Certified Seed/Stock - Available. PVP No. 8700084.

Preparer/Additional Information - Turf-Seed, Inc., P.O. Box 250, Hubbard, OR 97032, (503) 651-2130.

832

Developed by International Seeds Inc., Halsey, OR. Carried as accession no. ISI-832.

Source - Elka and selections from other commercial varieties.

Method of Breeding - Mass selection for several generations.

Intended Use - Permanent turf in all areas where improved turf-type perennial ryegrass is adapted. Also recommended for use in overseeding dormant bermudagrass turfs, including golf course greens, tees, and fairways. May be used in turfgrass blends and mixtures.

Description - Dark green turf-type variety that is quick to establish in both permanent turf and overseed situations. Shows good year-round vigor and growth which allows it to recover from wear and disease. Resistant to stem rust incited by *Puccinia graminis.* In seed production, it is medium-early for heading and shows good seed yield capacity.

Adapted to - LRR A, C, L, M, N, R, S, W, permanent turf. D, I, J, P, T, U, overseeding of dormant warm season turfs; PHZ 5, southern half of 4 east of the Rocky Mountains.

Released - 1990 by International Seeds Inc., Halsey, OR.

Breeder Seed/Stock - Maintained by International Seeds Inc.

Certified Seed/Stock - Certified and uncertified seed available from International Seeds Inc.

Preparer/Additional Information - Stephen W. Johnson, International Seeds Inc., P.O. Box 168, 820 W. First St., Halsey, OR 97348, (503) 369-2251.

856

Developed by International Seeds Inc., Halsey, OR. Carried as accession no. ISI-856.

Source - The ryegrass varieties Elka and Manhattan and an experimental breeding composite derived from selections out of commercial varieties.

Method of Breeding - Multiple cycles of mass selection in spaced plant nurseries.

Intended Use - Permanent turf in areas where perennial ryegrass is adapted. Also recommended for use in fall overseeding of warm-season grass turfs in the southern U.S.. May be sown alone or in turfgrass blends and mixtures.

Description - Medium-early maturing turf-type perennial with a high degree of resistance to stem rust and improved seed yield potential. Capable of producing quality turf across a wide range of environments.

Adapted to - LRR A, C, L, M, N, R, S, W, permanent turf. C, D, I, J, P, T, V, overseeding dormant warm season turf; PHZ 5 plus southern half of 4 east of Rocky Mountains.

Released - 1988 by Mid Valley Agricultural Products, Inc., Corvallis, OR.

Breeder Seed/Stock - Maintained by International Seeds Inc.

Certified Seed/Stock - Available from Mid Valley Agricultural Products, Inc. PVP No. 890047.

Preparer/Additional Information - Stephen W. Johnson, International Seeds Inc., P.O. Box 168, 820 W. First St., Halsey, OR 97348, (503) 369-2251.

Accolade

Source - Old turf in New Jersey pollinated with source material from Premier perennial ryegrass; germplasm developed at Rutgers University.

Method of Breeding - Synthetic variety.

Intended Use - Permanent turf; cool-season and fall overseeding of dormant warm-season turf. May be seeded alone or as a component in blends or mixtures.

Description - Early maturing, dark green, good mowing qualities with high endophyte level.

Adapted to - PHZ permanent turf - 4, 5, 6; winter overseeding dormant warm-season turf 7, 8, 9, 10.

Released - 1988, O.M. Scott & Sons Co.

Breeder Seed/Stock - O.M. Scott & Sons Co.

Certified Seed/Stock - Available.

Preparer/Additional Information - Virgil Meier, O.M. Scott & Sons Co., Marysville, OH 43041, (513) 644-0011.

All*Star (Reg. No. 108)

Developed cooperatively by the New Jersey AES and International Seeds Inc., and J & L Adikes. Carried as accession no. PI 509070.

Source - Maternal plants from old turf areas in Baltimore and College Park, MD; pollen sources were from a diverse polycross of plant material from California, France, Maryland, and New Jersey.

Method of Breeding - Seedlings of the parental accessions were transplanted into spaced-plant nursery. This crossing block was rogued, harvested, re-evaluated, multiplied, and released. Polycross progeny evaluated in turf trials. Synthetic variety developed by mass selection.

Intended Use - Permanent turf in all areas where improved turf-type perennial ryegrasses are adapted. Also suited for use in overseeding dormant warm-season turfs. May be seeded alone or used as a component in blends and mixtures.

Description - Leafy, persistent, turf-type capable of producing an attractive, medium-fine textured turf with a bright, medium green color. Exhibits good wear tolerance and winter-hardiness along with improved summer performance. Good resistance to brown patch disease. In seed production fields, All*Star is late maturing variety. Many lots of All*Star contain a high percentage of the endophyte fungus, associated with resistance to many turf insects including chinch bugs and sod webworms.

Adapted to - LRR A, C, E, K, M, L, R, S, T, N; PHZ 2, 3, 4, 5, 6.

Released - August 1983 by International Seeds, Inc. and Jacklin Seed Company, Post Falls, ID. U.S. PVP No. 8300059.

Breeder Seed/Stock - Maintained by Jacklin Seed Co.

Certified Seed/Stock - Available through J&L Adikes, International Seeds, and Jacklin Seed Co.

Preparer/Additional Information - Mark Sellmann, Jacklin Seed Co., W. 5300 Riverbend Ave., Post Falls, ID 83854, (208) 773-7581; or International Seeds Inc., P.O. Box 168, 820 W. First St., Halsey, OR 97348, (503) 369-2251.

Allaire I

KWS-AG, Einbeck, Germany.

Source - Collections from old pastures and turf areas in southeastern Europe plus selected clones of Manhattan II and Tando.

Method of Breeding - Clonal progeny evaluation, polycross breeding, recurrent mass selection of most desirable types. Reselected out of Allaire for more disease resistance to leaf spot and crown rust.

Intended Use - Permanent ornamental, recreational, and athletic turf; winter overseeding of dormant bermudagrass.

Description - Semi-upright, fine textured, moderately dark green; forms a dense sward with excellent mowing characteristics; early in maturity.

Adapted to - Throughout the northern cool, humid region; PHZ 4, 5, 6, (7, 8 winter overseeding).

Released - 1986, by KWS-AG.

Breeder Seed/Stock - KWS-AG.

Certified Seed/Stock - Distributed by TMI, Tangent, OR.

Preparer/Additional Information - Fred B. Ledeboer, Ph.D., TMI, 22068 Case Road NE, Aurora, OR 97002, (503) 678-2597.

Aquarius

KWS-AG, Einbeck, West Germany.

Source - Selected clones from old turf in Idaho and clonal selections out of an unidentified Manhattan II source, and Tando.

Method of Breeding - Crossing of the Idaho clones with Manhattan II clones and back-crossing to four Manhattan II clones to improve color; progeny testing in single-plant progeny turf plots, polycrossing and two cycles of recurrent mass selection for most desirable types.

Intended Use - Primarily for closely mown ornamental turf in the cool areas of the world with cool nights and moderately low humidity; also well suited for winter overseeding of sub-tropical grasses on golf courses, bowling greens, and home lawns.

Description - Very soft, semi-upright with very fine texture and very high sward density; very dark color, tolerant of lower mowing heights than most other turf type perennial ryegrasses, medium late in maturity.

Adapted to - Throughout the northern cool, humid region; PHZ 4, 5, 6, (7, 8 winter overseeding).

Released - 1988, KWS-AG.

Breeder Seed/Stock - KWS-AG.

Certified Seed/Stock - Distributed by TMI, Tangent, OR. This variety is extremely fine textured and should be maintained at a lower than average cutting height at moderately high fertility to obtain optimal turf performance. It is not well suited for blending and/or mixing with coarser textured grasses.

Preparer/Additional Information - Fred B. Ledeboer, Ph.D., TMI, 22068 Case Road NE, Aurora, OR 97002, (503) 678-2597.

Aquarius II

Source - Reselection of superior plant types out of Aquarius and three clones out of Allaire parent material.

Method of Breeding - Selection of desired plant types out of extensive space-plant nurseries, phenotypic matched mating for four generations for desired type.

Intended Use - Turf.

Description - Very fine texture (finer than any other perennial ryegrass variety), very dense turf sward production, outstandingly fine turf in winter overseeding of southern golf courses and lawns.

Adapted to - Throughout the zone of most ryegrasses.

Released - KWS-AG, Einbeck, Germany, in 1991.

Breeder Seed/Stock - KWS Seed, Inc. and TMI.

Certified Seed/Stock - Available.

Preparer/Additional Information - Fred B. Ledeboer, Ph.D., TMI, 22068 Case Road NE, Aurora, OR 97002, (503) 678-2597.

Assure

Source - In 1961, parental ryegrass material was collected in Maryland, Pennsylvania, New Jersey, and New York. Eight cycles of recurrent selection were conducted to incorporate stem and crown rust resistance along with improved turf qualities. The cycles of selection were conducted in space plantings and in mowed turf trials subjected to severe stresses of heat, drought, and important turf grass diseases.

Method of Breeding - Assure is an advanced generation synthetic cultivar selected from the progenies of 18 clones.

Intended Use - Athletic fields, golf course fairways, tees and roughs, home lawns, parks, institutional and commercial turf where a fine textured, wear-tolerant, quick recovery turf is desired. Assure is an excellent companion with bluegrass in a mixture because of its slower growth rate and lower growth habits.

Description - Assure is an improved turf-type perennial ryegrass with a dark green color, fine leaf texture, very good density, and improved mowing quality. Improved tolerance to heat and drought conditions. Exhibits quick germination and produces an excellent dense turf. Has a slower vertical growth rate and an overall lower growth habit as compared to many other turf-type perennial ryegrasses. Improved resistance to leaf spot, brown patch, and stem and crown rust. Has also shown improved tolerance to heat and drought. This variety contains a high level of the fungal endophyte that provides resistance to above ground feeding insects.

Adapted to - PHZ 3, 4, 5, 6, 7.

Released - 1989, by Rutgers University and Pure Seed Testing, Inc.

Breeder Seed/Stock - Rutgers University and Pure Seed Testing, Inc.

Certified Seed/Stock - Available.

Preparer/Additional Information - Art Wick, LESCO, Inc., 20005 Lake Road, Rocky River, OH 44116, (216) 333-9250.

Barball

Source - Ecotypes from Austria.

Method of Breeding - In 1974, selection of space plants, clones and families under turf conditions. Eight clones form the base of this variety.

Intended Use - Cool-season turfgrass for sportfields and parks.

Description - Good resistance to crown rust and snowmold; good seed yield.

Adapted to - Overseeding PHZ 6, 7, 8, 9, 10, 11.

Released - Barenbrug.

Breeder Seed/Stock - Maintained by Barenbrug Holland, Oosterhout, The Netherlands.

Certified Seed/Stock - Available in quantity from Barenbrug USA, Tangent, OR. U.S. PVP No. 8700027.

Preparer/Additional Information - Barenbrug USA, P.O. Box 239, Tangent, OR 97389, (800) 547-4101.

Barclay

Source - Dutch ecotypes.

Method of Breeding - In 1974, selection of space plants in a turf trial; five clones were selected to produce a synthetic variety.

Intended Use - Cool-season turfgrass suitable for mixtures, blends, lawns, and sportfields.

Description - It is distinguishable because of its creeping habit. Its stolon-like spruits are generative shoots, which do not produce heads and form roots on the nodes.

Released - Barenbrug.

Breeder Seed/Stock - Maintained by Barenbrug Holland, Oosterhout, The Netherlands.

Certified Seed/Stock - Available in quantity from Barenbrug USA, Tangent, OR. U.S. PVP No. 8300016.

Preparer/Additional Information - Barenbrug USA, P.O. Box 239, Tangent, OR 97389, (800) 547-4101.

Baristra

Source - Selection in the variety Bastion

Method of Breeding - Selected on persistency in a sward of cocksfoot. Breeders seed is produced of 16 highly identical plants for the production of the first clonal generation.

Intended Use - Permanent pasture and mixtures for early pastures.

Description - Tetraploid, very early variety. Very persistent and winter-hardy. Good crownrust resistance.

Adapted to - Central and northeastern U.S. PHZ 1, 2, 3, 4, 5.

Released - Barenbrug USA.

Breeder Seed/Stock - Barenbrug Holland, Oosterhout, The Netherlands.

Certified Seed/Stock - Available in quantity from Barenbrug USA.

Preparer/Additional Information - Barenbrug USA, 32080 Old Highway 34, P.O. Box 239, Tangent, OR 97389, (800) 547-4101.

Barlano

Developed by Barenbrug Holland. Tested as Bar Lp 77-2.

Source - Collected material from old pastures.

Method of Breeding - Selection was started in 1971 from plants collected in old Dutch pastures. These plants were cloned and tested in a mowing trial over three years. The selection criteria were yield, growing rhythm, and persistence. The seven clones finally selected were established in a polycross nursery and this seed was used for Syn-1 seed production.

Intended Use - Hay and pasture production.

Description - Barlano is a diploid variety. Growth habit is semi-erect; ear emergence is fairly early; ear length is short and stem length is medium long. The resulting sod is very dense with a high persistence. It has adequate winter-hardiness, good resistance to lodging, and spring growth begins very early. Slightly susceptible to crown rust. A good yielding forage type perennial ryegrass variety; outyielded Norlea in B.C. Forage Crops Committee Trials, 1979-82.

Adapted to - Central and northeastern U.S. PHZ 1, 2, 3, 4, 5.

Released - Barenbrug Holland. Registered in Canada in 1984 with registration no. 2389.

Breeder Seed/Stock - Barenbrug Holland.

Preparer/Additional Information - Agriculture Canada, Food Production and Protection Branch, Plant Products Division, Ottawa, Ontario K1A 0C6. Seed distributed by Richardson Seed Co., 4055 McConnell Dr., Burnaby, British Columbia V6A 3A7 and Barenbrug USA, 32080 Old Highway 34, P.O. Box 239, Tangent, OR 97389 (800) 547-4101.

Barlatra

Source - Selection from the Foundation for Plantbreeding, Wageningen, The Netherlands.

Method of Breeding - After tetraploidizing, selection of single plants, clones, and families.

Intended Use - Pastures and meadows.

Description - Tetraploid. Medium heading with a long culm. Good palatability and drought resistance. Quick initial growth makes it highly recommended as green manure.

Adapted to - Central and northeastern U.S. PHZ 1, 2, 3, 4, 5.

Released - Barenbrug.

Breeder Seed/Stock - Barenbrug Holland, Oosterhout, The Netherlands.

Certified Seed/Stock - Barenbrug USA.

Preparer/Additional Information - Barenbrug USA, 32080 Old Highway 34, P.O. Box 239, Tangent, OR 97389, (800) 547-4101.

Barmaco

Source - Ecotypes.

Method of Breeding - Selection of clones.

Intended Use - Pastures and especially meadows.

Description - Late heading variety. Very winter-hardy and persistent. Very good wear tolerance in pastures. Good digestibility over the years.

Adapted to - Central and northeastern U.S. PHZ 1, 2, 3, 4, 5.

Released - Barenbrug.

Breeder Seed/Stock - Barenbrug Holland, Oosterhout, The Netherlands.

Certified Seed/Stock - Barenbrug USA.

Preparer/Additional Information - Barenbrug USA, 32080 Old Highway 34, P.O. Box 239, Tangent, OR 97389, (800) 547-4101.

Barrage

Source - Ecotypes.

Method of Breeding - In 1976, selection of space plants, clones, and families under turf conditions, with a good disease resistance.

Intended Use - Cool-season turfgrass. Very suitable for sportfields and parks.

Description - The variety has an early heading date and a good resistance to crown rust and snow mold; it has a good wear-resistance.

Adapted to - Permanent PHZ 1, 2, 3, 4, 5, 6.

Released - Barenbrug.

Breeder Seed/Stock - Barenbrug Holland, Oosterhout, The Netherlands.

Certified Seed/Stock - Available in quantity from Barenbrug USA. U.S. PVP No. 8900084.

Preparer/Additional Information - Barenbrug USA, P.O. Box 239, Tangent, OR 97389, (503) 926-5801.

Barvestra

Source - Selection from the Foundation for Plantbreeding, Wageningen, The Netherlands.

Method of Breeding - Selection and evaluation of individual plants, clones, and families.

Intended Use - Permanent pastures and meadows.

Description - Tetraploid. Early maturing. Rather large leaves. Medium to half prostrate growth habit. Very palatable. Good disease resistance.

Adapted to - Central and northeastern U.S. PHZ 1, 2, 3, 4, 5.

Released - Barenbrug.

Breeder Seed/Stock - Barenbrug Holland, Oosterhout, The Netherlands.

Certified Seed/Stock - Barenbrug USA.

Preparer/Additional Information - Barenbrug USA, 32080 Old Highway 34, P.O. Box 239, Tangent, OR 97389, (800) 547-4101.

Bastion

Source - Mommersteeg International B.V., Vlijmen, The Netherlands.

Method of Breeding - Material selected from early diploid material was treated with colchicine in 1965. In the C_4 generation, plants and clones were selected. Selection was also carried out in C_5 at which time seven clones were selected and polycrossed to form the parental material of Bastion.

Intended Use - Hay and pasture production.

Description - This is a tetraploid variety. It is 10 cm shorter and 10 days earlier heading than Fantoom; European data show good resistance to rust and snow mold. Bastion is a forage-type perennial ryegrass with good forage production and acceptable winter-hardiness. Bastion consistently outyielded Norlea in British Columbia trials.

Adapted to - Recommended in Canada for use in British Columbia.

Released - Mommersteeg International, Vlijmen, The Netherlands. Registered in Canada in 1986 with Registration No. 2655.

Breeder Seed/Stock - Mommersteeg International, Vlijmen, The Netherlands.

Certified Seed/Stock - Available from Oseco Inc.

Preparer/Additional Information - Agriculture Canada, Food Production and Protection Branch, Plant Products Division, Ottawa, Ontario K1A 0C6. Seed distributed by Oseco Inc., Box 219, Brampton, Ontario L6V 2L2.

Belfort

Source - Mommersteeg International B.V., Vlijmen, The Netherlands.

Method of Breeding - From selected plants of the diploid variety Vigor. These selected plants were treated with colchicine in 1965 and the material was multiplied until the C_4 generation when plant selection was carried out. This was followed by clonal selection in the C_5 generation. The selection criteria were: heading date, growth habit, leaf color, mature plant height, panicle length, and flag leaf length and width. Seven plants were selected to form the synthetic population Mom Lp 175. This synthetic variety was evaluated for fresh yield, dry matter yield, disease resistance, persistence, winter hardiness, and sward density. Syn 1 seed was first produced in 1975 by harvesting seed from the seven parental plants. This variety was developed by Mommersteeg International B.V., Vlijmen, The Netherlands, and was tested as Mom Lp 175.

Intended Use - Hay and pasture production.

Description - This is a tetraploid variety; growth habit is intermediate; leaves are dark green; plant height is tall; and maturity is late to very late. In forage trials in British Columbia, Belfort consistently outyielded the check variety, Norlea.

Adapted to - Recommended in Canada for use in British Columbia.

Released - Mommersteeg International, B.V., Vlijmen, The Netherlands. Registered in Canada in 1987 with Registration No. 2800.

Breeder Seed/Stock - Mommersteeg International, B.V., Vlijmen, The Netherlands.

Certified Seed/Stock - Available from Richardson Seed Co. Ltd.

Preparer/Additional Information - Agriculture Canada, Food Production and Protection Branch, Plant Products Division, Ottawa, Ontario K1A 0C6. Seed distributed by Richardson Seed Co. Ltd., 4055 McConnell Dr., Burnaby, British Columbia V6A 3A7.

Blazer II

Source - Selected out of a source population located in NJ.

Method of Breeding - Synthetic cultivar development via polycross and progeny testing.

Intended Use - As a turfgrass for general use and as an overseeding grass for dormant warm-season species.

Description - It has been selected for lower and slower vertical growth habit as compared to other perennial ryegrass cultivars.

Adapted to - PHZ 7, 8, and eastern parts of 5 and 6.

Released - August 1988 by Pickseed West Inc.

Breeder Seed/Stock - Pickseed West Inc.

Certified Seed/Stock - Available.

Preparer/Additional Information - Pickseed West Inc., P.O. Box 888, Tangent, OR 97389, (503) 926-8886.

Bonita

Source - Van der Have B.V., The Netherlands.

Method of Breeding - Diploid ecotypes were collected in The Netherlands in 1966 and were evaluated in forage yield trials. The best ecotypes were treated with colchicine to induce tetraploidy and were re-evaluated for forage performance. The best 10 C_3 families formed the parental material of this variety. Syn 1 seed was first produced in 1975.

Intended Use - Hay and pasture production.

Description - This is a tetraploid variety. Leaves are rolled in young shoots, medium green, anthocyanin present in lower leaf sheath; maturity is medium, later than Pennfine. U.S. data indicate that Bonita is resistant to crown rust, leaf spot, and red thread and moderately resistant to powdery mildew. Bonita is a very good yielding variety.

Adapted to - Recommended in Canada for use in British Columbia.

Released - Van der Have, B.V., The Netherlands. Registered in Canada in 1987 with registration no. 2750.

Breeder Seed/Stock - Van der Have, B.V., The Netherlands.

Certified Seed/Stock - Available from Dawson Seed Co. Ltd.

Preparer/Additional Information - Agriculture Canada, Food Production and Protection Branch, Plant Products Division, Ottawa, Ontario K1A 0C6. Seed distributed by Dawson Seed Co. Ltd., P.O. Box 91204, North Vancouver, British Columbia V7J 2E7.

Brightstar

Source - Collections from New York, New Jersey, Pennsylvania, and Maryland, with selections from Loretta and Elka.

Method of Breeding - Advanced generation synthetic selected from the maternal progenies of 17 clones containing endophyte. Selected for darker green color, lower growth profile, and improved resistance to many diseases, insects, and environmental stresses in the mid-Atlantic U.S.

Intended Use - Cool-season turfgrass also for overseeding of dormant warm-season turfs in southern U.S.

Description - Excellent turf quality, improved density, very dark green color, reduced vertical growth.

Adapted to - PHZ 3, 4, 5, 6, 7, 8, 9, 10.

Released - Turf-Seed, Inc.

Breeder Seed/Stock - Pure Seed Testing, Inc.

Certified Seed/Stock - Available. PVP No. 9200202.

Preparer/Additional Information - Turf-Seed, Inc., P.O. Box 250, Hubbard, OR 97032, (503) 651-2130.

Calypso

Calypso was developed jointly by the O.M. Scott and Sons Company and the New Jersey AES - Virgil Meyer and Cyril R. Funk, Jr. The variety was purchased by O.M. Scott and Sons Company in 1983 and is licensed to Roberts Seed Company of Tangent, OR.

Source - Calypso perennial ryegrass is a composite of plants originating from tillers selected from 76 turf plots showing high resistance to sod webworms pollinated by several stem rust resistant clones. The resulting progeny were allowed to interpollinate for one generation. The progeny of this cross was planted and rogued of any plants showing late maturity or high rust levels. Germplasm developed at Rutgers University.

Method of Breeding - Composite variety with mass selections for maturity and rust resistance.

Intended Use - Calypso is intended as turf in cool zones on golf courses, grounds, and home lawns and for winter overseeding in warm zones where it outperforms annual ryegrasses used for this purpose.

Description - Calypso shows good resistance to many important turf diseases, including leaf spot, brown patch, and dollar spot. It has a fine leaf texture and a dark green color, establishes quickly, and has good wear tolerance. Early-mid maturity combined with attractive appearance and good mowing qualities.

Adapted to - LRR A, B, northern C & D, E, F, G, H, K, L, M, R & S year-round and all regions as winter overseed; PHZ 2, 3, 4, 5, 6, and Pacific coast year round, 7, 8, 9, 10, winter overseed.

Released - 1987 by Roberts Seed Co.

Breeder Seed/Stock - O. M. Scott and Sons Company.

Certified Seed/Stock - Available from Roberts Seed Co.

Preparer/Additional Information - Robert Simerly, Roberts Seed Co., 33095 Highway 99E, P.O. Box 206, Tangent, OR 97389, (503) 926-8891; or Virgil Meier, O.M. Scott & Sons Co., Marysville, OH 43041, (513) 644-0011.

Caravelle

Source - Original material from France developed by Mommersteeg International, The Netherlands.

Method of Breeding - Synthetic variety based on 13 clones.

Intended Use - Fall overseeding of dormant warm-season turf. Nurse crop for permanent turf in cool-season area.

Description - Very dark green plants with poor head and poor cold tolerance. Excellent spring transition to warm-season turf in winter overseeding area. In cool season area, excellent nurse crop for Kentucky bluegrass over a two- to three-year period.

Adapted to - PHZ winter overseeding - 7, 8, 9, 10; nurse crop - 4, 5, 6.

Released - 1977 by O.M. Scott & Sons Co.

Breeder Seed/Stock - Mommersteeg International B.V., The Netherlands.

Certified Seed/Stock - Not available.

Preparer/Additional Information - Virgil Meier, O.M. Scott & Sons Co., Marysville, OH 43041, (513) 644-0011.

Charger

Source - Selections from crosses with Citation II and Master and collections from old turfs in New Jersey.

Method of Breeding - Advanced generation synthetic of 25 clones selected for improved winter color in mowed turf trials and improved crown and stem rust resistance.

Intended Use - Cool-season turfgrass also for overseeding of dormant warm-season turfs in southern U.S.

Description - Has improved winter color and turf performance.

Adapted to - PHZ 3, 4, 5, 6, 7, 8, 9, 10.

Released - Turf-Seed, Inc.

Breeder Seed/Stock - Pure Seed Testing, Inc.

Certified Seed/Stock - Available. PVP No. 8900143.

Preparer/Additional Information - Turf-Seed, Inc., P.O. Box 250, Hubbard, OR 97032, (503) 651-2130.

Citadel

Source - Van der Have of Holland.

Method of Breeding - Citadel forage-type perennial ryegrass originates from colchicine treatment of selected diploid material performed in 1965 by Van der Have, B.V., The Netherlands. This treated material was multiplied until the C_4 generation when selection was carried out. Further plant and clone selection was carried out in C_5. Synthetic lines were then created and evaluated; one seven-family synthetic tested as Mom Lp166.

Intended Use - Hay and pasture production.

Description - This is a tetraploid variety. Growth habit is intermediate; leaf color is medium-dark green; stem length is medium to tall; heading has little tendency to head in year of sowing, intermediate heading date. Citadel consistently outyielded Norlea perennial ryegrass during three years of forage testing in British Columbia.

Adapted to - Recommended in Canada for use in British Columbia.

Released - Van der Have of Holland. Registered in Canada in 1986 with registration no. 2605.

Breeder Seed/Stock - Van der Have of Holland.

Certified Seed/Stock - Available from Buckerfield's Limited.

Preparer/Additional Information - Agriculture Canada, Food Production and Protection Branch, Plant Products Division, Ottawa, Ontario K1A 0C6. Seed distributed by Topnotch Nutri Ltd., P.O. Box 1030, Abbotsford, British Columbia B2S 5B5.

Citation II

Source - Missouri and Washington, DC, and selection from Citation.

Method of Breeding - Advanced generation synthetic resulting from five cycles of recurrent selection. Stem rust resistant sources were used in a modified backcrossing program with Citation selections.

Intended Use - Cool-season turfgrass also for overseeding of dormant warm-season turfs in southern U.S.

Description - Good stem rust resistance; improved leaf rust and crown rust resistance over Citation.

Adapted to - PHZ 3, 4, 5, 6, 7, 8, 9, 10.

Released - Turf-Seed, Inc.

Breeder Seed/Stock - Pure Seed Testing, Inc.

Certified Seed/Stock - Available. PVP No. 8400142.

Preparer/Additional Information - Turf-Seed, Inc., P.O. Box 250, Hubbard, OR 97032, (503) 651-2130.

Commander

Source - Stem rust-resistant progenies selected form Citation II were used in greenhouse crosses with other selected turf-type perennials. Following four cycles of selection, plants exhibiting superior turf qualities were chosen as the parents of Commander.

Method of Breeding - Commander is an advanced-generation synthetic variety resulting from four cycles of recurrent selection for improved stem rust and leaf spot resistance and bright green color.

Intended Use - Athletic fields, golf course fairways, tees, and roughs, home lawns, sod blends, parks, and institutional and commercial turf where wearability and rapid recovery is desired. Commander is particularly suited for slit seed renovating of fairways, roughs, and other turf areas where disease pressure has reduced the turf density and where *Poa annua* L. invasion is a problem. Commander is useful as an overseeded turf in dormant warm-season grasses in southern U.S.

Description - Commander is an improved turf-type perennial ryegrass with a medium dark green color. It produces a fine textured dense turf with very good heat tolerance ratings and has performed well over a range of soil types. It has performed well in both northern turf and as an overseeding grass for dormant bermudagrass. Excellent seedling vigor and very good resistance to net blotch, brown patch, and stem rust. Has shown good heat and cold tolerance, as well as a very high level of fungal endophyte that provides resistance to above ground feeding insects.

Adapted to - PHZ 3, 4, 5, 6, 7.

Released - 1987, Pure Seed Testing, Inc.

Breeder Seed/Stock - Pure Seed Testing, Inc.

Certified Seed/Stock - Available.

Preparer/Additional Information - Art Wick, LESCO, Inc., 20005 Lake Rd., Rocky River, OH 44116, (216) 333-9250.

Competitor

Developed by Pure Seed Testing.

Source - Old turf areas in New Jersey and Kentucky and selection from Manhattan II and Citation II.

Method of Breeding - Four cycles of recurrent selection.

Intended Use - Turf, overseeding.

Description - Low growing, dark green, rust resistant.

Adapted to - PHZ 4, 5.

Released - 1985 by Pure Seed Testing.

Breeder Seed/Stock - Pure Seed Testing.

Certified Seed/Stock - Available.

Preparer/Additional Information - Mike McCarthy, E.F. Burlingham & Sons, P.O. Box 217, Forest Grove, OR 97116, (503) 359-9368.

Competitor II

Developed by McCarthy-Burlingham Research.

Source - Materials originated using Competitor crossed with selected low-growing dark green clones from the Oregon Research Farm.

Method of Breeding - Polycross and reselection for uniform, short, dense, dark green plants. Materials were tested in turf trials for more refined performance selection.

Intended Use - Turf, overseeding of winter-dormant, warm-season grasses.

Description - Very low-growing, dwarf type with dark attractive green color. Resistance to rust and leaf spot pathogens. Quick to germinate and establish.

Adapted to - PHZ 7, 8, 9, 10 and overseeding in 11 - cool-season climates.

Released - 1992 by McCarthy-Burlingham Research.

Breeder Seed/Stock - McCarthy-Burlingham Research.

Certified Seed/Stock - Available in 1994.

Preparer/Additional Information - Mike McCarthy, E.F. Burlingham & Sons, P.O. Box 217, Forest Grove, OR 97116, (503) 359-9368.

Condesa

Source - D.J. Van der Have B.V., Kapelle, The Netherlands.

Method of Breeding - Diploid sources from the Netherlands were treated with colchicine, creating tetraploid families. The C_3 generation was grown in a spaced plant field on sandy soil and individual plant selection was conducted. Five clones were selected to form the parental material of Condesa.

Intended Use - Hay and pasture production.

Description - This is tetraploid. Growth habit is medium; leaf color is dark green; stem is medium length; inflorescence has little or no tendency to head in the seeding year, late emerging, long; plant height is short to medium; winter-hardiness is good; heading is late; and moderate resistance to crown rust. Condesa consistently produced higher forage yields than Melle and Belfort over three years of testing in British Columbia.

Adapted to - Recommended in Canada for use in British Columbia.

Released - D.J. Van der Have, Kapelle, The Netherlands. Registered in Canada in 1988 as registration no. 3017.

Breeder Seed/Stock - D.J. Van der Have, Kapelle, The Netherlands.

Certified Seed/Stock - Available from Oseco Inc.

Preparer/Additional Information - Agriculture Canada, Food Production and Protection Branch, Plant Products Division, Ottawa, Ontario K1A 0C6. Seed distributed by Oseco Inc., P.O. Box 219, Brampton, Ontario L6V 2L2.

Cowboy (Reg. No. 106)

Developed through the co-operative efforts of Pure Seed Testing of Hubbard, OR; Lofts Seed, Bound Brook, NJ; and the NJ AES. Its experimental name is Pure Seed 2EE.

Source - An eight-clone synthetic cultivar from two sources: St. Louis, MO, and Washington, DC.

Method of Breeding - Donor parents from sources used in modified back-crossing program with Birdie perennial as recurrent parents in five cycles and tested in turf trials.

Intended Use - Athletic fields, lawns, and golf course tees and fairways.

Description - Very early maturing turf-type ryegrass producing an attractive, persistent, medium dense turf with bright green color.

Adapted to - LRR all except D, I, J, O, P, T, U, as a perennial; PHZ as a perennial 4, 5, 6, 7.

Released - 1984, by Lofts Seed, Inc.

Breeder Seed/Stock - Pure-Seed Testing.

Certified Seed/Stock - Available.

Preparer/Additional Information - Lofts Seed Inc., Chimney Rock Rd, P.O. Box 146, Bound Brook, NJ 08805, (908) 560-1590.

Cutless

Developed by International Seeds Inc., Halsey, OR.

Source - Elka X Rust Resistant selections.

Method of Breeding - Three cycles of recurrent mass selection with emphasis on short growth habit, dense tillering, dark green color, fine leaf blades, and resistance to stem rust.

Intended Use - High-quality permanent turf in areas where perennial ryegrass is adapted. Also well suited for overseeding dormant bermudagrass and other warm-season turfs. May be used in blends with other ryegrasses and in poly-species turfgrass mixtures.

Description - Heavy tillering, fine leafed and dark green with an extremely short, compact growth habit; capable of producing a very dense, handsome turf; late flowering; stem rust resistant variety. Spaced plants of this variety have a unique pin cushion appearance.

Adapted to - A, C, L, M, N, R, S, W, permanent turf. C, D, I, J, P, T, U, V, overseeding of dormant warm season turf; PHZ 5, southern half of 4 east of the Rocky Mountains.

Released - 1990, by International Seeds Inc.

Breeder Seed/Stock - Maintained by International Seeds Inc.

Certified Seed/Stock - Available from International Seeds Inc. U.S. PVP applied for no. 8900195.

Preparer/Additional Information - Stephen W. Johnson, International Seeds Inc., P.O. Box 168, 820 W. First St., Halsey, OR 97348, (503) 369-2251.

Dandy (Reg. No. 128)

Developed by R.H. Bailey Seed using germplasm from the New Jersey AES.

Source - Synthetic cultivar from progeny of 70 clones. Parental germplasm mainly from the Mid-Atlantic but about 7% from European germplasm.

Method of Breeding - Modified backcrossing, phenotypic recurrent selection, single-plant progeny trials. Selection focused on improved resistance to crown rust, net blotch, and brown patch. Selection was also made for medium-fine leaf texture and density, good heat and drought tolerance, and improved mowing quality. Final selection was for high seed yield, uniformity, and disease resistance in seed production.

Intended Use - Dandy was developed for use on lawns, sports fields, parks, playgrounds, and golf courses in all regions where perennial ryegrasses are adapted. Dandy has performed exceptionally in dormant bermudagrass overseeding trials throughout the southern U.S.. Dandy performs well in seed mixtures and blends as well as alone in turfgrass. Excellent seedling vigor makes it useful for land reclamation where quick establishment is needed. Dandy is an excellent choice for renovating worn out lawns and sports turfs.

Description - Dandy is a dark green, leafy, dwarf cultivar with good winter-hardiness and summer stress tolerance which has excellent seedling vigor and good wear tolerance. Dandy is enhanced with Acremonium endophytes to provide improved resistance to billbugs, chinch bugs, and sod webworms. Dandy has also shown improved resistance to several pathogens, including crown rust, net blotch, and brown patch. Dandy has improved resistance to stem rust in seed production.

Adapted to - LRR A, B, C, D, E, F, G, H, I, J, K, L, M, N, O, P, Q, R, S, T, U, V; PHZ 3, 4, 5, 6, 7, plus Virginia, Maryland and coastal California.

Released - August 1988 by R. H. Bailey Seed, Salem, OR.

Breeder Seed/Stock - R.H. Bailey Seed, Salem, OR.

Certified Seed/Stock - Available through Northrup King Co., Minneapolis, MN, and R.H. Bailey Seed, Salem, OR. PVP No. 8800224.

Preparer/Additional Information - Eric K. Nelson, NK Lawn & Garden, P.O. Box 300, 33731 Highway 99E, Tangent, OR 97389, (503) 928-2393.

Dasher II

Source - Selections forming the base of the variety came from old turf plots located at New Brunswick, NJ.

Method of Breeding - Synthetic cultivar development via polycross and progeny testing.

Intended Use - As a turfgrass for general use and as an overseeding grass for dormant warm-season species.

Description - Dasher II is a lower and slower growing cultivar than many other types.

Adapted to - PHZ 7, 8, and eastern parts of 5 and 6.

Released - July 1987 by Pickseed West Inc.

Breeder Seed/Stock - Pickseed West Inc.

Certified Seed/Stock - Available.

Preparer/Additional Information - Pickseed West Inc., P.O. Box 888, Tangent, OR 97389, (503) 926-8886.

Derby (Reg. No. 77)

Developed by International Seeds Inc., Halsey, OR.

Source - Population of turf-type perennial ryegrasses assembled at Brookston, IN, in 1971. Population included derivatives of plants selected from old turf stands in New Jersey, New York City, and Baltimore, MD.

Method of Breeding - Selections made on basis of color, fine leaf texture, and lack of leaf disease. Twelve parent clones selected to produce synthetic variety.

Intended Use - Cool-season turfgrass; also used for fall overseeding of dormant warm season turfs in the southern

U.S.. May be seeded alone or as a component in turfgrass blends and mixtures.

Description - Moderately early flowering turf-type; produces an attractive turf with good mowing qualities.

Adapted to - LRR A, C, L, M, N, R, S, W, permanent turf. C, D, I, J, P, T, U, overseeding of dormant warm-season turfs; PHZ 5, southern half of 4.

Released - 1974 by International Seeds Inc.

Breeder Seed/Stock - Maintained by International Seeds Inc., Halsey, OR.

Certified Seed/Stock - Available in quantity from International Seeds Inc., Halsey, OR. Probably the most widely used of all the turf-type perennial ryegrass cultivars. U.S. PVP No. 7500009.

Preparer/Additional Information - Stephen W. Johnson, International Seeds Inc., P.O. Box 168, 820 W. First St., Halsey, OR 97348, (503) 369-2251.

Derby Supreme

Developed by International Seeds Inc., Halsey, OR. Tested as ISI-852.

Source - Derby perennial ryegrass and experimental breeding composites derived from crossing Elka perennial ryegrass with rust-resistant selections.

Method of Breeding - Multiple cycles of mass selection for desired plant types.

Intended Use - Establishment of new turf and the repair of old turf in areas where perennial ryegrass is adapted. Also well suited for overseeding dormant warm-season grass turfs, such as golf courses and athletic fields. May be sown alone or as a component in turfgrass blends and mixtures.

Description - Aggressively tillering, medium maturing, rust-resistant perennial; holds its green color during high temperatures and other stress periods. Establishes more quickly than many other ryegrasses; capable of rapidly forming a dense attractive turf. Well adapted for southern overseeding.

Adapted to - LRR A, C, L, M, N, R, S, W, permanent turf. D, I, J, P, T, U, overseeding of dormant warm-season turf; PHZ 5, southern half of 4.

Released - 1990, by International Seeds Inc.

Breeder Seed/Stock - Maintained by International Seeds Inc.

Certified Seed/Stock - Available from International Seeds Inc., uncertified seed also available. U.S. PVP applied for no. 9100064.

Preparer/Additional Information - Stephen W. Johnson, International Seeds Inc., P.O. Box 168, 820 W. First St., Halsey, OR 97348, (503) 369-2251.

Dillon

Developed by International Seeds Inc., Halsey, OR.

Source - Population derived from intercrossing selections out of commercial varieties.

Method of Breeding - Multiple cycles of mass selection.

Intended Use - Permanent turf in areas where perennial ryegrass is adapted. Also used for overseeding of dormant warm season turfs.

Description - Moderately dark green turf-type variety with good seedling vigor and resistance to crown rust. In production fields, Dillon is a late heading variety with good seed yield capabilities and superior resistance to stem rust.

Adapted to - A, C, L, M, N, R, S, W, permanent turf. C, D, I, J, P, T, U, overseeding of dormant warm-season turfs; PHZ 5, southern half of 4 east of the Rocky Mountains.

Released - 1988, by Mid Valley Agricultural Products, Corvallis, OR.

Breeder Seed/Stock - Maintained by International Seeds Inc.

Certified Seed/Stock - Available from Mid Valley Agricultural Products. US PVP no. 8900138.

Preparer/Additional Information - Stephen W. Johnson, International Seeds Inc., P.O. Box 168, 820 W. First St., Halsey, OR 97348, (503) 369-2251.

Dimension

Source - U.S.

Method of Breeding - Master x Manhattan II with three cycles of recurrent selection.

Intended Use - Lawns, playgrounds, parks, athletic fields, and golf courses.

Description - Good density, good leaf spot resistance, good stem rust resistance, and high tiller density.

Adapted to - PHZ 3, 4, 5, 6.

Released - Pure Seed Testing.

Breeder Seed/Stock - D.J. Van der Have, B.V., The Netherlands.

Certified Seed/Stock - Available from Advanta Seeds West, Inc.

Preparer/Additional Information - Kenneth Hignight, Advanta Seeds West, Inc., 33725 Columbus Street S.E., P.O. Box 1496, Albany, OR 97321-0452, (503) 967-8923.

Duet

Source - Selected from U.S. germplasm at Ceres Research Station, Christchurch, New Zealand, by Dr. Alan Stewart, Pyne Gould Guiness, Ltd.

Method of Breeding - Polycross of 10 parental clones selected on the basis of turf performance, seed yield, drought recovery, endophyte presence, and resistance to stem and crown rust. The breeding program commenced in 1984 and was completed in 1989. The germplasm in Duet traces back to Manhattan II (nine clones) and another un-named germplasm line.

Intended Use - Wherever cool-season grasses are adapted for turfgrass areas.

Description - Medium early variety, nine days later than the standard Coronet. Medium short in height, 8 cm shorter than Coronet in stem length, medium dark green, with flower spikelet being 5 cm shorter than Coronet. Contains more than 70% endophyte.

Adapted to - Useful in blends of turfgrass throughout the area where other cool season grasses are adapted.

Released - Pyne Gould Guiness, Ltd., Christchurch, New Zealand.

Breeder Seed/Stock - Pyne Gould Guiness, Ltd.

Certified Seed/Stock - Cascade International Seed Co.

Preparer/Additional Information - Irvin H. Jacob, Cascade International Seed Company, 8483 West Stayton Road, Aumsville, OR 97325, (503) 749-1822.

Edge

Source - Selected from source populations of perennial ryegrass in NJ.

Method of Breeding - Synthetic cultivar development via polycross and progeny testing.

Intended Use - Turf.

Description - The variety has a high level of fungal endophyte. It is an early maturing perennial ryegrass with good stem rust tolerance.

Adapted to - PHZ 7, 8, and eastern parts of 5 and 6.

Released - September 1990 by Pickseed West Inc.

Breeder Seed/Stock - Pickseed West Inc.

Certified Seed/Stock - Available

Preparer/Additional Information - Pickseed West Inc., P.O. Box 888, Tangent, OR 97389, (503) 926-8886.

Elka

Developed by Cebeco Zaden, Vlijmen, The Netherlands.

Source - Collections from various sites in The Netherlands.

Method of Breeding - Clonal evaluation and progeny testing for persistency and tillering when subjected to low mowing pressure. High-performing clones were selected to form synthetic variety.

Intended Use - Permanent turf in areas where perennial ryegrass is adapted; well adapted for sod farming. Also well suited for use as a low-growing ground cover for roadsides, orchards, and other low-maintenance areas either alone or in combination with creeping red fescue.

Description - Late heading, very fine leafed, heavy tillering turf-type that forms a dense bright green turf. Mature plant height is very short, significantly less than most other perennial ryegrass varieties. Highly resistant to crown rust.

Adapted to - LRR A, C, L, M, N, R, S, W; PHZ 5, southern half of 4 east of the Rocky Mountains.

Released - The Netherlands by Cebeco Zaden; in the U.S., 1980 by International Seeds Inc.

Breeder Seed/Stock - Cebeco Zaden, Vlijmen, The Netherlands.

Certified Seed/Stock - Available in quantity from International Seeds Inc. U. S. PVP No. 8100018.

Preparer/Additional Information - International Seeds Inc., P.O. Box 168, 820 W. First St., Halsey, OR 97348, (503) 369-2251.

Envy

Source - Collections from Maryland, Pennsylvania, New Jersey, and New York. Selections from Elka, Prelude, Loretta, Manhattan II, and Gator.

Method of Breeding - Advanced generation synthetic from the maternal progenies of 72 clones selected from a turf trial in New Jersey for resistance to pink snow mold.

Intended Use - Cool-season turfgrass and for overseeding of dormant warm-season turfs in southern U.S.

Description - Has good resistance to pink snow mold and good turf performance with reduced steminess in the spring.

Adapted to - PHZ 3, 4, 5, 6, 7, 8, 9, 10.

Released - Zajac Performance Seeds.

Breeder Seed/Stock - Zajac Performance Seeds.

Certified Seed/Stock - Available. PVP No. 9100081 pending.

Preparer/Additional Information - Zajac Performance Seeds, 33 Sicomac Rd., North Haledon, NJ 07508, (201) 423-1660.

Evening Shade

Source - Selections out of Allaire, Aquarius, and Patriot.

Method of Breeding - Selection of superior dark plants out of space-plant nurseries with phenotypic matched mating over three generations with specific goal towards very dark, fine-textured plants.

Intended Use - Primarily for overseeding of southern golf courses and lawns.

Description - Semi-upright growth, very dark green color with fine texture.

Adapted to - zone for ryegrass.

Released - 1992 by KWS-AG Einbeck, Germany.

Breeder Seed/Stock - KWS Seeds, Inc. and TMI.

Certified Seed/Stock - Fred B. Ledeboer, Ph.D., TMI 22068 Case Road NE, Aurora, OR 97002, (503) 678-2597.

Express

Source - Western Oregon.

Method of Breeding - Composite variety formed by blending equal amounts of seed from two closely related elite breeding populations.

Intended Use - Cool-season turfgrass.

Description - Preliminary results from the latest ryegrass variety trial shows the cultivar to have good tolerance to dollar spot and red thread diseases.

Adapted to - PHZ 7, 8, eastern parts of 5 and 6.

Released - September 1991 by Pickseed West Inc.

Breeder Seed/Stock - Pickseed West Inc.

Certified Seed/Stock - Available.

Preparer/Additional Information - Pickseed West Inc., P.O. Box 888, Tangent, OR 97389, (503) 926-8886.

Fantoom

Source - Mommersteeg International, Vlijmen, The Netherlands.

Method of Breeding - Fantoom forage type perennial ryegrass originates from selected diploid material treated with colchicine in 1965. This treated material was multiplied to the C4 generation, checking the chromosome number in the C_1 and C_2 generations. Plant selection in the C_4 generation was followed by plant and clone selection in the C_5 generation. Subsequently, progeny testing in mowing trials, single spaced plant trials and row trials were conducted. One synthetic out of this program was Mom Lp 165, later called Fantoom, a variety based on ten families.

Intended Use - Hay and pasture production.

Description - Fantoom is tetraploid. Leaves are folded in shoot, blue green, all plants have anthocyanin in lower leaf sheath; plant height is 79 cm.; maturity is medium early. U.S. data indicate that Fantoom is resistant to crown rust, leaf spot, and powdery mildew. Fantoom had a total yield approximately 20% greater than Norlea over 12 station-years of testing in British Columbia.

Adapted to - Recommended in Canada for use in British Columbia.

Released - Mommersteeg International, Vlijmen, The Netherlands. Registered in Canada with registration no. 2607.

Breeder Seed/Stock - Mommersteeg International, Vlijmen, The Netherlands.

Certified Seed/Stock - Available from Richardson Seed Company Ltd.

Preparer/Additional Information - Agriculture Canada, Food Production and Protection Branch, Plant Products Division, Ottawa, Ontario K1A 0C6. Seed distributed by Richardson Seed Co. Ltd., 4055 McConnell Dr., Burnaby, British Columbia V6A 3A7.

Fiesta II

Source - Selected out of source populations of perennial ryegrass in New Jersey.

Method of Breeding - Synthetic cultivar development via polycross and progeny testing.

Intended Use - As a turfgrass for general use and as an overseeding grass for dormant warm-season species.

Description - Fiesta II has been selected for its improved tolerance to several important turf diseases.

Adapted to - PHZ 7, 8, and eastern parts of 5 and 6.

Released - Pickseed West Inc.

Breeder Seed/Stock - Pickseed West Inc.

Certified Seed/Stock - Available.

Preparer/Additional Information - Pickseed West Inc., P.O. Box 888, Tangent, OR 97389, (503) 926-8886.

Frances

Source - Van der Have B.V., The Netherlands.

Method of Breeding - Frances forage-type perennial ryegrass is a synthetic variety based on the three best clones from a breeding nursery as determined by a top-cross progeny test.

Intended Use - Hay and pasture production.

Description - This is a diploid variety. Growth habit is intermediate; leaves are folded in shoot, medium green; stem length is medium to tall; heading is early with few heads in year of sowing; winter-hardiness is good. Data from The Netherlands indicate good rust resistance. Frances is an early flowering variety suitable for grazing and haymaking. It has high first cut yields and had total yields approximately 20% greater than Norlea over 12 station-years of testing in British Columbia.

Adapted to - Recommended in Canada for use in British Columbia.

Released - Van der Have B.V., The Netherlands. Registered in Canada in 1986 with registration no. 2606.

Breeder Seed/Stock - Van der Have B.V., The Netherlands.

Certified Seed/Stock - Available from Topnotch Nutri Ltd.

Preparer/Additional Information - Agriculture Canada, Food Production and Protection Branch, Plant Products Division, Ottawa, Ontario K1A 0C6. Seed distributed by Topnotch Nutri Ltd., P.O. Box 1030, Abbotsford, British Columbia B2S 5B5.

Friend

Snow Brand Seed Company, Daini-Hokkai Bldg, 3-8 Higashi-Nihonbashi 3 Chome, Chuo-Ku, Tokyo 103, Japan.

Source - Source nursery composed of five European tetraploid perennial ryegrasses (Petra 28, Reveille 25, Taptor 10, Tetraploid - 4 BV 4, and Tetraploid - 48L 11). Seventy-eight plants having relatively good winter-hardiness in Hokkaido, Japan, were selected.

Method of Breeding - The 78 plants were used to produce polycross seed, with 2,400 plants from the polycrossed seed being examined for three years for plant characteristics and productivity. Thirty-two clones were selected and used to produce polycross seed, which is the breeders seed of Friend.

Intended Use - For production of forage, either alone or in mixtures, for cutting or grazing.

Description - Friend most closely resembles Petra and Reveille and can be distinguished by being eight days earlier than Petra and 15 days later than Reveille. Friend is more erect-growing, darker green, has greater winter-hardiness, and more resistance to crown rust and leaf spot than Petra or Reveille.

Adapted to - LRR A, B, C, D, E, H, J, L, M, N, O, P, R, S, T; PHZ 4, 5, 6, 7, 8. *Released* - 1981 in Japan.

Breeder Seed/Stock - Maintained at the Hokkaido Research Station of Snow Brand Seed Company, Ltd.

Certified Seed/Stock - Available at Organization for Economics Corporation and Development.

Preparer/Additional Information - Susan H. Samudio, Jacklin Seed Company, W. 5300 Riverbend Ave., Post Falls, ID 83854, (208) 773-7581.

Gator (Reg. No. 90)

Developed by International Seeds Inc., Halsey, OR, using germplasm obtained from the New Jersey AES.

Source - Loretta perennial ryegrass and breeding composites developed from plants collected from old turfs in the northeastern U.S.

Method of Breeding - Clonal evaluation for disease resistance and progeny testing in turf of plants resulting from Loretta x breeding composites cross. Fifty-six elite clones selected to form synthetic variety.

Intended Use - Cool-season turfgrass; also recommended for overseeding of dormant warm-season turfs in the southernU.S. May be seeded alone or used as a component in perennial ryegrass blends and poly-species mixtures.

Description - Leafy, persistent, turf-type perennial capable of producing an attractive, dense, moderately low-growing turf with a bright, medium dark green color. Has good winter-hardiness and improved summer performance. Mowing qualities are superior to many currently available ryegrasses. Resistant to crown rusot and leaf spot. Small seeded variety.

Adapted to - LRR A, C, L, M, N, R, S, W, permanent turf. C, D, I, J, P, T, U, overseeding of dormant warm-season turf; PHZ 5, southern half of 4 east of Rocky Mountains.

Released - 1983 by International Seeds Inc.

Breeder Seed/Stock - Maintained by International Seeds Inc.

Certified Seed/Stock - Available in quantity from International Seeds Inc. U.S. PVP NO. 8300179.

Preparer/Additional Information - Stephen W. Johnson, International Seeds Inc., P.O. Box 168, 820 W. First St., Halsey, OR 97348, (503) 369-2251.

Gettysburg

Developed for Cascade International Seed Co., Aumsville, OR, and Jonathan Green and Sons, Inc. - Dr. C. Reed Funk, New Jersey AES, New Brunswick, and Dr. Kevin McVeigh, Willamette Valley Plant Breeders, Brownsville, OR.

Source - Germplasm obtained from the New Jersey AES was used to develop this cultivar. The parental clones of Gettysburg trace their maternal lineage to plants selected from old turfs in Maryland, Pennsylvania, New York, and

New Jersey and to plants selected from or related to Manhattan II and Loretta.

Method of Breeding - An advanced generation synthetic cultivar selected from the progenies of 131 clones immediately prior to anthesis in the late spring of 1987 and moved to an isolated nursery for interpollination. The 36 plants showing good floret fertility, the highest seed yields, and the presence of an endophyte were subsequently selected as the maternal parents of Gettysburg. Seed from each clone was used to establish single-plant progeny turf trials at North Brunswick, NJ, and also an isolated space-plant nursery near Brownsville, OR, during the late summer and fall of 1987. Syn-1 seed was harvested from this nursery from the most attractive and disease-free plants.

Intended Use - Improved turfgrass areas throughout the region of adaptation.

Description - A new and distinct, very dark green, dwarf plant that produces a fine bladed, attractive turf with reduced crown height and a reduced rate of vertical growth. It is most closely related to Pinnacle but is four days earlier at 50% anthesis. It is shorter in mature space plant height, 41 cm versus 55 cm for Pinnacle. It is finer in flag leaf width than Pinnacle, 3.2 mm versus 3.6 mm. It has a shorter flower spikelet than Pinnacle, 123 mm versus 168 mm.

Adapted to - It can be used throughout northern U.S. and southern Canada in turf seed mixtures or by itself. It has good persistence when used to overseed bermudagrass in southern U.S. markets.

Released - 1988 to Cascade International Seed Co. by Dr. Kevin McVeigh.

Breeder Seed/Stock - Dr. Kevin McVeigh, Willamette Valley Plant Breeders, Brownsville, OR.

Certified Seed/Stock - Cascade International Seed Co.

Preparer/Additional Information - Irvin H. Jacob, Cascade International Seed Company, 8483 West Stayton Road, Aumsville, OR 97325, (503) 749-1822.

Goalie

Developed by Northrup King Co. at research stations in Atmore, AL, and Minneapolis, MN.

Source - Four-clone synthetic.

Method of Breeding - Mass selection program using germplasm from Rutgers University and Penn State University. Major emphasis on heat tolerance, winter-hardiness, and disease resistance.

Intended Use - Goalie was developed to extend the adaptation of perennial ryegrasses to more northern latitudes where winter-hardiness of other perennial ryegrasses is a problem. Goalie is especially useful for sports turf and other areas receiving intense traffic. Goalie has exhibited excellent performance in lower maintenance turfs such as parks and lawns. Goalie is also compatible in blends of other ryegrasses and in mixture with Kentucky bluegrass and fine fescues.

Description - Goalie is a dark green, leafy perennial ryegrass distinguished by its low, subset crown which contributes to its excellent winter-hardiness, traffic tolerance, and recuperation from injury. Goalie exhibits excellent density and high percentage of living ground cover from early spring until fall. It has shown resistance to leaf spot, dollar spot, brown patch, and redthread diseases. Additionally, Goalie has shown inherent resistance to both chinch bugs and white grub incidence.

Adapted to - LRR A, B, C, D, E, F, G, H, I, J, K, L, M, N, O, P, Q, R, S, T, U, V, W; PHZ 3, 4, 5, 6, 7.

Released - Northrup King Co., now known as NK Lawn and Garden, Minneapolis, MN.

Breeder Seed/Stock - Maintained by NK Lawn and Garden, Minneapolis, MN, and Tangent, OR.

Certified Seed/Stock - Available through NK Lawn and Garden, Minneapolis, MN, and Tangent, OR.

Preparer/Additional Information - Eric K. Nelson, P.O. Box 300, 33731 Highway 99E, Tangent, OR 97389, (503) 928-2393.

Gunne

Source - Developed by Svalof AB, Svalof, Sweden.

Method of Breeding - A large spaced-plant population of Australian material was established. After two years of observation, 100 plants were selected in 1970 on the basis of winter-hardiness, improved regrowth, and yield. They were replanted in isolation to produce breeder seed. The population has been refined through repeated mass selection.

Intended Use - Hay and pasture production.

Description - Growth habit is semi-erect to erect; stems are stiff; heading is intermediate, earlier than Norlea; head is medium length; plant height is medium, similar to Norlea; winter-hardiness is similar to Norlea. Limited data indicate that Gunne is moderately susceptible to crown rust. Gunne is a good-yielding variety with good winter-hardiness as demonstrated in Atlantic trials.

Adapted to - Recommended in Canada for use in the Alantic region of Canada.

Released - Svalof AB, Svalof, Sweden. Registered in Canada in 1989 with Registration No. 3178.

Breeder Seed/Stock - Svalof AB, Svalof, Sweden.

Certified Seed/Stock - Available from Bonin and Co.

Preparer/Additional Information - Agriculture Canada, Food Production and Protection Branch, Plant Products Division, Ottawa, Ontario K1A 0C6.

Legacy

Source - In 1961, parental ryegrass material was collected in Maryland, Pennsylvania, New Jersey, and New York. Ten cycles of recurrent selection were conducted to incorporate stem rust resistance along with improved turf qualities. The cycles of selection were conducted in space plantings and in mowed turf trials subjected to severe stresses of heat, drought, and important turf grass diseases. The parents selected were dwarf, fine-textured plants with a very dark green color.

Method of Breeding - Legacy is an advanced generation synthetic cultivar selected from the progenies of 18 clones.

Intended Use - Athletic fields, golf course fairways, tees and roughs, home lawns, parks, and institutional and commercial turf where a fine textured, heat tolerant, quick recovery turf is desired. Legacy is an excellent companion with bluegrass in a mixture due to its slower growth rate and lower growth habit.

Description - Legacy is an improved turf-type perennial ryegrass with a dark green color, fine leaf texture, very good density, dwarf growth habit, and improved mowing quality. This variety has improved tolerance to heat and drought conditions. Exhibits quick germination and produces an excellent dense turf. Has a slower vertical growth rate and an overall lower growth habit as compared to many other turf-type perennial ryegrasses. Legacy perennial ryegrass has improved resistance to leaf spot, brown patch, and stem rust. It has also shown improved tolerance to heat and drought, This variety contains a very high level of the fungal endophyte that provides resistance to above ground feeding insects.

Adapted to - PHZ 3, 4, 5, 6, 7.

Released - 1990, Rutgers University and Pure Seed Testing, Inc.

Breeder Seed/Stock - Rutgers University and Pure Seed Testing, Inc.

Certified Seed/Stock - Available.

Preparer/Additional Information - Art Wick, LESCO Inc., 20005 Lake Road, Rocky River, OH 44116, (216) 333-9250.

Lindsay

Developed by International Seeds Inc., Halsey, OR. Tested as ISI-851.

Source - Elka, Manhattan, and a breeding composite derived from selections from commercial varieties.

Method of Breeding - Mass selection for resistance to rust, fine leaf blades, and dark green color.

Intended Use - Permanent cool-season turf; also recommended for overseeding dormant warm-season turfs including golf greens, tees, and fairways. May be seeded alone or as component in turfgrass blends and mixtures.

Description - Fine leafed, dark green genetic color; produces an attractive moderately dense turf; resistant to crown rust and stem rust .

Adapted to - LRR A, C, L, M, N, R, S, W, permanent turf. C, D, I, J, P, T, U, overseeding of dormant warm-season turf; PHZ 5, southern half of 4 east of Rocky Mountains.

Released - 1988 by International Seeds Inc.

Breeder Seed/Stock - Maintained by International Seeds Inc.

Certified Seed/Stock - Available in quantity through International Seed, Inc. U.S. PVP No. 8700176.

Preparer/Additional Information - Stephen W. Johnson, International Seeds Inc., P.O. Box 168, 820 W. First St., Halsey, OR 97348, (503) 369-2251.

Linn

Selected at Oregon AES, Corvallis, OR.

Source - Introduced from New Zealand in 1928. Original introduction grown on one farm for four years; seed harvested to establish plantings on two farms where this particular source has been grown since 1932.

Method of Breeding - Source identified and selected for certification on basis of field inspections and comparative tests. Linn 1 and Linn 2 refer to two experimental seed lots subsequently bulked for seed production because they were comparable in seed producing potential.

Description - Representative of best Oregon perennial types. Good seed yield and typical perennial characteristics. Increased on limited generation basis with two classes of seed registered and certified.

Adapted to - LRR A, B, E; PHZ 4, 5, 6.

Released - 1961, by Oregon AES.

Breeder Seed/Stock - Oregon AES (registered seed class produced on approved farms.)

Certified Seed/Stock - Available.

Preparer/Additional Information - Department of Crop and Soil Sciences, Crop Sciences Building 107, Oregon State Univ., Corvallis, OR 97331-3002, (503) 737-4513.

Loretta

Source - Original material from Austria. Developed by Saatzucht Steinach, Germany.

Method of Breeding - Selective multiplication of original material.

Intended Use - Permanent turf in cool-season region. Fall overseeding of dormant warm-season turf.

Description - Medium light green color, good wear tolerance, late maturity, and small seed size.

Adapted to - PHZ permanent turf - 4, 5, 6; winter overseeding - 7, 8, 9, 10.

Released - 1975, O.M. Scott & Sons Co.

Breeder Seed/Stock - Saatzucht Steinach, Germany.

Certified Seed/Stock - Available from O.M. Scott & Sons Co.

Preparer/Additional Information - Virgil Meier, O.M. Scott & Sons Co., Marysville, OH 43041, (513) 644-0011.

Magella

Developed by Cebeco Zaden B.V., Vlijmen, The Netherlands.

Source - Experimental lines and breeding composites developed by the Dutch Foundation for Plant Breeding at Wageningen, from ecotypes collected in The Netherlands. Carried as experimental no. EER 89 B.

Method of Breeding - Recombination of the best-producing clones under unusual summer stress conditions in The Netherlands, followed by two cycles of phenotypic mass selection.

Intended Use - High-quality pasture and hay production for cattle, sheep, and horses. Used with alfalfa as a companion or nurse crop. Also used for renovating old alfalfa stands.

Description - Intermediate maturity, moderately tall, moderately broad-leafed, vigorous; very cold-tolerant and productive, with improved persistence. Life of stand is three to five years depending upon environment and management. Magella combines high productivity with a good resistance to crown rust and a good persistency. Diploid (2n=2x=14).

Adapted to - LRR A, K, L, M, N, P, R, S, T, W; PHZ 3.

Released - 1983 in The Netherlands.

Breeder Seed/Stock - Cebeco Zaden B.V.

Certified Seed/Stock - Available.

Preparer/Additional Information - Craig W. Edminster, International Seeds Inc., P.O. Box 168, 830 W. First Street, Halsey, OR 97348, (503) 369-2251.

Manhattan (Reg. No. 18)

Selected at New Jersey AES, New Brunswick - C.R. Funk, R.E. Engel, and P.M. Halisky.

Source - Primarily plants collected in Central Park, New York City, NY.

Method of Breeding - Polycross progenies of clones selected from old turf areas and spaced plant nurseries evaluated under turf maintenance. Sixteen plants selected to produce synthetic variety.

Intended Use - Fine lawns and play fields.

Description - Leafy, late flowering, persistent, turf-type that produces an attractive, moderately dark green turf of finer texture and greater density than common perennial ryegrass.

Adapted to - LRR S, T, P, M; PHZ 5, 6, 7. *Released* - 1967, by New Jersey AES.

Breeder Seed/Stock - New Jersey AES.

Certified Seed/Stock - Not available in U.S., but sold widely in Western Europe.

Preparer/Additional Information - C. Reed Funk, NJ AES, Rutgers Univ., Crop Science Department, P.O. Box 231, New Brunswick, NJ 08903, (908) 932-9480.

Manhattan II (E)

Source - P.I. 197, 270 (Finland), Sprinter and collections from old turfs in New Jersey and Maryland.

Method of Breeding - Advanced generation synthetic of 22 parental clones. Selected for stem and crown rust resistance.

Intended Use - Cool-season turfgrass and for overseeding of dormant warm-season turfs in southern U.S.

Description - Forms a persistent turf with good density, good summer performance, and excellent seedling vigor. Improved resistance to stem and crown rust, leaf spot, and brown patch. Good resistance to above ground feeding insects.

Adapted to - PHZ 3, 4, 5, 6, 7, 8, 9, 10.

Released - Turf-Seed, Inc.

Breeder Seed/Stock - Pure Seed Testing, Inc.

Certified Seed/Stock - Available. PVP No. 8200154.

Preparer/Additional Information - Turf-Seed, Inc., P.O. Box 250, Hubbard, OR 97032, (503) 651-2130.

Melle

Source - Rijksstation Voor Plantenveredeling (the Government Plant Breeding Station), Merelbeke, Belgium.

Method of Breeding - Melle, tested as Vigor, originated from material collected in old natural pastures in Belgium. Melle was bred at Rijksstation Voor Plantenveredeling, Merelbeke, Belgium, in 1948. Until 1968, the composition of the variety changed regularly. Today it is a stable variety based on 22 clones selected by the polycross method. This variety is sold in Europe as Vigor R.V.P. and Melle.

Intended Use - Hay and pasture production.

Description - It is diploid; growth habit is semi-prostrate and very leafy; leaves are 200 mm in length and are medium green; heads are medium to long; plant height is 100 cm; good spring growth; heading date is very late; and winter-hardiness is good to very good. It is moderately resistant to crown rust. Yields well over several cuts; total yields were approximately 18% greater than Norlea over 12 station-years of testing in British Columbia.

Adapted to - Recommended in Canada for use in British Columbia.

Released - Rijksstation Voor Plantenveredeling, Merelbeke, Belgium. Registered in Canada in 1986 with Registration No. 2614.

Breeder Seed/Stock - Rijksstation Voor Plantenveredeling, Merelbeke, Belgium.

Certified Seed/Stock - Available from Oseco Inc.

Preparer/Additional Information - Agriculture Canada, Food Production and Protection Branch, Plant Products Division, Ottawa, Ontario K1A 0C6. Seed distributed by Oseco Inc., P.O. Box 219, Brampton, Ontario L6V 2L2.

Mondial

Source - U.S.

Method of Breeding - Loretta x Experimentals, with four cycles of selection based on 12 clones.

Intended Use - Parks, playgrounds, athletic fields, and golf courses.

Description - Highly wear tolerant. Good resistance to crown rust, red thread, and leaf spot.

Adapted to - PHZ 3, 4, 5, 6.

Released - Mommersteeg International B.V., The Netherlands.

Breeder Seed/Stock - Mommersteeg International.

Certified Seed/Stock - Available from Advanta Seeds West Inc.

Preparer/Additional Information - Kenneth Hignight, Advanta Seeds West, Inc., 33725 Columbus Street S.E., P.O. Box 1496, Albany, OR 97321-0452, (503) 967-8923.

Morning Star (Syn P)

Source - Selected plants that form the basis of the variety were collected from old turf plots at Rutgers University in New Jersey.

Method of Breeding - Mass selection and bulked seed harvest. Variety was progeny tested for uniformity utilizing spaced-plant nurseries.

Intended Use - Turf.

Description - Based on preliminary NTEP results, the cultivar exhibits good summer and fall density and a high tolerance to brown patch disease.

Adapted to - Wide adaptation wherever perennial ryegrasses are used.

Released - 1992, Pickseed West, Inc.

Breeder Seed/Stock - Pickseed West, Inc.

Certified Seed/Stock - Available.

Preparer/Additional Information - Pennington Seed Inc. of Oregon, P.O. Box 386, Lebanon, OR 97355, (503) 451-5261.

Navajo

Source - Australia, Italy, Illinois, Missouri, Washington, DC, New Jersey, Washington, and Oregon.

Method of Breeding - Selected for medium late maturity, dark color, fine leaves, low growth habit, crown rust and leaf spot resistance, and good turf quality. Twelve clones were selected as the parents to produce a synthetic after three cycles of selection.

Intended Use - Cool-season turfgrass also for overseeding of dormant warm-season turfs in southern U.S.

Description - Improved crown rust resistance and reduced vertical growth; produces a dark, dense fine-leaved turf.

Adapted to - PHZ 3, 4, 5, 6, 7, 8, 9, 10.

Released - Turf-Seed, Inc.

Breeder Seed/Stock - Pure Seed Testing, Inc.

Certified Seed/Stock - Available. PVP No. pending.

Preparer/Additional Information - Turf-Seed, Inc., P.O. Box 250, Hubbard, OR 97032, (503) 651-2130.

Nomad

Source - Selected material out of Barelay, Manhattan II, and parent lines of Allaire and Patriot.

Method of Breeding - Plant evaluations for prostrate growth habit in space plant nurseries, phenotypic matched mating for four generations for prostrate types and aggressive growth patterns.

Intended Use - Well adapted to fairway turf because of aggressive prostrate growth.

Description - Very prostrate growth habit, aggressive spreading type growth, medium to late maturity.

Adapted to - Through adaptation zone for ryegrass.

Released - 1991, KWS-AG, Einbeck, Germany.

Breeder Seed/Stock - KWS Seed, Inc. and TMI.

Certified Seed/Stock - Available. *Preparer/Additional Information* - Fred B. Ledeboer, Ph.D., TMI, 22068 Case Road NE, Aurora, OR 97002, (503) 678-2597.

Norlea

Selected at Canada Department of Agriculture Research Station, Ottawa, Ontario - R.M. MacVicar.

Source - Worldwide collection of seed lots.

Method of Breeding - Repeated selection and progeny evaluation through six generations; in final synthesis, 12 proven clones included in synthetic variety.

Intended Use - Hay and pasture plantings.

Description - Sufficient hardiness to survive and to be productive in areas where species had been of little or no value. Leafy, somewhat later in maturity than short-leaf ryegrass strains. Susceptible to leaf rust in some areas, but susceptibility does not appear to affect yield, since it consistently out-yielded other varieties in forage and seed production. Useful as a turf species.

Adapted to - LRR F, K, R; PHZ 3, 4.

Released - 1958, by Canada Department of Agriculture.

Breeder Seed/Stock - Canada Department of Agriculture Research Station, Ottawa, Ontario.

Certified Seed/Stock - Limited availability.

Preparer/Additional Information - Canada Seed Grower's Association, P.O. Box 8445, Ottawa, Ontario K1G 3T1.

Nova

Developed by International Seeds Inc., Halsey, OR.

Source - Experimental breeding composite and the varieties Elka and Manhattan.

Method of Breeding - Individual plants screened in spaced-plant nurseries for early maturity and resistance to stem rust. Forty individual plants selected to develop synthetic variety.

Intended Use - Permanent turf in areas where perennial ryegrass is adapted. Also recommended for overseeding dormant bermudagrass turfs. May be seeded alone, in blends with other perennial ryegrass, or in mixtures with other turfgrass species.

Description - Dark green, leafy, early maturing variety. Produces a handsome, moderately dense turf. Nova has exhibited resistance to snow mold and stem rust.

Adapted to - LRR A, C, L, M, N, R, S, W - permanent turf C, D, I, J, P, T, U - overseeding of dormant warm-season turfs; PHZ 5 plus southern half of 4 east of the Rocky Mountains.

Released - 1989 by Seed Research of Oregon, Corvallis, OR.

Breeder Seed/Stock - International Seeds, Inc.

Certified Seed/Stock - Available. U.S. PVP No. 8900046.

Preparer/Additional Information - International Seeds, Inc., P.O. Box 168, Halsey, OR 97348, (503) 369-2251.

Omega II

Source - Missouri, Washington, DC, and Omega.

Method of Breeding - Advanced-generation synthetic resulting from three cycles of recurrent selection. Rust resistant sources were used as donor parents in a modified backcrossing program.

Intended Use - Cool-season turfgrass also for overseeding of dormant warm-season turfs in southern U.S.

Description - Excellent seedling vigor and medium maturity, producing a dense turf which does not become stemmy in the spring.

Adapted to - PHZ 3, 4, 5, 6, 7, 8, 9, 10.

Released - Zajac Performance Seeds.

Breeder Seed/Stock - Pure Seed Testing, Inc.

Certified Seed/Stock - Available. PVP No. 8400141.

Preparer/Additional Information - Zajac Performance Seeds, 33 Sicomac Rd., North Haledon, NJ 07508, (201) 423-1660.

Ovation

Source - Cross between Loretta and Pennfine developed by Mommersteeg International, The Netherlands.

Method of Breeding - Synthetic variety - 10 clones.

Intended Use - Permanent cool-season turf and for fall overseeding of dormant warm-season turf.

Description - Medium green, medium maturity, dense with good mowing qualities.

Adapted to - PHZ permanent turf - 4, 5, 6; winter overseeding - 7, 8, 9, 10.

Released - 1982 by O.M. Scott & Sons Co.

Breeder Seed/Stock - Mommersteeg International B.V.

Certified Seed/Stock - Available from O.M. Scott & Sons Co.

Preparer/Additional Information - Virgil Meier, O.M. Scott & Sons Co., Marysville, OH 43041, (513) 644-0011.

Palmer (Reg. No. 85)

Developed and released by Lofts Seed, Inc., using germplasm obtained from the New Jersey AES. Named in honor of professional golfer, Arnold Palmer. Carried as accession no PI GT-I, Greektown.

Source - Breeding material developed from PI 231597, a plant from Greece and plant material from the New Jersey AES.

Method of Breeding - An advanced-generation synthetic cultivar selected from the progenies of 36 clones.

Intended Use - Athletic fields, home lawns, golf courses, playgrounds, and winter overseeding of dormant warm-season turf grasses in the southern U.S..

Description - A leafy, turf-type ryegrass. Produces dense, medium-low growing, fine-textured turf with a bright, dark green color. Performs well in full sun to moderate shade, heat tolerant and winter-hardy.

Adapted to - LRR all except D, I, J, O, P, T, U, as a perennial; PHZ 4, 5, 6, 7.

Released - 1982, Lofts Seed, Inc., in cooperation with the New Jersey AES.

Breeder Seed/Stock - Lofts Seed, Inc., with the cooperation of the New Jersey AES.

Certified Seed/Stock - Available.

Preparer/Additional Information - Lofts Seed, Inc., Chimney Rock Rd., P.O. Box 146, Bound Brook, NJ 08805, (908) 560-1590.

Palmer II

Source - Palmer II is an advanced generation synthetic cultivar selected from the maternal progenies of 40 clones. Most of the germplasm traces back to plants selected from old turfs in Maryland, Pennsylvania, New York, and New Jersey.

Method of Breeding - Phenotypic and genotypic recurrent selection. Screening of the seedlings was done for disease resistance and quality under close mowing.

Intended Use - Cool-season turfgrass, can also be used for fall overseeding of dormant warm-season grasses. Performs well when seeded alone or in mixtures.

Description - Palmer II is a medium-early maturing, leafy, turf-type perennial ryegrass capable of producing a very persistent, dense, attractive, low growing, fine textured turf with a bright, very dark green color. It has improved mowing qualities, good summer performance, and good winter-hardiness on well-drained soils. Palmer II is also endophyte enhanced.

Adapted to - Wherever cool-season grasses are found.

Released - Lofts Seed, Inc., 1992.

Breeder Seed/Stock - Lofts Seed, Inc. and Pure Seed Testing, and the New Jersey AES.

Certified Seed/Stock - Available. *Preparer/Additional Information* - Marie Pompei, Lofts Seed, Inc., P.O. Box 146, Bound Brook, NJ 08805, (908) 560-1590.

Patriot

KWS-AG, Einbeck, Germany.

Source - Collections from old pastures and turf areas in southeastern Europe plus selected clones of Tando and Corona.

Method of Breeding - Clonal progeny evaluation, in space-plant nurseries, turf performance, poly-crossbreeding, recurrent mass selection of most desirable types.

Intended Use - Primarily for winter overseeding of dormant sub-tropical turf areas, but also well suited for permanent ornamental, recreational, and athletic turf throughout the cool, humid regions of the world.

Description - Strongly upright, medium-fine textured, early maturing with moderately dark green color, producing a sward of moderate density, medium-early in maturity.

Adapted to - The northern cool, humid region; PHZ 4, 5, 6, (7, 8 winter overseeding).

Released - 1986, KWS-AG.

Breeder Seed/Stock - KWS-AG.

Certified Seed/Stock - Distributed by TMI, Tangent, OR.

Preparer/Additional Information - Fred B. Ledeboer, Ph.D., TMI, 22068 Case Road NE, Aurora, OR 97002, (503) 678-2597.

Patriot II

Source - Reselection of superior plant types out of Patriot.

Method of Breeding - Superior plant types used in poly-crossing nurseries over two synthetic generations.

Intended Use - Turf and winter overseeding of southern golf courses and lawns.

Description - Dark green, upright, outstanding turf characteristics, medium maturity, and moderate to good resistance to crown rust and leaf spot.

Adapted to - The adaptation zone for ryegrasses.

Released - KWS-AG, Einbeck, Germany.

Breeder Seed/Stock - KWS-Seeds, Inc. and TMI.

Certified Seed/Stock - Available.

Preparer/Additional Information - Fred B. Ledeboer, Ph.D., TMI, 22068 Case Road NE, Aurora, OR 97002, (503) 678-2597.

Pebble Beach

Source - Germplasm from Manhattan II, Citation II, two old-turf areas in Kentucky, one old-turf area in New Jersey, and one shaded-turf area in Oregon.

Method of Breeding - Four cycles of recurrent selection to produce this synthetic variety.

Intended Use - Turf blends, overseeding dormant warm-season turfgrasses.

Description - Dark, low-growing, with excellent mowing qualities.

Adapted to - PHZ 4, 5, 6, 7, 8, 9 - east to west coast.

Released - 1986, Pure Seed Testing, Inc.

Breeder Seed/Stock - Pennington Seed Inc.

Certified Seed/Stock - Available.

Preparer/Additional Information - Pennington Seed Inc. of Oregon, P.O. Box 386, Lebanon, OR 97355, (503) 451-5261.

Pennant

Developed by Rutgers University.

Source - Selections from old turf in Maryland.

Method of Breeding - Sixty-five clone synthetic polycross.

Intended Use - Turf, winter overseeding of dormant Bermudagrass.

Description - Attractive, moderately dark green turf, endophyte enhanced, rapid germination.

Adapted to - PHZ 4, 5.

Released - 1978, Rutgers University.

Breeder Seed/Stock - Rutgers University.

Certified Seed/Stock - Available. PVP # 8000141.

Preparer/Additional Information - Mike McCarthy, E.F. Burlingham & Sons, P.O. Box 217, Forest Grove, OR 97116, (503) 359-9368.

Pennant II

Developed by McCarthy-Burlingham Research Farm.

Source - Selections were made from old-turf plots in New Jersey, Massachusetts, and Oregon.

Method of Breeding - Cycles of recurrent selections crossed with top-performing selections derived from Pennant to produce a synthetic polycross variety.

Intended Use - Turf, overseeding of dormant bermudagrass in southern areas.

Description - Very dark green, attractive, endophyte-enhanced turf. Low-growing and requiring less mowing than many other cultivars. Resistant to leaf spot. Has good mowing characteristics, and responds quickly to fertilizer and irrigation.

Adapted to - PHZ 7, 8, 9, 10 and overseeding in 11 - cool-season climates.

Released - 1992, McCarthy-Burlingham Research.

Breeder Seed/Stock - McCarthy-Burlingham Research.

Certified Seed/Stock - Available in 1995.

Preparer/Additional Information - Mike McCarthy, E.F. Burlingham & Sons, P.O. Box 217, Forest Grove, OR 97116, (503) 359-9368.

Pennfine (Reg. No. 26)

Selected at Pennsylvania AES, University Park - J.M. Duich, A.T. Perkins, and H. Cole.

Source - Plants collected from recreational areas.

Method of Breeding - Clonal evaluation and progeny testing. Three superior turf-type clones selected for synthetic variety.

Intended Use - Turf.

Description - Fine textured, comparatively dense and persistent, with less leaf shredding after mowing than available varieties. At 25 mm mowing height, leaf width about 2.5 mm., in comparison with 2.7 mm. for Manhattan and 3.8 mm. for Linn; vertical seedling growth 50% less than Linn. Resistant to leaf spot, head scab, dollar spot, and leaf rust; and moderate resistance to red thread and snow mold. Adapted for use as component in cool-season turf grass mixtures and for overseeding damaged turf-grass areas, and shows promise for overseeding bermudagrass putting greens.

Adapted to - Cool temperate regions in the eastern U.S.

Released - 1969, by Pennsylvania AES.

Breeder Seed/Stock - Not available.

Certified Seed/Stock - Available in quantity.

Preparer/Additional Information - Pennsylvania State Univ., Agronomy Department, University Park, PA 16802, (814) 865-6541.

Phoenix

Developed by Cebeco Zaden B.V., Vlijmen, The Netherlands.

Source - Colchicine-treated experimental lines and breeding composites derived from plants collected from old farms throughout The Netherlands. Carried as experimental number EER 400.

Method of Breeding - Recombination of the 29 most persistent and best dry matter-producing clones followed by two cycles of phenotypic mass selection.

Intended Use - High-quality pasture and hay production for cattle, sheep, and horses. Used with alfalfa as a companion or nurse crop. Also used for renovating old alfalfa stands.

Description - Intermediate maturity, moderately tall, broad-leafed, vigorous; very cold-tolerant and productive, with superior persistence. Life of stand is three to seven years depending upon environment and management. Tetraploid (2n=4x=28).

Adapted to - LRR A, K, L, M, N, P, R, S, T, W; PHZ 3.

Released - 1987 in The Netherlands.

Breeder Seed/Stock - Available from Cebeco Zaden B.V.

Certified Seed/Stock - Available.

Preparer/Additional Information - Craig W. Edminster, International Seeds, Inc, P.O. Box 168, 830 W. First Street, Halsey, OR 97348, (503) 369-2251.

Pinnacle

Source - Fourteen clones selected from Manhattan II, citation II, Jazz, and plants selected from old turfs in the northeastern U.S.

Method of Breeding - Advanced generation synthetic cultivar originally developed by the New Jersey AES, Rutgers University with Dr. C. Reed Funk.

Intended Use - For turf, either alone or in blends.

Description - Dark green, fine leaved, dwarf growth characteristics, dense turf, improved disease resistance, high endophyte for insect resistance, excellent low mowing.

Adapted to - All zones.

Released - 1987 by Barenbrug USA/Normarc.

Breeder Seed/Stock - Pickseed West and Barenbrug USA.

Certified Seed/Stock - Available from Barenbrug USA.

Preparer/Additional Information - Barenbrug USA, P.O. Box 239, Tangent, OR 97389, (503) 926-5801.

PR 8820

Developed by International Seed, Inc., Halsey, OR.

Source - Populations derived from crosses of Elka with selections from various American varieties.

Method of Breeding - Multiple cycles of mass selection for fine leaf texture, tillering, and resistance to stem rust and crown rust. Synthetic derived from progenies of 14 fine-bladed, rust-resistant plants.

Intended Use - Permanent turf in all areas where turf-type perennial ryegrass is adapted. Well suited for golf course tees and fairways; also may be used for overseeding dormant warm-season turfs in the southern U.S.

Description - Fine bladed, aggressively tillering medium-late maturing variety. Produces a dense, moderately dark green turf with improved tolerance to close mowing. Has good resistance to crown and stem rusts.

Adapted to - LRR A, C, L, M, N, R, S, W - permanent turf; C, D, I, J, P, T, U - overseeding of dormant warm-season turfs.

Released - Expected 1994 by International Seeds, Inc.

Breeder Seed/Stock - International Seeds, Inc.

Certified Seed/Stock - Not yet available.

Preparer/Additional Information - Stephen W. Johnson, International Seeds, Inc., P.O. Box 168, Halsey, OR 97348, (503) 369-2251.

Prelude

Developed and released by Lofts Seed, Inc., using germplasm obtained from the New Jersey AES. Carried as accession no. PI R-40.

Source - A 205-clone synthetic cultivar developed from plants collected from old turfs in Maryland, Pennsylvania, New York, and New Jersey and European origin.

Method of Breeding - Screening of more than 250,000; subsequently, 25,000 clones were evaluated in spaced-plant nurseries. Single-plant progenies in seed turf trials subjected to close mowing.

Intended Use - Home lawns, parks, athletic fields, and golf courses. Has performed well for fall and winter overseeding of dormant warm-season grasses in the southern U.S.

Description - Prelude is an early maturing, leafy, turf-type perennial ryegrass capable of producing a moderately dense, attractive, low-growing, fine textured turf with a bright green color. It has exhibited good heat tolerance and good summer performance characteristics.

Adapted to - LRR all except D, I, J, O, P, T, U, as a perennial; PHZ as a perennial 4, 5, 6, 7.

Released - 1982, Lofts Seed, Inc.

Breeder Seed/Stock - Lofts Seed with co-operation of the New Jersey AES.

Certified Seed/Stock - Available.

Preparer/Additional Information - Lofts Seed Inc., Chimney Rock Road, P.O. Box 146, Bound Brook, NJ 08805, (908) 560-1590.

Prelude II

Source - Prelude II is an advanced-generation synthetic cultivar selected from the maternal progenies of 20 clones.

Method of Breeding - Plants related to Prelude were crossed with other elite perennial ryegrass germplasm to initiate a program of phenotypic and genotypic recurrent selection for greater stress tolerance, improved turf performance, increased disease resistance, lower growth profile, and darker green color.

Intended Use - Cool-season turfgrass; also useful for fall overseeding of dormant warm-season grasses.

Description - Compared to Prelude, Prelude II has a darker green color, slightly finer leaves, a lower growth profile, and increased resistance to stem rust, net blotch, and scald. Prelude II is also endophyte enhanced.

Adapted to - Wherever cool-season grasses are found.

Released - Lofts Seed, Inc. with the cooperation of Pure Seed Testing and the New Jersey AES.

Breeder Seed/Stock - Lofts Seed, Inc.

Certified Seed/Stock - Available.

Preparer/Additional Information - Marie Pompei, Lofts Seed, Inc., P.O. Box 146, Bound Brook, NJ 08805, (908) 560-1590.

Premier

Source - Nine clones originating from plots at North Brunswick, NJ.

Method of Breeding - Synthetic cultivar developed by the New Jersey AES with C. Reed Funk.

Intended Use - For turf, either alone or in seed mixtures.

Description - Dark green, medium-fine texture, dense turf, good disease resistance, high endophyte for insect resistance, excellent mowing characteristics, and aggressive establishment.

Adapted to - All zones.

Released - Barenbrug USA/Normarc, 1984.

Breeder Seed/Stock - Barenbrug USA.

Certified Seed/Stock - Available, Barenbrug USA.

Preparer/Additional Information - Barenbrug USA, P.O. Box 239, Tangent, OR 97389, (503) 926 5801.

Prizm

Source - New York, New Jersey, Pennsylvania, Maryland, and Loretta.

Method of Breeding - Recurrent phenotypic selection for endophyte, a lower growth habit, dark green color, stress tolerance, and seed production.

Intended Use - Cool-season turfgrass also for overseeding of dormant warm-season turfs in southern U.S.

Description - Low growth habit, reduced clippings, and good density.

Adapted to - PHZ 3, 4, 5, 6, 7, 8, 9, 10.

Released - Zajac Performance Seeds.

Breeder Seed/Stock - Zajac Performance Seeds.

Certified Seed/Stock - Available. PVP pending.

Preparer/Additional Information - Zajac Performance Seeds, 33 Sicomac Rd., North Haledon, NJ 07508, (201) 423-1660.

Profit

Developed by Cebeco Zaden, B.V., Vlijmen, The Netherlands.

Source - Cross of elite plants from the famous old Dutch variety Ferma with selected clones from ecotypes collected in the Dutch polder "Beemster." Carried as experimental no. ELP 15.

Method of Breeding - Recombination of the 10 highest dry matter-producing clones, followed by two cycles of phenotypic mass selection to a variety with broad climactic adaptation.

Intended Use - High-quality pasture and hay production for cattle, sheep, and horses. Used with alfalfa as a companion or nurse crop. Also used for renovating old alfalfa stands.

Description - Late maturity, moderately broad-leafed, vigorous; very cold tolerant and productive, with superior persistence. Life of stand is three to seven years depending upon environment and management. Diploid (2n=2x=14).

Adapted to - LRR A, K, L, M, N, P, R, S, T, W; PHZ 3.

Released - 1986 in The Netherlands.

Breeder Seed/Stock - Cebeco Zaden B.V.

Certified Seed/Stock - Available.

Preparer/Additional Information - Craig W. Edminster, International Seeds, Inc, P.O. Box 168, Halsey, OR 97348, (503) 369-2251.

Quickstart

Source - Maryland, Pennsylvania, New Jersey, and New York.

Method of Breeding - Advanced generation synthetic resulting from eight cycles of recurrent selection for disease resistance, the presence of endophyte, and improved turf quality and seed yield.

Intended Use - Cool-season turfgrass; also for overseeding of dormant warm-season turfs in southern U.S.

Description - Quick establishment, improved turf quality, good leaf spot resistance, and fusarium resistance.

Adapted to - PHZ 3, 4, 5, 6, 7, 8, 9, 10.

Released - Turf-Seed, Inc.

Breeder Seed/Stock - Pure Seed Testing, Inc.

Certified Seed/Stock - Available. PVP No. 9100214.

Preparer/Additional Information - Turf-Seed, Inc., P.O. Box 250, Hubbard, OR 97032, (503) 651-2130.

Regal (Reg. No. 76)

Developed by International Seeds Inc., Halsey, OR.

Source - Population of turf-type perennial ryegrass plants assembled at Albany, OR, including derivatives of plants selected from old turfs in the mid-Atlantic U.S.

Method of Breeding - Polycross progeny evaluated under turf maintenance. Three plants selected to produce synthetic variety.

Intended Use - Commonly used for cool-season turf. Also well suited for overseeding dormant warm-season turfs in the southern U.S. It is used alone, in blends with other cultivars, and in mixtures with other turfgrass species.

Description - Open crowned, early heading cultivar; forms an exceptionally dark, moderately dense turf. Good resistance to red thread. Regal contains high levels of the endophytic fungus which is associated with increased resistance to insect damage.

Adapted to - LRR A, C, L, M, N, R, S, W, permanent turf. D, I, J, P, T, U, overseeding of dormant warm-season turfs; PHZ 5, southern half of 4 east of the Rocky Mountains.

Released - 1977 by International Seeds Inc.

Breeder Seed/Stock - Maintained by International Seeds Inc., Halsey, OR.

Certified Seed/Stock - Available from International Seeds Inc., Halsey, OR. U.S. PVP No. 7700110.

Preparer/Additional Information - Stephen W. Johnson, International Seeds Inc., P.O. Box 168, 820 W. First St., Halsey, OR 97348, (503) 369-2251.

Regency

Source - Crosses were made between selected plants from old Citation perennial ryegrass and selected plants from old-turf areas in New Jersey. The seedlings from 17 of these crosses were put in a space-plant nursery to initiate the cycles of selection. Selection was based on improved leaf spot resistance and vigor.

Method of Breeding - Regency is an advanced-generation synthetic variety resulting from four cycles of recurrent selection.

Intended Use - Athletic fields, golf course fairways, tees and roughs, home lawns, parks, and institutional and commercial turf. Especially useful as an overseeded turf in dormant warm-season grasses.

Description - Regency perennial ryegrass is a turf-type with bright dark green color, producing a low-growing turf with medium fine texture and medium density. Good cold, heat, and wear tolerance, as well as good resistance to brown patch and fusarium blight and moderate resistance to red thread and dollar spot. Very good mowing qualities, excellent seedling vigor and moderately good resistance to leaf spot and crown rust. It has a moderate level of the fungal endophyte for enhanced insect resistance.

Adapted to - PHZ 3, 4, 5, 6, 7.

Released - 1986, Pure Seed Testing, Inc.

Breeder Seed/Stock - Pure Seed Testing, Inc.

Certified Seed/Stock - Available.

Preparer/Additional Information - Art Wick, LESCO, Inc., 20005 Lake Rd., Rocky River, OH 44116, (216) 333-2950.

Repell (Reg. No. 93)

Developed by Lofts Seed, Inc., using germplasm obtained from the New Jersey AES. Carried as accession no. PI GT-11.

Source - Maternal plant from Central Park, New York City; paternal plant was PI 231,597 from Greece (resistant to crown rust). Recurrent parents include Manhattan, Citation, Pennfine, and others selected from old turfs in Maryland, New Jersey, Pennsylvania, and New York.

Method of Breeding - Repell is an advanced-generation synthetic cultivar selected from the progenies of 27 clones. Each clone contained a lolium endophyte which enhances resistance to a number of insect pests.

Intended Use - Parks, playgrounds, lawns, and winter overseeding of dormant warm-season turfs.

Description - A leafy, turf-type ryegrass producing persistent, dense, low-growing turf of dark green color. Seed of Repell contains high levels of endophyte, which provides enhanced pest resistance. Newly harvested seed or seed maintained in cold storage should be used in propagation to maintain endophyte effectiveness. This seed should be avoided in pastures and foragelands.

Adapted to - LRR all except D, I, J, O, P, T, U, as a perennial; PHZ as a perennial 4, 5, 6, 7.

Released - 1983, Lofts Seed, Inc.

Breeder Seed/Stock - Lofts Seed, Inc.

Certified Seed/Stock - Available.

Preparer/Additional Information - Lofts Seed Inc., Chimney Rock Road, P.O. Box 146, Bound Brook, NJ 08805, (908) 560-1590.

Repell II

Source - An advanced-generation synthetic cultivar selected from the maternal progenies of 17 clones. Plants were selected that had persisted and thrived in old lawn-type turfs under the environmental stresses, diseases, and insect pests common to the mid-Atlantic region.

Method of Breeding - Following varying cycles of phenotypic and genotypic recurrent selection, the 17 maternal clones of Repell II were selected from 11 separate breeding composites. Selection was based on attractiveness, rich dark green color, medium-low growth habit, good seed yield, and freedom from disease.

Intended Use - For cool-season turf; can also be used for fall overseeding of dormant warm-season grasses.

Description - Repell II is an attractive, persistent, lower-growing, turf-type cultivar capable of producing a turf with medium-fine texture, medium-high density, and a reduced rate of vertical growth. It has a darker green color than most ryegrasses in commercial production. Repell II is endophyte enhanced.

Adapted to - Wherever cool-season grasses are found.

Released - 1991, Lofts Seed, Inc.

Breeder Seed/Stock - Lofts Seed, Inc.

Certified Seed/Stock - Available.

Preparer/Additional Information - Marie Pompei, Lofts Seed, Inc., P.O. Box 146, Bound Brook, NJ 08805, (908) 560-1590.

Rodeo II

Source - Reselection of superior plant types out of Rodeo.

Method of Breeding Phenotypic matched mating in extensive polycross for three synthetic generations.

Intended Use - Turf and winter overseeding of southern golf courses and lawns.

Description - Contains endophyte, excellent clean cutting, medium dark green, semi-upright growth habit, moderate resistance to crown rust and leaf spot, and compact growth habit.

Adapted to - Zone for ryegrasses.

Released - 1990 by KWS-AG Einbeck, Germany.

Breeder Seed/Stock - KWS-Seeds, Inc. and TMI.

Certified Seed/Stock - Available.

Preparer/Additional Information - Dr. Fred B. Ledeboer, TMI, 22068 Case Road NE, Aurora, OR 97002, (503) 678-2597.

Saturn

Source - Selections from the following crosses Citation II x Manhattan II, Citation II x Prelude, Jazz x Manhattan II and Jazz x Prelude.

Method of Breeding - Advanced generation resulting from two cycles of selection for stem and crown rust resistance, improved seed yield, and turf quality.

Intended Use - Cool-season turfgrass; also for overseeding of dormant warm-season turfs in southern U.S.

Description - Good resistance to net blotch, pink snow mold, brown patch, and stem rust; good turf performance across the U.S.

Adapted to - PHZ 3, 4, 5, 6, 7, 8, 9, 10.

Released - Zajac Performance Seeds.

Breeder Seed/Stock - Zajac Performance Seeds.

Certified Seed/Stock - Available. PVP No. 8800114.

Preparer/Additional Information - Zajac Performance Seeds, 33 Sicomac Rd., North Haledon, NJ 07508, (201) 423-1660.

Seville PE 8

Developed by R.J. Peterson Enterprises Inc., Hillsboro, OR 97123, using germplasm from the New Jersey AES, New Brunswick, NJ.

Source - Plants selected from old turfs located in Maryland, New Jersey, New York, and Pennsylvania.

Method of Breeding - An advanced-generation synthetic cultivar selected from progenies of 14 clones. Selection was directed toward plants showing a lower growth profile during the short days of fall and early spring.

Intended Use - Turf use on lawns and sports fields in temperate zone and for overseeding of dormant warm-season turfs.

Description - A leafy persistent, turf-type perennial ryegrass. It is capable of producing an attractive turf with a rich dark green color, medium-high density, medium-fine leaf texture, improved mowing qualities, and a moderate reduction in rate of vertical growth.

Adapted to - Temperate Zone.

Released - 1989 by R. J. Peterson Enterprises, Inc., Forest Grove, OR.

Breeder Seed/Stock - R. J. Peterson Enterprises, Inc. with the assistance of the New Jersey AES.

Certified Seed/Stock - Available.

Preparer/Additional Information - Robert J. Peterson, R.J. Peterson Enterprises, Inc, P.O. Box 312, Forest Grove, OR 97116-1132, (503) 357-4336.

Sherwood

Developed by Cascade International Seed Co in cooperation with Jonathan Green and Sons, Inc., Hubbard Seed and Supply Co., Pickseed West Inc., and the New Jersey AES.

Source - The seven breeding composites originated from material collected by Dr. Reed Funk and others at the New Jersey AES in the spring of 1962. Final selection of the 105 parental clones of Sherwood was made during a period of heat and drought stress during early August of 1985.

Method of Breeding - Seedlings from seven breeding composites were established in a large spaced-plant nursery at Adelphia, NJ. Propagules of 105 selected parental clones of Sherwood were established in a replicated, randomized, isolated crossing block for the production of Syn-1 breeder seed. Seed was harvested from the 95 clones that contained endophytic fungus.

Intended Use - Turfgrass in areas of adaptation. Because of its diverse composition, it may be seeded alone or in mixtures and blends with other turfgrasses.

Description - Exhibits a medium-early heading date, a moderately low-growing mature plant height, and a moderately good level of stem rust resistance. Sherwood most closely resembles Citation II, from which it can be distinguished by a significantly longer flag leaf (153 mm vs. 116 mm), less resistance to stem rust; turf quality scores of Sherwood, however, are significantly higher than Citation II.

Adapted to - Extending from southern regions of Canada to southern regions of the U.S. where perennial ryegrass is used to overseed dormant bermudagrass. It has been tested in Europe and exhibits improved cold tolerance.

Released - 1988, Jonathan Green and Sons, Inc., Cascade International Seed Co., and Hubbard Seed and Supply Co.

Breeder Seed/Stock - Pickseed West, Inc.

Certified Seed/Stock - Available.

Preparer/Additional Information - Irvin H. Jacob, Cascade International Seed Company, 8483 West Stayton Road, Aumsville, OR 97325, (503) 749-1822.

SR 4000 (Reg. No. 115)

Developed by Pure Seed Testing, Inc., Seed Research of Oregon, and New Jersey AES - W.A. Meyer, C.A. Rose-Fricker, M.A. Robinson and C.R. Funk. Carried as accession no. PI 525459.

Source - Germplasm collected from old turfs in the eastern U.S.

Method of Breeding - Cycles of population improvement including phenotypic recurrent selection for resistance to stem rust and winter net blotch, attractive dark green color, medium-early maturity and presence of the Lolium endophyte. Progeny testing was in closely mowed trials receiving severe summer stress and incorporation of genes for stem rust resistance by modified backcrossing. From this, 11 clones were selected for uniformity, seed yield potential, and disease resistance and incorporated into a polycross progeny.

Intended Use - Home lawns, parks, sports turf, school grounds, and play areas. Winter overseeding of dormant warm-season turfgrasses.

Description - Attractive, leafy, persistent turf-type perennial ryegrass with a medium-fine texture, medium-high density, and a bright dark green color. Produces a moderately low-growing turf with improved mowing qualities, good winter-hardiness, good summer performance, and excellent wear tolerance. It shows good resistance to current races of crown rust, stem rust, winter net blotch, and large brown patch. It has good seedling vigor and the ability to produce high seed yields. Seed lots containing high percentages of Lolium endophyte can be expected to produce a turf resistant to many insects, including billbugs and sod webworms.

Adapted to - LRR A, B, C, D, E, F, G, H, K, L, M, N, P, R, S (I, J, O, T, U, V for overseeding); PHZ 3, 4, 5, 6, 7, 8, 10 (8, 9, 10 for overseeding).

Released - 1986 by Seed Research of Oregon.

Breeder Seed/Stock - Pure Seed Testing of Hubbard, OR, using precautions to maintain a high percentage of plants containing Lolium endophyte.

Certified Seed/Stock - Available.

Preparer/Additional Information - Dr. Leah A. Brilman, Seed Research of Oregon, Inc., P.O. Box 1416, Corvallis, OR 97339, (800) 253-5766.

SR 4100 (Reg. No. 112)

Seed Research of Oregon; Pure-Seed Testing, and New Jersey AES - M.F. Robinson, W.A. Meyer, R.F. Bara, W.K. Dickson and C.R. Funk. Carried as accession no. PI 518662.

Source - Breeding composite MC, derived from Citation II and M382 was used as the female parent with breeding composite BM, derived 90% from selections from old turfs in Maryland, Pennsylvania, and New Jersey and 10% from the European cultivar, used as the male parent.

Method of Breeding - Advanced generation synthetic from progenies of crosses of breeding composite MC and breeding composite BM. Some 4,656 seedlings were evaluated in a spaced-plant nursery for turf-type growth, bright dark green color, resistance to foliar diseases, medium-early maturity, and high seed yield. The 1,156 plants showing least damage from stem rust and highest floret fertility were harvested. Composite MC plants were selected based on progeny trials in turf and presence of Lolium endophyte. Composite BM plants were selected through cycles of selection based on screening for crown rust resistance and spaced-plant and turf trials.

Intended Use - Home lawns, parks, golf courses, athletic fields, schools, playgrounds, and industrial sites. Winter overseeding of dormant, warm-season turfgrasses.

Description - Leafy, turf-type perennial ryegrass producing an aggressive, persistent, dense, attractive, fine-textured, medium-low growing turf with a bright, dark green color. Good mowing qualities, winter-hardiness, heat tolerance, and summer performance. Excellent seedling vigor and good wear tolerance. Good resistance to crown rust, winter net blotch, and large brown patch; moderate resistance to stem rust. Seed lots with a high percentage of Lolium endophyte produce turf with resistance to billbugs and sod webworms.

Adapted to - LRR A, B, C, D, E, F, G, H, K, L, M, N, P, R, S (I, J, O, T, U, V for overseeding); PHZ 3, 4, 5, 6, 7, 8, 10 (portions of 8, 9, 10 for overseeding).

Released - 1986 by Seed Research of Oregon.

Breeder Seed/Stock - Seed Research of Oregon - Corvallis, OR with cooperation of Pure-Seed Testing and New Jersey AES under conditions to maintain high levels of Lolium endophyte.

Certified Seed/Stock - Available.

Preparer/Additional Information - Dr. Leah A. Brilman, Seed Research of Oregon, Inc., P.O. Box 1416, Corvallis, OR 97339, (800) 253-5766.

SR 4200

Seed Research of Oregon and Crop Science Dept. of Rutgers Univ. - M.F. Robinson, Leah A. Brilman, B.A. Adams and C.R. Funk.

Source - Plants collected from old turfs in Maryland, New Jersey, New York, and Pennsylvania from 1962 to 1977.

Method of Breeding - Superior plants selected from closely mowed turf plots and spaced plants. Intercrosses were employed to initiate multiple cycles of population improvement including phenotypic recurrent selection in spaced-plant nurseries, disease screening, and progeny evaluation in closely mowed turf trials. From this, 90 parental clones were selected from a spaced-plant nursery containing 8,000 plants based on low growth, attractiveness, medium-early maturing, and high seed yield potential. Syn-I seed was produced in isolation and 4,550 progeny from the 90 clones were established at Corvallis, OR. From these, 1,214 progeny were selected based on disease resistance, seed yield, and uniformity with breeder seed produced. Plants contained high levels of Lolium endophyte.

Intended Use - Home lawns, school play areas, parks, sports fields, and golf course tees, fairways and cart paths where adapted. Fall and winter overseeding of dormant warm-season turf where its lower growth profile makes it particularly useful.

Description - Lower-growing turf-type ryegrass of medium-early maturity and improved mowing qualities. High tillering produces fine-textured, leafy, dense turf that is less likely to become stemmy. Bright dark-green color and good tolerance of temperature extremes. Excellent seedling vigor, wear tolerance, and recuperative ability. Medium-high resistance to net blotch, and improved resistance to brown patch and moderate resistance to stem rust. Presence of the Lolium endophyte for insect resistance and endophyte-enhanced performance.

Adapted to - LRR A, B, C, D, E, F, G, H, K, L, M, N, P, R, S (I, J, O, T, U, V for overseeding); PHZ 3, 4, 5, 6, 7, 8, 10 (portions of 8, 9, 10 for overseeding).

Released - 1990 by Seed Research of Oregon, Corvallis, OR.

Breeder Seed/Stock - Seed Research of Oregon, Corvallis, OR.

Certified Seed/Stock - Available.

Preparer/Additional Information - Dr. Leah A. Brilman, Seed Research of Oregon, Inc., P.O. Box 1416, Corvallis, OR 97339, (800) 253-5766.

Stallion

KWS-AG, Einbeck, Germany.

Source - Progenies from collections from old pastures in southeastern Europe and selected clones out of Tando.

Method of Breeding - Clonal evaluation, phenotypically matched mating, individual progeny evaluation on the basis of turf performance, polycrossing of vegetative propagules, and two cycles of recurrent mass selection of upright, dark-colored types.

Intended Use - Dormant winter overseeding of sub-tropical grasses, permanent turf throughout the cool humid regions of the world.

Description - Strongly upright, moderately dark green color, quite coarse as individual plants but still forming a moderately dense sward with medium coarse texture; maturity is medium early.

Adapted to - The northern cool, humid regions; PHZ 4, 5, 6 (7, 8 winter overseeding).

Released - 1988, KWS-AG.

Breeder Seed/Stock - KWS-AG.

Certified Seed/Stock - Distributed by TMI, Tangent, OR.

Preparer/Additional Information - Fred B. Ledeboer, Ph.D., TMI, 22068 Case Road NE, Aurora, OR 97002, (503) 678-2597.

Stallion Select

Source - Selection from stem and crown resistant perennial ryegrass varieties.

Method of Breeding - Recurring phenotypic selection.

Intended Use - Turf.

Description - Early maturing, dark green color, semi-dwarf, multiple pest resistance to foliage diseases.

Adapted to - PHZ 2, 3, 4, 5, 6, 7, 8, 9, 10 - all areas where perennial ryegrass grows.

Released - 1990, Kevin McVeigh, Willamette Valley Plant Breeders.

Breeder Seed/Stock - Willamette Valley Plant Breeders.

Certified Seed/Stock - Available.

Preparer/Additional Information - David Lundell, Fine Lawn Research, Inc., P.O. Box 1051, Lake Oswego, OR 97034, (503) 636-2600.

Sunrise 246

Source - Rust-resistant progeny of Citation II, Manhattan II, and selections from old-turf areas in New Jersey.

Method of Breeding - Recurrent selection for dark green color and resistance to stem rust and leaf spot.

Intended Use - Turf blends and overseeding dormant warm-season turfgrasses.

Description - Low growing, excellent mowing qualities.

Adapted to - PHZ 4, 5, 6, 7, 8, 9 - east to west coast.

Released - 1986, Pure Seed Testing, Inc.

Breeder Seed/Stock - Pure Seed Testing, Inc.

Certified Seed/Stock - Available.

Preparer/Additional Information - Pennington Seed Inc. of Oregon, P.O. Box 386, Lebanon, OR 97355, (503) 451-5261.

Surrey

Selected at Florida AES - G.M. Prine. Carried as accession no. PI FL X 1986 LR.

Source - Marshall annual ryegrass.

Method of Breeding - Recurrent mass selection from gridded nurseries.

Intended Use - Pasture, hay, and silage production; ground cover of every kind.

Description - Late-maturing, crown-rust-resistant diploid ryegrass similar to Marshall in forage and seed yields, adaptation and cold tolerance, but slightly earlier maturity in most environments. Does not have non-vernalizing plants during warm growing seasons. This cultivar was developed to replace crown-rust-susceptible cultivars such as Marshall that were contributing to the higher incidence of ryegrass crown rust in the southeast. PVP has been applied for.

Adapted to - LRR A, J, N, O, P, T, U; PHZ 7, 8, 9, 10.

Released - 1989 by Florida AES.

Breeder Seed/Stock - Agronomy Department, University of Florida, Gainesville, FL.

Certified Seed/Stock - Available from International Seeds Inc., and Smith Seed Services, Inc., Halsey, OR.

Preparer/Additional Information - Gordon M. Prine, University of Florida, Agronomy Department, 304 Newell Hall, Gainesville, FL 32611, (904) 392-1811.

Tara

Source - Germplasm obtained from the New Jersey AES was used in the development of Tara (BT-1), Rutgers University, Cook College.

Method of Breeding - Selections were made for lower growth, color, disease resistance, finer texture, mowability, and overall turf performance. The selections were directed toward a mid-to-later maturity, and the variety to have no endophyte infection.

Intended Use - For a quality turf performance in low- to high-maintenance conditions in both full sun and moderate shade on golf courses, lawns, parks, school grounds, sports fields, and sod farms.

Description - Tara was top-rated in the 1983 national perennial ryegrass trial. It is a leafy-type perennial ryegrass, producing a persistent, dense, very attractive low-growing, fine-textured turf of a bright dark-green color. Resistant to brown patch, leaf spot, crown rust, and winter net blotch disease. Excellent seedling vigor and the ability to establish and grow on a wide range of soils. Displays improved mowing qualities, good heat tolerance, summer performance, and winter-hardiness. Performs well for winter overseeding of bermudagrass turfs throughout the southern U.S.

Adapted to - LRR A, L, M, N, R, S, W - Permanent turf; C, D, I, J, P, T, U - Overseed dormant warm-season turfs; PHZ 5, southern half 4.

Released - 1984, Hubbard Seed & Supply Co.

Breeder Seed/Stock - Hubbard Seed & Supply Co., Hubbard, OR.

Certified Seed/Stock - Available.

Preparer/Additional Information - Gordon W. & Sharon K. Jones, Hubbard Seed & Supply Co., P.O. Box 310, 13346 Whiskey Hill Road NE, Hubbard, OR 97032, (503) 981-4136.

Troubadour

Developed by Van Engelen Zaden B.V., Vlijmen, The Netherlands.

Source - The ryegrass varieties Elka and Manhattan.

Method of Breeding - Selections made from progeny of Elka by Manhattan cross.

Intended Use - Turf for sports fields, parks, home lawns, and commercial areas. Also planted as ground cover on roadsides and in orchards. May be used for overseeding warm-season grass, especially when quick spring transition is desired.

Description - Late flowering perennial with very fine, bright medium green leaves. Exhibits a reduced rate of vertical growth and requires less mowing than other ryegrass varieties. Exceptional wear tolerance and good winter-hardiness. Resistant to crown rust and red thread.

Adapted to - LRR A, C, L, M, N, R, S, W, permanent turf. D, I, J, P, T, U, overseeding of dormant warm season turf; PHZ 5, 4 east of the Rocky Mountains.

Released - The Netherlands by Van Engelen Zaden; in the U.S., 1988 by International Seeds Inc.

Breeder Seed/Stock - Maintained by Cebeco Zaden B.V.

Certified Seed/Stock - Available through International Seeds Inc. The suitability of Troubadour for use on sports fields has been established through extensive testing.

Preparer/Additional Information - Stephen W. Johnson, International Seeds Inc., P.O. Box 168, 820 W. First St., Halsey, OR 97348, (503) 369-2251.

Vantage

Developed by International Seeds Inc., Halsey, OR.

Source - Experimental breeding composite along with selections from Elka and other commercial cultivars.

Method of Breeding - Multiple cycles of mass selections.

Intended Use - Permanent turf in areas of the cool season and transition zones where perennial ryegrass is adapted. Also, it is well suited for winter overseeding of bermudagrass sod in the southwestern U.S.

Description - Leafy, moderately late maturing variety with excellent resistance to stem rust. Has exhibited good turf quality at multiple U. S. locations.

Adapted to - LRR A, C, L, M, N, R, S, W - permanent turf. C, D, I, J, P, T, U - overseeding of dormant warm turfs; PHZ 5, southern half 4 east of the Rocky Mountains.

Released - 1989, by Proprietary Seeds, Inc. of Salem, OR.

Breeder Seed/Stock - Maintained by International Seeds Inc.

Certified Seed/Stock - Available from Proprietary Seeds, Inc. U.S. PVP has been applied for No. 8900265.

Preparer/Additional Information - Stephen W. Johnson, International Seeds Inc., P.O. Box 168, 820 W. First St., Halsey, OR 97348, (503) 369-2251.

Vintage

Source - Parental material consisted of three sources of stem rust-resistance collected from old-turf areas of St. Louis and Washington, DC. This rust-resistant material was them crossed with improved turf-type perennial ryegrass clones selected from old turf-areas of the northeastern U.S. The seedlings from these crosses were then moved to space- plant nurseries to initiate the cycles of phenotypic recurrent selection from stem rust and leaf spot resistance and attractive appearance.

Method of Breeding - An advanced-generation synthetic variety resulting from four cycles or recurrent selection.

Intended Use - Overseeding golf courses and other turf areas in the southern U.S., parks, and institutional and commercial turf areas. Excellent companion with Regency and Commander in turf-type perennial ryegrass blends.

Description - A dark green colored variety capable of producing a dense, medium-fine textured turf. It has shown good heat and cold tolerance. It can produce a good northern turf and is well adapted for overseeding dormant bermudagrass in southern U.S. Vintage exhibits good mowing qualities, as well as good seedling vigor and wear tolerance.

Adapted to - PHZ 3, 4, 5, 6, 7.

Released - 1985 by Pure Seed Testing, Inc.

Breeder Seed/Stock - Pure Seed Testing, Inc.

Certified Seed/Stock - Available.

Preparer/Additional Information - Art Wick, LESCO, Inc., 20005 Lake Road, Rocky River, OH 44116, (216) 333-2950.

Wizzard

Developed by McCarthy-Burlingham Research Farm.

Source - Parent clones were selected from large space-plant nurseries on the basis of color, growth habit, disease resistance, and endophyte content.

Method of Breeding - Several lines of parents selected were polycrossed and cycled to obtain uniformity and consistent performance in turf trials.

Intended Use - Turf, overseeding of winter-dormant, warm-season grasses.

Description - Exceptional dark green color and dwarf characteristics, good mowing quality, dense and fine leafed, erect leaf blade growth habit, leaf spot and rust resistant.

Adapted to - PHZ 7, 8, 9, 10 and overseeding in 11 - Cool season climates.

Released - McCarthy-Burlingham Research in 1992.

Breeder Seed/Stock - McCarthy-Burlingham Research.

Certified Seed/Stock - Available in 1994.

Preparer/Additional Information - Mike McCarthy, E.F. Burlingham & Sons, P.O. Box 217, Forest Grove, OR 97116, (503) 359-9368.

Yatsyn 1

Source - Four-parent polycross synthetic variety from Auckland, New Zealand.

Method of Breeding - Selection of clones for increased summer-autumn yield and persistence. Method of maintenance by preservation of clones and seed.

Intended Use - Permanent pasture with winter growth and bulk production in early spring.

Description - Diploid, very early variety. Longest stem is very long. Some winter growth and very early spring growth. Very resistent to rust. Stock with endophytes is available.

Adapted to - Central and northeastern U.S. PHZ 1, 2, 3, 4, 5.

Released - Agriseeds, New Zealand.

Breeder Seed/Stock - New Zealand Agriseeds, Christchurch, New Zealand.

Certified Seed/Stock - Available in quantity from Barenbrug USA.

Preparer/Additional Information - Barenbrug USA, 32080 Old Highway 34, P.O. Box 239, Tangent, OR 97389, (800) 547-4101.

Yorktown III

Source - Yorktown III is an advanced-generation synthetic cultivar selected from the maternal progenies of 13 clones. The parental germplasm were collected from Central Park, New York City, Riverside and Paterson Parks in Baltimore, MD, and old lawns, parks, and sports fields in New Jersey and southwestern Pennsylvania.

Method of Breeding - Following varying cycles of phenotypic and genotypic recurrent selection, the 13 maternal clones of Yorktown III were selected from five separate breeding composites. Selection was based on attractive appearance, fine leaves, dark green color, medium-low growth profile, good seed yield, and relative freedom from disease.

Intended Use - Cool-season turfgrass; also for fall overfeeding of dormant warm-season grasses.

Description - An attractive, persistent, lower growing turf-type cultivar capable of producing a turf with a fine texture, medium-high density, and a reduced rate of vertical growth. It has darker green color than most commercial ryegrass varieties. It is also endophyte enhanced.

Adapted to - Wherever cool-season grasses are found.

Released - Lofts Seed, Inc. 1991.

Breeder Seed/Stock - Lofts Seed, Inc.

Certified Seed/Stock - Available.

Preparer/Additional Information - Marie Pompei, Lofts Seed, Inc., P.O. Box 146, Bound Brook, NJ 08805, (908) 560-1590.

Lolium perenne L. ssp. *multiflorum* (Lam.) Husnot - Italian ryegrass

Major cool-season annual from Europe. Grown principally in Pacific coast states west of the Cascades and as a winter annual in southern U.S.. Used for pasture, hay, silage, cover crop, temporary lawns, and overseeding warm-season turfgrasses in southern U.S.. Yields well on productive soils; palatable; tolerates heavy grazing. Subject to winterkilling in northern states. Some common seed sources represent genetic mixtures of Italian and perennial ryegrass; hybrids between these two species used in developing improved varieties, which behave as short-lived perennials.

Ace

Snow Brand Seed Company, Chiba, Japan.

Source - Base population of 2,000 plants of tetraploid Italian ryegrass from a Snow Brand variety and foreign varieties. One hundred plants were selected and transplanted to isolation plot to obtain F_1 seed.

Method of Breeding - The above nursery was subjected to two cycles of selection and remultiplication in isolated polycross blocks. The F_2 seed was used to propagate 5,000 spaced plants in 1970. The best 1,000 plants were used to create a third crossing block. F_3 seed was mass harvested and designated as breeder seed in 1972.

Intended Use - Forage.

Description - Ace is most similar to Tetrone and can be differentiated from it by being a little earlier and having better summer survival and higher dry-matter yields. The stalk length is longer (81.4 cm versus 73.6 cm for Tetrone) and 1,000 seeds weigh less (36.2 mg versus 44.53 mg for Tetrone). Ace is very resistant to snow blight and snow mold.

Adapted to - LRR A, B, C, D, E, H, I, J, K, L, M, N, O, P, R, S, T; PHZ 4, 5, 6, 7, 8.

Released - 1984, Japan.

Breeder Seed/Stock - Maintained by Snow Brand Seed Company at Chiba, Japan.

Certified Seed/Stock - Ace is multiplied by Jacklin Seed Company, Albany, OR.

Preparer/Additional Information - Susan H. Samudio, Jacklin Seed Company, W. 5300 Riverbend Ave., Post Falls, ID 83854, (208) 773-7581.

Barmultra

Developed by Barenbrug Holland N.V., Arnhem, The Netherlands.

Source - Collections from lawns, pastures, and other uncultivated areas.

Method of Breeding - Selection and evaluation of individual plants, clones, and families.

Intended Use - Forage.

Description - Late-heading tetraploid; extensive and fibrous root system; quick recovery after grazing or cutting; resistant to mildew and various races of rust. High digestibility and high soluble carbohydrate content.

Adapted to - Throughout the U.S., extensively used in the south.

Released - 1963, Agency for U.S. and Canada, Steven J.R. Frohlich & Co., Princeton, NJ.

Breeder Seed/Stock - Barenbrug Holland N.V.

Certified Seed/Stock - This line is available and is eligible for certification.

Preparer/Additional Information - Barenbrug USA, P.O. Box 239, Tangent, OR 97389, (503) 926-5801.

Billiken

Goro Miura of Snow Brand Seed Company was the breeder.

Source - Tetraploid Westerwold ryegrass.

Method of Breeding - Developed by repeated mass selection from population of tetraploid Westerwold ryegrass which was induced by colchicine treatment in Sapporo, a northern district of Japan, for the purpose of having early growth, good regrowth, and erect type habit.

Intended Use - Pasture, silage, hay.

Description - Billiken is most similar to Grasslands Tama. Billiken can be differentiated from Tama by better early growth vigor, formation of more heads, lighter color, better regrowth, and plant height (78.4 cm versus 75.8 cm for Tama). Billiken has fair resistance to crown rust and resistance to snow blight, typula blight, and dollar spot.

Adapted to - LRR A, B, C, D, E, H, I, J, K, L, M, N, O, P, R, S, T; PHZ 4, 5, 6, 7, 8.

Released - 1985, Jacklin Seed Company.

Breeder Seed/Stock - Maintained by Snow Brand Seed Company.

Certified Seed/Stock - Available.

Preparer/Additional Information - Susan H. Samudio, Jacklin Seed Company, W. 5300 Riverbend Ave., Post Falls, ID 83854, (208) 773-7581.

Bulldog

Developed by Lofts Seed, Inc. Also known as L-FAR-1.

Source - Some of the germplasm used in the development of Bulldog can be traced back to plants selected out of Marshall, Magnolia, and Florida 80.

Method of Breeding - Recurrent selection, synthetic variety.

Intended Use - Forage use in southern U.S.; winter overseeding on warm-season grass pasture.

Description - A high-yielding, forage-type ryegrass.

Adapted to - LRR T, O, P, U, J; PHZ 7, 8, 9, 10.

Released - 1987.

Breeder Seed/Stock - Lofts Seed, Inc., 1986 in Rickreall, OR.

Certified Seed/Stock - Not available.

Preparer/Additional Information - Lofts Seed Inc., Chimney Rock Road, P.O. Box 146, Bound Brook, NJ 08805, (908) 560-1590.

Florida 80 (Reg. No. 102)

Selected at University of Florida Institute of Food and Agricultural Sciences, Agronomy Department, Gainesville, FL - G.M. Prine. Carried as accession no. FL 1977 B.

Source - Reseeding populations of Kinderlou (local reseeding ecotype in Lowndes County, GA), Florida Rust Resistant, Magnolia, Gulf, and other unknown reseeding ecotypes.

Method of Breeding - Four cycles of mass selection from gridded nurseries.

Intended Use - Pasture, hay, and ground cover.

Description - Early-maturing, rapid-developing, crown rust-resistant diploid ryegrass which grows well in mixture with clovers. Earlier in maturity than Gulf. Excellent reseeding, if properly managed. Superior yields December to March.

Adapted to - A, P, T, U; PHZ 8, 9, 10.

Released - 1982 by Florida Agricultural Experiment Stations.

Breeder Seed/Stock - Agronomy Department, University of Florida, Gainesville, FL.

Certified Seed/Stock - Available.

Preparer/Additional Information - Pennington Seed Inc. of Oregon, P.O. Box 386, Lebanon, OR 97355, (503) 451-5261.

Florida Rust Resistant

Selected at North Florida Experiment Station, Quincy - T.E. Webb and W.H. Chapman.

Source - Selections from domestic ryegrass and introductions.

Method of Breeding - Mass selection.

Intended Use - Pasture and hay production and various ground cover uses.

Description - Rapid developing. Rust resistant, with about 90% of plants highly resistant to rust. Earlier in maturity than Gulf and much earlier than common ryegrass. Good seed producer, possessing slender seedheads that are unusually weak, as is typical of many improved varieties.

Adapted to - LRR A, P, T, U; PHZ 8, 9, 10.

Released - 1965, by Florida AES.

Breeder Seed/Stock - Agronomy Dept., Univ. of Florida, Gainesville.

Certified Seed/Stock - Available. Normarc Seed Co., Tangent, OR.

Preparer/Additional Information - Gordon M. Prine, University of Florida, Agronomy Department, 304 Newell Hall, Gainesville, FL 32611, (904) 392-1811.

Georgia Reseeding (Reg. No. GP-56)

ARS, Coastal Plain Experiment Station, University of Georgia, Tifton, GA - W.W. Hanna, W.G. Monson, and P.R. Utley. Carried as accession no. PI 517948.

Source - An adapted reseeding of an early maturing ecotype collected from a pasture near Valdosta, GA.

Method of Breeding - Two cycles of mass selection for vigorous leafy plants and crown rust resistance.

Intended Use - Especially suited for overseeding on sods of warm-season perennial grasses such as bermudagrass and bahiagrass.

Description - An early maturing annual ryegrass with up to 98% of its plants flowering by the first week of May compared to less than 10% of the plants of Marshall or Common at Tifton, GA. Shows crown rust and stem rust resistance. Dry-matter yields equaled the yields of the best commercial cultivars in 29 of 37 tests over 12 years. It is a diploid.

Adapted to - LRR J, O, P, T; PHZ 7, 8.

Released - April 1987 by ARS and the University of Georgia, Coastal Plain Experiment Station at Tifton, GA.

Breeder Seed/Stock - ARS, Tifton, GA.

Certified Seed/Stock - Not available.

Preparer/Additional Information - Wayne Hanna, ARS Coastal Plain Experiment Station, P.O. Box 748, Tifton, GA 31793, (912) 386-3353.

Gulf (Reg. No. 8)

Increased at Texas A&M University Agricultural Research and Extension Center, Beaumont, Texas, ARS cooperating - R.M. Weihing. Carried as accession no. PI 193145.

Source - PI 193145, introduction of La Estanzuela 284 received from Uruguay.

Method of Breeding - Comparative tests.

Intended Use - As a winter forage utilized by grazing cattle.

Description - High forage production and quality and crown rust resistance. Good seed yields in Oregon, but susceptible to winter killing below -12°C.

Adapted to - LRR J, O, P, T, U; PHZ 8, 9.

Released - 1958, cooperatively by Texas AES, College Station, and Plant Science Research Division, ARS.

Breeder Seed/Stock - Foundation seed, Texas A&M University, College Station, TX.

Certified Seed/Stock - Not available.

Preparer/Additional Information - Dr. Lloyd R. Nelson, Texas A&M Research Center, Drawer E, Overton, TX 75684, (214) 834-6191.

Jackson (Reg. No. 135)

Mississippi Agricultural and Forestry Experiment Station, Mississippi State, MS - C.E. Watson, Jr., S.D. McLean, and N.C. Edwards, Jr. Carried as accession no. MSR-86-1.

Source - Selections from Marshall annual ryegrass.

Method of Breeding - Three cycles of phenotypic recurrent selection for crown rust resistance utilizing artificial inoculation techniques in the greenhouse.

Intended Use - Winter annual pasture and cover crop.

Description - Intermediate to late maturing diploid cultivar; flowers approximately one week later than Gulf; tall, erect-growing type with wide leaves; good forage yield potential; forage yields similar to Gulf and Marshall; highly cold tolerant, only slightly less than that of Marshall; highly resistant to crown rust.

Adapted to - LRR A, H, J, N, O, P, T, U; PHZ 6, 7, 8, 9, 10.

Released - 1989 by Mississippi Agricultural and Forestry Experiment Station.

Breeder Seed/Stock - Mississippi Agricultural and Forestry Experiment Station.

Certified Seed/Stock - Available.

Preparer/Additional Information - C.E. Watson, Jr., Mississippi State Univ., Agronomy Department, Box 5248, Mississippi State, MS 39762, (601) 325-2732.

Magnolia (Reg. No. 15)

Selected at Delta Branch Experiment Station and Mississippi Agricultural and Forestry Experiment Station, State College, MS, ARS cooperating - H.W. Bennett and H.W. Johnson.

Source - PI 194395 introduced from Uruguay (contained 35% rust-resistant plants when random population of seedlings inoculated in greenhouse). PI 201980, introduction of La Estanzuela 284, obtained from Uruguay (contained 51% rust-resistant plants). Other accessions were PI 194394 and 193145, and T.O. 1882.

Method of Breeding - Selfing and progeny testing after artificial inoculation with crown rust was conducted at Stoneville and State College, MS. Seedlings that remained rust-free were transplanted to the field, and the cycle continued with selections resistant to rust and leaf spot. Several rust-resistant experimental synthetics were developed at both locations. State College No. 7 and Stoneville No. 3 were selected on the basis of yield, comparable maturity, and rust resistance, and equal quantities of seed blended for production of breeder seed. State College No. 7 and Stoneville No. 3 are maintained as separate entities, and equal quantities of seed are blended to provide new plantings for breeder-seed production.

Intended Use - Winter annual pasture and cover crop.

Description - Similar in appearance to common annual ryegrass and comparable in maturity; high yielding and highly resistant to rust; well adapted in MS and adjacent states. Roots of one-quarter to one-half of the seedlings give negative reaction when tested for fluorescence; this suggests inclusion in the original introductions of natural crosses between annual and perennial ryegrass.

Adapted to - LRR H, J, N, O, P, T, U, A; PHZ 6, 7, 8, 9, 10.

Released - 1965, cooperatively by Mississippi AES and Plant Science Division, ARS.

Breeder Seed/Stock - Mississippi AES.

Certified Seed/Stock - Available.

Preparer/Additional Information - C.E. Watson, Jr., Mississippi State U., Agronomy Department, Box 5248, Mississippi State, MS 39762, (601) 325-2732.

Marshall (Reg. No. 72)

Mississippi Agricultural and Forestry Experiment Station, Mississippi State, MS - B.L. Arnold, C.E. Watson, Jr., and N.C. Edwards, Jr. Carried as North Mississippi Reseeding.

Source - Seed harvested from a pasture at the North Mississippi Branch Experiment Station, Holly Springs, MS, which had been planted to common annual ryegrass and allowed to reseed itself under grazing conditions for 29 years.

Method of Breeding - Ecotype selection resulting from natural selection.

Intended Use - Winter annual pasture and cover crop.

Description - Late maturing diploid cultivar; flowers approximately two to three weeks later than Gulf; tall, erect-growing type with wide leaves; highly cold tolerant; good late and early spring forage production; good forage yield potential, particularly in areas subject to cold winters; good seed producer.

Adapted to - LRR A, H, J, N, O, P, T, U; PHZ 6, 7, 8, 9, 10.

Released - 1980, by Mississippi Agricultural and Forestry Experiment Station.

Breeder Seed/Stock - Mississippi Agricultural and Forestry Experiment Station.

Certified Seed/Stock - Available.

Preparer/Additional Information - C.E. Watson, Jr., Mississippi State Univ., Agronomy Department, Box 5248, Mississippi State, MS 39762, (601) 325-2732.

Sakura-Wase

Snow Brand Seed Company, Daini-Hokkai Bldg, 3-8 Higashi-Nihonbashi 3 Chome, Chuo-Ku, Tokyo 103, Japan.

Source - In 1974, 4,000 plants of Tottori annual ryegrass and 28 additional lines were selected for early maturity and erect growth type. Selected plants were transplanted in an isolation plot for F_1 seed production by natural crossing.

Method of Breeding - Six lines were selected out of F_1 seed and transplanted to isolation plots for F_2 seed production. This produced 30 lines of which eight lines (54 plants) were selected and transplanted to isolation to produce F_3 seed. Seeds from above 54 plants used to establish mass production plot of 1,000 plants;, offtypes (10%) were rogued. This produced F_4 seed which became breeders seed.

Intended Use - For short season, early forage.

Description - Sakura-Wase most closely resembles Gulf and can bc differentiated by Sakura-Wase's plant height being 12 cm shorter and flagleaf size being 13 cm longer and 3.5 cm wider. Sakura-Wase also has an earlier heading date than Gulf.

Adapted to - LRR A, B, C, D, E, H, I, J, K, L, M, N, O, P, R, S, T; PHZ 4, 5, 6, 7, 8.

Released - Applied to Japanese Plant Variety Protection Certificate (MAFF) and was accepted October 21, 1982.

Breeder Seed/Stock - Maintained by Snow Brand Seed Company, Japan.

Certified Seed/Stock - Available.

Preparer/Additional Information - Susan H. Samudio, Jacklin Seed Company, W. 5300 Riverbend Ave., Post Falls, ID 83854, (208) 773-7581.

Tachiwase

Snow Brand Seed Company, Daini-Hokkai Bldg, 3-8 Higashi-Nihonbashi 3 Chome, Chuo-Ku, Tokyo 103, Japan.

Source - In 1979-80, started with 5,500 plants each from Tottori, Miyazaki, and Waseaoba and selected 130 plants from each variety for early maturity and erect growth. Three separate 130-plant isolation fields were started from the transplants to get F_1 seed.

Method of Breeding - F_1 was space planted; 28 plants were selected for lodging resistance and erect type which were transplanted and isolated to get F_2 seed. F_2 seed was screened at two locations for erect and vigorous plants which were transplanted to isolation fields for F_3 seed production. This was repeated until F_4 seed was put into mass production fields and rogued for offtypes. F_5 seed became breeder seed.

Intended Use - Areas which need early maturing forage.

Description - Tachiwase closely resembles Gulf annual ryegrass and can be differentiated from Gulf in that Tachiwase exhibits better spring vigor, earlier heading date, spike length of 25.7 cm versus 30.6 cm for Gulf, longer awns, darker green leaf color and an extra erect growth habit and upright leaves (not drooping).

Adapted to - LRR A, B, C, D, E, H, I, J, K, L, M, N, O, P, R, S, T; PHZ 4, 5, 6, 7, 8.

Released - 1987 by Snow Bran Seed Company in Japan.

Breeder Seed/Stock - Maintained by Snow Brand Seed Company.

Certified Seed/Stock - Available.

Preparer/Additional Information - Susan H. Samudio, Jacklin Seed Company, W. 5300 Riverbend Ave., Post Falls, ID 83854, (208) 773-7581.

TAM 90

Source - Texas AES at Overton. Carried as accession no. TX-R-85-2.

Method of Breeding - Polycross involving three parents which were Gulf, Marshall, and TX-R-78-2.

Intended Use - Winter forage crop.

Description - TAM-90 was developed for improved forage yield potential and winter-hardiness. It is a robust diploid annual ryegrass with wide leaves that make it look like some tetraploids. It was selected at Amarillo and Overton, TX, for winter-hardiness.

Adapted to - LRR I, J, P, T, A; PHZ 7, 8, 9.

Released - 1991.

Breeder Seed/Stock - The Texas AES will maintain breeders seed. Contact person is: Dr. L. R. Nelson, P.O. Drawer E, Overton, TX 75684.

Certified Seed/Stock - TAM-90 will be protected under PVP, and be available through several seed companies in Texas and the southeastern U.S.

Preparer/Additional Information - Dr. Lloyd R. Nelson, Professor of Plant Breeding, Texas A&M Research Center, Drawer E, Overton, TX 75684, (903) 834-6191.

Wimmera 62 (Reg. No. 11)

Selected at Plant Materials Center, SCS, Pleasanton, CA - H.W. Miller and O.K. Hoglund. Carried as accession no. PI P-11419.

Source - Naturalized in Wimmera-Mallee areas of Victoria, Australia. Possibly hybrid between *Lolium rigidum* Gaudin and *L. perenne* ssp. *multiflorum* (Lam.) Husnot. Seed obtained from F.H. Brunning, Ltd., Melbourne, Australia.

Method of Breeding - Natural selection and roguing for 13 generations at Plant Materials Center, SCS, Pleasanton, CA. Increased at Plant Materials Center, SCS, Pleasanton, CA, and SCS Nursery, San Fernando, CA, as P-11419.

Intended Use - Cover crop, revegetation of disturbed areas, and wildfire burn rehabilitation.

Description - Awnless, deep bright green, erect, early-maturing annual. Tends to lodge first year in brush-burn seedings. Uniform in appearance, leafy, with numerous culms. Provides rapid temporary cover for erosion control. Useful as self-perpetuating cover crop in irrigated orchards and vineyards. No advantage over common ryegrass in areas of high humidity, on fertile soils where rainfall exceeds 300 mm annually, or above 610 m in elevation.

Adapted to - LRR A, C; PHZ 9.

Released - 1962, cooperatively by California AES, Davis, CA, and Plant Materials Center, SCS, Pleasanton, CA.

Breeder Seed/Stock - Plant Materials Center, SCS, Lockeford, CA.

Certified Seed/Stock - Limited availability.

Preparer/Additional Information - R. Slayback, SCS, 2121-C, 2nd St., Davis, CA 95616, (916) 757-8257.

***Muhlenbergia wrightii* Vasey ex Coult. - spike muhly**

Warm-season, tufted native grass that has short rootstalks. Found in southwestern Colorado, Arizona, New Mexico, and Mexico. Common throughout the region, but abundant only in local areas. Reasonably productive and palatable. Adapted to widely varying moisture conditions and valued for soil-binding qualities.

El Vado

Increased at Plant Materials Center, SCS, Los Lunas, NM, and New Mexico AES, Las Cruces, cooperating - G.C. Niner and J.E. Anderson. Carried as accession no. PI 476996, P 15618, NM199.

Source - Collected in 1958, 16 k west of Park View, NM, at an elevation of 2,230 m and 400 mm annual precipitation.

Method of Breeding - Increase of original collection under isolation, as NM-199.

Intended Use - Revegetation of disturbed areas and range reseeding.

Description - Fairly upright as compared with average of other collections tested. Average culm height 580 mm, leaf length 400 mm, basal leaves dense and wider than average tested, plants uniform. Seed production very good. Free of gall (caused by wheat curl mite), which has been found in varying degrees on seed heads of all other strains tested. This gall can reduce or eliminate seed production in certain years.

Adapted to - LRR D, G, H; PHZ 4.

Released - 1973, cooperatively by SCS-Los Lunas and New Mexico and Colorado AESs.

Breeder Seed/Stock - Plant Materials Center, SCS, Los Lunas, NM.

Certified Seed/Stock - Available.

Preparer/Additional Information - Plant Materials Center, SCS, 1036 Miller St. SW, Los Lunas, NM 87031, (505) 865-4684.

Nassella viridula **(Trin.) Barkworth, green needlegrass - feather bunchgrass**

Stipa viridula **Trin.**

Cool-season, native bunchgrass found from Wisconsin to Montana and south to New Mexico. Valuable in northern Great Plains, Wyoming, and Colorado. Palatable and nutritious. Starts growth relatively early and remains green until late in season.

Lodorm (Reg. No. 19)

Selected at Northern Great Plains Research Laboratory, ARS, Mandan, ND - H.M. Schaaf and G.A. Rogler. Carried as accession no. Mandan 2611.

Source - Bulk collection made in native stand north of Bismarck, ND, in 1935.

Method of Breeding - Three cycles of recurrent selection for low seed dormancy were practiced within progenies of selected parents of this highly self-pollinated species.

Intended Use - Lodorm has rapid recovery growth and high palatability and has been used in monoculture and in mixtures with other grasses and alfalfa to provide complementary pasture for native rangelands in the northern Great Plains.

Description - Except for low seed dormancy, Lodorm has no other traits that distinguish it from other cultivars and strains of green needlegrass. Superior to green stipagrass in level of dormancy exhibited by newly harvested seed and comparable in forage and seed yields - Morphologically indistinguishable from green stipagrass

Adapted to - LRR E, F, G. *Released* - 1970, by ARS in cooperation with State Agricultural Experiment Stations of North Dakota, South Dakota, and Montana.

Breeder Seed/Stock - U.S. Northern Great Plains Research Laboratory, Mandan, ND.

Certified Seed/Stock - Available.

Preparer/Additional Information - John Berdahl, ARS, Northern Great Plains Research Laboratory, P.O. Box 459, Mandan, ND 58554, (701) 663-6445.

SD-93

Selected at the Plant Materials Center, SCS, Bismarck, ND - John McDermand, Erling T. Jacobson, and Russell J. Haas. Carried as accession no. PI 478007.

Source - A commercial harvest of seed in 1962 from native grasslands in western South Dakota (Custer County), near Fairburn. Seed was purchased from Bober Seed Company, Rapid City, SD.

Method of Breeding - Mass selection and increase of an individual ecotype.

Intended Use - For use in mixtures for range seeding, critical area stabilization, surface mine revegetation, rural beautification, and other plantings where the establishment of native vegetation is the objective. Forage produced is of high quality, nutritious, and well liked by livestock.

Description - This accession is a leafy, cool-season bunch grass with fine stems. It has good drought resistance, is winter-hardy, and can tolerate mildly alkaline conditions. Since green needlegrass is a predominantly self-pollinating species, the accession is fairly uniform. The average height is 75 cm and it tends to have better stand persistence than the standard varieties.

Adapted to - LRR F, G; PHZ 3a - 4b.

Released - No. (Seed available for field testing.)

Breeder Seed/Stock - Plant Materials Center, SCS, Bismarck, ND.

Certified Seed/Stock - Not available.

Preparer/Additional Information - Russell J. Haas, SCS, P.O. Box 1458, Bismarck, ND 58502, (701) 250-4425.

Oryzopsis hymenoides **(Roemer & J.A. Schultes) Ricker ex Piper - Indian ricegrass**

Valuable cool-season, native bunchgrass. Widely distributed as mixture with other native grasses over western U.S. from North Dakota to Washington and south to California and Texas. Adapted to dry, sandy soils; drought resistant. Nutritious, palatable; good as standing winter seed. High level of seed dormancy restricts use.

CSU-3

Department of Agronomy, Colorado State University, Fort Collins, CO - Robin L. Cuany, Gary L. Thor, and R.S. Zemetra.

Source - Plants of Paloma, which resulted from germinating six-month old seed with no scarification or other dormancy breaking treatment.

Method of Breeding - Recurrent selection for unstimulated six-month germination of hand-harvested mature seed set on plants in isolated blocks which were selected as seedlings.

CSU-3 is derived from 120 plants in the third cycle. This generation has shown up to 59% germination compared to 17% for control Paloma, both measured in greenhouse soil rows with no dormancy breaking treatment.

Intended Use - For dryland, range, especially sandy soils and blowout areas, to stop further wind erosion. Should be adapted wherever Paloma is adapted but will come up more reliably because of lower dormancy. Grower should be able to save on seeding rate.

Description - Dryland, cool-season bunchgrass of good palatability and adapted to sandy areas. Nutritious forage cures well and has special value on winter ranges and at elevations of 600-2,500 m in semi-arid, plains, foothills, and intermountain plateaus, including dry southwest aspects. In 1987 test of 1,986 seeds, CSU-3 germinated 34% compared to 4% for Paloma. The absolute levels change somewhat from year to year, but CSU-3 is always more non-dormant than the source cultivar Paloma.

Adapted to - LRR D, E, G, H; PHZ 3, 4, 5.

Released - Not yet released; in advanced trials.

Breeder Seed/Stock - Department of Agronomy, Colorado State University, Fort Collins, CO.

Certified Seed/Stock - Not available. A foundation seed plot is growing at the Plant Materials Center, SCS, Los Lunas, NM.

Preparer/Additional Information - Robin L. Cuany, Dept. of Agronomy, Colorado State Univ., Ft Collins, CO 80523, (303) 491-6832.

CSU-10

Department of Agronomy, Colorado State University, Fort Collins, CO - Robin L. Cuany, Gary L. Thor, and R.S. Zemetra.

Source - Ten parent plants from: Warner farm (two), Paloma (one), NM-168 (three), M-700 (one), Breeders (North Dakota) (two), and P-15597 (one).

Method of Breeding - Phenotypic and progeny-test selection of clones in a source nursery of 15 accessions led to 34 clones with acceptable seed size and seed yield, both greater than average. Progenies were tested in the Piceance Basin of northwest Colorado on disturbed soils for emergence, vigor, and survival to third growing season. Parental clones of the 10 best progenies were put into a crossing-block for composite or synthetic production.

Intended Use - For dryland ranges, especially sandy soils and blowout areas, to stop further wind erosion. Tested for adaptation and good emergence in loamy soils of intermountain plateau region of northwest Colorado.

Description - Dryland cool season bunchgrass of good palatability and adapted to sandy and loamy soils. Nutritious forage, cures well and has special value on winter ranges and at elevations of 600-2,500 m in rolling hills and plains, foothills and intermountain plateaus including dry southwest aspects. As a broad-based composite, it should have some adaptation over a wide range from Montana to New Mexico and Idaho to Utah.

Adapted to - LRR D, E, G, H; PHZ 3, 4, 5.

Released - Not yet released; in advanced trials.

Breeder Seed/Stock - Maintained by Department of Agronomy, CSU, Fort Collins, CO.

Certified Seed/Stock - Not available. A foundation seed plot is growing at Plant Materials Center, SCS, Los Lunas, NM.

Preparer/Additional Information - Robin L. Cuany, Dept. of Agronomy, Colorado State Univ., Ft Collins, CO 80523, (303) 491-6832.

Nezpar

Selected at Plant Materials Center, SCS, Pullman, WA - J.L. Schwendiman. Carried as accession no. P-2575 and PI 469230.

Source - Native collection made in 1935, eight kilometers south of White Bird, ID.

Method of Breeding - Selected from among 152 accessions for its good vegetative characteristics and low hard-seed content. Selection repeated through several generations before initial increase.

Intended Use - Forage for livestock and wildlife and for soil stabilization, particularly on sandy soils.

Description - Large, erect plant type. Robust stems; broad, flat, abundant leaves; medium-small, dark, almost naked, elongate seeds. Excellent seedling vigor, averaging less than 50% of hard seeds.

Adapted to - LRR B, D, E; PHZ 4, 5.

Released - 1978 by the Plant Materials Center, SCS, Aberdeen, ID.

Breeder Seed/Stock - Plant Materials Center, SCS, Aberdeen, ID.

Certified Seed/Stock - Available.

Preparer/Additional Information - Gary Young, Plant Materials Center, SCS, Box 296, Aberdeen, ID 83210-0296, (208) 397-4133.

Paloma

Selected at Plant Materials Center, SCS, Los Lunas, NM, and New Mexico AES, Las Cruces; cooperatively - G.C. Niner, and J.E. Anderson. Carried as accession nos. PI 476997, C-42, BN-15909.

Source - Collected 3.6 kg in 1957 west of Pueblo, CO, by Bob Searway, SCS, at an elevation of 1,530 m and precipitation of 250-300 mm. Soils were medium textured. Growing in association with blue grama and western wheatgrass.

Method of Breeding - Bulk increase of multi-plant source material. Selected in comparison rows with 62 other strains in Los Lunas, NM; Las Cruces, NM; and Tucson, AZ. Stands, vigor, seed yields, and forage production were consistently higher for Paloma.

Intended Use - Paloma was the first named cultivar of Indian ricegrass and is intended for use in soil stabilization, range revegetation, and mined land reclamation. Paloma is excellent forage grass and considered highly palatable to livestock and wildlife.

Description - Besides its superiority in stand establishment, vigor, and forage production, Paloma is very drought tolerant. In one trial, Paloma was superior in resistance to root rot damage.

Adapted to - LRR D, G, E, H; PHZ 4.

Released - 1974, cooperatively with SCS and New Mexico and Colorado AESs and New Mexico State Highway Department.

Breeder Seed/Stock - Maintained by Plant Materials Center, SCS, Los Lunas, NM.

Certified Seed/Stock - Available.

Preparer/Additional Information - Plant Materials Center, SCS, 1036 Miller St. SW, Los Lunas, NM 87031, (505) 865-4684.

Panicum amarum Elliot - coastal panicgrass

***Panicum amarulum* A.S. Hitchc. & Chase**

Occurring along the tall, robust, slightly rhizomatous, warm-season perennial grass coastal regions from Massachusetts to Texas. Well suited for direct seeding on coastal sand dunes and other critical sites; additional value as standup winter cover for wildlife, and grass barriers for wind erosion control.

Atlantic (Reg. No. 82)

USDA-SCS and NJ AES - C.R. Belcher, W. Curtis Sharp, R.W. Duell. Carried as accession no. PI 421136.

Source - USDI, Back Bay Wildlife Refuge, Princess Anne, VA.

Method of Breeding - Bulk selection of original seed collected in 1955.*Intended Use* - Stabilization of disturbed areas, including sand dunes and inland sandy, gravelly sites.

Description - Tall, robust, native warm-season perennial grass. Upright calms grow to a height of 2 m. It is bunchgrass that spreads from ascending rhizomes. Bluish green, leafy plant. Resembles switchgrass.

Adapted to - LRR N, R, S, T; PHZ 5a, 8b.

Released - 1981, SCS, NJ AES.

Breeder Seed/Stock - SCS, Cape May Plant Materials Center, Cape May Court House, NJ.

Certified Seed/Stock - Available.

Preparer/Additional Information - Christopher Miller, SCS, 1370 Hamilton St., Somerset, NJ 08873, (908) 246-4110.

Panicum amarum var. *amarulum* (A.S. Hitchc. & Chase) P.G. Palmer, bitter panicgrass

Decumbent warm-season perennial occurring on coastal dunes from New Jersey to Texas. Reproduction is by vigorous rhizomes which spread in the sand dunes; produces very little viable seed; propagation is by culms and rhizomes. Used for stabilizing sand dunes.

Northpa

Plant Materials Center, SCS, Brooksville, FL. Carried as PI No. 421957.

Source - Native collection from near Wanchese, Dare County, NC, by Dr. S. Broome, (North Carolina State University, Raleigh, NC), 1972.

Method of Breeding - Selected from a collection from North Carolina State University. Evaluated at NC Experiment Station, Clayton, and transferred to Brooksville Plant Materials Center where it was additionally evaluated. In field plantings in combination with more southerly accessions, it was sole survivor.

Intended Use - Coastal dune erosion control plantings in area of the Carolinas and northward.

Description - Sub-erect to slightly decumbent growth habit.

Foliage dies back in winter regardless of geographic location planted. Strong stems fall over, trap sand, and produce a root-stem network. Established only vegetatively at present.

Adapted to - LRR T; PHZ Coastal 8.

Released - 1992 by SCS, Plant Materials Center, Brooksville, FL.

Breeder Seed/Stock - Plant Materials Center, SCS, Brooksville, FL.

Certified Seed/Stock - Not available.

Preparer/Additional Information - Plant Materials Center, SCS, 14119 Broad St., Brooksville, FL 34601, (904) 796-9600.

Southpa

Plant Materials Center, SCS, Brooksville, FL. Carried as PI no. 561721.

Source - Collected from a native stand in Palm Beach County, FL, in 1977 by R.E. Somer.

Method of Breeding - Single accession selected from a collection. Selection made on basis of comparative growth and survival in various geographic areas.

Intended Use - Coastal dune vegetation extending south from NC to and including the Gulf coast.

Description - Upright growth, excellent for sand trapping. Vegetatively propagated.

Adapted to - LRR T, U; PHZ southern 8, 9, 10.

Released - 1992 Plant Materials Center, SCS, Brooksville, FL.

Breeder Seed/Stock - Plant Materials Center, SCS, Brooksville, FL.

Certified Seed/Stock - Not available.

Preparer/Additional Information - Plant Materials Center, SCS, 14119 Broad St., Brooksville, FL 34601, (904) 796-9600.

Panicum antidotale **Retz. - blue panicgrass**

Warm-season, sod-forming grass native to India and introduced from Australia in 1912. Important in parts of southwestern U.S. for dryland and irrigated pastures and erosion control. Not winter-hardy in much of the U.S. Coarse, vigorous, extensive root system. High forage and seed yields. Grows best on fertile, well-drained soils; responds to nitrogen.

A-130

Increased at SCS Nursery, Tucson, AZ. Carried as accession no. PI 469224.

Source - Australia.

Method of Breeding - Increase of accession.

Intended Use - Dryland/irrigated pasture and wildlife applications.

Description - Original increase of blue panicgrass widely used in Texas and in southwestern U.S.

Adapted to - LRR D; PHZ 7, 8, 9, 10.

Released - 1950, cooperatively by Arizona AES and SCS Nursery, Tucson.

Breeder Seed/Stock - Plant Materials Center, SCS, Tucson, AZ.

Certified Seed/Stock - Available.

Preparer/Additional Information - Plant Materials Center, SCS, 3241 N. Romero Rd., Tucson, AZ 85705, (602) 241-2966.

Panicum coloratum **L. - kleingrass**

Warm-season species complex introduced from Africa. Includes bunchgrasses and spreading types used for hay, pasture, and silage, primarily in southern Texas. Adapted to moist, heavy soils. Withstands considerable drought; not cold tolerant. Seed subject to shattering.

OKPC-1

Selected at the Oklahoma State University, Stillwater, OK - Oklahoma AES and USDA-ARS cooperating. C.M. Taliaferro and R.M. Ahring.

Source - PI's 142284, 166400, 196363, and 196364.

Method of Breeding - In 1954, equal amounts of seed of the four plant introductions were bulked and planted for increase on the Southwestern Livestock and Forage Research Station (SWLFRS), El Reno, OK (35¯ 31' N. latitude). During the ensuing 26 years, the population was advanced through 10 generations using seed from plants that survived one or more winters. All of the generational advances were made on the SWLFRS except the last, which was made on the North Central Research Station near Lahoma, OK (36¯ 25' N. latitude).

Intended Use - Germplasm source of increased cold tolerance.

Description - No artificial selection was imposed during the 10 generations of advance and OKPC-1 is highly heterozygous producing heterogeneous plant populations. Morphological characters for which variability has been observed include foliage color, leaf/stem ratio, pubescence, growth habit, and inflorescence traits such as color and compactness. Comparison of OKPC-1 and Selection 75 in small plot trials and space plant nurseries in Oklahoma indicated the former to have substantially better winter-hardiness as indicated by spring recovery and growth. The enhanced cold tolerance of OKPC-1 presumably resulted from natural selection during the generations of advance. However, OKPC-1 does not have sufficient winter-hardiness for reliable pasture use north of about 34° N. latitude.

Adapted to - PHZ 6, marginally 7.

Released - 1982, Oklahoma AES.

Breeder Seed/Stock - Oklahoma AES.

Certified Seed/Stock - Available.

Preparer/Additional Information - C.M. Taliaferro, Agronomy Department, 368 Ag Hall, Oklahoma State Univ., Stillwater, OK 74078, (405) 744-6410.

Selection 75

Selected at the SCS Nursery, San Antonio, TX, Texas AES cooperating - J. E. Smith, Jr. Exp. No. T-20275, PMT-10. Carried as accession no. PI 166400.

Source - Introduced from Kimberley, Union of South Africa. Received March 1952 as PI 166400, BN-5225.

Method of Breeding - Selected as best in forage production from large number of similar African accessions. Evaluated in pasture and range planting for persistence and animal response. Increased for testing as T-20275.

Intended Use - Pastureland, range reseeding, hay production, wildlife food, and cover (birds).

Description - Produces abundant, hard seed. Nutrition forage readily grazed by all classes of livestock; whitetail deer graze young growth. Moderate salinity tolerance. Drought hardy. Responds well to fertilizer and/or irrigation. Cures for good winter forage in drier regions. Growth starts late winter or early spring and continues to late fall.

Adapted to - LRR D, H, I, J, M, N, P, T; PHZ 7, 8.

Released - 1957, informally by SCS. Formally in 1969, cooperatively by Texas AES and Plant Sciences Division, SCS.

Breeder Seed/Stock - Plant Materials Center, SCS, Knox City, TX.

Certified Seed/Stock - Available.

Preparer/Additional Information - Plant Materials Center, SCS, Rt 1 Box 155, Knox City, TX 79529, (817) 658-3922.

TEM-LD1

Carried as accession no. PI 559908.

Source - Base population from Selection 75 (75%) and OKPC-1 germplasm (25%).

Method of Breeding - Six cycles of recurrent selection for rapid germination immediately following harvest.

Intended Use - A source of kleingrass germplasm with low seed dormancy.

Description - Germination of freshly harvested TEM-LD1 was 89%, that of the base population was 19%. Winter survival of TEM-LD1 was less than Selection-75. Forage yields of TEM-LD1 were similar to Selection-75.

Adapted to - Not widely tested.

Released - October 1991, USDA-ARS and Texas AES.

Breeder Seed/Stock - Grassland, Soil & Water Research Lab., Temple, TX.

Certified Seed/Stock - Not available. Germplasm.

Preparer/Additional Information - B.A. Young, USDA-ARS, 808 E. Blackland Rd., Temple, TX 76504, (817) 770-6524.

TEM-SR1

Carried as accession no. PI 564168.

Source - Open-pollinated seed from PI 410177 grown in a large and diverse kleingrass nursery.

Method of Breeding - Two cycles of recurrent selection for resistance to seed shattering.

Intended Use - A source of resistance to seed shattering in *Panicum coloratum* L.

Description - TEM-SR1 has over twice the seed-retaining capability of the cultivar Selection-75. Dry-matter yields significantly lower than Selection-75 and Verde.

Adapted to - Not widely tested.

Released - October 1992, USDA-ARS.

Breeder Seed/Stock - Grassland, Soil and Water Research Lab., Temple, TX 76502.

Certified Seed/Stock - Not available. Germplasm.

Preparer/Additional Information - B.A. Young, USDA-ARS, 808 E. Blackland Road, Temple, TX 76502, (817) 770-6524.

Verde

Carried as PI-483444.

Source - Developed under the experimental designations of 73-23, 73-24, and 77-28. African introductions from which

Kleingrass 75 originated and others obtained after that time provided the germplasm from which Verde was developed. Selection for increased seed size (weight) resulted in approximately 70 plants selected from two parent clones and 38 O.P. progeny.

Method of Breeding - Heritability estimate for seed weight was 68.2% with expected advance of 22.8% based on selection of the top 10% of the parent plants. Two polycross blocks; one based on 11 parent clones (74-23) and one based on the top 15 O.P. progeny (74024) in the parent-O.P. progeny nursery. The parents of both crosses were combined in a new cross (77-28) after eliminating six of the 26 parent plants based on progeny performance.

Intended Use - Pasture and range reseeding.

Description - Verde does not differ appreciably from Kleingrass 75, except in seed size. The cultivar is described as follows: A fibrous root system; the plant spreads by tillers, which are initiated throughout the growing season, and/or short rhizomes. Fine stems, both erect and geniculate, 1-1.25 m at maturity. Culms simple or branched, ranging from glabrous to minutely pubescent. Geniculate stems will root at nodes when in contact with moist soil. Leaves generally medium green to occasionally glaucous.

Adapted to - LRR D, H, I, J, M, N, P, T; PHZ 7, 8.

Released - 1982, Texas AES and SCS.

Breeder Seed/Stock - Texas AES.

Certified Seed/Stock - Available.

Preparer/Additional Information - Texas A&M University, AES, Dept. of Range Science, College Station, TX 77843, (409) 845-5579.

Panicum hemitomon J.A. Schultes - maidencane

Spread by extensive creeping rhizomes, often producing numerous sterile shoots; culms 50-150 cm tall; occurs on moist soil along rivers and ditches, borders of lakes, ponds, often in the water, sometimes a weed; occurs naturally in coastal regions from New Jersey to Florida and Texas; Tennessee.

Halifax

Selected at Plant Materials Center, SCS, Coffeeville, MS. Mississippi Agricultural and Forestry Experiment Station cooperating. Carried as accession nos. PI 434171, MS 2108.

Source - Collected from a native stand near U.S. Highway 1 south of Halifax, NC, by Karl E. Graetz.

Method of Breeding - Vegetative increase from original collection.

Intended Use - Shoreline and stream bank erosion control in the southern U.S.

Description - More vigorous, robust, cold hardy, and spreads by rhizomes faster than material from native stand.

Adapted to - LRR T, P, O, U; PHZ 7.

Released - 1974, cooperatively by Mississippi AES and Plant Materials Center, SCS, Coffeeville, MS.

Breeder Seed/Stock - None. Established from rhizomes.

Certified Seed/Stock - None. Limited amount of vegetative material being produced.

Preparer/Additional Information - David Lane, SCS, Jamie L. Whitten Plant Materials Center, Rte 3, Box 215A, Coffeeville, MS 38922, (601) 675-2588.

Panicum maximum Jacq. - guineagrass

Warm-season bunch grass from Africa. Used primarily for pasture in Hawaii and the Pacific Basin. Used to limited extent for pasture and silage in Florida and parts of southern Texas and California. Tall, rather coarse. Adapted in moist regions, but tolerates some drought; not cold tolerant; grows well on fertile soils. Nutritive value high when leafy and green. Established vegetatively and from seed. Several varieties available in tropical regions. Slender guineagrass (*Urochloa maxima* (Jacq.) R. Webster, formally *Panicum maximum* var. *publiglume* K. Schum), small, slender variety with finer leaves than common, and purple top *(P. maximum* var. *coloratum),* low, coarse variety; both grown in Queensland, Australia. Brazilian varieties include ordinary robust type of common and Sempre-Verde and fine-leaved, drought-resistant type. A major limiting factor is indeterminate seed ripening and seed shattering, reducing its value.

HA-696

Selected plants increased at Plant Materials Center, SCS, Wailuku, HI - E.A. Lewis. Carried as accession no. PI 156074.

Source - PI 156074.

Method of Breeding - Selected plants from early observational planting on Oahu used to vegetatively establish observational planting at Plant Materials Center, Wailuku, in 1957. Seed harvested for further increase.

Intended Use - Forage and grazing.

Description - Fine-stemmed, fine-leaved type that regenerates in approximately 40 days; produces good yields of forage and seed; stands up well under heavy grazing. In Hawaii is adapted to annual rainfall of 600-2,500 mm and from sea level to 760 m elevation.

Adapted to - LRR I, T, U, V, Z; PHZ 9.

Released - No. Distributed for field-scale plantings throughout Hawaii.

Breeder Seed/Stock - Plant Materials Center, SCS, Hoolehua, Molokai, HI.

Certified Seed/Stock - Not available.

Preparer/Additional Information - Plant Materials Center, SCS, P.O. Box 236, Hoolehua, HI 96729, (808) 567-6378.

Panicum miliaceum L. - proso millet

Warm-season annual introduced from Asia. Grown as grain crop since prehistoric times. Grown in China, former Soviet Union, the Balkans, northwestern Europe, and northern Great Plains in the U.S. It is a short-season plant often requiring only 60-65 days from seeding to maturity. Hay is coarse and forage yield comparatively low. In the U.S. used as feed grain, in birdseed mixtures, and as food crop for game birds.

Cerise

Developed at the University of Nebraska, Panhandle Station.

Source - The initial seed source was obtained by selecting the red seeds from a predominantly white seeded line, PI 170603. This selection was tested as R.B. 170603 Red in 1972 and 1973.

Method of Breeding - The line was further purified by selecting 150 heads of the most uniform plants growing in the plots in 1972. These selections were grown head to row in the greenhouse for two generations to eliminate selections with color, panicle, or height variations.

Intended Use - Grain production.

Description - Cerise is a red-seeded proso with an open panicle. It was evaluated at six locations in 1972 and four locations in 1973. It resembles Turghai in appearance, with similar panicle type, seed color, and height. Cerise heads slightly earlier than Turghai and has yields equal to or slightly better than Turghai. The seed is used primarily for caged bird and wild bird feed but also can be used for human food and livestock feed.

Adapted to - Nebraska.

Released - University of Nebraska, Panhandle Station, Mitchell, NE.

Breeder Seed/Stock - University of Nebraska, Panhandle Station, Mitchell, NE.

Certified Seed/Stock - Available.

Preparer/Additional Information - David D. Baltensperger, Panhandle Research & Extension Center, 4502 Avenue I, Scottsbluff, NE 69361, (308) 632-1261.

Dawn

Developed at the University of Nebraska, Panhandle Station.

Source - Tested under the experimental number IPm 1108 which was in a proso collection including PI 260053 from the former USSR.

Method of Breeding - The PI line was purified through head selection to give a plant type which has short stature, compact panicle, and a uniform white seed color.

Intended Use - Grain production.

Description - Dawn was evaluated from 1970 to 1975 at several locations. It has a yield record slightly lower and more erratic than Panhandle in the 1.2 m wide plots but performed better in wider plots. In 1974 and 1975, Dawn was allowed to stand in the field until the moisture was at a safe storage level. Under the weather conditions of those two years, Dawn showed good promise for direct combining. The seed is used primarily for caged bird and wild bird feed but also can be human food and livestock feed.

Adapted to - Nebraska.

Released - 1976 by University of Nebraska, Panhandle Station.

Breeder Seed/Stock - University of Nebraska, Nebraska AES, Scottsbluff.

Certified Seed/Stock - Available.

Preparer/Additional Information - David D. Baltensperger, Panhandle Research & Extension Center, 4502 Avenue I, Scottsbluff, NE 69361, (308) 632-1261.

Dove

Selected at the Plant Materials Center, SCS, Americus, GA; Georgia AES, Athens, cooperating. Carried as accession no. PI 196292.

Source - PI 196292 received from India.

Method of Breeding - Best vigor and grain yield among 35 accessions evaluated at Americus, GA. Seed multiplied at Plant Materials Center, SCS, Coffeeville, MS.

Intended Use - Wildlife food and cover, primarily doves.

Description - In contrast to most other varieties, it is adapted to southern latitudes. Quick maturing, upright, loose-drooping panicle, and seed of a light straw color. Pale or yellow green, lax leaves and comparatively heavy stems. Useful food plant for upland game birds.

Adapted to - LRR T, P; PHZ 8.

Released - 1977 by SCS and University of Georgia.

Breeder Seed/Stock - Plant Materials Center, SCS, Americus, GA.

Certified Seed/Stock - Not available. Common seed is available.

Preparer/Additional Information - Plant Materials Center, SCS, 295 Morris Dr., Americus, GA 31709, (912) 924-2286.

Panhandle

Selected at Nebraska AES, Lincoln.

Source - Common white seed stock included in testing program.

Method of Breeding - Twelve plants selected, with seed from individual heads planted in single progeny rows. Seed from four most similar rows bulked and rogued for offtype plants for several generations.

Intended Use - Forage and grain production.

Description - Medium in height, relatively early, only fair straw strength, heads semicompact, and seed shattering average. Seed creamy white with high test weight. In Nebraska, produces higher grain yields than other varieties tested.

Adapted to - Widely adapted in Nebraska.

Released - 1967, by Nebraska AES.

Breeder Seed/Stock - Nebraska AES.

Certified Seed/Stock - Available.

Preparer/Additional Information - David D. Baltensperger, Panhandle Research & Extension Center, 4502 Avenue I, Scottsbluff, NE 69361, (308) 632-1261.

Rise (Reg. No. 89)

Developed at the University of Nebraska, Panhandle Station. It was tested under the experimental number 76004-3-8.

Source - Derived from a cross of Dawn x Minn. 402 made at the Nebraska AES in 1976.

Method of Breeding - A single plant selection was made in the F_2 generation for simply inherited traits such as panicle type, height, and seed color. The head row planted from the F_2 selection was again selected in the F_3 generation for lodging resistance and to more precisely determine height, seed color, and panicle type. The seed produced in the F_4 generation was bulked to obtain enough seed to begin testing the F_5 generation in 1980.

Intended Use - Grain production.

Description - Rise was tested for three years in six yield trials per year in western Nebraska. Rise is about 12-15 cm taller than Dawn and about 5-8 cm shorter than Panhandle. It has a heading date midway between Cope and Dawn, similar to Minco. It had grain yields that exceeded Cope, Dawn, and Minco during all three years of testing. It has a compactum panicle type similar to Dawn, although the seed is smaller than Dawn. It has lodging resistance similar to Dawn. Planting rates and dates for Rise will be similar to other cultivars of medium maturity.

Released - Nebraska AES.

Breeder Seed/Stock - Nebraska AES.

Certified Seed/Stock - Available.

Preparer/Additional Information - David D. Baltensperger, Panhandle Research & Extension Center, 4502 Avenue I, Scottsbluff, NE 69361, (308) 632-1261.

Sunup (Reg. No. 124)

Developed by the University of Nebraska AES.

Source - Sunup (79012-9-B-8) is an increase of an F_4 derived proso line from the cross Rise x Dawn made in 1979.

Method of Breeding - Identified as a line in 1981, grown in an observation block in 1982, and entered in preliminary yield trials in 1983. From 1984 to 1988 it was tested in yield trials in western Nebraska. Current breeder seed originated from a block grown in Sidney in 1987 and increased at Mead in 1988.

Intended Use - Grain production.

Description - A white seeded proso millet with seed size intermediate between Dawn and Rise. It has a head type that is compactum, but not as compact as Dawn. Sunup is 5 cm taller than Rise, but is not as tall as Panhandle. In five years of testing involving 25 locations, Sunup has yielded 100 kg ha-1 more than Rise and 300 kg ha-1 more than Minco and Cope. Sunup has been as resistant to lodging as Dawn or Rise in spite of its taller height.

Sunup has performed well throughout western Nebraska and has a good yield record on black fallow, notill, early, and late seeding. Sunup has a seed size larger than Rise, and matures two days earlier than Rise.

Adapted to - It has not been tested outside of Nebraska, but should be adapted to areas north of Nebraska.

Released - 1989, University of Nebraska, AES, Lincoln, NE.

Breeder Seed/Stock - Foundation Seed Division, University of Nebraska, Lincoln.

Certified Seed/Stock - Available.

Preparer/Additional Information - David D. Baltensperger, Panhandle Research & Extension Center, 4502 Avenue I, Scottsbluff, NE 69361, (308) 632-1261.

***Panicum virgatum* L. - switchgrass**

Important warm-season, native, sod-forming grass. Occurs throughout most of U.S.. Especially valuable for forage, pasture, and erosion control in central and southern parts of Great Plains. Most abundant on relatively moist, fertile areas. Coarse stemmed, vigorous root system; short rhizomes. High yields of seed and forage. Good seedling vigor. Quality acceptable during periods of rapid growth, but low as standing winter feed.

421138

Increased at SCS, Plant Materials Centers, Beltsville, MD, and Cape May Court House, NJ. Carried as accession no. PI 421138.

Source - Single clone collected vegetatively by K. E. Graetz in 1957 near Carthage, NC.

Method of Breeding - Clone multiplied vegetatively in isolation. Open-pollinated seed from this isolation constituted initial material for multiplication. Tested under numbers BN-8624, SC-56-32, AM-77, and NJ 50.

Intended Use - Grazing, hay, wildlife strips, critical areas, streambanks, dams, and spoil-disposal areas.

Description - Leafy, better than average spread, high nutrient value and early spring recovery. Seed production fair. Growth about 1.5 m tall. Adapted to NC, TN, AR, and northward, but northern limits undetermined. Suggested for use as grazing, hay, wildlife strips, and for planting streambanks, dams, and spoil-disposal areas.

Adapted to - LRR S, P, T, N; PHZ 7, 8.

Released - No. Distributed for testing.

Breeder Seed/Stock - Plant Materials Center, SCS, Cape May, NJ.

Certified Seed/Stock - Not available.

Preparer/Additional Information - D.W. Hamer, Plant Materials Center, SCS, 1536, Rte 9 No, Cape May Court House, NJ 08210, (609) 465-5901.

Alamo

Selected in 1975 at James E. "Bud" Smith Plant Materials Center, Knox City, TX, and increased for field planting as T-788. Carried as accession no. PI 422006.

Source - Original seed collection made by Laramie McIntire of the SCS near George West, TX, in 1964.

Method of Breeding - Selected over similar accessions as better forage and seed producer and to extend range south of previous cultivars.

Intended Use - Range reseeding, pastureland, hay production, wildlife food and cover (birds), revegetation, shoreline stabilization, and erosion control.

Description - Longer, wider leaves than Blackwell; taller and much greater forage producer. Adapted south of area where Blackwell and Kanlow perform well, to the Rio Grande Valley of Texas. Flowers in fall one to two months later than Blackwell. Moderate salinity tolerance. Grows on all soil types with 630 mm or more annual precipitation. Responds well to fertilizer and/or irrigation.

Adapted to - LRR H, I, J, M, N, O, P, T, U; PHZ 6.

Released - 1978 - SCS, Texas AES.

Breeder Seed/Stock - Plant Materials Center, SCS, Knox City, TX.

Certified Seed/Stock - Not available.

Preparer/Additional Information - Plant Materials Center, SCS, Rte 1 Box 155, Knox City, TX 79529, (817) 658-3922.

Blackwell

Selected at Plant Materials Center, SCS, Manhattan, KS. Carried as accession nos. 421520 and KG-208.

Source - Seed harvested in 1934 from single plant growing in native prairie near Blackwell, OK.

Method of Breeding - Single plant selected in comparison with many other collections at SCS Nursery, Manhattan. Tested as KG-208.

Intended Use - Range reseeding, pasture plantings, waterways, and revegetation of disturbed areas.

Description - Upland-type switchgrass of medium height, with rather large stems. Ranked high in leafiness, total forage produced, and resistance to rust and other diseases. Good seedling vigor. Wide adaptation in range seedings, pasture plantings, waterways, and other permanent plantings in Kansas, Oklahoma, southern Nebraska, and northern Texas in areas of 500 mm or more of annual precipitation. Will grow on sandy ranges and favorable lowland sites in area of 380-760 mm of annual precipitation.

Adapted to - LRR D, G, H, J, L, M, N, O, P, R, S; PHZ 5a.
Released - 1944, cooperatively by Plant Materials Center, SCS, Manhattan, KS, and Kansas AES.

Breeder Seed/Stock - Plant Materials Center, SCS, Manhattan, KS.

Certified Seed/Stock - Available.

Preparer/Additional Information - SCS, 760 S. Broadway, Salina, KS 67401, (913) 823-4541.

Caddo (Reg. No. 4)

Selected at Oklahoma AES, Stillwater, ARS cooperating - H.W. Staten, W.C. Elder, R.A. Chessmore, and J.R. Harlan.

Source - Field collections from southern Great Plains, especially central Oklahoma.

Method of Breeding - Mass selection in space-planted nurseries, with elimination of undesirable types. Process repeated, using most promising lines; seed from selected plants used to establish rows. Five rows selected for uniformity and superior production; seed bulked to form experimental strain 4200.

Intended Use - Forage production for grazing and haying.

Description - Tall, robust, upland switchgrass generally characteristic of central Oklahoma. Leafy, productive, considerable rust resistance, rather uniform when seeded in rows for seed production. Gives heavy yield of seed under favorable conditions. Forage yield under irrigation outstanding for native grass; recovers well after mowing. No special features distinguish it positively from other varieties, but tends to be greener and contains less red pigment in stems and heads than many other varieties.

Adapted to - LRR H; PHZ 7.

Released - 1955, cooperatively by Oklahoma AES and Plant Science Research Division, ARS.

Breeder Seed/Stock - Oklahoma AES.

Certified Seed/Stock - Not available.

Preparer/Additional Information - C.M. Taliaferro, Oklahoma AES, OSU, Agronomy Department, 368 Ag Hall, Stillwater, OK 74078, (406) 744-6421.

Cave-In-Rock

Plant Materials Center, SCS in cooperation with the Missouri AES. Carried as accession no. PI 469228.

Source - Selected from a native stand of grass at Cave-In-Rock, IL, in 1958 by Virgil B. Hawk and R.K. Lawson, Agronomist. Seed of this accession was planted in the Plant Materials Center rod rows for testing. Plants proved superior to other accessions including Blackwell. These plants were then put out in a two-row breeders block in 1962. Seed from this block was used to establish a foundation field in 1967.

Method of Breeding - Cross-pollination.

Intended Use - Pasture, hay production, and range reseedings.

Description - Greater seedling vigor, more resistance to dampening off or leaf spot, higher seed yields, resistance to lodging, lowland type of switchgrass. Tolerant to flooding, will withstand droughty soils but is better suited to moderately wet soils.

Adapted to - LRR H, M, N, O; PHZ 4b.

Released - 1973.

Breeder Seed/Stock - Plant Materials Center, SCS, Elsberry, MO.

Certified Seed/Stock - Available.

Preparer/Additional Information - Plant Materials Center, SCS, RR 1, Box 9, Elsberry, MO 63343, (314) 898-2012.

Dacotah (Reg. No. CU-132)

Collected by the ARS, Northern Great Plains Research Laboratory, Mandan, ND, and Plant Materials Center, SCS, Bismarck, ND - George Rogler, Reed E. Barker, John McDermand, Erling T. Jacobson, and Russell J. Haas. Carried as accession nos. 478002, NDG-965-98.

Source - Original plants collected from a native stand near Breien, Morten County, ND.

Method of Breeding - Collected plants were grown in comparison with other collections. After three generations in open pollinated nurseries, 10 plants were selected for uniform plant type with good leafiness, high plant vigor, seed yields, winter hardiness, and uniform green color. Phenology, persistence, and forage yield have been extensively evaluated by SCS in comparative field evaluation studies and on farm field plantings in North Dakota, South Dakota, and Minnesota.

Intended Use - Erosion control, critical area, wildlife habitat, surface mine revegetation, range, pasture, and natural area seedings.

Description - Dacotah is 27 days earlier in anthesis than Forestburg and 45-50 days earlier than Blackwell, Summer Cave-in-Rock, Pathfinder, and Nebraska-28. Tends to be shorter in mature height and have less rank growth than southern cultivars. Appears to have increased drought tolerance for this species. Chromosome number 2n=4x=36

Adapted to - LRR F, G, K; PHZ 3a - 4a.

Released - Cooperatively in 1989 by the ARS; Plant Materials Center, SCS, Bismarck, ND; and the North Dakota and Minnesota Agricultural Experiment Stations.

Breeder Seed/Stock - ARS, Northern Great Plains Research Laboratory, Mandan, ND.

Certified Seed/Stock - Available in quantity.

Preparer/Additional Information - Russell J. Haas, SCS, P.O. Box 1458, Bismarck, ND 58502, (701) 250-4425.

Forestburg (Reg. No. 110)

Selected at the Plant Materials Center, SCS, Bismarck, ND, and ARS, Northern Great Plains Research Laboratory, Mandan, ND, cooperating - John McDermand, Erling T. Jacobson, Russell J. Haas, and Reed E. Barker. Carried as accession nos. 478001, PM-SD-149.

Source - Composite of four accessions collected from native stands in Sanborn County near Forestburg, SD.

Method of Breeding - Selection based on performance in comparison with many accessions in initial evaluation studies. The phenology, persistence, forage yield and quality, animal performance, and wildlife habitat potential have been extensively tested in comparative field evaluation studies and field plantings in North Dakota, South Dakota, and Minnesota.

Intended Use - Range and pasture seedings, wildlife habitat, natural areas, surface mine revegetation, and critical area seedings.

Description - Superior winter-hardiness and persistence, seed production ability, and earlier maturity than other accessions. Forage production at northern latitudes exceeds that of Dacotah and is equal to/greater than Nebraska 28. Forestburg is similar in performance and adaptation to Sunburst. Chromosome number 2n=6x=36. Average daily gains of yearling steers were slightly higher for Forestburg than Pathfinder in grazing studies at Morris, MN.

Adapted to - LRR F, G, K, M; PHZ 3b - 4b.

Released - Released cooperatively in 1987 by the Plant Materials Center, SCS, Bismarck, ND; ARS; and the North Dakota, South Dakota, and Minnesota Agricultural Experiment Stations.

Breeder Seed/Stock - ARS, Northern Great Plains Research Laboratory, Mandan, ND.

Certified Seed/Stock - Available.

Preparer/Additional Information - Russell Haas, SCS, P.O. Box 1458, Bismarck, ND 58502, (701) 250-4425.

Grenville

Increased at former SCS Nursery, Albuquerque, NM. Carried as accession nos. PI 414066 & A-5669.

Source - Collection near Grenville, NM, at elevation of 1,800 m and annual precipitation of 400 mm.

Method of Breeding - Bulk increase of source material.

Intended Use - Range reseeding and revegetation of disturbed areas.

Description - Intermediate type between northern and southern geographic strains. Plants uniform, leafy, fine stemmed, and remain green well into fall. Height at maturity 1-1.2 m. Medium maturity date. No rust or other diseases observed.

Adapted to - LRR G, H; PHZ 5.

Released - Informally in 1940 - Spur, TX - Plant Materials Center, SCS.

Breeder Seed/Stock - Plant Materials Center, SCS, Los Lunas, NM.

Certified Seed/Stock - No.

Preparer/Additional Information - Plant Materials Center, SCS, 1036 Miller St. SW, Los Lunas, NM 87031, (505) 865-4684.

Kanlow

Developed at Kansas AES, Manhattan, ARS cooperating - F.L. Barnett and K.L. Anderson. Carried as accession no. 421521.

Source - SCS collection from lowland site near Wetumka, OK, in 1957.

Method of Breeding - Collection planted at Manhattan in spring of 1958; 200 plants selected for leafiness, vigor, and retention of green late in season; selections isolated at Ashland Farm near Manhattan.

Intended Use - A lowland type switchgrass for soil conservation in poorly drained or frequently flooded sites. Can be utilized as hay or pasture.

Description - Tall, coarse, productive, especially adapted to lowlands where flooding, high water table, or other excess water problems occur, but performs well on upland where soils are not too thin or droughty. Not intended to replace upland varieties, such as Caddo and Blackwell, but to supplement them because of adaptation to wet locations.

Adapted to - LRR H; PHZ 5.

Released - 1963, cooperatively by Kansas AES and Plant Science Research Division, ARS.

Breeder Seed/Stock - Kansas AES.

Certified Seed/Stock - Available in limited quantity.

Preparer/Additional Information - SCS, 760 S. Broadway, Salina, KS 67401, (913) 823-4541.

KY-1625 (Reg. No. GP-0057)

Selected at the SCS Plant Materials Center, Quicksand, KY - Donald S. Henry and Charles F. Gilbert. Carried as accession no. PI 431575.

Source - Collected as KY-584 from Raleigh County, WV. Clonal selection of PI 431575 (KY-1625) made at the Quicksand Plant Materials Center in 1970.

Method of Breeding - Increase of seed from selected clones.

Intended Use - Cover and forage plant on marginal low-fertility hillside pasture. Selected to meet the need of mid-summer supplemental grazing during summer slump period of cool-season grass pastures.

Description - A late maturing, leafy, fine stem, perennial rhizomatous native warm season grass, that responds well to small increments of nitrogen. It has a high leaf-stem ratio and has shown higher protein and digestibility when compared to other switchgrasses. Poor seed quality and seedling vigor may be a limiting factor. Preliminary cytogenetic studies of the University of KY indicate chromosomes numbers of 2N=36, 54, 72.

Adapted to - LRR F, G, H, I, J, K, L, M, N, O, P, R, S; PHZ 4, 5, 6, 7, 8.

Released - 1987 as germplasm by the Plant Materials Center, SCS, Quicksand, KY, and KY AES.

Breeder Seed/Stock - Plant Materials Center, SCS, Quicksand, KY.

Certified Seed/Stock - Not available.

Preparer/Additional Information - Laura Ray, Plant Materials Center, SCS, 175 Robinson Dr., Quicksand, KY 41363-9008, (606) 666-5069.

Nebraska 28

Developed at Nebraska AES, Lincoln, ARS and SCS cooperating - L.C. Newell. Carried as accession no. PI 477003.

Source - Native stand of switchgrass collected in Holt County, NE, in 1935.

Method of Breeding - Spaced plants grown at this experiment station from original collection selected for type and allowed to cross-pollinate in isolation. Resulting seed bulked and increased.

Intended Use - Forage production and soil conservation purposes such as seeded waterways.

Description - Relatively early maturing strain of switchgrass, representative of Nebraska sandhill types. Average plants semi-decumbent, with fine stems of moderate height, bluish green, and leafy; but considerable variation in plant type exists. Well adapted to diverse soils and used successfully for pasturage and soil conservation purposes, such as seeded waterways in pure stands or mixtures. Matures seed in mid-August to early September. In areas with longer growing seasons is susceptible to rust, which is likely to be a serious factor in production.

Adapted to - LRR F, G, M; PHZ 4, 5.

Released - 1949, cooperatively by Nebraska AES; USDA-ARS; and Nursery Division, SCS.

Breeder Seed/Stock - Nebraska AES and USDA-ARS, Univ. of Nebraska, Lincoln.

Certified Seed/Stock - Available.

Preparer/Additional Information - K.P. Vogel, ARS-USDA, 344 Keim Hall, Univ. of Nebraska or Jeff Pederson, Dept. of Agronomy, Univ. of Nebraska, Lincoln, NE 68583, (402) 472-1564, (402) 472-2811.

Pathfinder (Reg. No. 17)

Selected at Nebraska AES, Lincoln, ARS cooperating - L.C. Newell.

Source - Domestic collections in 1953 from Nebraska and Kansas.

Method of Breeding - Clones selected as "Type f" from space-planted nurseries of collections were polycrossed and progeny tested. Twelve superior clones of selected type polycrossed in isolation. Within 12 progenies, 192 plants selected and intercrossed in isolation to produce breeder seed.

Intended Use - Forage production for late spring and summer grazing.

Description - Winter-hardy, vigorous, leafy, late maturing, and rust resistant in region of adaptation. Good stand establishment and forage production for late-spring and summer grazing. Used in pure stands or in mixtures with other warm-season prairie grasses. Tests indicate its adaptation in Nebraska and adjacent areas. Most favorable area for seed production is in eastern third of Nebraska south of Platte River.

Adapted to - LRR northern H, eastern G, M; PHZ 4, 5.

Released - 1967, cooperatively by Nebraska AES and Plant Science Research Division, ARS.

Breeder Seed/Stock - Nebraska AES and USDA-ARS, Univ. of Nebraska, Lincoln.

Certified Seed/Stock - Available.

Preparer/Additional Information - K.P. Vogel, ARS-USDA, 344 Keim Hall, Univ. of Nebraska or Jeff Pederson, Dept. of Agronomy, Univ. of Nebraska, Lincoln, NE 68583, (402) 472-1564, (402) 472-2811.

Shelter

Selected at Big Flats Plant Materials Center, SCS, Corning, NY. Carried as accession no. NY 4006, PI 430240.

Source - Frank Glover collected switchgrass seed from a stand located south of St. Mary's, WV, in 1956.

Method of Breeding - The original accession was selected for upright form and stiff stems and was advanced under isolation for two generations. Sixty plants were then selected for large number of stems, early maturity, large stem diameter, and leafiness and moved to an isolated cross-pollination block.

Intended Use - Spring nesting cover for ground nesting birds, and escape cover for wildlife. Can be used as a component of warm-season grass mixtures on droughty sites.

Description - Shelter has thicker stems and fewer leaves than other released varieties with the exception of Kanlow. It is 5-40 mm taller than Blackwell after the second growing season, but exhibits less seedling vigor during the establishment year. Shelter reaches full anthesis 7-10 days earlier than Blackwell.

Adapted to - LRR L, R, S, M, N, P, T; PHZ 4.

Released - 1986 - cooperatively by SCS, Cornell University, New York Department of Environmental Conservation Division of Fish & Wildlife, and the Pennsylvania Game Commission. Previously a germplasm release as NY-4006.

Breeder Seed/Stock - Big Flats Plant Materials Center, SCS, Corning, NY.

Certified Seed/Stock - Available.

Preparer/Additional Information - Big Flats Plant Materials Center, SCS, RD1, Rte 352, Box 360A, Corning, NY 14830, (607) 562-8404.

Summer

Selected at South Dakota AES, Brookings - J.G. Ross.

Source - Native collection, PI 214759, made by W.L. Tolstead and L.C. Newell south of Nebraska City, NE, in 1953.

Method of Breeding - Collection found superior when grown at Brookings. Mass selection for earliness, leafiness, and rust resistance. Selections from replicated nursery of progenies from these plants made and placed in polycross. Since high degree of uniformity of desirable type was present, foundation field was established from seed harvested from this nursery.

Intended Use - Range and pasture seedings.

Description - Tall, upright, with abundant, somewhat coarse leaves. Starts growth after June 1 and matures seed in mid-September. Produces high yield of forage and seed.

Adapted to - LRR M; PHZ 4.

Released - 1963, by South Dakota AES.

Breeder Seed/Stock - South Dakota AES.

Certified Seed/Stock - Available.

Preparer/Additional Information - South Dakota AES, Brookings, SD 57007, (605) 688-5125.

Trailblazer

Developed by USDA-ARS (L.C. Newell) and Nebraska Agricultural Research Division, Dept. of Agronomy, Univ. of Nebraska. Carried as PI 549094; reg. no. CV-146.

Source - Collections from natural grasslands in Nebraska and Kansas.

Method of Breeding - The result of a genetic study designed to determine if the forage quality of switchgrass could be improved by restricted phenotypic selection and to obtain heritability estimates for in vitro dry-matter digestibility (IVDMD). Two Nebraska experimental strains, ey and ff, were used; both are similar to Pathfinder in maturity, origin, and breeding history. In 1974, approximately 800 plants that seemed to be superior for IVDMD were selected from a 1973 nursery planting of 2,200+ plants. Single-plant selections for high and low IVDMD were made in each of the nursery rows. In 1976, ramets from 25 of the high-IVDMD selections and from 25 of the low-IVDMD were transplanted to establish isolated high- and low-IVDMD polycross nurseries. Syn-1 seed was harvested from these nurseries in 1977 and from the same open-pollinated clones in the selection nursery in 1976 was used in 1978 to establish a replicated, seeded sward nursery in which the following strains were compared: high-IVDMD PC (Trailblazer), high-IVDMD OP, and low-IVDMD PC. Strains did not differ significantly for forage yield in any of three years; thus, the genetic gain in IVDMD was achieved without forage yield loss.

Intended Use - A warm-season pasture grass.

Description - A 25-clone synthetic that is similar to Pathfinder in maturity, appearance, and area of adaptation. Higher IVDMD than Pathfinder. Twelve of the 25 clones are from the Nebraska strain ff, while the remainder are from the ey strain. Hexaploid (2n=6x=54).

Adapted to - Central Great Plains and adjacent midwestern states.

Released - 1984, by USDA-ARS and Nebraska AES.

Breeder Seed/Stock - Dept. of Agronomy, Nebraska Agricultural Research Division, Lincoln.

Certified Seed/Stock - Available.

Preparer/Additional Information - K.P. Vogel, ARS-USDA, 344 Keim Hall, Univ. of Nebraska or Jeff Pederson, Dept. of Agronomy, Univ. of Nebraska, Lincoln, NE 68583, (402) 472-1564, (402) 472-2811.

***Pascopyrum smithii* (Rydb.) A. Love - western wheatgrass**

***Agropyron smithii* Rydb.**
***Elytrigia smithii* (Rydb.) Nevski**
***Elymus smithii* (Rydb.) Gould**

Important cool-season, sod-forming grass. Widely distributed from Wisconsin to central Washington and south into New Mexico and Texas panhandle. Major range grass in northern and central Great Plains, where used for pasture, hay, and erosion control. Develops slowly from seed, is drought resistant, and has moderate alkali tolerance.

9007213

Increased at Plant Materials Center, SCS, Pullman, WA - J.L. Schwendiman.

Source - Collected from a naturalized stand near George, WA. Carried as accession P-727.

Method of Breeding - Bulk increase of original collection after comparison with other accessions collected throughout the western states.

Intended Use - Rangeland reseeding.

Description - A distinctly vigorous, semi-coarse, robust, broad-leaved, open-sodding type. A good seed producer, which rapidly spreads vegetatively more so than other strains. Chromosome number is 2n=56.

Adapted to - LRR B; PHZ 5.

Released - Not released.

Breeder Seed/Stock - Plant Materials Center, SCS, Pullman, WA.

Certified Seed/Stock - Not available.

Preparer/Additional Information - Plant Materials Center, SCS, Room 104, Hulbert Agricultural Sciences Building, Washington State Univ., Pullman, WA 99164-6211, (509) 335-7376.

Ariba

Plant Materials Center, SCS, Los Lunas, NM - J.A. Downs. Carried as accession nos. PI 432402 & P-15614.

Source - Collected in 1957 near Flagler, CO, from High Plains site at elevation of 1,530 m and annual precipitation of about 400 mm.

Method of Breeding - Bulk increase from initial planting. Increased through five successive generations to improve seed-production potential.

Intended Use - Range reseeding and revegetation of disturbed areas.

Description - Rapid germination and good seedling establishment. Dense, dark-green, medium-height foliage; aggressive rhizomes. Superior to other accessions tested in seed production. Production block at Los Lunas yielded 74 kg of seed per acre in 1963.

Adapted to - LRR D, G, H, F; PHZ 4.

Released - 1973, cooperatively by SCS-Los Lunas, NM, Colorado AES, and New Mexico State Highway Department.

Breeder Seed/Stock - Plant Materials Center, SCS, Los Lunas, NM.

Certified Seed/Stock - Available.

Preparer/Additional Information - SCS, 517 Gold Ave. SW, Room 3301, Albuquerque, NM 87102-3157, (505) 766-3277.

Barton

Increased at Plant Materials Center, SCS, Manhattan, KS, with Kansas AES and ARS, cooperating - R.D. Lippert and H.L. Hackerott. Carried as accession nos. PM-K-402, PM-K-27, KG-2036.

Source - Seed collected in 1947 from natural grassland on clay bottomland along Walnut Creek near Heizer in Barton County, KS.

Method of Breeding - Field seed collection increased and tested as PM-K-402.

Intended Use - Range reseeding, revegetation of disturbed areas, stabilization of earth structures, and revegetation of saline areas.

Description - A strongly rhizomatous leafy ecotype intermediate in growth between northern and southern types;

shows little evidence of rust in western KS and relatively free of rust at Manhattan. In plot evaluations at Manhattan, superior in forage production and disease resistance to accessions representing areas where western wheatgrass seed is frequently harvested for commercial use. At Hays Experiment Station, Barton western wheatgrass ranked first in seed culm development and forage yield in comparison with 16 other accessions.

Adapted to - LRR F, G, H; PHZ 4b.

Released - 1970, cooperatively by Plant Materials Center, SCS, Manhattan, KS, AES, and Plant Science Research Division, ARS.

Breeder Seed/Stock - Plant Materials Center, SCS, Manhattan, KS.

Certified Seed/Stock - Available.

Preparer/Additional Information - SCS, 760 S. Broadway, Salina, KS 67401, (913) 823-4541.

Flintlock

Selected at Nebraska AES, Lincoln, ARS, cooperating - L.C. Newell.

Source - Plant selections from 30 accessions collected in 1957 from native grasslands of central and southwestern Nebraska and northwestern Kansas.

Method of Breeding - A base population of 30 accessions was subjected to two cycles of recurrent selection for forage and seed yields and rhizomatous spread. One hundred parent clones were then selected and intermated to produce a synthetic with a broad genetic base.

Intended Use - Flintlock has been grown in monoculture or in mixtures with other cool-season grasses in the central Great Plains and portions of adjacent regions for conservation purposes, roadside and park plantings, dryland hay production, and early season pasture that complements the grazing of summer grasses on rangeland.

Description - Flintlock spreads aggressively by rhizomes, has relatively coarse culms, and relatively soft leaves compared with other strains of western wheatgrass. Fully developed spikes are robust with double spikelets frequent on lower nodes.

Adapted to - LRR E, G; PHZ 3, 4, 5.

Released - 1975, cooperatively by Nebraska AES and ARS.

Breeder Seed/Stock - Nebraska AES and USDA-ARS, Univ. of Nebraska, Lincoln.

Certified Seed/Stock - Available.

Preparer/Additional Information - K.P. Vogel, ARS-USDA, 344 Keim Hall, Univ. of Nebraska or Jeff Pederson, Dept. of Agronomy, Univ. of Nebraska, Lincoln, NE 68583, (402) 472-1564, (402) 472-2811.

ND-WWG931

Source - Base population consisted of 5,140 genotypes that were vegetatively collected in 1977 from 1,028 sites in western North and South Dakota. Population was assembled and evaluated at Mandan, ND.

Method of Breeding - This material was subjected to two cycles of phenotypic recurrent selection for plant vigor, rhizomatous spread, density of foliage cover, and seed yield. Selection intensities were 8% and 20% (i.e., 400 selected parents) in Cycles 0 and 1, respectively.

Intended Use - To serve as a source of improved cultivars to reseed soils, abandoned cropland, and rangeland in the northern Great Plains.

Description - ND-WWG931 has a very broad genetic base. It produces forage yields equal to that of the cultivars Flintlock and Walsh. It is more digestible, but has produced less forage than Rodan.

Adapted to - Northern Great Plains and the prairie provinces of Canada.

Released - USDA-ARS, USDA-SCS, North Dakota AES. Scheduled for release in 1993.

Breeder Seed/Stock - USDA-ARS, P.O. Box 459, Mandan, ND 58554.

Certified Seed/Stock - No.

Preparer/Additional Information - Ian M. Ray, ARS Northern Great Plains Research Lab, P.O. Box 459, Mandan, ND 58554, (701) 667-3025.

ND-WWG932

Source - Base population consisted of open-pollinated seed collected from 468 genotypes growing on rangeland sites in Alberta and Saskatchewan, Canada. Population was evaluated at Mandan, ND.

Method of Breeding - Material was subjected to one cycle of phenotypic recurrent selection (selection intensity of 5%, i.e. 234 selected parents) for plant vigor, rhizomatous spread, and density of foliage cover.

Intended Use - To serve as a source of improved cultivars to reseed erosive soils, abandoned cropland, and rangeland in the northern Great Plains.

Description - ND-WWG932 is similar to ND-WWG931 in plant vigor and foliage density, but matures four days earlier. Rate of rhizome production of ND-WWG932 is 10% less than that of ND-WWG931.

Adapted to - Northern Great Plains and the prairie provinces of Canada.

Released - USDA-ARS, USDA-SCS, North Dakota AES. Scheduled for release in 1993.

Breeder Seed/Stock - USDA-ARS, P.O. Box 459, Mandan, ND 58554.

Certified Seed/Stock - No.

Preparer/Additional Information - Ian M. Ray, ARS Northern Great Plains Research Lab, P.O. Box 459, Mandan, ND 58554, (701) 667-3025.

Rodan (Reg. No. 14)

Selected at Northern Great Plains Research Center, ARS, Mandan, ND - G.A. Rogler. Carried as accession no. PI477993.

Source - Selections by G.A. Rogler in 1936 from a commercial seed field of unknown origin grown in the Missouri River bottoms near Mandan, ND.

Method of Breeding - Thirteen half-sibs from a 40-plant progeny row were selected for density of foliage cover, leafiness, and resistance to stem rust. These 13 selections were intermated in isolation, and the bulked seed was in turn grown in isolation for seven open-pollinated seed increase generations. Low intensity selection was practiced during this time by roguing plants with obvious defects. Natural selection favored upland types with drought tolerance. Three hundred ramets were selected at random from the seventh-generation seed increase field and used to establish an isolation plot for production of breeder seed.

Intended Use - Rodan is adapted for grazing and hay in the northern Great Plains region of the U.S. and the Canadian prairie region. Rodan has had extensive use in revegetation of disturbed lands. In its area of adaptation, Rodan has had higher forage production on coarse-textured soils than other current western wheatgrass cultivars.

Description - Vigorous, leafy, rust resistant.

Adapted to - LRR F, G; PHZ 2, 3, 4.

Released - 1983, by ARS, SCS, and the North Dakota AES.

Breeder Seed/Stock - ARS, Northern Great Plains Research Laboratory, Mandan, ND.

Certified Seed/Stock - Available.

Preparer/Additional Information - Ian M. Ray, ARS, Northern Great Plains Research Laboratory, P.O. Box 459, Mandan, ND 58554, (701) 663-6445.

Rosana

Increased at Plant Materials Center, SCS, Bridger, MT - J.L. McWilliams and A.A. Thornburg. Carried as accession nos. M-23, P-15582, PI 469236.

Source - Commercial harvest from native meadows along Porcupine Creek, northwest of Forsyth, MT, 1959.

Method of Breeding - Direct increase of field collection after comparison with approximately 60 other accessions representing native collections from Montana and Wyoming. Tested as M-23.

Intended Use - Rosana was developed for irrigated hay or pasture in short water supply areas and overflow sites, or reseeding dryland range sites. Used in native mixtures for reclamation of drastically disturbed areas.

Description - Excellent seedling vigor and strong rhizomes. Plants blue-green, leafy, with moderately fine stems. Good forage and seed production. Produces the tightest sod of any accession tested.

Adapted to - LRR B, D, E, F, G; PHZ 3.

Released - 1972, cooperatively by Plant Materials Center, SCS, Bridger, MT, and Montana AES.

Breeder Seed/Stock - Plant Materials Center, SCS, Bridger, MT.

Certified Seed/Stock - Available.

Preparer/Additional Information - Plant Materials Center, SCS, RR 1, Box 1189, Bridger, MT 59014, (406) 662-3579.

Walsh (Reg. No. 15)

Selected at Agriculture Canada Research Station, Lethbridge, Alberta - S. Smoliak and A. Johnston. Carried as accession no. L2381.

Source - The 20 clones included in Walsh trace to 468 different western wheatgrass ecotypes collected in 1968 and 1969 from indigenous stands in southern Alberta and southwestern Saskatchewan.

Method of Breeding - The 20 clones constituting this synthetic cultivar were selected for high forage and seed yields, vigor and leafiness, freedom from disease, aggressive rhizomatous spread, and uniformity in plant height and maturity. Final selection of the 20 clones was based on data from clonal and polycross progeny tests.

Intended Use - Walsh is adapted for pasture and hay in rangeland seedings and is used extensively in revegetation of disturbed lands. Primary use has been for hay production on heavy clay soils that flood periodically, but the cultivar also is adapted to medium textured soils. Moderately tolerant of saline soils, survives drought, and is responsive to added moisture and nutrients.

Description - A hardy perennial with moderately aggressive creeping root system and drought tolerance. Strongly rhizomatous and forms a dense blue-green sward under dryland conditions. Noted for its tolerance to salinity and its ability to grow and yield well on spring-flooded clay sites. Plants have erect leaves 25 cm or more long. Similar to Rosana in visual appearance.

Adapted to - LRR F, G; PHZ 2, 3, 4.

Released - 1982, by Agriculture Canada.

Breeder Seed/Stock - Agriculture Canada, Research Station, Lethbridge, Alberta T1J 4B1.

Certified Seed/Stock - Available. SeCan Association, 885 Meadowlands Dr., Suite 512, Ottawa, Ontario K2C 3N2.

Preparer/Additional Information - Surya N. Acharya, Agriculture Canada, Research Station, P.O. Box 3000, Lethbridge, Alberta T1J 4B1, (403) 327-4561.

Paspalum dilatatum **Poir. - dallisgrass**

Major warm-season, slightly spreading bunchgrass introduced from Argentina or Uruguay in mid-1800s. Used for pasture throughout much of Cotton Belt wherever annual rainfall is as much as 760 mm. Palatable, nutritious. Tolerates moderately close grazing, and should be grazed to prevent accumulation of dead leaves and stalks. Serious weed in lawns.

Prostrate

Selected at Georgia Coastal Plain Experiment Station, Tifton, ARS cooperating - G.W. Burton.

Source - Obtained from B. Smith, North Carolina AES, Raleigh, who received it from Bernardo Rosengurt of Montevideo, Uruguay.

Method of Breeding - Seeds from several progenies that appeared to be similar in type and 100% apomictic were blended and increased to furnish seed released in regional tests.

Intended Use - Breeding germplasm and possible cultivar if seed production can be improved.

Description - More prostrate, more resistant to foliage diseases, more persistent, maintaining good stands much longer than common dallisgrass. Outyielded common dallisgrass in clipping tests at Tifton. Very susceptible to ergot, very irregular in meiosis, and poor in seed production.

Adapted to - LRR and PHZ not known, but very persistent at Tifton, GA.

Released - No. Included in regional testing program.

Breeder Seed/Stock - Georgia Coastal Plain Experiment Station.

Certified Seed/Stock - Limited breeders seed only.

Preparer/Additional Information - ARS, Coastal Plain Experiment Station, P.O. Box 748, Tifton, GA 31793, (912) 386-3353.

Paspalum hieronymi **Hack. - paspalum**

Its natural range extends to both hemispheres and it is found growing on the seacoasts from Australia to southern Spain and from Argentina and Chile to Baja California and North Carolina.

Tropic Lalo

SCS and Hawaii Institute of Tropical Agriculture and Human Resources, Department of Agronomy and Soil Science - Robert J. Joy and Peter P. Rotar. Carried as accession no. PI 310108.

Source - Brazil. Seed received from National Plant Materials Center, SCS, Beltsville, MD.

Method of Breeding - Vegetatively propagated from original observational planting which was grown from seed.

Intended Use - Primarily as a ground cover for erosion control in orchards, waterways, roadsides and other erosion-prone areas.

Description - Rapidly spreading, low-growing, relatively low-maintenance with a dense, matlike growth habit. Stolons are tough and somewhat coarse; hence it will tolerate fairly heavy use from wheeled equipment and foot traffic.

Adapted to - LRR I, T, U, V, Z; PHZ 9.

Released - 1984, U.S. Department of Agriculture, SCS, and the University of Hawaii, Hawaii Institute of Tropical Agriculture and Human Resources.

Breeder Seed/Stock - Plant Materials Center, SCS, Hoolehua, Molokai, HI, is responsible for the maintenance of vegetative material.

Certified Seed/Stock - Not available.

Preparer/Additional Information - Plant Materials Center, SCS, P.O. Box 236, Hoolehua, HI 96729, (808) 567-6378.

Paspalum nicorae Parodi - brunswickgrass

Warm-season grass, rhizomatous, slender, erect to suberect culms. Produces dense sod. Introduced from Uruguay and southern Brazil.

PI 202044

Increased at SCS, Plant Materials Centers at Arcadia, FL, Americus, GA, and Coffeeville, MS. Carried as accession no. 202044.

Source - PI 202044 from Argentina.

Method of Breeding - Compared with other accessions for rate of spread, height, sod density, seed-producing potential, and overall vigor. Identified for testing as MS-906 and F-18888. It is a tetraploid apomict.

Intended Use - Cover crop in waterways and for pasture use.

Description - Typical of species. At maturity, averages about 760 mm tall. Sod has bluish or glaucous cast; seed yield and quality good. Adapted in the southern coastal plain, silty uplands of southern Mississippi Valley, and blackland prairie of Alabama and Mississippi. Rated as a preferred grazing plant in South America. Has potential for grazing, hay, and cover plant in waterways and for seeding eroded areas.

Adapted to - LRR P, T; PHZ 8.

Released - No. Distributed for testing.

Breeder Seed/Stock - Plant Materials Center, SCS, Americus, GA.

Certified Seed/Stock - No.

Preparer/Additional Information - Plant Materials Center, SCS, 295 Morris Dr., Americus, GA 31709, (912) 924-2286.

PI 310131

Plant Materials Center, SCS, Americus, GA. Carried as accession no. PI 310131.

Source - Collected from Brazil.

Method of Breeding - Direct increase of PI 310131.

Intended Use - Cover crop for grassed waterways.

Description - The ability to germinate and provide protection for waterways in conservation systems more quickly than other tested varieties.

Adapted to - LRR U, T, P; PHZ 8.

Released - Projected 1993, SCS.

Breeder Seed/Stock - Plant Materials Center, SCS, Americus, GA.

Certified Seed/Stock - Not available.

Preparer/Additional Information - Plant Materials Center, SCS, 295 Morris Dr., Americus, GA 31709, (912) 924-2286.

Paspalum notatum Flugge - bahiagrass

Major warm-season grass that spreads slowly by short, stout rhizomes. One of the first USDA introductions from Brazil in 1914. Common bahiagrass adapted in Florida and lower coastal plain. Rated as more palatable but less winter-hardy than Pensacola. Aggressive species that spreads rapidly from seed. Well suited for pasture use on sandy soils of low fertility or where good fertilizer programs are not maintained.

Paraguay 22

Increased at Georgia Coastal Plain Experiment Station, Tifton, ARS cooperating - J.L. Stephens. Carried as accession no. 158822.

Source - PI 158822 collected by J.L. Stephens in Paraguay in 1947.

Method of Breeding - Selection of one plant in source nursery. Progeny tests indicated that selection was true breeding and probably 100% apomictic.

Intended Use - Grazing and hay.

Description - A common bahiagrass, except more cold-hardy, narrower blades, smaller seed, and more responsive to fertilization. Seed germination excellent, with full stands and ground cover in 8-12 weeks. Adapted throughout Florida.

Adapted to - LRR T, U; PHZ 9.

Released - No. Included in regional testing program as Tifton bahiagrass.

Breeder Seed/Stock - Georgia Coastal Plain Experiment Station.

Certified Seed/Stock - Not available. Some commercial production in Florida.

Preparer/Additional Information - ARS, Coastal Plains Experiment Station, P.O. Box 748, Tifton, GA 31793, (912) 386-3354.

Pensacola

Plant Materials Center, SCS, Americus, GA. Carried as accession no. PI 422024.

Source - Plants growing along docks and railroad tracks at Pensacola. Thought to have arrived by fruit boat from Central or South America.

Method of Breeding - Comparative tests conducted at several experiment stations. Experimental pastures and plots for forage yield and chemical composition planted at Gainesville, FL, in 1942.

Intended Use - Pasture, cover crop.

Description - Similar to common bahiagrass, except more cold-hardy, narrower blades, smaller seed, and more responsive to fertilization. Seed germination excellent, with full stands and ground cover in 8-12 weeks. Adapted throughout southeastern coastal plain area and to all Florida.

Adapted to - LRR T, P, N, U; PHZ 8.

Released - Approved as superior forage by Florida AES, Gainesville in 1944. Seed distributed by SCS Nursery, in Americus, 1942, with first large-scale distribution in 1944.

Breeder Seed/Stock - Plant Materials Center, SCS, Americus, GA.

Certified Seed/Stock - Available.

Preparer/Additional Information - Plant Materials Center, SCS, 295 Morris Dr., Americus, GA 31709, (912) 924-2286.

RCP

Selected at Fort Lauderdale Research and Education Center, University of Florida - Philip Busey.

Source - Population derived from polycross selection.

Method of Breeding - Mass selection was performed on naturalized southeastern U.S. highway accessions which were reproduced asexually and evaluated for coverage at four locations in southern Florida. Selected clones (29) were hybridized through two generations of random mating to produce a heterogeneous population. Superior combining F_1s (15) were selected based on rapid coverage of F_2 progeny in a non-irrigated, non-fertilized test site. First generation seed from selected F_1s was described as RCP-1 (Rapid Coverage Polycross); second generation bulk increase as RCP-2.

Intended Use - Highway rights-of-way and conservation areas.

Description - Similar to Pensacola, but slightly taller foliage. Significantly faster lateral growth rate than Argentine and Pensacola, resulting in higher establishment ratings in the first year after seeding. The rapid coverage advantage of RCP is greatest at low seeding rates.

Adapted to - LRR J, I; PHZ 8.

Released - Proposed for commercial release. Available for experimental use.

Breeder Seed/Stock - Fort Lauderdale Research and Education Center, University of Florida.

Certified Seed/Stock - Not available.

Preparer/Additional Information - Fort Lauderdale Research and Education Center, University of Florida, 3205 College Ave., Fort Lauderdale, FL 33314, (305) 475-8990.

Tifton 9 (Reg. No. 118)

ARS in cooperation with the University of Georgia Coastal Plain Experiment Station, Tifton, GA. Carried as accession no. PI 531086.

Source - Selected for long leaves and seed production potential.

Method of Breeding - Ninth cycle of recurrent restricted phenotypic selection (RRPS) of Pensacola cultivar. RRPS that required three years per cycle has been improved to permit one cycle per year with the same increase in yield per cycle. Each cycle screens the most vigorous seedlings from a population of about 20,000 rows in flats in the greenhouse and the best 200 plants in 1,000 spaced plants in the field. Field selection of the best five plants in each 25-plant block is made visually. Three culms from each plant ready to flower the next day are placed in gallon jars of water and grouped together under a one-meter diameter paper tent where they are thoroughly intermated each morning as they shed pollen. Seeds that mature on these culms, 125 per selection, are planted in rows in flats of steam-sterilized soil in the greenhouse in December to start the next cycle of RRPS.

Intended Use - Grazing and hay.

Description - Longer leafed, more vigorous seedling stage, more succulent and 40% more productive of forage than Pensacola.

Adapted to - LRR P, T, U; PHZ 8, 9.

Released - March 1987, University of Georgia and ARS.

Breeder Seed/Stock - ARS, Coastal Plain Experiment Station, Tifton, GA.

Certified Seed/Stock - Available in quantity.

Preparer/Additional Information - ARS, Coastal Plain Experiment Station, P.O. Box 748, Tifton, GA 31793, (912) 386-3353.

Wilmington

Increased at SCS Nursery, Rock Hill, SC. Carried as accession no. PI 434189.

Source - Collected in 1940 from naturalized stand near Wilmington, NC, by Paul Tabor.

Method of Breeding - Increase of original collection. Tested as SC 20-338, AM-1284, and MS-131.

Intended Use - Pasture and critical area erosion control.

Description - Narrow leaf, cold-hardy bahiagrass; makes dense sod; plants of medium size. Only bahiagrass not injured by cold at Chapel Hill, NC, between 1941 and 1953. Seed production poorer than Pensacola but adequate for multiplication. Seed about 30% larger than Pensacola. Adapted for forage and conservation use in the southern U.S. and of greatest value north of area in which Pensacola is adapted.

Adapted to - LRR O, P; PHZ 7.

Released - 1971, cooperatively by Mississippi AES and Plant Materials Center, SCS, Coffeeville, MS.

Breeder Seed/Stock - Plant Materials Center, SCS, Coffeeville, MS.

Certified Seed/Stock - Not available.

Preparer/Additional Information - SCS, Jamie L. Whitten Plant Materials Center, Rte 3, Box 215A, Coffeeville, MS 38922, (601) 675-2588.

Paspalum vaginatum **Swartz - seashore paspalum**

Seashore paspalum grows in saltwater coastal marshes and coastal mud and sand flats in the Hawaiian Islands, American Samoa, Caroline Islands, Guam, and the Commonwealth of the Northern Marianas Islands. It is one of the most salt-tolerant grasses known and has been reported to grow with water containing total soluble salts of more than 10,000 parts per million. It will also grow in fresh water.

Tropic Shore

SCS and Hawaii Institute of Tropical Agriculture and Human Resources, Department of Agronomy and Soil Science - Robert J. Joy and Peter P. Rotar. Carried as accession no. PI 543854.

Source - Collected by D.N. Palmer along the seashore about 1.2 km SE of the Kawainui Canal outlet in Kailua, Oahu, HI.

Method of Breeding - Asexual propagation of original material.

Intended Use - Primarily for stabilizing the shoreline and banks of aqua culture ponds, canals, and streams having brackish or salty water.

Description - Rapidly spreading and low-growing. Stolons contain numerous nodes that root to form a dense, sod-like cover. Provides good protection from strong waves. Very salt-tolerant. Less aggressive with regard to growing out in the water than other grass strains tested. This is important to aqua culture harvesting operations.

Adapted to - LRR I, T, U, V, Z; PHZ 9.

Released - 1988, SCS and the University of Hawaii, Hawaii Institute of Tropical Agriculture and Human Resources.

Breeder Seed/Stock - SCS, Plant Material Center, Hoolehua, Molokai, HI.

Certified Seed/Stock - Not available.

Preparer/Additional Information - Plant Materials Center, SCS, P.O. Box 236, Hoolehua, HI 96729, (808) 567-6378.

Pennisetum americanum **(L.), Leeke - pearl millet**

Pennisetum tyhoides auct. non (Burm.) Stapf & C.E. Hubbard

Important warm-season annual from India and Africa, where grown primarily for grain. Used for pasture, soiling, and silage from Maryland to Florida and west to Texas. Well adapted in coastal plain. Grows best in moist, warm locations; grows on poor sandy soils; responds well to fertilizer. Highly nutritious and palatable. Good regrowth under proper management.

Gahi 3

ARS and the University of Georgia Coastal Plain Experiment Station at Tifton, GA.

Source - Selected from other hybrids for ease of planting with grain drill and later maturity for longer grazing period.

Method of Breeding - Gahi 3 is a first-generation hybrid between Tift 23DA and Tift 186. The short Tift 23DA is used in the commercial production of Gahi 3 to facilitate cross-pollination and seed harvest.

Intended Use - Grazing, hay, and silage.

Description - Very uniform, medium-stemmed, leafy hybrid with smooth leaves and stems.

Adapted to - LRR J, N, O, P, T, U; PHZ 7, 8, 9.

Released - 1972, ARS and University of Georgia, Coastal Plain Experiment Station, Tifton, GA.

Breeder Seed/Stock - University of Georgia, Coastal Plain Experiment Station, Tifton, GA.

Certified Seed/Stock - Limited supplies available.

Preparer/Additional Information - ARS, Coastal Plain Experiment Station, P.O. Box 748, Tifton, GA 31793, (912) 386-3353.

Tiflate

Selected at Georgia Coastal Plain Experiment Station, Tifton, ARS cooperating - G.W. Burton.

Source - Fifty-four introductions from Nigeria and Upper Volta that bred true for short-day photoperiod sensitivity.

Method of Breeding - Fifty-four introductions from Nigeria and Upper Volta that bred true for less than 12-hour day photoperiod sensitivity were allowed to cross pollinate, isolated in one green-house section, in winter of 1963-64. Seed was increased in two succeeding years under isolation in winter plantings in Puerto Rico.

Intended Use - Grazing and hay.

Description - Highly heterozygous, uniform only to extent that all plants remain vegetative until day length is 12 hours or shorter. At latitude of Tifton, GA, plantings made from April to August will not flower until late October or early November. When not grazed or cut, April and August plantings may reach respective heights of 4.6 and 1.5 m when mature. Most plants have pubescent leaves; seeds variable in size and considerably larger than those of Gahi 1 and Starr. Photoperiod sensitivity keeps it in a vegetative condition for a longer period in summer, gives better seasonal distribution of forage, more succulent, leafier, and more digestible forage, increases length of growing season, and increases ease of management. Although it produces less dry matter per acre under most cutting regimes, two years of grazing data indicate that it will be equal, and perhaps superior, to Gahi 1 under grazing. More resistant to leaf spot than a number of other millets, including common, Starr, and Gahi 1. Like other pearl millets, it contains no prussic acid.

Adapted to - LRR J, O, P, T, U; PHZ 7, 8, 9.

Released - 1969, cooperatively by Georgia Coastal Plain Experiment Station and Plant Science Research Division, ARS.

Breeder Seed/Stock - Georgia Coastal Plain Experiment Station.

Certified Seed/Stock - Has been approved for certification in Texas.

Preparer/Additional Information - ARS, Coastal Plain Experiment Station, P.O. Box 748, Tifton, GA 31793, (912) 386-3353.

Tifleaf 2 (Reg. No. 1)

ARS and University of Georgia Coastal Plain Experiment Station, Tifton, GA. Carried as accession no. PI 518646.

Source - Selected for disease resistance in southern U.S.

Method of Breeding - A uniform first generation dwarf leafy forage hybrid between dwarf cytoplasmic male sterile, Tift 85D2A1, and dwarf pollinator, Tift 383 (2). It has the same male parent as Tifleaf 1 (Tift 383) and its female parent (Tift 85D2A1) is genetically similar to Tift 23D2A1, the female parent of Tifleaf 1. Tift 85DB (maintainer for Tift 85D2A1, female parent for Tifleaf 2) was developed by backcrossing Tift 23D2B1 (maintainer line of Tift 23D2A1) to a highly rust and leaf spot resistant introduction from Senegal for five generations followed by nine genertions of selfing and selection. The resistance of Tifleaf 2 to both diseases is controlled by separate major dominant genes in Tift 85D2A1. Tifleaf 2 sheds pollen and is male and female fertile. It will reach a height of 1.8 m if not defoliated.

Intended Use - Grazing, hay, and silage.

Description - Similar to and cannot be distinguished from Tifleaf 1 in disease-free situations. Highly resistant to rust and leaf spot.

Adapted to - LRR J, N, O, P, T, U; PHZ 7, 8.

Released - 1987, ARS and University of Georgia Coastal Plain Experiment Station, Tifton, GA.

Breeder Seed/Stock - ARS Coastal Plain Experiment Station, Tifton, GA.

Certified Seed/Stock - Available.

Preparer/Additional Information - ARS, Coastal Plain Experiment Station, P.O. Box 748, Tifton, GA 31793, (912) 386-3353.

Pennisetum ciliare **(L.) Link - buffelgrass**
Cenchrus ciliaris **L.**

Warm-season grass from Union of South Africa. Includes bunch and spreading types. Used for pasture in southern Texas and to limited extent in parts of gulf coast states. Adapted to lighter sandy soils; responds to fertilizer. Good seedling vigor. Drought resistant; not cold tolerant; withstands fairly heavy grazing. Nutritious.

B-1S (Reg. No. GP1)

Selected at Texas AES, College Station, ARS cooperating - E.C. Bashaw.

Source - A variant plant on Pat Higgins' ranch, Southerland Springs, TX. Presumed to be a sexual mutant of apomictic Blue buffelgrass.

Method of Breeding - Vegetative increase of original plant.

Intended Use - Range seedings.

Description - A vigorous, rhizomatous sexual plant heterozygous for method of reproduction. For use as female parent in crosses with a apomictic strains or production of segregating selfed progeny. Crosses readily with other buffelgrass or birdwoodgrass *(C. setigerus).*

Adapted to - LRR J, I, T; PHZ 9.

Released - 1966, cooperatively by Texas AES and Plant Science Research Division, ARS.

Breeder Seed/Stock - Selfed seed and vegetative material, Texas AES.

Certified Seed/Stock - Not available.

Preparer/Additional Information - Texas A&M University, AES, Department of Range Science, College Station, TX 77843, (409) 845-5579.

Blue

Selected at SCS Nursery, San Antonio, TX - James E. Smith, Jr. Carried as accession no. PI 133898.

Source - Pretoria, Union of South Africa, PI 133898. Received March 1940.

Method of Breeding - Selected as most vigorous and productive of 21 similar accessions from Union of South Africa. Increased for testing as T-3782.

Intended Use - Range reseeding.

Description - Selected over other similar accessions because of early spring growth recovery (about three weeks ahead of T-4464 buffelgrass), vigorous summer growth, high forage production, rapid spread by means of short rhizomes, drought tolerance, resistance to injury by leafhoppers and aphids, and tolerance to light frost (active growth continues in fall about three weeks longer than for T-4464 buffelgrass). Best adapted to clay soils in Texas from Sonora eastward and Waco southward. Relatively low seed producer; limited in use within its area of adaptation by chronic shortage of commercial seed supplies. Both green and cured forage readily eaten by cattle.

Adapted to - LRR I, T; PHZ 9.

Released - Informally by SCS in 1952.

Breeder Seed/Stock - Not available.

Certified Seed/Stock - Not available, common is sold.

Preparer/Additional Information - Plant Materials Center, SCS, Rte 1 Box 155, Knox City, TX 79529, (817) 658-3922.

Higgins (Reg. No. 14)

Selected at Texas AES, College Station, ARS cooperating - E.C. Bashaw.

Source - A single sexual plant found at Southerland Springs, TX, and identified as B-1s.

Method of Breeding - Selection made in first generation selfed progeny of B-1s. Progeny from selections evaluated for agronomic characteristics in rows and solid seedlings.

Intended Use - Range and pasture seedings.

Description - Green foliage, brownish-wine inflorescence, and a rhizomatous root system. Typical involucres of the inflorescence contain a single spikelet, but basal members may have one to four spikelets. Resembles the T-4464 (common) variety in foliage and inflorescence color but may be identified by presence of rhizomes and more compact inflorescence. Distinct from Blue buffelgrass that has bluish foliage and tan-colored inflorescence. Produces somewhat less forage than Blue buffelgrass but far superior in seed production. Adapted to southern Texas.

Adapted to - LRR I, T; PHZ 9.

Released - 1968, cooperatively by Texas AES and Plant Science Research Division, ARS.

Breeder Seed/Stock - Texas AES.

Certified Seed/Stock - Available.

Preparer/Additional Information - Texas A&M University, AES, Department of Range Science, College Station, TX 77843, (409) 845-5579.

Llano

Source - An apomictic F hybrid derived from the cross TAM-CRD B-1s sexual clone (Reg. No. GPI) x 1/a rhizomatous Blue-type introduction from Africa

Method of Breeding - Llano is an apomictic F hybrid derived from the cross TAM-CRD B-1s sexual clone (Reg. No. GPI_ x 1/a rhizomatous Blue-type introduction from Africa. Llano (experimental hybrid #331) reproduces by obligate apospory and breeds true and thus no further selection was required to assure uniformity. Mode of reproduction was determined by cytological study of embryo sac development of the F_1 and confirmed by progeny tests through four generations.

Intended Use - Range and pasture plantings.

Description - Llano advantages over presently available cultivars are superior cold tolerance, significantly better forage production, earlier spring production, and excellent persistence due to an extensive rhizomatous root system. Stands are often severely damaged by freezing temperatures up to 125 km south of the line from San Antonio to Uvalde, Texas. Well established stands of Llano have survived up to 167 km further north than T-4464 and Higgins. Llano is adapted some 87 km further north than Nueces and also provides an earlier highly productive cultivar for areas further south.

Adapted to - LRR I, J, T; PHZ 9.

Released - 1977, cooperatively by ARS Plant Science Research Division, Texas AES and SCS.

Breeder Seed/Stock - Soil and Crop Science Department, Texas AES.

Certified Seed/Stock - Available.

Preparer/Additional Information - Texas A&M University, AES, Department of Range Science, College Station, TX 77843, (409) 845-5579.

Nueces

Carried as accession no. PI 476989.

Source - Nueces is an apomictic F_1 hybrid derived from the cross TAM-Crd B-1s sexual clone (Reg. No. GPI) x a rhizomatous Blue-type introduction from Africa.

Method of Breeding - Nueces (experimental hybrid #2-1) reproduces by obligate apospory and breeds true; thus, no further selection was required to assure uniformity. Mode of reproduction was determined by cytological study of embryo sac development of the F_1 and confirmed by progeny tests through six generations.

Intended Use - Range and pasture plantings.

Description - Advantages over presently available cultivars are improved cold tolerance, superior forage production, and higher digestibility, based upon in vitro studies. Present buffelgrass cultivars, T-4464 (common) and Higgins are generally considered adapted south of a line from San Antonio to Uvalde, TX. However, stands are often severely damaged by freezing temperatures up to 116 km south of this line. Nueces has proven winter hardy 125 km further north than present cultivars and provides a buffelgrass for an entirely new area.

Adapted to - LRR I, J, T; PHZ 9.

Released - 1977, cooperatively by ARS Plant Science Research Division, Texas AES, and SCS.

Breeder Seed/Stock - Soil and Crop Science Department, Texas AES.

Certified Seed/Stock - Available.

Preparer/Additional Information - Texas A&M University, AES, Department of Range Science, College Station, TX 77843, (409) 845-5579.

T-4464 (Common)

Selected at SCS Nursery, San Antonio, TX, and distributed to ranchers as T-4464. Carried as accession no. PI 153671.

Source - Turkana Desert of northern Africa in 1946.

Method of Breeding - Selected for drought tolerance and forage production.

Intended Use - Range reseeding and pasture planting.

Description - Rapid growth from very early spring through late summer. Growth continues during hottest part of summer with brief showers. Fast recovery from grazing and drought.

Adapted to - LRR I, J, T; PHZ 9.

Released - Informally by SCS in 1949.

Breeder Seed/Stock - Not available.

Certified Seed/Stock - Not available, common on market.

Preparer/Additional Information - Plant Materials Center, SCS, Rte 1 Box 155, Knox City, TX 79529, (817) 658-3922.

***Pennisetum clandestinum* Hochst. ex Chiov. - kikuyugrass**

Warm-season, sod-forming grass from Africa. Recommended in Hawaii. Aggressive pasture species where adapted, but can be serious weed in lawns and cultivated crops. Not winter-hardy; unadapted in southeastern U.S. Propagated vegetatively and by seed.

Hosaka

Source - University of Hawaii.

Method of Breeding - Unknown.

Intended Use - Pasture and erosion control.

Description - A strong, long-lived grass. It grows to a height of approximately 0.6 m. The stout stolons are much-branched and root freely at the nodes. The plant forms a dense mat. The stems are horizontal to somewhat upright and produce abundant leaves. The stems and stolons of Hosaka are thicker than cv. Breakwell, Whittet or common kikuyu in Hawaii. The leaves are covered with soft, short hairs. They are narrow, 130-380 mm long and spread outward stiffly at the ends of the stems. The light-colored sheaths are more hairy than the leaves and the ligule is a row of hairs. Flowering stems have three or four flowers which bloom on short side shoots. The side shoots are almost entirely or partially enclosed in the terminal leaf sheath. The topmost flower is on a short stalk while the others are without flower-stalks.

Adapted to - Areas in Hawaii with more than 900 mm of annual rainfall and at elevations from sea level to 1,830 m.

Released - By Dr. U. Urata of the Hawaii AES, University of Hawaii.

Breeder Seed/Stock - Vegetative material is available from the University of Hawaii and the SCS Hawaii Plant Materials Center.

Certified Seed/Stock - Not available.

Preparer/Additional Information - Robert Joy, Plant Materials Center, SCS, P.O. Box 236, Hoolehua, HI 96729, (808) 567-6378.

Pennisetum flaccidum **Griseb. - flaccidgrass**

Flaccidgrass is tetraploid (2n=36) apomicts and is adapted to areas in the southeastern U.S. that are in the transitional belt of warm-season, cool-season perennial grasses. Flaccidgrass is a montane grass of central Asia and is found at 1,600-4,300 m elevation from western Nepal, Tibet, and China through Kashmir to Afghanistan. Flaccidgrass is tall (100-200 cm), rhizomatous, deep-rooted perennials have high dry-matter production of 12,000 kg/ha, desirable dry-metter seasonal distribution, and adequate nutritive value; and are particularly low in fiber composition relative to other warm-season perennial grasses.

Carostan

Dept. of Crop Science, North Carolina State University - David H. Timothy. Carried as experimental no. Sel. 169.

Source - Traces to PI220606 collected by E.E. Smith on 6-17-54 near Girdah, Kataghan, Afghanistan. Received at Beltsville ARS on 8-24-54.

Method of Breeding - Two cycles of mass selection for most vigorous seedlings and most vigorous plants.

Intended Use - Grazing and hay.

Description - Carostan is taller than most accessions, growing to 1.5-2 m tall with ascending leaves that arch at maturity. It is apomictic, with 2n=36. Nodes are purple, lightly pubescent, with glabrous sheaths having sparsely ciliate margins. Blades are sparsely pilose with pronounced white mid-veins. Florets have purple stigmas, and exerted anthers are light saddle tan. Rapid and continuous growth from early spring until frost. Vigorous rhizomes. *In vitro* dry matter disappearance (IVDMD) is usually 60-65% when harvested at heights below 50 cm. Continuous grazing alters normally erect growth habit by stimulating profuse basal tiller and rhizome growth to form a moderately dense solid stand, even when grazed to a 2-cm stubble height. Carostan should perform well in a broad arc extending south from central Pennsylvania through North Carolina and west into eastern Oklahoma encompassing the lower mountains, piedmont, and upper coastal plain areas of that crescent.

Adapted to - LRR most of N, northern P, western S; PHZ most of 7, eastern 6, northern 8.

Released - May 25, 1987, North Carolina AES.

Breeder Seed/Stock - N.C. Foundation Seed Producers, Inc.

Certified Seed/Stock - Not available. Protected under PVP. Seed available from Aaron's Engineering, Rte 1 Box 26, Fargo, OK 73840, (405) 698-2613.

Preparer/Additional Information - Crop Science Dept., Box 7620, NC State Univ., Raleigh, NC 27695-7620, (919) 737-2650.

PI 315868

Increased at Plant Materials Center, SCS, Los Lunas, NM. Carried as accession no. PI 315868.

Source - The original source was from Czechoslovakia in 1969 by the Plant Introduction Center, Experiment, GA. The original material was mis-identified as *Cenchrus ciliaris* L. now *Pennisetum ciliare* (L.) Link.

Method of Breeding - Original accession was field planted, evaluated, and selected in plot comparison studies. Seed was bulk harvested from surviving plants and increased.

Intended Use - As a warm-season perennial forage grass, 315868 has potential in irrigated and dryland pastures in the southwest and higher moisture (300 mm) range seedings.

Description - The accession was originally selected for cold tolerance in comparison with buffelgrass (*Cenchrus ciliaris* L., now *Pennisetum ciliare* (L.) Link. Cold tolerance to -19⁻C have been measured. PI 315868 is very resistant to drought. It appears earlier maturing than other strains.

Adapted to - LRR D, G, H; PHZ 5.

Released - Not released, experimental seed available.

Breeder Seed/Stock - Plant Materials Center, SCS, Los Lunas, NM.

Certified Seed/Stock - Not available.

Preparer/Additional Information - Plant Materials Center, SCS, 1036 Miller St. SW, Los Lunas, NM 87031, (505) 865-4684.

Pennisetum purpureum **Schumach. - napiergrass**

Warm-season, slightly spreading bunchgrass introduced from Africa in 1913. Limited use for green feed, silage, and rotational grazing in Florida, gulf coast, and parts of southwest U.S. Coarse; forms large clumps. Must be cultivated to maintain high-yield potential. Grows best in moist, fertile soil. Fair drought resistance; moderate frost resistance. Propagated vegetatively. Considerable variation among geographical races. In Hawaii, clonal varieties developed from selected seedlings.

Banagrass

Hawaiian Sugar Planters' Association. Carried as accession no. 9037869.

Source - Australia (Bureau of Sugarcane Experiment Stations). Brought to Hawaii through quarantine at Beltsville in 1976.

Method of Breeding - Asexual propagation of original plant material.

Intended Use - Windbreak grass and for crossing with pearl millet for production of pearl millet x napiergrass (PMN) hybrids. Also has potential for fiber and energy production.

Description - An erect cultivar of napiergrass. Chromosome number confirmed as 28. Weed potential is moderate to high; should double chromosomes then cross to produce a sterile triploid.

Adapted to - LRR I, T, U, V, Z; PHZ 9.

Released - Not officially, although being used as windbreak material.

Breeder Seed/Stock - Vegetative material will be maintained by the Hawaiian Sugar Planters' Association, Aiea, HI, and the Plant Materials Center, SCS, Hoolehua, HI.

Certified Seed/Stock - Not available.

Preparer/Additional Information - Hawaiian Sugar Planters' Association, 99-193 Aiea Heights Dr., Aiea, HI 96701, (808) 487-5561.

Merkeron

Selected at Georgia Coastal Plain Experiment Station, Tifton, ARS cooperating - G.W. Burton.

Source - F_1 hybrid between two selections carrying numbers 1 and 208.

Method of Breeding - From 1936-41 selection within open pollinated seedling progenies of local types and introductions practiced. In fall of 1941, several of these selections hybridized to combine desirable characteristics, including resistance to *Helminthosporium* Link eyespot. One of these crosses involving selection 1, vigorous common type, and selection 208, plant with very short internodes and many tillers, gave plants yielding 35% more than checks and best common napiergrass hybrids. In 1944, several of best of these hybrids sent to Rio Piedras, PR, for testing. Best of this cross between selection 1 and 208 released under name Merkeron in 1955 by Velez Fortuno, head of plant breeding at Experiment Station, Rio Piedras.

Intended Use - Grazing, hay, biomass.

Description - Leafy, many tillered, late maturing F_1 hybrid resistant to *Helminthosporium* eye-spot.

Adapted to - LRR P, T, U; PHZ 8, 9, 10.

Released - 1955, by Experiment Station, Rio Piedras, PR.

Breeder Seed/Stock - Experiment Station, Rio Piedras and Coastal Plain Experiment Station, Tifton, GA.

Certified Seed/Stock - Not available.

Preparer/Additional Information - ARS, Coastal Plain Experiment Station, P.O. Box 748, Tifton, GA 31793, (912) 386-3353.

Mott

Selected at Georgia Coastal Plain Experiment Station - W.W. Hanna.

Source - Tift N75 dwarf napiergrass germplasm released in 1986 by Georgia Coastal Plain Experiment Station.

Method of Breeding - Selfed progeny of Merkeron napiergrass, a tall hybrid selected from a dwarf x tall napiergras cross made in 1941 by G.W. Burton. Tift N75 is heterozygous and does not breed true from seed, so it has been vegetatively maintained in the field since 1977.

Intended Use - Mott dwarf elephantgrass has use as a pasture grass in tropic and subtropic areas. The leaf-to-stem ratio is high and provides good grazing material. It has a high fertility requirement and persists for many years if managed properly.

Description - Maintains high forage quality over a wider range of maturities than is characteristic of most tropical grasses. Excellent drought tolerance producing new growth longer into the dry season than other grasses. Vegetatively propagated using stem cuttings. Establishment cost may be high due to expense of planting material and intense labor needs.

Adapted to - LRR U; PHZ southern 9, 10.

Released - 1988, jointly by University of Florida, Institute of Food and Agricultural Sciences, Gainesville, FL and ARS, Coastal Plain Experiment Station, Tifton, GA.

Breeder Seed/Stock - Florida Foundation Seed Producers, Inc., Greenwood, FL, and ARS at Georgia Coastal Plain Experiment Station.

Certified Seed/Stock - Not available.

Preparer/Additional Information - University of Florida, Institute of Food and Agricultural Science, Agronomy Department, Gainesville, FL 32611, (904) 392-1814.

Tift N75

ARS and University of Georgia Coastal Plain Experiment Station, Tifton, GA. Carried as experimental no. GP-55 and accession no. PI 517947.

Source - Selected in 1977 from a selfed progeny of Merkeron napiergrass for shorter height.

Method of Breeding - Selfed progeny of Merkeron napiergrass, a tall hybrid selected from a dwarf x tall napiergrass cross made in 1941 by G.W. Burton. Tift N75 is heterozygous and does not breed true from seed, so it has been vegetatively maintained in the field since 1977.

Intended Use - Grazing and hay.

Description - A low-management cultivar that produces high quality forage. Dwarf growth habit attains an uncut height of five feet.

Adapted to - LRR T, U; PHZ 8, 9, 10.

Released - March 1986, ARS and University of Georgia Coastal Plain Experiment Station, Tifton, GA.

Breeder Seed/Stock - Vegetatively distributed by ARS, Coastal Plain Experiment Station, Tifton, GA.

Certified Seed/Stock - Limited supply available.

Preparer/Additional Information - ARS Coastal Plain Experiment Station, P.O. Box 748, Tifton, GA 31793, (912) 386-3353.

Phalaris aquatica L. - hardinggrass

Phalaris tuberosa L. var. _stenoptera_ (Hack.) A.S. Hitchc.
Phalaris tuberosa L. var. _hirtiglumis_ Batt. and Trabut

Species now contains hardingrass and koleagrass. Hardinggrass is a cool-season, slightly-spreading grass from Union of South Africa; indigenous to northern Africa. Used for pasture and erosion control in California and sparingly in other parts of the southwestern U.S. Adapted to subtropical winter rainfall climate; best suited to heavy soils; drought resistant; palatable; withstands heavy grazing. Koleagrass was received from AES, Rabat, Morocco, in 1955. California accession number T.O. 2143. Coarse bunchgrass. Resembles hardinggrass in general appearance, but has round, bulb-like enlargements at base of culm somewhat similar to *Phalaris coerulescens* Desf. Some indication that it requires higher rainfall than hardinggrass.

Au Oasis (Reg No. 84)

Developed by Alabama AES Auburn University, AL. Carried as accession no. AP-2.

Source - PI 240280, PI 236482, PI 240284, PI 219636, PI 240242, PI 207960, plus one selection from an old Phalaris nursery established at Tallassee, AL.

Method of Breeding - Clonal evaluation for vigor, winter growth, regrowth potential, and disease resistance, plus progeny testing of selected clones for forage yield distribution. Eight high-performing clones were selected for synthetic variety.

Intended Use - Forage grass, grown alone or in combination with legumes for pasture and hay. Also used in wetlands mixtures and for wildlife food and cover.

Description - Early maturing, aggressively tillering variety with improved cold tolerance; exhibits a more even seasonal forage distribution than that of many other *Phalaris* cultivars.

Adapted to - LRR A, C, P, T, I; PHZ 8.

Released - 1981, Auburn University, AL.

Breeder Seed/Stock - Auburn University, AL.

Certified Seed/Stock - OECD certification. Available through International Seeds Inc. PVP No. 8900076

Preparer/Additional Information - International Seeds Inc., P.O. Box 168, 820 W. First St., Halsey, OR 97348, (503) 369-2251.

Perla

Selected at Plant Materials Center, SCS, Pleasanton, CA - H.W. Miller and O.K. Hoglund. Carried as accession no. PI 202480.

Source - Introduced from Morocco. Received as *Phalaris tuberosa* var. *stenoptera* (Hack.) A.S. Hitchc. (PI 202480) and assigned accession number P-14529. Later reidentified as *P. tuberosa* L. var. *hirtiglumis* Batt. & Trabut, now *P. aquatica* L.

Method of Breeding - Increased under isolation with minor phenotypic selection to eliminate off-types.

Intended Use - Range reseeding and wildlife cover.

Description - Tall, robust, rapid-developing bunchgrass, with short rhizomes. Resembles hardinggrass in general appearance, but has much stronger seedling vigor and hairy glumes. Easy to establish, grows well during cold winter months, and produces good seed crops. Stands frequently improve through natural reseeding, which differs from hardinggrass. Range livestock graze it readily; reaches range readiness about three weeks before hardinggrass. In dry years produces more forage because of early growth. Adapted to soils with restricting layer in Mediterranean climatic zone wherever average annual rainfall is 400 mm or more.

Adapted to - LRR A, C; PHZ 8b, 9.

Released - 1970, cooperatively by California AES and Plant Materials Center, SCS, Pleasanton, CA.

Breeder Seed/Stock - Plant Materials Center, SCS, Lockeford, CA.

Certified Seed/Stock - Limited availability.

Preparer/Additional Information - Robert Slayback, SCS, 2121-C, 2nd St., Davis, CA 95616, (916) 757-8257.

Wintergreen

Selected at the Texas A&M University Agricultural Research Center, McGregor - M.J. Norris.

Source - Surviving plants from two plant introductions (PI 193056 and PI 196338) originally planted in 1952. Surviving plants noted following severe drought in late 1954.

Method of Breeding - Increase of three plants that survived extended drought in 1954, and tested under similar conditions in 1963, 1966, and 1967.

Intended Use - Grazing.

Description - Cool-season, perennial bunchgrass somewhat more erect than commercial hardinggrass; distinguishable mainly by summer survival in area with extended dry periods and high temperatures.

Adapted to - Southeastern U.S., severely limited by lack of cold-hardiness and its susceptibility to nematodes.

Released - 1969, by Texas AES.

Breeder Seed/Stock - Texas A&M University, Agricultural Research Center, McGregor.

Certified Seed/Stock - Available.

Preparer/Additional Information - Texas A&M University, AES, Department of Range Science, College Station, TX 77843, (409) 845-5579.

Phalaris arundinacea L. - reed canarygrass

Cool-season, sod-forming grass indigenous to North America, Europe, and Asia. Locally important for hay, pasture, silage, and erosion control, especially in the north central states and on the west coast from northern California to Washington, and to a limited extent in the northeastern, southern, and intermountain regions. Well adapted to poorly drained soils subject to flooding, but can be grown on drier upland soils.

Castor

Developed by Agriculture Canada, Beaverlodge Research Station, Alberta - S.G. Bonis.

Source - Castor is a four-clone synthetic variety. Of the four clones, two originated from a seed accession from the former USSR; one from a commercial seed lot from the Ramy Seed Company, Minneapolis, MN; and one from a breeders seed lot from Dr. R.P. Knowles, Agriculture Canada, Saskatoon Research Station, Saskatoon.

Method of Breeding - Mass selection.

Intended Use - Hay.

Description - Castor is a high seed-retaining variety with characteristics of forage productivity, maturity, quality, and vigor similar to Frontier. Its seed yield is approximately double that of Frontier. Castor has performed well in all areas of Canada where the species is grown, being particularly well suited to irrigated culture, peat soils, and heavy soils where moisture is adequate.

Adapted to - LRR A, B, E, C, F, G, K, L, R, S, M, N, P; PHZ 2.

Released - 1972, Agriculture Canada, Beverlodge Research Station.

Breeder Seed/Stock - Maintained at the Beaverlodge, Agriculture Canada Research Station.

Certified Seed/Stock - Available. Distributed by Otto Pick and Sons Seeds Ltd., Richmond Hill, Ontario.

Preparer/Additional Information - Agriculture Canada, Sainte-Foy Research Station, 2560 Hochelaga Blvd., Sainte-Foy, Quebec G1V 2J3, (418) 648-7980.

Grove

Selected at Canada Department of Agriculture Research Station, Ottawa, Ontario - R.M. MacVicar and D.R. Gibson.

Source - Native collections.

Method of Breeding - Large number of selections evaluated in clonal and outcross progeny tests. Four superior clones identified and intercrossed in isolation to provide basic seed.

Intended Use - Hay production.

Description - Very leafy; some 7-10 days later in maturity than Frontier. First growth is not equal to Frontier, but good second growth makes it equal in crude-protein production. Main attributes are late maturity and ability to give high yields of very leafy herbage at high levels of soil fertility. May be more suitable than common reed canarygrass for haylage production.

Adapted to - All areas of the northern U.S. and southern Canada where reed canarygrass does well; PHZ 3, 4, 5, 6.

Released - 1970, by Canada Department of Agriculture.

Breeder Seed/Stock - Canada Department of Agriculture Research Station, Ottawa.

Certified Seed/Stock - Available.

Preparer/Additional Information - Dr. A. R. McElroy, Forge Bldg. #12, Plant Research Center, Ottawa, Ontario, (613) 995-3700.

Ioreed

Selected at Iowa AES, Ames, SCS cooperating - H.D. Hughes and C.P. Wilsie.

Source - Parental clones selected from German Steenacker 1; German Rodowbrooker 18; Oregon commercial; Minnesota J18, J15C, J15A, J20B; U.S. Department of Agriculture 55009 and 55018; and old Iowa strain 503.

Method of Breeding - Ten clones from above sources selected on basis of forage, seed-yielding ability, and forage quality; saved and recombined. Iowa clone represented about one-third and other nine sources about 7% each of seed recombined to form Ioreed. Synthetic 1 seed obtained in 1945.

Intended Use - Hay production.

Description - Hardy, vigorous, moderately productive, with good leaf-disease resistance. Mid-early in maturity, fair in seed production, rather susceptible to seed shattering. Appears similar to commercial types from long-time stands in Iowa and Minnesota.

Adapted to - LRR K, M, N, L, R; PHZ 3, 4, 5, 6.

Released - 1946, cooperatively by Iowa AES and Nursery Division, SCS.

Breeder Seed/Stock - Iowa AES.

Certified Seed/Stock - Not available, common is available.

Preparer/Additional Information - SCS, Plant Materials Center, RR1, Box 9, Elsberry, MO 63343 (314) 898-2012.

Palaton

Land O'Lakes, Inc. - R.R. Kalton, J. Shields, P.A. Richardson. Carried as accession no. PI 531088, Reg No. 120, Experimental No. P.S.3.

Source - An eight-clone synthetic variety. Parent clones derived from Flare (three), Vantage (two), Rise (one), a polycross progeny (one), and a germplasm collection (one).

Method of Breeding - Parent clones were selected from a large source nursery and evaluated for several years in clonal and polycross progeny trials for forage and seed yield. Parental clones also were evaluated several times for animal palatability, alkaloid content, and seed shattering.

Intended Use - Palaton should be suitable for hay, pasture, greenchop, and haylage use as well as for wildlife food, cover, and conservation use on a wide variety of soil types both well and poorly-drained.

Description - Palaton is superior to such similar varieties as Rise, Vantage, and Flare in seed yield and seed holding capacity. It also is lower in alkaloid content; contains only gramine and no tryptamine and carboline which cause digestive upsets in cattle and sheep; and promises to give increased palatability and animal gains from its forage. It is similar in forage yield and leaf disease resistance to Venture, Rise, Vantage, and Flare; exhibits excellent winter-hardiness and persistence; and shows good seedling vigor and stand establishment ability because of its improved shattering resistance. Palaton is a high seed- and forage-producing cultivar with improved palatability and good drought and flood tolerance. It is licensed in Canada and in tests in northern Europe, Japan, Korea, and northern China.

Adapted to - All areas of the northern U.S. and southern Canada where reed canarygrass does well; PHZ 2, 3, 4, 5, 6.

Released - Limited release in 1985 with more extensive commercial production in 1986. Released by Land O' Lakes, Inc.

Breeder Seed/Stock - Breeders and foundation seed produced in Iowa by Land O' Lakes, Inc.

Certified Seed/Stock - Available.

Preparer/Additional Information - Robert R. Kalton, Land O' Lakes Research Farm, 1025 190th St., Webster City, IA 50595, (515) 543-4852.

Rival

Developed at the University of Manitoba.

Source - Rival is derived from nine clones of diverse origin (including Ottawa Synthetic C and the cultivar Grove), selected on the basis of being tryptamine alkaloid-free and relatively low in total alkaloid content.

Method of Breeding - Mass selection for the alkaloid selection. Selection criteria were uniformity of type of forage production, flowering date, and general combining ability with respect to seed production potential. Selected clones within Ottawa Synthetic C and the variety Grove were crossed with tryptamine-free clones of the Saskatoon seed retention line S6982. The progeny were sown and further selections made on the basis of forage yield and seed yield.

Intended Use - Hay.

Description - Rival has an erect growth habit with leaves extending up both vegetative and floral culms. It has an early flowering and is more winter-hardy than Frontier. The seeds are grey, with approximately 5% yellow seeds. There is no evidence of susceptibility to leaf diseases.

Adapted to - LRR A, B, C, E, F, G, K, L, R, S, M, N, P; PHZ 2.

Released - 1985, University of Manitoba.

Breeder Seed/Stock - Maintained at the University of Manitoba.

Certified Seed/Stock - Available, distributed by Secan Association.

Preparer/Additional Information - Agriculture Canada, Sainte-Foy Research Station, 2560 Hochelasa Blvd., Sainte-Foy, Quebec G1V 2J3, (418) 648-7980.

Vantage

Selected at Iowa Agriculture and Home Economics Experiment Station - I.T. Carlson. Carried as accession no. RC-1.

Source - Seed collections made in Iowa and southern Minnesota in 1954.

Method of Breeding - Developed from six clones selected on basis of individual plant performance, clonal evaluation, and topcross progeny performance. Selection mainly for high seed yield and seed retention. Winter-hardiness, maturity, forage yield, disease reaction, and palatability also were considered.

Intended Use - Best suited for pasture, but it also can be used for hay and silage.

Description - It has a unique combination of early heading, high seed yield, good seed retention, high 100-seed weight, high forage yield, and the forage is free of tryptamine-carboline types of indole alkaloids. Its main advantage is superior seed retention.

Adapted to - LRR A, B, E, F, G, K, L, M, N, R, S; PHZ 3.

Released - 1972, Iowa Agriculture and Home Economics Experiment Station.

Breeder Seed/Stock - Iowa Agriculture and Home Economics Experiment Station, Ames, IA.

Certified Seed/Stock - Available.

Preparer/Additional Information - Irving T. Carlson, Agronomy Department, Iowa State Univ., Ames, IA 50011, (515) 294-9653.

Venture (Reg. No. 121)

Land O'Lakes, Inc. - R.R. Kalton, P.A. Richardson, J. Shields. Carried as accession no. PI 531089, Experimental No. P.S.2A.

Source - A seven-clone synthetic variety. Parental clones derived from Flare (two), Vantage (two), polycross progeny (one) an elite seeding clone (one), and a field collection from northern Iowa (one).

Method of Breeding - Clones originally selected for animal palatability in 7,500 plant source nursery. Clones evaluated for seed yield, shattering resistance, alkaloid content, and palatability in clonal trials and for forage and seed yield in polycross progeny trials.

Intended Use - All-purpose forage use (hay, pasture, greenchop and haylage), conservation use, and wild life food and cover on both fertile, well-drained soils as well as on poorly drained soils.

Description - Venture is a tall-growing, rhizomatous, low-alkaloid cultivar of reed canarygrass. It exhibits improved shattering resistance, palatability, and seed yield compared with most available varieties. It is very winter-hardy, persistent, and flood tolerant and contains no tryptamines or carboline alkaloids which cause digestive upsets in cattle and sheep. Venture is a highly productive, palatable cultivar with good management and exhibits good drought as well as flooding tolerance compared with many other cool-season grasses such as smooth bromegrass, orchardgrass, tall fescue, timothy, etc.

Adapted to - Semi-humid and humid areas in northern U.S. and southern Canada; PHZ 2, 3, 4, 5, 6.

Released - Limited release in 1985, first substantial commercial crop in 1986. Research Seeds, Inc., and Peterson Seed Company.

Breeder Seed/Stock - Land O' Lakes, Inc., Webster City, IA. Breeders and foundation seed produced in Iowa.

Certified Seed/Stock - Available.

Preparer/Additional Information - Robert R. Kalton, Research Seeds, Inc., 1025 190th St., Webster City, IA 50595, (515) 543-4852.

Phalaris canariensis L. - annual canarygrass

A robust annual that furnished the canary seed of commerce. Introduced from the Mediterranean region. Generally found across North America and south through Mexico and into South America. Generally found in moist wasteplaces and on wet streambanks.

Elias

Source - Developed at the Minnesota AES.

Method of Breeding - Limited to four generations of select seed multiplication to form foundation seed in Canada. Elias breeder seed was increased from a single plant selection of USDA PI 170622.

Intended Use - Hay and pasture production.

Description - Inflorescence is 16 mm wide, 28 mm long, compact and oval, larger than Keet; plant height is taller than Keet; lodging resistance is similar to Keet; shattering resistance is good; and maturity is similar to Keet. Elias is used in birdseed mixtures. It has a higher test weight than Keet. It is well adapted to the dark brown and black soil zones of western Canada.

Adapted to - Recommended in Canada for use in Saskatchewan.

Released - 1983 at the Minnesota AES. Registered in Canada in 1988 with registration no. 3030.

Breeder Seed/Stock - University of Saskatchewan.

Certified Seed/Stock - Available.

Preparer/Additional Information - Agriculture Canada, Food Production and Protection Branch, Plant Products Division, Ottawa, Ontario K1A 0C6.

Keet

Source - Minnesota AES, St. Paul.

Method of Breeding - Developed from a single plant selection (PI 250741) from Iran. Seed of this selection was increased and approximately 100 plants were selected for early and uniform maturity, lodging resistance, test weight, and seed yield. Progeny rows were selected and bulked to form breeder seed of Keet.

Intended Use - Hay and pasture production.

Description - Leaves are slightly narrower and shorter than those of Alden (average 4 mm x 103 mm); inflorescence is slightly narrower and shorter than those of Alden (average 17 mm x 34 mm); plant height is slightly shorter than Alden; and maturity is similar to Alden. Keet is a good-yielding variety with adequate seed yields in Saskatchewan and good yields in Manitoba. It has a slightly higher test weight than Alden and may be slightly more resistant to lodging. Keet is adapted to dark brown and black soil zones of Western Canada. Its use is as birdseed or in birdseed mixtures.

Released - 1983, Minnesota AES. Registered in Canada in 1983 with registration no. 2345.

Breeder Seed/Stock - Minnesota AES, St. Paul.

Certified Seed/Stock - Available.

Preparer/Additional Information - Agriculture Canada, Food Production and Protection Branch, Plant Products Division, Ottawa, Ontario K1A 0C6.

Phleum pratense L. - timothy

Major cool-season bunchgrass from Europe. Used for hay, pasture, and silage throughout humid sections of northern U.S.. Long lived in cool, humid regions; winter-hardy, but not resistant to close, continuous grazing. Palatable and nutritious; valuable hay grass. Poor recovery with limited moisture; does not tolerate drought or high temperatures.

Alexander

Developed by Svalof AB, Sweden.

Source - Three Swedish populations.

Method of Breeding - Mass selection, seven plant synthetics.

Intended Use - Hay.

Description - Semi-erect growth, leaves are larger and medium green; flag leaf is semi-erect, larger size and color. Maturity similar to Climax.

Adapted to - LRR K, L, R, S, M, N, P; PHZ 2.

Released - 1987.

Breeder Seed/Stock - Svalof AB, Svalof, Sweden.

Certified Seed/Stock - Available. Speare Seeds, CP 171, Harriston, Ontario NO6 120.

Preparer/Additional Information - Agriculture Canada, Sainte-Foy Research Station, 2560 Hochelasa Blvd., Sainte-Foy, Quebec G1V 2J3, (418) 657-7980.

Alma

Source - Agriculture Research Institute, Jokoiinen, Finland.

Method of Breeding - A four-clone synthetic originating from the cross between the European varieties Tarmo and Bodin made in 1961. Several cycles of mass selection for improved forage and seed yield without loss of winter-hardiness were conducted. Superior lines were further evaluated for increased regrowth yield and seed yield; based on progeny test results, four lines were finally selected. Breeder seed was first bulked in 1983. Tested as JO 0166.

Intended Use - Hay and pasture production.

Description - Growth habit is semi-erect; ploidy is hexaploid, leaves are medium green; uppermost internode is long; plant height is tall; maturity is slightly earlier than Bottnia II; similar to Climax under Alberta conditions. Alma is a winter-hardy variety, giving good first-cut yields. It has good spring vigor and the regrowth after first cutting is intermediate. Seed yields are excellent. In Ontario, Alma produced lower forage yields than Climax or Timfor, but demonstrated good yields in western Canada.

Adapted to - Recommended in Canada for use in British Columbia.

Released - Agricultural Research Institute, Jokoiinen. Registered in Canada in 1990 as registration no. 3222.

Breeder Seed/Stock - Agricultural Research Institute, Jokoiinen.

Certified Seed/Stock - Available.

Preparer/Additional Information - Agriculture Canada, Food Production and Protection Branch, Plant Products Division, Ottawa, Ontario K1A 0C6. Seed is distributed by Oseco Inc., P.O. Box 219, Brampton, Ontario L6V 2L2.

Argus

Source - Weibullsholm Plant Breeding Institute, Landskrona, Sweden.

Method of Breeding - Developed from the cross I3 Kampe II-213 x I2 Climax-332. These parents were selected on the basis of superior performance during inbreeding. Tested as WWT100.

Intended Use - Hay and pasture production.

Description - Leaves are dark in early growth, lighter in regrowth; narrow, upright; heading is one day later than Champ; good resistance to leaf spot diseases. Argus has high yield especially in the first cut.

Adapted to - Recommended in Canada for use in Ontario.

Released - Weibullsholm Plant Breeding Institute. Registered in Canada in 1988 as registration no. 2914.

Breeder Seed/Stock - Weibullsholm Plant Breeding Institute.

Certified Seed/Stock - Available.

Preparer/Additional Information - Agriculture Canada, Food Production and Protection Branch, Plant Products Division, Ottawa, Ontario K1A 0C6. Seed is distributed by Oseco Inc., P.O. Box 219, Brampton, Ontario L6V 2L2.

Barmoti

Source - Dutch ecotypes.

Method of Breeding - Selection of space plants.

Intended Use - Cool-season turfgrass. May be seeded as a component in mixtures.

Description - It is an early flowering variety.

Adapted to - Central & northeastern U.S.; PHZ 1, 2, 3, 4, 5.

Released - Barenbrug.

Breeder Seed/Stock - Barenbrug Holland, Oosterhout, The Netherlands.

Certified Seed/Stock - Available from Barenbrug USA.

Preparer/Additional Information - Barenbrug USA, P.O. Box 239, Tangent, OR 97389, (503) 926-5801.

Barvanti

Source - Dutch ecotypes.

Method of Breeding - Selection in space plants, and after that clones in turf conditions.

Intended Use - Cool-season turfgrass. Only to be used in turfgrass mixtures.

Description - A very good turf quality with a very prostrate growth habit; very fine leaved.

Adapted to - Central and northeastern U.S.; PHZ 1, 2, 3, 4, 5.

Released - Barenbrug.

Breeder Seed/Stock - Barenbrug Holland, Oosterhout, The Netherlands.

Certified Seed/Stock - Available from Barenbrug USA.

Preparer/Additional Information - Barenbrug USA, P.O. Box 239, Tangent, OR 97389, (503) 926-5801.

Basho

Selected at the Plant Research Center, Canada, Dept. of Agriculture Research Station, Ottowa, Ontario - W.R. Childers.

Source - Farmers' fields under grazing pressure for over 15 years.

Method of Breeding - Single-clone synthetic from one genotype isolated within the variety Champ.

Intended Use - Hay.

Description - Grows 100-150 mm taller than Champ, long penicules, good regrowth with a maturity similar to Champ.

Adapted to - LRR K, L, R, S, M, N, P; PHZ 2.

Released - 1974, Canada Department of Agriculture.

Breeder Seed/Stock - Agriculture Canada, Plant Research Center, Ottawa.

Certified Seed/Stock - Available. Public variety.

Preparer/Additional Information - Agriculture Canada, Sainte-Foy Research Station, 2560 Hochelasa Blvd., Sainte-Foy, Quebec G1V 2J3, (418) 657-7980.

Bottnia II

Developed by Svalof AB, Sweden.

Source - Old field near Bottnia in northern Sweden.

Method of Breeding - Mass selection.

Intended Use - Hay.

Description - Erect growth habit, with long leaves, average width, and medium green. In Alberta, Canada, it reached maturity two days before Climax.

Adapted to - LRR K, L, R, S, M, N, P; PHZ 2.

Released - 1985.

Breeder Seed/Stock - Svalof AB, Svalof, Sweden.

Certified Seed/Stock - Available. Canadian Distributor: Henri Malon Ltd, 15 Woodland Hts., Toronto, Ontario M65 2W5.

Preparer/Additional Information - Agriculture Canada, Sainte-Foy Research Station, 2560 Hochelasa Blvd., Sainte-Foy, Quebec G1V 2J3, (418) 657-7980.

Bounty

Selected at Canada Department of Agriculture Research Station, Ottawa, Ontario - W.R. Childers.

Source - Original clones selected from material supplied by Morgan Evans, Wooster, OH, and from variety Otofte.

Method of Breeding - Superior late-maturing clones selected and included in open pollinated progeny test. Five highest yielding progenies noted, and their maternal clones established vegetatively in isolated increase block.

Intended Use - Pasture and hay production.

Description - Tall, with leaves held high on the stem; maturity 7-10 days later than Climax. Stems larger and leaves broader than Climax. Under higher summer temperatures, the lateness differential becomes shorter.

Released - 1966, Canada Department of Agricultural Research Station, Ottawa.

Breeder Seed/Stock - Canada Department of Agricultural Research Station, Ottawa.

Certified Seed/Stock - Available. Phillips Seed Farm Ltd., Box 249, Tisdale, Saskatchewan S0E 1T0.

Preparer/Additional Information - Canada Seed Growers' Association, P.O. Box 8445, Ottawa, Ontario K1G 3T1.

Carola

Source - Weibullsholm Plant Breeding Institute, Landskrona, Sweden. Tested as WWT118.

Method of Breeding - This variety originated from a cross between the varieties Climax and Vanadis in the early 1970s. The parent clones were inbred for one to three generations. Following several generations of selection for homogeneity and productivity and after subsequent progeny testing, nine superior clones were selected in 1978 on the basis of dry-matter yield potential and persistence. Syn 1 seed was produced from the nine parent clones established in a polycross nursery in 1979. Breeder seed was produced in 1981.

Intended Use - Hay and pasture production.

Description - Growth habit is erect; leaves are light green, medium-length flag leaf; top internode length is short to medium; and maturity is similar to Climax. Carola, a leafy hay-type timothy, has yielded well in Ontario trials. It has good winter-hardiness.

Adapted to - Recommended in Canada for use in Ontario.

Released - Weibullsholm Plant Breeding Institute, Landskrona, Sweden. Registered in Canada in 1990 with registration no. 3197.

Breeder Seed/Stock - Weibullsholm Plant Breeding Institute, Landskrona, Sweden.

Certified Seed/Stock - Available.

Preparer/Additional Information - Agriculture Canada, Food Production and Protection Branch, Plant Products Division, Ottawa, Ontario K1A 0C6. Seed is distributed by Oseco Inc., P.O. Box 219, Brampton, Ontario L6V 2L2.

Champ

Selected at Canada Department of Agriculture Research Station, Ottawa, Ontario - W.R. Childers and L.P. Folkins.

Source - Fifty-six plants selected from a local farmer's field that had been closely grazed for 15 years.

Method of Breeding - Clones increased and tested in greenhouse for dry-matter production and rapidity of new culm formation. Top 10 clones selected for field testing. Replicated polycross progeny tests and yield trials established throughout Canada. Top four clones selected for synthetic.

Intended Use - Pasture and hay production.

Description - Excellent seedling vigor; seedlings stool rapidly to produce thick stand; stems finer than most varieties. Matures five to seven days earlier than Climax; flowering heads shorter, more dense than most varieties. Most important characteristic is quick recovery and good aftermath production. Seed yields have been 15-20% less than Climax.

Released - 1967, Canada Department of Agriculture.

Breeder Seed/Stock - Canada Department of Agriculture Research Station, Ottawa.

Certified Seed/Stock - Available in quantity.

Preparer/Additional Information - Canada Seed Growers' Association, P.O. Box 8445, Ottawa, Ontario K1G 3T1.

Clair (Reg. No. 3)

Selected by the Kentucky AES and the ARS.

Source - Farm of Clair Andrew near Vevay, IN.

Method of Breeding - Mass selection.

Intended Use - Hay production.

Description - Vigorous, early-maturing variety, with good aftermath. Stem appears to be somewhat larger and leafed broader than those of other timothy.

Adapted to - LRR K, L, R, S, M, N, P; PHZ 2.

Released - 1971.

Breeder Seed/Stock - Kentucky AES.

Certified Seed/Stock - Not available. Public variety.

Preparer/Additional Information - Agriculture Canada, Sainte-Foy Research Station, 2560 Hochelasa Blvd., Sainte-Foy, Quebec G1V 2J3, (418) 657-7980.

Climax

Selected at Canada Department of Agriculture Research Station, Ottawa, Ontario - R.M. MacVicar.

Source - Wide collection of seed lots.

Method of Breeding - Synthetic variety developed by combining several progeny-tested clones.

Intended Use - Hay production.

Description - Tall, fine stemmed. Characterized by marked leafiness; leaves carried high on stems. Under conditions of good fertility, aftermath growth excellent. Highly resistant to rust. Seven to ten days later in maturity than common.

Adapted to - LRR K, L, R, S, M, N, P; PHZ 2.

Released - 1947, Canada Department of Agriculture.

Breeder Seed/Stock - 1947, Canada Department of Agriculture Research Station, Ottawa.

Certified Seed/Stock - Available.

Preparer/Additional Information - Canada Seed Growers' Association, P.O. Box 8445, Ottawa, Ontario K1G 3T1.

Drummond

Selected at Macdonald College, Quebec, Canada - J.N. Bird. Licensed in 1940.

Source - Strain from Northern Europe, S-48 and S-51 from Wales, and F.C. 15150 from ARS and Ohio AES, Wooster, all introduced during 1930-33.

Method of Breeding - Maternal line selection, with space-planted progeny tests.

Intended Use - Pasture and hay production.

Description - Reaches flowering and seed stage about 10-14 days later than common timothy at Macdonald College. Winter-hardy, with appreciable amount of rust resistance. Slightly inferior to Climax in midsummer aftermath yield.

Adapted to - PHZ 4.

Released - Macdonald College, Quebec.

Breeder Seed/Stock - Macdonald College of McGill Univ., Quebec.

Certified Seed/Stock - Available.

Preparer/Additional Information - B. Coulman, Macdonald College of McGill Univ., 21,111 Lakeshore Rd., Ste. Anne de Bellevue, Quebec H9X 3V9, (514) 398-7872.

Dynasty

Source - FFR Cooperative, West Lafayette, IN. Tested as Syn S.

Method of Breeding - A seven-clone synthetic developed by progeny line selection, and at least two generations of phenotypic recurrent selection. The seven clones were selected in 1979 for their vigor, leafiness and similarity in maturity. Breeder seed (Syn 1) was produced from vegetative portions of the original seven clones of Syn S, and was bulked in 1981, 1982, 1986, and 1987.

Intended Use - Hay and pasture production.

Description - Erect growth habit; plant height is similar to Climax; leaves have a length and width similar to Climax; heads are similar to Climax, dense, and spike-like; maturity is later than Climax; seed width is similar to Climax and Clair. In 14 station-years of trials in Ontario, Dynasty produced higher forage yields than Climax and Clair.

Adapted to - Recommended in Canada for use in Ontario.

Released - FFR Cooperative, West Lafayette, IN, and registered in Canada in 1991 with Registration No. 3353.

Breeder Seed/Stock - FFR Cooperative, West Lafayette, IN.

Certified Seed/Stock - Available.

Preparer/Additional Information - Agriculture Canada, Food Production and Protection Branch, Plant Products Division, Ottawa, Ontario K1A 0C6. Seed is distributed by United Cooperative of Ontario, Box 527, Station A, Mississauga, Ontario L5A 3A4.

Farol

Selected by Cebeco-Handelsraad in The Netherlands.

Source - A six-clone synthetic developed from individual plant selection from the variety Lofar (unlicensed) in 1959.

Method of Breeding - Mass selection for drought tolerance.

Intended Use - Hay.

Description - Semi-erect to semi-prostrate; medium-green leaves; flag leaf is medium to long, medium width, semi-erect; medium to long stem; late maturity.

Adapted to - LRR K, L, R, S, M, N, P; PHZ 2.

Released - 1985.

Breeder Seed/Stock - Cebeco-Handelsraad, 3000 AD Rotterdam, The Netherlands.

Certified Seed/Stock - Available.

Preparer/Additional Information - Agriculture Canada, Sainte-Foy Research Station, 2560 Hochelasa Blvd, Sainte-Foy, Quebec G1V 2J3, (418) 657-7980.

Glenmor

Source - Northrup King Co., Minneapolis, MN. Tested as K4-215 and K215.

Method of Breeding - This variety originated from three clones of Climax and one clone of Lorain. These clones were progeny-tested for yield; open-pollinated progeny were recombined in 1972 and underwent recurrent selection for vigor, leafiness, disease resistance, and uniformity in 1974. Two hundred plants were recombined to form the parental material of Glenmor.

Intended Use - Hay and pasture production.

Description - Growth habit is upright with leaves extending well up on the stems of most plants; leaves have a similar width to Climax; maturity is later than Climax and Timfor; and it is moderately resistant to stem rust. Glenmor is a late maturing variety which produces acceptable forage yields.

Adapted to - Recommended in Canada for use in British Columbia and Ontario.

Released - Northrup King Co., Minneapolis, MN. Registered in Canada in 1988 with registration no. 3002.

Breeder Seed/Stock - Northrup King Co., Minneapolis, MN.

Certified Seed/Stock - Available.

Preparer/Additional Information - Agriculture Canada, Food Production and Protection Branch, Plant Products Division, Ottawa, Ontario K1A 0C6. Seed is distributed by Northrup King Seeds Ltd., P.O. Box 1207, Cambridge, Ontario N1R 6C9.

Hokuo

Developed by Snow Brand Seed Ltd., Sapporo, Japan.

Source - Selected from 150 individual plants introduced from the Ukraine.

Method of Breeding - Mass selection four-clone synthetic.

Intended Use - Hay.

Description - Darker green than Climax, with smaller flag leaves which are erect; an early variety with good winter-hardiness and intermediate height. Moderately susceptible to *Cladosporium phlei* (C.T. Gregory) G.A. De Vries and moderately resistant to *Scolecorichum graminis* (Fuckel) Deighton.

Adapted to - LRR K, L, R, S, M, N, P; PHZ 2.

Released - 1982.

Breeder Seed/Stock - Snow Brand Seed Co., Ltd., Sapporo, Japan.

Certified Seed/Stock - Available. Landis Seed Canada, CP 217, Lindsay, Ontario K9V 4SI.

Preparer/Additional Information - Agriculture Canada, Sainte-Foy Research Station, 2560 Hochelasa Blvd., Sainte-Foy, Quebec G1V 2J3, (418) 657-7980.

Itasca (Reg. No. 1)

Developed by the Minnesota Experimental Station, St.Paul, MN.

Source - Traced to seven inbred lines, one from T7, two from Cornell 1620, three from Cornell 1777, and one from Minnesota-grown common timothy.

Method of Breeding - Composed of seven inbred lines from following sources: One from Minnesota commercial seed, one from T7, two from Cornell 1620, and three from Cornell 1777. Synthetic tested as Minnesota 1630.

Intended Use - Hay.

Description - Medium maturity one to two days earlier than Climax and 10 days later than Clair in Minnesota upright growth.

Adapted to - LRR K, L, R, S, M, N, P; PHZ 2.

Released - 1972, Minnesota AES.

Breeder Seed/Stock - Minnesota AES.

Certified Seed/Stock - Available.

Preparer/Additional Information - OSECO Limited, Box 219, Brampton, Ontario; or Agriculture Canada, Sainte-Foy Research Station, 2560 Hochelasa Blvd., Sainte-Foy, Quebec, (418) 657-7980.

Korpa

Source - Rannsoknastofnun Lanbunadarins Agricultural Research Institute, Iceland.

Method of Breeding - This strain was collected from native stands of grass in old, permanent Icelandic hayfields in 1952. Many of the selected plants had been chosen for their freedom from injury caused by severe winters. Originally, 600 ecotypes were collected and compared in a randomized field trial. Based on the criterion of good agronomic performance, 30 clones were selected. This was narrowed down to 18 clones, mostly on the basis of seed production. Finally, from that group, seven clones were chosen according to vigor, leafiness, winter-hardiness, height, and seed setting. The breeder seed was first produced in 1962.

Intended Use - Hay and pasture production.

Description - Hexaploid; erect growth habit; height is 1.1 m, slightly taller than Bottnia II; length of longest stem is medium, slightly longer than Bottnia II; leaves are dark green, anthocyanin coloration present at maturity; ascending carriage, 0.9 cm in width, 17 cm in length; 72% are flat, 4% are folded, and 24% are rolled. Head: 8% are cylindrical, 56% conical, and 36% fusiform; anthocyanin coloration is present; at mid-length, width is 0.75 cm; medium to long. Flowering is slow speed of emergence in the year of sowing, medium time of emergence; persistence is 4-10 years; maturity is medium to late, similar to Climax; regrowth after first cutting is slow. European data indicate susceptibility to ergot and leaf blotch. In western Canadian trials, Korpa produced higher seed yields and forage yields than Climax. Korpa performed poorly under Ontario conditions.

Adapted to - Recommended in Canada for use in British Columbia.

Released - Agricultural Research Institute, Iceland. Registered in Canada in 1990 with registration no. 3216.

Breeder Seed/Stock - Agricultural Research Institute, Iceland.

Certified Seed/Stock - Available.

Preparer/Additional Information - Agriculture Canada, Food Production and Protection Branch, Plant Products Division, Ottawa, Ontario K1A 0C6. Seed is distributed by Oseco Inc., P.O. Box 219, Brampton, Ontario L6V 2L2.

Mariposa

Developed by Maple Leaf Mills, Ltd., Georgetown, Ontario.

Source - Variety of Richmond. Carried as accession no. MLM 17018.

Method of Breeding - Nine-clone synthetic chosen. The parental clones were selected for high yield and earliness on the basis of progeny tests.

Intended Use - Hay.

Description - Earlier than Climax, but similar to Salvo and Richmond in maturity. Rapid aftermath with leafiness and height similar to Richmond. Panicles longer than Climax and Richmond.

Adapted to - LRR K, L, R, S, M, N, P; PHZ 2.

Released - 1984.

Breeder Seed/Stock - Otto Pick & Sons Seed, Ltd., CP 126, Richmond Hill, Ontario L4C 4X9.

Certified Seed/Stock - Available.

Preparer/Additional Information - Agriculture Canada, Sainte-Foy Research Station, 2560 Hochelasa Blvd., Sainte-Foy, Quebec G1V 2J3, (418) 657-7980.

Mohawk (Reg. No. 63)

Selected at FFR Cooperative - S.J. Baluch, S.D. Stratton, and R.J. Buker. Carried as accession no. Syn P.

Source - Timothy clones were selected from public cultivars in the fall of 1967 for vigor and leaf disease resistance at West Lafayette, IN.

Method of Breeding - Nine-clone synthetic selections out of spaced-plants nursery, which had good clonal evaluations and polycross progeny yield data.

Intended Use - Released for hay production.

Description - High-yielding timothy variety with good stand persistence. Mohawk has shown more resistance to fall leaf diseases than Clair in FR testing. It is generally later than Clair and similar in maturity to Climax.

Adapted to - LRR A, F, H, K, L, M, N, P, R, S, T; PHZ 2, 3, 4, 5, 6, 7.

Released - FFR Cooperative, West Lafayette, IN.

Breeder Seed/Stock - FFR Cooperative, West Lafayette, IN.

Certified Seed/Stock - Available.

Preparer/Additional Information - FFR Cooperative, 4112 E. State Road 225, West Lafayette, IN 47906, (317) 567-2115.

N7-126

Developed by Northrup, King & Co., Minneapolis, MN.

Source - Itasca, Drummond, Lorain, Climax, and common.

Method of Breeding - Individual plants selected for leaf-disease resistance, leafiness, and aftermath recovery. Seventeen clones combined in synthetic.

Intended Use - Hay.

Description - Leafy hay-type with leaves well up the culms; resistant to stem rust and good tolerance to leaf diseases. Maturity similar to Climax.

Adapted to - LRR K, L, R, S, M, N, P; PHZ 2.

Released - No. Under consideration for release.

Breeder Seed/Stock - Northrup, King & Co.

Certified Seed/Stock - Available. Canadian Distributor: National NK Seeds Ltd., C.P. 485, Kitcheno, Ontario N264A7.

Preparer/Additional Information - Agriculture Canada, Sainte-Foy Research Station, 2560 Hochelasa Blvd., Sainte-Foy, Quebec G1V 2J3, (418) 657-7980.

Nike

Source - Otto Pick and Sons Seeds Ltd., Blenheim, Ontario.

Method of Breeding - The variety traces back to Climax (7%), Champ (7%), and unknown wide background material. Open-pollinated progeny was produced in 1979 and underwent selection for seed yield in 1981. Fourteen plants of medium maturity were planted in isolation to produce prebreeder seed. Breeder seed (Syn 2) was produced in 1982.

Intended Use - Hay and pasture production.

Description - Growth habit is erect; ploidy is hexaploid; plant height is tall; flowering similar to Champ; no anthocyanin present; long panicle; leaves are medium green; medium length, narrow; semi-erect attitude; regrowth is similar to Climax. Nike is a medium maturing variety which produces good seed and forage yields.

Adapted to - Recommended in Canada for use in Ontario.

Released - Otto Pick and Sons Seeds Ltd. Registered in Canada in 1989 as registration no. 3153.

Breeder Seed/Stock - Otto Pick and Sons Seeds Ltd.

Certified Seed/Stock - Available.

Preparer/Additional Information - Agriculture Canada, Food Production and Protection Branch, Plant Products Division, Ottawa, Ontario K1A 0C6. Seed is distributed by Otto Pick and Sons Seeds Ltd.

Richmond

Developed by Maple Leaf Mills, Georgetown, Ontario. Carried as accession no. MLM 17011.

Source - Old hay field near Richmond, Ontario.

Method of Breeding - Mass selection.

Intended Use - Hay.

Description - Early maturity, leafy, erect growth.

Adapted to - LRR K, L, R, S, M, N, P; PHZ 2.

Released - 1976.

Breeder Seed/Stock - Maple Leaf Mills, Ltd., Oakwood, Ontario.

Certified Seed/Stock - Available. Maple Leaf Mills Ltd., Oakwood, Ontario.

Preparer/Additional Information - Agriculture Canada, Sainte-Foy Research Station, 2560 Hochelasa Blvd., Sainte-Foy, Quebec G1V 2J3, (418) 657-7980.

Salvo

Developed at the Plant Research Center, Agriculture Canada, Ottawa. Carried as accession no. T009 (Ottawa).

Source - Open-pollinated Champ.

Method of Breeding - Forty-one plants mass selected from an open-pollinated field were selected for upright growth, leafiness, and large, long heads. Seed from the 12 earliest progeny was used as Syn-1.

Intended Use - Hay.

Description - Very early maturity, good aftermath, rapidly producing new culms.

Adapted to - LRR K, L, R, S, M, N, P; PHZ 2.

Released - 1980.

Breeder Seed/Stock - Plant Research Center, Agriculture Canada, Ottawa, Ontario.

Certified Seed/Stock - Available. Association Secan, 512-885 Promenade Meadowlands, Ottawa.

Preparer/Additional Information - Agriculture Canada, Sainte-Foy Research Station, 2560 Hochelasa Blvd., Sainte-Foy, Quebec G1V 2J3, (418) 657-7980.

Tiiti

Developed by Hankkija Plant Breeding Institute, Helsinki, Finland. Carried as accession no. Hja 1160.

Source - Forty-one individuals.

Method of Breeding - Polycross bred for adaptation, five best families from 41-family polycross. From the five-family progensis, 700 individual were selected and used for the testing.

Intended Use - Hay.

Description - Erect, pale green with slightly curved leaf tips, are inflorescent or pale with no anthocyanin coloration. Maturity is similar to or slightly earlier than Climax.

Adapted to - LRR K, L, R, S, M, N, P; PHZ 2.

Released - 1985.

Breeder Seed/Stock - Hankkija Plant Breeding Institute, Helsinki, Finland.

Certified Seed/Stock - Available. Henri Malon Ltd, 15 Woodland Heights, Toronto, Ontario M65 2W5.

Preparer/Additional Information - Agriculture Canada, Sainte-Foy Research Station, 2560 Hochelasa Blvd., Sainte-Foy, Quebec G1V 2J3, (418) 657-7980.

Tiller

Developed by Van Der Have B.V., The Netherlands. Carried as accession no. HT 34.

Source - Various sources.

Method of Breeding - Several progenies were selected for good agronomic performance and matching type.

Intended Use - Hay.

Description - Very early maturity, morphological description according to 4PDV guidelines TG 34/6 of 07-11-1984 is as follows: Leaf color - 4; leaf width - 6; growth habit - 2; heading date - 1, flagleaf length - 4; flagleaf width - 5; stem length - 6; ear length - 5; and culm length - 5.

Adapted to - LRR K, L, R, S, M, N, P; PHZ 2.

Released - 1989.

Breeder Seed/Stock - Van Der Have B. V., The Netherlands.

Certified Seed/Stock - Available. OSECO, Inc., P.O. Box 219, Brampton, Ontario LGV 2L2.

Preparer/Additional Information - Agriculture Canada, Sainte-Foy Research Station, 2560 Hochelasa Blvd., Sainte-Foy, Quebec G1V 2J3, (418) 657-7980.

Timfor

Developed by Northcup King and Company of Minneapolis, MN. Carried as accession no. N7-128 (Ontario).

Source - Fifteen clones from Itasca, Lorain, MN common, and commercial sources.

Method of Breeding - Selected for earliness, lack of leaf disease and stem rust, and leafiness. Mass selection followed by polycross progeny testing.

Intended Use - Hay.

Description - Semi-early, very leafy, hay-type, relatively free of leaf diseases.

Adapted to - LRR K, L, R, S, M, N, P; PHZ 2.

Released - 1975.

Breeder Seed/Stock - Northrup, King & Co.

Certified Seed/Stock - Not available.

Preparer/Additional Information - Agriculture Canada, Sainte-Foy Research Station, 2560 Hochelasa Blvd., Sainte-Foy, Quebec G1V 2J3, (418) 657-7980.

TM8601

Selected at FFR Cooperative - S.J. Baluch, S.D. Stratton and B.L. Winsett.

Source - Selections from PI 419862, progeny selection rows, Mohawk, Clair x Mohawk crosses.

Method of Breeding - Eight-clone synthetic. Selections out of source material based on maturity and phenotypic evaluations such as spring vigor and desirability.

Intended Use - Will be released as a hay-use cultivar.

Description - High-yielding cultivar with good resistance to fall leaf diseases. It is slightly later in maturity than Clair but significantly earlier than Mohawk and Climax.

Adapted to - LRR A, F, H, K, L, M, N, P, R, S, T; PHZ 2, 3, 4, 5, 6, 7.

Released - Has not been released at this time. Scheduled for release in 1994.

Breeder Seed/Stock - FFR Cooperative, West Lafayette, IN.

Certified Seed/Stock - Not yet available.

Preparer/Additional Information - FFR Cooperative, 4112 E. State Road 225, West Lafayette, IN 47906, (317) 567-2115.

Toro

Selected at the Institute Sperimentale Per Le Colture Foraggere in Lodi, Italy.

Source - Local population of the Po Valley.

Method of Breeding - Mass selection in a local population of the Po Valley.

Intended Use - Hay type.

Description - Equal to Champ in maturity; yield slightly more than Climax in Ontario. Erect growth habit; early; taller with larger stems and shorter leaves than Champ.

Adapted to - LRR K, L, R, S, M, N, P; PHZ 2.

Released - 1972.

Breeder Seed/Stock - Institute Sperimentale Per Le Colture Foragggere, Lodi, Italy.

Certified Seed/Stock - Available. OSECO, Inc., CP 219, Brampton, Ontario LGV 2L2.

Preparer/Additional Information - Agriculture Canada, Sainte-Foy Research Station, 2560 Hochelasa Blvd., Sainte-Foy, Quebec G1V 2J3, (418) 657-7980.

Winmor

Developed by Northrup King Company at Norwood, MN. Carried as accession no. K4-216 and K-216.

Source - Cross between two experimental varieties whose parents were selected for disease resistance, vigor, leafiness, and wide leaves.

Method of Breeding - The 2,000-3,000-plant nursery of each experimental variety were rogued for non-vigorous, non-wide-leafed, and rusted or diseased plants. The remaining 400-500 plants were allowed to recombine to form the Syn-1.

Intended Use - Hay.

Description - Medium-early maturity, later than Timfor and Climax. Wider leaves than Climax or Timfor. Majority of leaves have a leaf angle of less than 30%.

Adapted to - LRR K, L, R, S, M, N, P; PHZ 2.

Released - 1984.

Breeder Seed/Stock - Northrup King Company, Minneapolis, MN.

Certified Seed/Stock - Available. Northrup King Seeds, Ltd.

Preparer/Additional Information - Agriculture Canada, Sainte-Foy Research Center, 2560 Hochelasa Blvd., Sainte-Foy, Quebec G1V 2J3, (418) 657-7980.

***Phleum pratense* L. ssp. *nodosum* (L.) Arcang., timothy**

***Phleum nodosum* L.**

Leafy, persistent, low-growing species. Some forms are almost prostrate and spread by creeping stolons. No varieties.

***Phragmites australis* (Cav.) Trin. ex Steud. - common reed**

Perennial reeds with broad, flat, linear blades and large terminal panicles. Name, from the Greek, refers to its growin like a fence along streams. Culms erect 2-4 m tall, with stout creeping rhizomes and often also with stolons. Marshes, banks of lakes and streams, and around springs. Widely scattered throughout the U.S., West Indies and south to Chile and Argentina; Eurasia, Africa, and Australia.

Shoreline

Selected at the James E. "Bud" Smith Plant Materials Center, Knox City, TX, and increased for field testing as PMT-2376. Carried as accession no. PI 434204.

Source - Vegetatively collected in 1970 by Arnold Davis, SCS employee, at Lawrence, TX.

Method of Breeding - Selected from 10 similar accessions for shoreline stabilization.

Intended Use - Shoreline stabilization and erosion control.

Description - Outstanding for control of wave action erosion. Tolerates slight salinity.

Adapted to - LRR H, I, J, O, P, T, U; PHZ 7.

Released - 1978, SCS and TX Agricultural Extension Service.

Breeder Seed/Stock - Plant Materials Center, SCS, Knox City, TX (rhizomes).

Certified Seed/Stock - Not available.

Preparer/Additional Information - Plant Materials Center, SCS, Rte 1 Box 155, Knox City, TX 79529, (817) 658-3922.

***Piptatherum miliaceum* (L.) Coss. - smilograss**

***Oryzopsis miliacea* (L.) Benth. & Hook.f. ex Asch. & G. Schweinf**

From Mediterranean region, introduced in California, New Jersey, and Pennsylvania. Used for forage, as a sand binder, and for seeding burned areas. Grows to 1.5 m, stout, branching, erect, but base sometimes decumbent. Panicle to 30 cm long, loose.

Smilo

Source - Introduced from Mediterranean region; first tested at California AES in 1879.

Method of Breeding - Increased directly from initial evaluation rows by California AES personnel.

Intended Use - Critical area stabilization and range replantings, especially after wildfire burns.

Description - Drought tolerant, perennial bunchgrass performing better on droughty, lighter soils with about the same climatic adaptation as hardinggrass. Difficult to obtain stands except in the ash of brush burns or on very light soils. Needs scarification for best stands. Less palatable than veldtgrass or hardinggrass.

Adapted to - LRR C; PHZ 9, 10, 11.

Released - 1947, California Crop Improvement Association.

Breeder Seed/Stock - CA AES, and Plant Materials Center, SCS, Lockeford, CA.

Certified Seed/Stock - Not available. Common seed available commercially.

Preparer/Additional Information - Dave Dyer, Plant Materials Center, SCS, P.O. Box 68, 21001 N. Elliott Rd., Lockeford, CA 95237, (209) 727-5319.

Poa alpina **L. - alpine bluegrass**

Plant is erect, culms arising from a tight crown, 10-30 cm tall. The blades are short, 2-5 mm wide, the uppermost occurring about the middle of the culm. Occurs in mountain meadows, Arctic regions of the Northern Hemisphere, extending south to Quebec, northern Michigan, and the alpine summits of Colorado, Utah, Washington, and Oregon.

Gruening

Source - Mass selection from a six-meter row of PI 235491, originally received from the U.S. NPGS in 1977. Original material (a single plant) collected at a high elevation site near LaCure, Switzerland.

Method of Breeding - Open-pollinated increase from original selection.

Intended Use - Erosion control, reclamation, and restoration in arctic, subarctic, and boreal regions. Outcompetes other bluegrass cultivars on gravelly, alpine slopes.

Description - Low-growing, non-rhizomatous, perennial bunchgrass. Extremely early seed maturation and superb winter-hardiness. Forms a noticeable thatch with persistent leaves from previous year's growth. Gruening is the first named cultivar of this species.

Adapted to - LRR W (alpine areas), X, Y (northern half); PHZ 1, 2, 3.

Released - 1986, Alaska Department of Natural Resources and USDA-SCS.

Breeder Seed/Stock - Alaska Plant Materials Center.

Certified Seed/Stock - Available.

Preparer/Additional Information - Stoney J. Wright, Manager, Alaska Plant Materials Center, State of Alaska, HC 02, Box 7440, Palmer, AK 99645, (907) 745-4469.

Poa arida **Vasey - plains bluegrass**

Culms erect, 20-60 cm tall; blades mostly basal, firm, folded. Panicle narrow, 2-10 cm long. Prairies, plains, and alkali meadows, up to 3,000 m elevation. Manitoba to Alberta, south to western Iowa, Texas, and New Mexico.

PI 434231

Selected at Plant Materials Center, SCS, Bridger, MT - M.E. Majerus and J.G. Scheetz. Carried as accession no. WY-573.

Source - Collected from saline, sub-irrigated site five kilometers west of Lander, WY, by M.E. Majerus.

Method of Breeding - Direct increase of field collection.

Intended Use - Include in mixtures on saline-alkaline affected soils for grazing or cover and stabilization.

Description - Good seedling vigor, mildly rhizomatous, and has ability to establish excellent stands in saline-alkaline affected soils.

Adapted to - LRR B, D, E, F, G; PHZ 3.

Released - No. Distributed for limited field testing.

Breeder Seed/Stock - Plant Materials Center, SCS, Bridger, MT.

Certified Seed/Stock - Not available.

Preparer/Additional Information - Plant Materials Center, SCS, RR 1 Box 1189, Bridger, MT 59014, (406) 662-3579.

Poa bulbosa **L. - bulbous bluegrass**

Cool-season bunchgrass from Europe. Used for pasture and erosion control in parts of western U.S., including southwestern Idaho, Oregon, and northern California. Increased from bulblets that form in panicle. No varieties are available in the U.S.

Poa compressa **L. - Canada bluegrass**

Cool-season, sod-forming grass from Europe. Used for pasture and erosion control in humid parts of northern U.S.. Adapted to open, rather poor, dry soils, but does not withstand heavy grazing. Less desirable than Kentucky bluegrass for turf.

Canon

Selected at Ontario Agricultural College, University of Guelph, Guelph, Ontario.

Source - Collections from U.S. and Canada.

Method of Breeding - Mass selection for leafiness, disease resistance, and type.

Intended Use - Turf.

Description - Hardier and earlier in spring growth, flowering, and maturity than normal for species. Improved in leafiness, disease resistance, longevity, and yield. Comparatively pure as to type.

Released - 1944, by Ontario Agricultural College.

Breeder Seed/Stock - Available, Ontario Agricultural College.

Certified Seed/Stock - Not available.

Preparer/Additional Information - Canadian Seed Growers' Association, P.O. Box 8455, Ottawa, Ontario K1G 3T1.

Reubens

Developed at Jacklin Seed Company, Post Falls, ID - A.W. Jacklin.

Source - Phenotypic selection of plants from an established field of common Canada bluegrass near Reubens, ID.

Method of Breeding - The progeny of selected mother plants were row-planted, and rogued to remove common type material. Phenotypic selection was repeated in subsequent generations to improve trueness to type.

Intended Use - Low-maintenance turf and erosion control cover in areas of low fertility, irregular moisture supply, and where mowing is difficult.

Description - Low-growing apomictic, rhizomatous perennial. Spring leaf color is dark green, changing to bluegreen in late spring. At maturity, stems turn to light yellow-green. Under a low-maintenance mowing schedule, it becomes a denser turf than common Canada bluegrass but more open than Kentucky bluegrass. Leaf blades are short and stiff, tapering to a boat-shaped apex. Culms are distinctly compressed and sharply keeled.

Adapted to - LRR B, E, G, F, K, L, S, M, R, H; PHZ 4.

Released - 1976.

Breeder Seed/Stock - Jacklin Seed Company, Post Falls, ID.

Certified Seed/Stock - Available.

Preparer/Additional Information - Jacklin Seed Co, W. 5300 Riverbend Ave., Post Falls, ID 83854, (208) 773-7581.

Poa glauca Vahl ssp. *glaucantha* (Gauldin) Lindm. - glaucous or upland bluegrass

A perennial, loosely tufted bunchgrass that spreads by tillering; numerous compressed, fine, wiry culms, decumbent at the base; many flat, short, well-distributed, dark green leaf blades. Seed heads numerous, lax, becoming brownish, compact, and nodding at maturity. Seeds are small, lemmas lightly pubescent, and sparsely webbed at the base. Plants resemble Canada bluegrass but become sodbound less readily, lodge less, and produce more seed. Adapted to low-fertility soils for ground cover. Introduced in 1935 by Westover and Enlow from Chorsum, Turkey.

Draylar (Reg. No. 3)

Selected at Plant Materials Center, SCS, Pullman, WA - J.L. Schwendiman.

Source - Introduced as *Poa* spp., PI 109350, in 1935 by Westover-Enlow expedition from Chorsum, Turkey. Tested and propagated as P-410.

Method of Breeding - Aberrant plants removed from original introduction and remaining apomictic (2n=50) plants increased.

Intended Use - Soil erosion control on disturbed lands with a minimum annual precipitation of 45 cm.

Description - Numerous compressed, fine, wiry culms; decumbent at base. Many flat, short, dark-green, well-distributed leaves. Seedhead numerous, lax, becoming brownish, compact, and nodding at maturity. Seeds small; lemmas lightly pubescent and sparsely webbed at base. Plants resemble Canada bluegrass, but become sodbound less readily, lodge less, and produce more seed. Adapted to low-fertility soils for ground cover. Draylar is apomictic 2n=50 and has been used in the Carnegie Institution of Washington bluegrass hybrid studies.

Adapted to - LRR B, E; PHZ 5.

Released - 1951, cooperatively by Washington AES and Plant Materials Center, SCS, Pullman, WA. Named Draylar in 1963.

Breeder Seed/Stock - Plant Materials Center, SCS, Pullman, WA.

Certified Seed/Stock - Not available.

Preparer/Additional Information - Plant Materials Center, SCS, Room 104, Hulbert Agricultural Sciences Bldg., WSU, Pullman, WA 99164-6211, (509) 335-7376.

Tundra (Reg. No. 19)

Selected at the Palmer Research Center, University of Alaska Fairbanks Agricultural and Forestry Experiment Station, Palmer, AK - William W. Mitchell.

Source - Bulk seed collection from indigenous plants in arctic Alaska along the Sagavanirktok River about 116 km south of the northern coast at Prudhoe Bay.

Method of Breeding - Twenty-three plants selected from spaced-plant nursery for upright growth habit and persistence were established in isolated breeding nursery. Uniformity of 23 plants suggested a common origin and seed was bulked to be seed-propagated for breeder generation.

Intended Use - Revegetation purposes.

Description - First cultivar of species.

Adapted to - Arctic regions to northern fringes of boreal forest in Alaska and neighboring areas of Canada. Hardy across northern boreal and arctic regions; PHZ Y.

Released - 1976, by the Alaska Agricultural and Forestry Experiment Station.

Breeder Seed/Stock - Palmer Research Center, Agricultural and Forestry Experiment Station, Palmer, AK.

Certified Seed/Stock - Available. PVP No. 7700033.

Preparer/Additional Information - Palmer Research Center, Agricultural and Forestry Experiment Station, 533 E. Fireweed, Palmer, AK 99645, (907) 745-3257.

Poa nemoralis L. - wood meadowgrass

Introduced from Europe; occasional across Canada and south into the northern tier of states, particularly in New England and down to Wyoming.

Barnemo

Source - Dutch ecotypes.

Method of Breeding - Selection of space plants, clones, and families. Different families were selected to produce a synthetic variety.

Intended Use - Cool-season turfgrass. Particularly suitable for lawns in the shade, or low-maintenance lawns under trees.

Description - It has a good turf under low maintenance, an early ear emergence, and a fine leaf texture.

Adapted to - Pacific northwest, north central, northeastern U.S.; PHZ 1, 2, 3, 4, 5.

Released - Barenbrug.

Breeder Seed/Stock - Barenbrug Holland, Oosterhout, The Netherlands.

Certified Seed/Stock - Available from Barenbrug USA.

Preparer/Additional Information - Barenbrug USA, P.O. Box 239, Tangent, OR 97389, (503) 926-5801.

Poa pratensis L. - Kentucky bluegrass

Major cool-season, sod-forming grass from Europe. Extensively used for pasture, recreational turf, and erosion control in northeastern and north central states and southward in Appalachians to northern Georgia. Important lawn grass in Pacific northwest and throughout much of northern and central Great Plains and intermountain region. Best adapted to well-drained, productive soils of limestone origin.

4 Aces

Source - New York, Alabama, Tennessee, and New Jersey.

Method of Breeding - A composite of the progenies of four highly apomictic plants: a collection from New York, a progeny from the cross NJE H-1 x Baron, a progeny from the cross Brunswick x Baron, and a progeny from a collection from Tennessee.

Intended Use - Cool-season turf, sod, athletic fields, and golf courses.

Description - Improved leaf spot and stripe rust resistance, good seed yield, and widely adapted.

Adapted to - PHZ 2, 3, 4, 5, 6, 7, 8, 10.

Released - Turf-Seed, Inc.

Breeder Seed/Stock - Pure Seed Testing, Inc.

Certified Seed/Stock - Available. PVP No. 9100137.

Preparer/Additional Information - Turf-Seed, Inc., P.O. Box 250, Hubbard, OR 97032, (503) 651-2130.

A-34

Selected at Warren's Turf Nursery, Palos Park, IL - B.O. Warren.

Source - Collection of plants from old-turf areas, including golf fairways and cemeteries.

Method of Breeding - Spaced-plant testing and sod-plot evaluation under artificial shade (65%).

Intended Use - General turf use.

Description - Leaf medium in width and color; weaker than most bluegrasses; density above average. Superior shade tolerance in comparison with other Kentucky bluegrasses, fine fescues, and rough bluegrass. Resistance to leaf spot is medium; good resistance to stem rust, stripe smut, and powdery mildew. Above average vigor, and in open sunlight will tolerate 12-mm mowing. In shade, mowing at 3.5-5 mm is necessary. Level of apomixis above 90%.

Adapted to - Central U.S.

Released - 1964, by Warren's Turf Nursery.

Breeder Seed/Stock - Warren's Turf Nursery.

Certified Seed/Stock - Available.

Preparer/Additional Information - Warren's Turf Nursery, Box 459, Suisun City, CA 94585, (707) 422-5100.

Abbey

Source - O.M. Scott breeding program.

Method of Breeding - First-generation hybrid from the cross of Victa x Windsor with apomictic seed increase.

Intended Use - Permanent turf alone or in a mixture with other cool-season turf species.

Description - Excellent seed yield combined with moderately good turf performance.

Adapted to - PHZ 2, 3, 4, 5, 6, 7.

Released - 1986 by O.M. Scott & Sons Co.

Breeder Seed/Stock - O.M. Scott & Sons Co., Marysville, OH.

Certified Seed/Stock - Available.

Preparer/Additional Information - Virgil Meier, O.M. Scott & Sons Co., Marysville, OH 43041, (513) 644-0011.

Able 1

New Jersey AES - Bill Meyers and Reed Funk. Carried as accession no. N37.

Source - Developed as a cross between A25 K bluegrass and Nugget Kentucky bluegrass.

Method of Breeding - A25 was pollinated with Nugget seedlings and were planted from this cross and selections were made. The variety is highly apomictic and very stable in reproducing itself.

Intended Use - General turf use.

Description - Able 1 is a leafy, low-growing, turf-type Kentucky bluegrass capable of producing an attractive, dark green turf with high density and a medium fine texture. It performs very well in shady areas.

Adapted to - LRR A, B, E, G, K, L, M, R, S; PHZ 5.

Released - 1985.

Breeder Seed/Stock - Dr. William A. Meyer, Pure Seed Testing, Hubbard, OR.

Certified Seed/Stock - Available.

Preparer/Additional Information - Warren's Turf Nursery, Box 459, Suisun City, CA 94585, (707) 422-5100.

Adelphi

Selected at New Jersey AES, New Brunswick, and developed in cooperation with J.L. Adikes, Inc., Jamaica, NY - G.W. Pepin, R.A. Russell, and C.R. Funk.

Source - Apomictic F_1 hybrid from Bellevue x Belturf cross.

Method of Breeding - Single hybrid plant selected from progeny rows planted with seed from Bellevue x Belturf cross.

Intended Use - Turf - Athletic, park, playground, and home lawns.

Description - Moderately low growing, good vigor, medium leaf texture, producing dense turf with a rich, dark-green color. Maintains good color throughout growing season; good resistance to leaf spot, moderate resistance to stripe smut and leaf rust.

Adapted to - PHZ 2, 3, 4, 5, 6.

Released - Jacklin Seed Co. and J.L. Adikes.

Breeder Seed/Stock - Jacklin Seed Co.

Certified Seed/Stock - Available.

Preparer/Additional Information - Kim Peterson, Jacklin Seed Co., W. 5300 Riverbend Ave., Post Falls, ID 83854, (208) 773-7581.

Amazon

Svalof AB, Svalov Sweden. Carried as accession no. P-024.

Source - Tavored/The Netherlands, southwest Sweden.

Method of Breeding - Seed from collection area was planted in increase nursery; spaced-planted nursery was established and rogued to remove off types; 25 progeny plants were established from 90 space-nursery plants; 38 progeny rows were harvested and used to establish and maintain breeders seed.

Intended Use - General turf use.

Description - Medium green, very fine textured leaves; high quality turf in spring and early summer; outstanding drought tolerance and spring greenup; and seedling vigor, summer/fall density, and summer coverage are moderate.

Adapted to - LRR A, B, E, G, , K, L, M, R, S; PHZ 5.

Released - 1982 in Sweden by Svalof AB; 1983 in U.S. by Jacklin Seed Company, Post Falls, ID.

Breeder Seed/Stock - Svalof AB, Svalov, Sweden.

Certified Seed/Stock - Available.

Preparer/Additional Information - Kim Peterson, Jacklin Seed Co., W. 5300 Riverbend Ave., Post Falls, ID 83854, (208) 773-7581.

America

Source - Source material traces back to Syracuse, NY, and Beltsville, MD.

Method of Breeding - Originated as a single, highly apomictic plant. It was selected from the open-pollinated progeny of a highly sexual hybrid selected from the F_1 progeny of a single cross.

Intended Use - Cool-season turf.

Description - Maintains good winter color in protected areas. Compatible in blends with most other Kentucky bluegrass cultivars.

Adapted to - PHZ 5, 6, 7, and parts of 8.

Released - 1980 by Pickseed West Inc. and International Seeds, Inc.

Breeder Seed/Stock - International Seeds, Inc.

Certified Seed/Stock - Available.

Preparer/Additional Information - Pickseed West Inc., P.O. Box 888, Tangent, OR 97389, (503) 926-8886.

Apex

Source - Originated from an ecotype selection from the golf course of Kevinge near Stockholm.

Method of Breeding - Ecotype selection from golf course in Stockholm, Sweden.

Intended Use - Turf.

Description - Dark green, fine-leaved variety with high shoot density and excellent shade tolerance.

Adapted to - Northern and central U.S. and southern Canada.

Released - 1992, Peder Weibull.

Breeder Seed/Stock - Weibullsholm Plant Breeding Institute, Sweden.

Certified Seed/Stock - Available.

Preparer/Additional Information - Kim Peterson, Jacklin Seed Co., W. 5300 Riverbend Ave., Post Falls, ID 83854, (208) 773-7581.

Asset

Source - U.S. germplasm.

Method of Breeding - Clonal material increased by seed and trialed in The Netherlands and Germany. Superior performer was selected. Few off-types. Highly apomictic.

Intended Use - Turf for athletic fields, playgrounds, home lawns, and golf courses.

Description - Excellent dark color, very dense. Few seed heads produced in turf.

Adapted to - PHZ 2, 3, 4, 5, 6 - cool humid.

Released - D.J. Van der Have, B.V., The Netherlands.

Breeder Seed/Stock - D.J. Van der Have.

Certified Seed/Stock - Available from Advanta Seeds West, Inc.

Preparer/Additional Information - Kenneth Hignight, Advanta Seeds West, Inc., 33725 Columbus Street S.E., P.O. Box 1496, Albany, OR 97321-0452, (503) 967-8923.

Banff

Source - National Park at Banff, Alberta, Canada.

Method of Breeding - Source material originally collected was re-selected for several years until the degree of uniformity indicated that the final selection was based on a single apomictic line.

Intended Use - Cool-season turf.

Description - It is a dwarf turf-type Kentucky bluegrass that tolerates close mowing well.

Adapted to - PHZ 5, 6, 7, and parts of 8.

Released - University of Alberta, Canada.

Breeder Seed/Stock - Department of Plant Science, University of Alberta, Canada.

Certified Seed/Stock - Available.

Preparer/Additional Information - Pickseed West Inc., P.O. Box 888, Tangent, OR 97389, (503) 926-8886.

Banjo

Developed by the New Jersey AES and Jacklin Seed Co.

Source - Single plant selected in Huntsville, AL.

Method of Breeding - Ecotype selection and evaluation.

Intended Use - Turf for home lawns, golf course roughs, and commercial/industrial areas. Sown either alone or in blends with other bluegrasses or as part of polyspecies mixtures.

Description - Medium green variety with an upright growth habit; excellent heat tolerance and good brown patch resistance; has shown better persistence than other Kentucky bluegrasses in hot, humid environments.

Adapted to - LRR B, C, D, E, F, G, H, K, L, M, N, P, R, S; PHZ 3.

Released - 1991 by International Seeds, Inc.

Breeder Seed/Stock - Jacklin Seed Co.

Certified Seed/Stock - International Seeds, Inc.

Preparer/Additional Information - Stephen W. Johnson, International Seeds, Inc., P.O. Box 168, Halsey, OR 97348, (503) 369-2251.

Barblue

Source - A cross between the varieties Parade and Barbie.

Method of Breeding - High breed plants of the cross were selected in 1974 and seed was harvested in the same year. Turf plots were seeded in the fall of 1974 and Barblue appeared to be superior in turf quality.

Intended Use - Cool-season turfgrass. May be seeded in turfgrass blends or mixtures. Especially for the sod industry.

Description - Good winter color and very fine leaved.

Adapted to - Southwestern U.S.

Released - Barenbrug.

Breeder Seed/Stock - Barenbrug Holland, Oosterhout, The Netherlands.

Certified Seed/Stock - Available from Barenbrug USA.

Preparer/Additional Information - Barenbrug USA, P.O. Box 239, Tangent, OR 97389, (503) 926-5801.

Barmax

Source - Ecotypes.

Method of Breeding - Selection of space plants in a turf. The most important selection was for wear tolerance.

Intended Use - Cool-season turfgrass. Used for lawns, fairways, parks, and sportfields.

Description - The variety has very good wear tolerance.

Adapted to - Northeastern, central, mountain, and northwestern U.S.

Released - Barenbrug.

Breeder Seed/Stock - Barenbrug Holland, Oosterhout, The Netherlands.

Certified Seed/Stock - Available from Barenbrug USA.

Preparer/Additional Information - Barenbrug USA, P.O. Box 239, Tangent, OR 97389, (503) 926-5801.

Baron (Reg. No. 20)

Developed and released by Barenbrug Holland B. V., Arnhem, The Netherlands. Carried as accession no. BAR 64-1.

Source - Clonal selection from a meadow in eastern Holland.

Method of Breeding - Seed from clonal parent placed in evaluation plots. Plants from first generation evaluated and one plant was selected and increased vegetatively.

Intended Use - Parks, lawns, athletic fields, cemeteries, sod farms, and golf courses in cool-season grass regions.

Description - Dense, medium texture, dark green, persistent winter-hardy turf; semi-dwarf habit that tolerates close mowing. Excellent seed yield potential.

Adapted to - LRR A, B, C, E, F, G, H, K, L, M, N, R, S; PHZ 3, 4, 5, 6, 7.

Released - 1973 by Barenbrug, Arnhem, The Netherlands.

Breeder Seed/Stock - Great Western Seed Company, subsidiary of Lofts Pedigreed Seed in Albany, OR.

Certified Seed/Stock - Available.

Preparer/Additional Information - Lofts Seed Inc., Chimney Rock Road, P.O. Box 146, Bound Brook, NJ 08805, (908) 560-1590; and Barenbrug USA, 32080 Old Highway 34, P.O. Box 239, Tangent, OR 97389, (800) 547-4101.

Bartitia

Source - Ecotypes from Sweden.

Method of Breeding - Selection of space plants in a turf and selection on seed yield.

Intended Use - Sportfields, lawns, and parks. It is a cool-season turf grass.

Description - The variety is a late flowering turf type with good wear tolerance.

Adapted to - Southwestern U.S.

Released - Barenbrug.

Breeder Seed/Stock - Barenbrug Holland, Oosterhout, The Netherlands.

Certified Seed/Stock - Available from Barenbrug USA.

Preparer/Additional Information - Barenbrug USA, P.O. Box 239, Tangent, OR 97389, (503) 926-5801.

BAR VB 1184 (Barcelona)

Source - Ecotypes from Hungary.

Method of Breeding - Selection under turf conditions with artificial wear.

Intended Use - Cool-season turf grass for lawns, sportfields, and parks. May be used in blends or mixtures.

Description - The variety has a very good Drechsleria resistance, good drought tolerance and a very good wear tolerance.

Adapted to - Northeastern, central, mountain, and northwestern U.S.

Released - Barenbrug.

Breeder Seed/Stock - Barenbrug Holland, Oosterhout, The Netherlands.

Certified Seed/Stock - Available from Barenbrug USA.

Preparer/Additional Information - Barenbrug USA, P.O. Box 239, Tangent, OR 97389, (503) 926-5801.

Barzan

Source - Dutch ecotypes.

Method of Breeding - Selection in space plantings under turf conditions.

Intended Use - Cool-season turfgrass. May be used in mixtures and blends. Low maintenance.

Description - Low-maintenance quality is unique. Leaf sheath has anthocyanin coloration in early development.

Adapted to - Southwestern U.S.

Released - Barenbrug.

Breeder Seed/Stock - Barenbrug Holland, Oosterhout, The Netherlands.

Certified Seed/Stock - Available from Barenbrug USA.

Preparer/Additional Information - Barenbrug USA, P.O. Box 239, Tangent, OR 97389, (503) 926-5801.

Blacksburg

Source - Virginia.

Method of Breeding - A single apomictic ecotype collected from a putting green in Virginia.

Intended Use - Cool-season turf for sod, athletic fields, and golf courses.

Description - Dwarf growth habit, very dark blue-green color, and high density in turf.

Adapted to - PHZ 2, 3, 4, 5, 6, 7, 8, 10.

Released - Turf-Seed, Inc.

Breeder Seed/Stock - Pure Seed Testing, Inc.

Certified Seed/Stock - Available. PVP No. 8700140.

Preparer/Additional Information - Turf-Seed, Inc., P.O. Box 250, Hubbard, OR 97032, (503) 651-2130.

Bristol

Source - Germplasm from New Jersey AES.

Method of Breeding - First-generation hybrid from the cross of Bellevue x Anheuser Dwarf with apomictic seed increase.

Intended Use - Permanent turf alone or in mixtures with other cool-season turf species.

Description - Very high turf performance, with dark green color and excellent shade tolerance.

Adapted to - PHZ 2, 3, 4, 5, 6, 7.

Released - 1974 by O.M. Scott & Sons Co.

Breeder Seed/Stock - O.M. Scott & Sons Co.

Certified Seed/Stock - Available.

Preparer/Additional Information - Virgil Meier, O.M. Scott & Sons Co., Marysville, OH 43041, (513) 644-0011.

Challenger

Source - Old-turf areas in Washington, DC, and northern Kentucky.

Method of Breeding - A single highly apomictic plant selected from the progeny of the cross NJEP-123 x K106.

Intended Use - Cool-season turf for sod, athletic fields, and golf courses.

Description - Excellent spring and winter color; good resistance to leaf spot, leaf rust, stripe smut, red thread and stem rust.

Adapted to - PHZ 2, 3, 4, 5, 6, 7, 8, 10.

Released - Turf-Seed, Inc.

Breeder Seed/Stock - Pure Seed Testing, Inc.

Certified Seed/Stock - Available. PVP No. 8500082.

Preparer/Additional Information - Turf-Seed, Inc., P.O. Box 250, Hubbard, OR 97032, (503) 651-2130.

Chateau

Source - O.M. Scott breeding program.

Method of Breeding - First-generation hybrid cross of two selections. Apomictic seed increase.

Intended Use - Turf.

Description - Premier turf quality, semi-dwarf with excellent disease resistance and shade performance.

Adapted to - PHZ 2, 3, 4, 5, 6, 7 - Northeatern, central, and western U.S.

Released - O.M. Scott and Sons.

Breeder Seed/Stock - Fine Lawn Research, Inc.

Certified Seed/Stock - Available.

Preparer/Additional Information - Pennington Seed Inc. of Oregon, P.O. Box 386, Lebanon, OR 97355, (503) 451-5261.

Cheri

Source - Svalof Seed Co., Sweden. A selection from southern Sweden.

Method of Breeding - Ecotype selection from old lawn in southern Sweden.

Intended Use - Turf.

Description - Medium dark green Kentucky bluegrass with an intermediate winter color. Resistance to leaf spot, necrotic ring spot, and billbug are intermediate; excellent winter-hardiness.

Adapted to - Northern and central U.S. and southern Canada.

Released - 1975, Jacklin Seed Co.

Breeder Seed/Stock - Jacklin Seed Co.

Certified Seed/Stock - Available.

Preparer/Additional Information - Kim Peterson, Jacklin Seed Co., W. 5300 Riverbend Ave., Post Falls, ID 83854, (208) 773-7581.

Classic

Rutgers University, New Brunswick, NJ, and Jacklin Seed Company, Post Falls, ID - C.R. Funk and L.A. Brilman. Carried as accession no. H74-225, 225.

Source - Cross between NJE P-59 and Baron.

Method of Breeding - Maternal parent P-59 was hybridized with paternal parent Baron. Individual plants were selected from the space-planted progeny of this cross. Progeny of these selections were evaluated as turf, reselected, multiplied, and released.

Intended Use - General turfgrass use in blends with other Kentucky bluegrass or in turf mixtures with fescues and ryegrass.

Description - Bright, fine textured, medium dark green leaves; dense turf formation; good winter color and hardiness; early spring greenup.

Adapted to - LRR A, B, E, G, F, K, L, M, R, S; PHZ 5.

Released - 1983 by Jacklin Seed Company, Post Falls, ID.

Breeder Seed/Stock - Jacklin Seed Company, Post Falls, ID.

Certified Seed/Stock - Available.

Preparer/Additional Information - Jacklin Seed Co., W. 5300 Riverbend Ave., Post Falls, ID 83854, (208) 773-7581.

Cobalt

Source - Single plant selection from old turf in Maryland in 1985.

Method of Breeding - Spaced plants derived from open pollinated seed. Abberant types were rogued from plots. Based on results obtained in turf as well as in space-plant nurseries.

Intended Use - Turf for high- and medium-maintenance areas.

Description - Dark green with excellent turf quality. Upright growing with medium fine texture and good resistance to leaf spot.

Adapted to - PHZ 2, 3, 4, 5, 6. Northern U.S.

Released - KWS-AG, Einbeck, Germany, and Turf Merchants, Inc. 1990.

Breeder Seed/Stock - Turf Merchants, Inc.

Certified Seed/Stock - Available.

Preparer/Additional Information - Fred B. Ledeboer, Ph.D., Turf Merchants, Inc, 33390 Tangent Loop, Tangent, OR 97389, (800) 421-1735.

Coventry

Source - O.M. Scott breeding program.

Method of Breeding - First-generation hybrid from the cross Gnome x unreleased variety in O.M. Scott breeding program with apomictic seed increase.

Intended Use - Permanent turf alone or in a mixture with other cool-season turf species.

Description - Moderate seed yield combined with excellent turf performance and shade tolerance.

Adapted to - PHZ 2, 3, 4, 5, 6, 7.

Released - 1986, O.M. Scott & Sons Co.

Breeder Seed/Stock - O.M. Scott & Sons Co., Marysville, OH.

Certified Seed/Stock - Available.

Preparer/Additional Information - Virgil Meier, O.M. Scott & Sons Co., Marysville, OH 43041, (513) 644-0011.

Cynthia

Source - England.

Method of Breeding - Ecotype collected in England. Trialed in turf and then selected based on performance.

Intended Use - Athletic fields, home lawns, parks, playgrounds, and golf courses.

Description - Very fine leaves, deep rooted, and excellent density.

Adapted to - PHZ 2, 3, 4, 5, 6 - cool humid.

Released - Mommersteeg International, Vlijmen, The Netherlands.

Breeder Seed/Stock - Mommersteeg International.

Certified Seed/Stock - Available from Advanta Seeds West, Inc.

Preparer/Additional Information - Kenneth Hignight, Advanta Seeds West, Inc., 33725 Columbus Street S.E., P.O. Box 1496, Albany, OR 97321-0452, (503) 967-8923.

Dawn

Jacklin Seed Company, Post Falls, ID, and Rutgers University, New Brunswick, NJ - A.D. Brede and C.R. Funk. Carried as accession no. F-1328.

Source - A single plant selected from F_1 progeny of the cross NJE P120 x PSU K106 in 1963.

Method of Breeding - Progeny from greenhouse hybridization was space planted in nursery. A single-plant selection was multiplied, tested, and reselected through several cycles prior to multiplication and release.

Intended Use - General turf usage in blends with other bluegrasses or in mixtures with ryegrass, red fescues, or turf-type tall fescues. Home lawns; athletic fields; parks; sod production; golf course fairways, tees, and roughs; and institutional and commercial turf. Maximum durability if turf is allowed to establish before heavy use.

Description - Moderately low-growing variety with a dark green color and medium-fine textured leaf blade. It has good color retention in cool weather and an attractive early spring color. Provides a moderately aggressive turf or medium-fine texture and good density. Has demonstrated moderately good resistance to leaf spot, dollar spot, pink snow mold, and leaf rust and excellent resistance to powdery mildew. Good drought tolerance.

Adapted to - LRR A, B, E, G, F, K, L, M, R, S; PHZ 3, 4, 5, 6, 7.

Released - 1987 by Jacklin Seed Company, Post Falls, ID, and Rutgers University, New Brunswick, NJ.

Breeder Seed/Stock - Jacklin Seed Company, Post Falls, ID, and Rutgers University.

Certified Seed/Stock - Available.

Preparer/Additional Information - Kim Peterson, Jacklin Seed Co., W. 5300 Riverbend Ave., Post Falls, ID 83854, (208) 773-7581 or Art Wick, LESCO, Inc., 20005 Lake Road, Rocky River, OH 44116, (216) 333-9250.

Destiny

Jacklin Seed Company, Post Falls, ID, and Rutgers University, New Brunswick, NJ - A.D. Brede C.R. Funk. Carried as accession no. H74-222, 222.

Source - Hybridization between NJE P-59 and Baron Kentucky bluegrass.

Method of Breeding - Original selection was a single F_1 plant from space-planted nursery of progeny from a hybridization between NJE P-59 and Baron.

Intended Use - General use in blends and mixtures for general turf, home lawns, parks, athletic fields, institutional grounds, golf courses, and school play areas.

Description - Medium dark green leaves; moderately low-growing, and moderately aggressive turf. Medium texture and density. Tolerant of both winter and summer temperature stress.

Adapted to - LRR A, B, E, G, F, K, L, M, R, S; PHZ 5.

Released - 1987 by Jacklin Seed Company, Post Falls, ID.

Breeder Seed/Stock - Jacklin Seed Company, Post Falls, ID.

Certified Seed/Stock - Available.

Preparer/Additional Information - Kim Peterson, Jacklin Seed Co., W. 5300 Riverbend Ave., Post Falls, ID 83854, (208) 773-7581.

Eclipse

Source - Rutgers University.

Method of Breeding - Progeny of the cross 64-765-4 and Anheuser Dwarf. The female parent 64-765-4 was selected from the cross SP-1 x Belturf.

Intended Use - Turf.

Description - Low-growing, leafy, dark green turf that has good resistance to leaf spot, dollar spot, stripe smut, powdery mildew, and fusarium blight.

Adapted to - Northern and central U.S. and southern Canada.

Released - 1980, Jacklin Seed Co. and Zajac Performance Seeds.

Breeder Seed/Stock - Jacklin Seed Co.

Certified Seed/Stock - Available.

Preparer/Additional Information - Kim Peterson, Jacklin Seed Co., W. 5300 Riverbend Ave., Post Falls, ID 83854, (208) 773-7581.

Estate

Source - O.M. Scott & Sons breeding program.

Method of Breeding - First-generation hybrid from the cross of Gnome and unreleased selection in the Scott breeding program with apomictic seed increase.

Intended Use - Permanent turf alone or in a mixture with other cool-season turf species.

Description - Excellent turf performance in sun and shade, resistant to leaf spot and powdery mildew, and medium level of seed yield potential.

Adapted to - PHZ 2, 3, 4, 5, 6, 7.

Released - Roberts Seed Co., 1987.

Breeder Seed/Stock - O.M. Scott & Sons Co.

Certified Seed/Stock - Available from Roberts Seed Co.

Preparer/Additional Information - Virgil Meier, O.M. Scott & Sons Co., Marysville, OH 43041, (513) 644-0011.

Freedom

Jacklin Seed Company, Post Falls, ID, and Rutgers University, New Brunswick, NJ - A.D. Brede, C.R. Funk. Carried as accession no. F-1872.

Source - Single plant selected from F_1 progeny of a cross between NJE P-59 and Glade.

Method of Breeding - F_1 progeny was space planted in nursery. The single plant selected was harvested, multiplied, tested, and released.

Intended Use - General turf for home lawns, sports fields, institutional grounds, golf courses, and school play areas. Favorable cultivar for blends with other Kentucky bluegrasses and in mixtures with turf-type tall fescues, ryegrass, and fine-leafed red fescues.

Description - Moderately low-growing bluegrass with medium green color; good cool-weather color retention and spring color; produces an aggressive medium-fine textured turf; good tolerance to drought and shade.

Adapted to - LRR A, B, E, G, F, K, L, M, R, S; PHZ 5.

Released - 1988 Vaughans Seed Company, Downers Grove, IL, Vaughans Seed Company, Bound Brook, NJ; Arkansas Valley Seeds, Denver, CO; Beckman Turf Irrigation, Chesterfield, MO; Jacklin Seed Company, Post Falls, ID.

Breeder Seed/Stock - Jacklin Seed Company, Post Falls, ID.

Certified Seed/Stock - Available.

Preparer/Additional Information - Kim Peterson, Jacklin Seed Co., W. 5300 Riverbend Ave., Post Falls, ID 83854, (208) 773-7581.

Fylking

Developed at Swedish Seed Association, Svalof, Sweden. Carried as accession no. 0217.

Source - Single-plant collections in southern Sweden.

Method of Breeding - Selection in spaced-planted nurseries and plot evaluation.

Intended Use - As a component of high-quality turfgrass mixtures.

Description - Low-growing, semiprostrate, dense turf; short, medium-wide blades; good root system. Moderate to good resistance to leaf spot, rust, *Fusaruim roseum* Link: Fr., and stripe smut. Rate of growth during establishment not as rapid as Merion. Tolerates close mowing.

Adapted to - LRR A, B, E, G, F, K, L, M, R, S; PHZ 5.

Released - 1969, in U.S. by Jacklin Seed Company, Post Falls, ID. Swedish Seed Association. (Clone patented in U.S., but protection does not apply to seed.)

Breeder Seed/Stock - Plant Breeding Institute of Swedish Seed Association. (Proprietary.)

Certified Seed/Stock - Available. Distributed in U.S. by Jacklin Seed Company, Post Falls, ID.

Preparer/Additional Information - Kim Peterson, Jacklin Seed Co., W. 5300 Riverbend Ave., Post Falls, ID 83854, (208) 773-7581.

Georgetown

The University of Rhode Island AES - Dr. Fred Ledeboer. Carried as accession no. RI PpI.

Source - Original clone discovered in a farm yard in Oregon in 1959.

Method of Breeding - Seed harvested from a single apomictic clone of Kentucky bluegrass and evaluated.

Intended Use - For home lawns, sod fields, golf courses, and parks.

Description - A medium-textured Kentucky bluegrass variety that possesses an attractive medium-dark green color. Superior resistance to fusarium.

Adapted to - LRR A, B, C, E, F, G, H, K, L, M, N, R, S; PHZ 3, 4, 5, 6, 7.

Released - 1982 by Lofts Seed, Inc., Bound Brook, NJ.

Breeder Seed/Stock - Lofts Seed, Inc. Bound Brook, NJ .

Certified Seed/Stock - Available.

Preparer/Additional Information - Lofts Seed Inc., Chimney Rock Road, P.O. Box 146, Bound Brook, NJ 08805, (908) 560-1590.

Geronimo

Source - European germplasm.

Method of Breeding - Ecotype selections.

Intended Use - Turf for golf courses, home lawns, athletic fields, parks, and playgrounds.

Description - Fast establishment, good resistance to leaf spot and leaf rust, good wear tolerance.

Adapted to - PHZ 2, 3, 4, 5, 6.

Released - Mommersteeg International, Vlijmen, The Netherlands.

Breeder Seed/Stock - Mommersteeg International, Vlijmen, The Netherlands.

Certified Seed/Stock - Available from Advanta Seeds West, Inc.

Preparer/Additional Information - Kenneth Hignight, Advanta Seeds West, Inc., 33725 Columbus Street S.E., P.O. Box 1496, Albany, OR 97321-0452, (503) 967-8923.

Glade (Reg. No. 12)

Selected at New Jersey AES, New Brunswick, and developed jointly with Jacklin Seed Co., Inc., Dishman, WA - C.R. Funk, D.W. Jacklin, and W.T. Boyd. Carried as accession no. P-29.

Source - Single plant found in lawn at Albany, NY.

Method of Breeding - Six thousand plants collected from old-turf areas were screened in either spaced-plant nurseries or in clonal plots receiving turf maintenance. Promising selections subsequently evaluated in solid-seeded turf plots, spaced-plant progeny tests, and seed-production trials.

Intended Use - General turf use and in shaded areas.

Description - Attractive, leafy, moderately fine textured, dark green, turf type. Excellent resistance to stripe smut and leaf rust, good resistance to powdery mildew, and moderately good resistance to leaf spot. Produces dense aggressive turf. Good record of persistence, spread, and overall turf performance.

Adapted to - LRR A, B, E, G, F, K, L, M, R, S; PHZ 5.

Released - 1975 Jacklin Seed Company, Post Falls, ID.

Breeder Seed/Stock - Jacklin Seed Co., Inc.

Certified Seed/Stock - Available.

Preparer/Additional Information - Kim Peterson, Jacklin Seed Co., W. 5300 Riverbend Ave., Post Falls, ID 83854, (208) 773-7581.

Huntsville

Jacklin Seed Company, Post Falls, ID, and Rutgers University, New Brunswick, NJ - A.D. Brede C.R. Funk. Carried as accession no. H76-2499.

Source - Single-plant discovery from an old-turf area in northern Alabama.

Method of Breeding - The original plant was vegetatively multiplied in greenhouse and transferred to space-planted nursery. Progeny from seed was tested, multiplied, and released.

Intended Use - General turf use and specialty use in mixtures with turf-type tall fescues.

Description - Dark green, wide leaves; moderately tall, forming dense upright type of turf; good seedling vigor. May be adversely stressed by close mowing; morphology, color, growth style, and heat tolerance indicate good compatibility in turf mixtures with tall fescue.

Adapted to - LRR A, B, E, G, F, K, L, M, R, S; PHZ 5.

Released - 1986 Jacklin Seed Company, Post Falls, ID.

Breeder Seed/Stock - Jacklin Seed Company, Post Falls, ID.

Certified Seed/Stock - Available.

Preparer/Additional Information - Kim Peterson, Jacklin Seed Co., W. 5300 Riverbend Ave., Post Falls, ID 83854, (208) 773-7581.

Ikone

Max-Planck-Institute, Germany - W. Nitzsche. Carried as accession no. WRP 156.

Source - Cross between Poa pratensis L., cv. Ottos and irradiated Poa palustris L., cv. Huana.

Method of Breeding - Huana was irradiated to induce sexuality. Sexual plants were crossed with stored pollen derived from Otto. Hybrid plant progeny was selected for growth type, rhizomes, ligules, and hairiness.

Intended Use - Turfgrass, either as a component of blends with other Kentucky bluegrass cultivars or in mixtures with other turf type species.

Description - Moderately low-growing, medium dark green color, and a medium textured blade. Forms a moderately aggressive turf of medium texture and density.

Adapted to - LRR A, B, E, G, F, K, L, M, R, S; PHZ 5.

Released - 1985, L.C. Nungesser, Darmstadt, Germany, and Jacklin Seed Company, Post Falls, ID.

Breeder Seed/Stock - International Seed.

Certified Seed/Stock - Available.

Preparer/Additional Information - Kim Peterson, Jacklin Seed Co., W. 5300 Riverbend Ave., Post Falls, ID 83854, (208) 773-7581.

Julia

Van Engelen Zaden B.V., Vlijmen, The Netherlands. Carried as accession no. FVB 5584.

Source - Collection of plants from northern Germany. In 1974, the first large-scale multiplication was initiated, with the first commercial seed production in 1975.

Method of Breeding - Plant collection was space planted. Progeny of selected plants was reviewed for turf quality. Most promising mother plants were cloned and increased for more intensive evaluation.

Intended Use - Home lawns; athletic fields; parks; sod production; golf course fairways, tees, and roughs; and institutional and commercial turf. Highly recommended as a component of premium bluegrass mixtures and ryegrass-bluegrass mixtures. Maximum durability if turf is allowed to establish before heavy use.

Description - Moderately dark green variety with a medium-fine leaf texture and upright growth habit. Early spring greenup; maintains its color into the fall. Excellent density and wear tolerance even at a short cutting height. Little or no seedhead development occurs under turf conditions. Excellent resistance to leaf spot and moderate resistance to stem rust as well as good resistance to melting out.

Adapted to - LRR A, B, E, G, F, K, L, M, R, S; PHZ 3, 4, 5, 6, 7.

Released - 1975 Van Engelen Zaden, B.V., Vlijmen, The Netherlands.

Breeder Seed/Stock - Van Engelen Zaden, B.V.

Certified Seed/Stock - Available.

Preparer/Additional Information - Jacklin Seed Co., W. 5300 Riverbend Ave., Post Falls, ID 83854, (208) 773-7581; or Art Wick, LESCO, Inc., 20005 Lake Rd., Rocky River, OH 44116, (216) 333-9250.

Kenblue

Increased at Kentucky AES, Lexington - R.C. Buckner.

Source - Blend of seed from farms located in major seed-producing counties in central Kentucky.

Method of Breeding - Increase of blend of original collections.

Intended Use - Turf and pasture uses.

Description - Consistently superior in performance to all named varieties and to seed lots of foreign origin, in tests at Kentucky AES, Lexington. Superiority attributed to resistance to diseases and tolerance to sod webworn.

Adapted to - LRR M, N, S, P; PHZ 4, 5, 6, 7.

Released - 1967, by Kentucky AES.

Breeder Seed/Stock - Kentucky AES.

Certified Seed/Stock - Available.

Preparer/Additional Information - Dr. A.J. Powell, AES, Univ. of Kentucky, Lexington, KY 40506, (606) 257-9000.

Liberty

Jacklin Seed Company, Post Falls, ID, and Rutgers University, New Brunswick, NJ - A.D. Brede, L.A. Brilman, A.W. Jacklin, J.J. Zajac, and C.R. Funk. Carried as accession no. A-609 and/or JW4.

Source - Original plant selected from F_1 progeny of cross NJE P-123 x PSU K-106.

Method of Breeding - Progeny from greenhouse hybridization was space planted in nursery. Single plant selection was harvested, multiplied, tested, and reselected through several cycles.

Intended Use - General turf use in blends with other Kentucky bluegrass cultivars and in mixtures with turf-type tall fescue, perennial ryegrass, and red fescue.

Description - Dark green leaves; dwarf plants form a medium-textured quality turf with good density and good fall and winter color.

Adapted to - LRR A, B, E, G, F, K, L, M, R, S; PHZ 5.

Released - 1986, Zajac Performance Seeds, N. Haledon, NJ.

Breeder Seed/Stock - Jacklin Seed Company, Post Falls, ID.

Certified Seed/Stock - Available.

Preparer/Additional Information - Zajac Performance Seeds, 33 Sicomac Rd., N. Haledon, NJ 07508, (201) 423-1660.

Limousine

Deutsche Saatveredelung, Lippstadt-Bremen, Germany. Carried as accession no. 02941-17061.

Source - Vogelsborg region of Germany.

Method of Breeding - Original collection was cloned and space planted in nursery. Observation and selection resulted in a single clone selection for advanced testing, multiplication, and release.

Intended Use - Cultivar characteristics indicate superior performance for playing fields and sports turf when hard wear is expected.

Description - Dark green, very fine leaves; very dense, fine textured turf of medium-late maturity.

Adapted to - LRR A, B, E, G, F, K, L, M, R, S; PHZ 5.

Released - 1982.

Breeder Seed/Stock - Deutsche Saatveredlung Lippstadt-Bremen, Germany.

Certified Seed/Stock - Available.

Preparer/Additional Information - Kim Peterson, Jacklin Seed Co., W. 5300 Riverbend Ave., Post Falls, ID 83854, (208) 773-7581.

Livingston

Source - Washington, DC.

Method of Breeding - A single, highly apomictic plant selected from the progeny of a selection from an old lawn in Washington, DC, and crossed with Boron.

Intended Use - Cool-season turf for sod, athletic fields, and golf courses.

Description - Broad leaves, mixes well with tall fescues, good seed yield, and a dark green color.

Adapted to - PHZ 2, 3, 4, 5, 6, 7, 8, 10.

Released - Turf-Seed, Inc.

Breeder Seed/Stock - Pure Seed Testing, Inc.

Certified Seed/Stock - Available.

Preparer/Additional Information - Turf-Seed, Inc., P.O. Box 250, Hubbard, OR 97032, (503) 651-2130.

Lofts 1757

Developed and released by Lofts Seed, Inc., using germplasm obtained from the New Jersey AES. Carried as accession no. Lofts F-1757.

Source - The maternal source, Brunswick, was selected from an old lawn on the Cook College campus of Rutgers University in New Brunswick, NJ. Anheuser Dwarf was collected from old turf in St. Louis, MO.

Method of Breeding - A single, highly apomictic plant selected from the progeny of the cross Brunswick Kentucky bluegrass x Anheuser Dwarf Kentucky bluegrass.

Intended Use - Home lawns, parks, cemeteries, sod farms, and athletic fields.

Description - A moderately low-growing, turf-type cultivar with a medium dark green color. It produces a persistent turf of medium density, medium aggressiveness, and stiff wide leaves. It shows good resistance to leaf spot and melting-out disease.

Adapted to - LRR A, B, C, E, F, G, H, K, L, M, N, R, S; PHZ 3, 4, 5, 6, 7.

Released - 1987, Lofts Seed, Inc.

Breeder Seed/Stock - Lofts Seed, Inc.

Certified Seed/Stock - Available.

Preparer/Additional Information - Lofts Seed Inc., Chimney Rock Road, P.O. Box 146, Bound Brook, NJ 08805, (908) 560-1590.

Merion (Reg. No. 1)

Selected at Plant Industry Station, Beltsville, MD, by U. S. Golf Association Green Section, ARS cooperating - F.V. Grau.

Source - Single-plant selection made by Joseph Valentine of Merion Golf Club, Ardmore, PA, in 1936 and increased by John Monteith, Jr., former director, U.S. Golf Association Green Section.

Method of Breeding - Plant selection and apomictic seed progenies obtained through succeeding generations tested in cooperative turf research program of Plant Science Research Division, ARS, and U.S. Golf Association Green Section. Tested as B-27.

Intended Use - Home lawns, parks, and sod farms.

Description - Low-growing, short leaves, good color. High degree of resistance to leaf spot. More tolerant to close mowing than common Kentucky bluegrass. Susceptible to rust and stripe smut.

Adapted to - LRR K, L, M, N, P, R, S; PHZ 4, 5, 6, 7.

Released - 1947, cooperatively by Plant Science Research Division, ARS, and U.S. Golf Association Green Section.

Breeder Seed/Stock - Not available.

Certified Seed/Stock - Available.

Preparer/Additional Information - Pennsylvania State Univ., Agronomy Department, University Park, PA 16802, (814) 865-6541.

Midnight

Source - Washington, DC, and Glade.

Method of Breeding - A single, highly apomictic plant selected from the progeny of NJE-P-154 and a selection from old lawn in Washington, DC, crossed with Glade.

Intended Use - Cool-season turf for sod, athletic fields, and golf courses.

Description - Dark blue color, excellent turf quality across the U.S., with good stripe smut, dollar spot and leaf spot resistance, and very good heat and cold tolerance.

Adapted to - PHZ 2, 3, 4, 5, 6, 7, 8, 10.

Released - Turf-Seed, Inc.

Breeder Seed/Stock - Pure Seed Testing, Inc.

Certified Seed/Stock - Available. PVP No. 8200181.

Preparer/Additional Information - Turf-Seed, Inc., P.O. Box 250, Hubbard, OR 97032, (503) 651-2130.

Minstrel

Source - Hybrid between Fylking and Bristol.

Method of Breeding - Cross Fylking x Bristol. F_1 plants compared to Fylking. Deviating plants had seed harvested and evaluated in turf trials during 1975-81. Top performers were increased.

Intended Use - Turf for athletic fields, golf courses, home lawns, parks, and playgrounds.

Description - Widely adapted, dark green color, and dense turf.

Adapted to - PHZ 2, 3, 4, 5, 6.

Released - D.J. Van der Have, The Netherlands.

Breeder Seed/Stock - D.J. Van der Have.

Certified Seed/Stock - Available from Advanta Seeds West, Inc.

Preparer/Additional Information - Kenneth Hignight, Advanta Seeds West, Inc., 33725 Columbus St. S.E., P.O. Box 1496, Albany, OR 97321-0452, (503) 967-8923.

Mystic (Reg. No. 25)

Developed by the cooperative efforts of the U.S. Golf Association Green Section; Lofts Seed, Inc.; and the New Jersey AES. Carried as accession nos. P-140, P-141, Echo Lake P1h.

Source - A single, highly apomictic plant selected from Echo Lake Country Club in Westfield, NJ.

Method of Breeding - Field grown, spaced-plant progenies.

Intended Use - Golf course fairways and tees, sod farms.

Description - Moderately low-growing, fine-leaved, turf-type bluegrass showing bright green color. Provides dense turf that is highly aggressive, tolerates close mowing, and has good winter color.

Adapted to - LRR A, B, C, E, F, G, H, K, L, M, N, R, S; PHZ 3, 4, 5, 6, 7.

Released - 1980, Lofts Seed, Inc., U.S. Golf Association, Green Section, and the New Jersey AES.

Breeder Seed/Stock - Lofts Seed, Inc.

Certified Seed/Stock - Available.

Preparer/Additional Information - Lofts Seed Inc., Chimney Rock Road, P.O. Box 146, Bound Brook, NJ 08805, (908) 560-1590.

Nassau (Reg. No. C87-116R)

Developed and released by Jacklin Seed Company, Post Falls, ID and Lofts Seed Inc., Bound Brook, NJ - L.A. Brilman, A.W. Jacklin, R.H. Hurley, B.B. Clarke and C.R. Funk. Carried as accession no. 243.

Source - Progeny of a cross between NJE P-59 and Baron.

Method of Breeding - The maternal parent, NJE P-59, was crossed with the paternal parent, Baron. Nassau was developed from the progeny of a single plant from this hybridization.

Intended Use - Turfgrass in full sun and light shade.

Description - Moderately low-growing, turf-type Kentucky bluegrass, medium-wide leaves, medium-dark green color, medium density, and good winter color retention.

Adapted to - LRR A, B, E, G, F, K, L, M, R, S; PHZ 5.

Released - 1983 by Jacklin Seed Company, Post Falls, ID, and Lofts Seed, Inc., Bound Brook, NJ.

Breeder Seed/Stock - Maintained by Jacklin Seed Company.

Certified Seed/Stock - Available.

Preparer/Additional Information - Kim Peterson, Jacklin Seed Co., W. 5300 Riverbend Ave., Post Falls, ID 83854, (208) 773-7581.

NE 80-47

Developed at the University of Nebraska.

Source - Hybrid cross from selections found in a park at Falls City and selection 72-50.

Method of Breeding - Parents selected on the basis of stress tolerance, color, and quality.

Intended Use - Cool-season turfgrass adapted for the midwest and heat-stress areas of the U.S. Intended for use on golf courses and home lawns.

Description - Cool-season grass with moderately fine leaves. Selected for its very dark green color and normal growth habit. Very attractive under low or high maintenance.

Adapted to - PHZ 3, 4, 5 are areas of best adaptation and will perform well under heat stress as well as ideal management.

Released - Pending.

Breeder Seed/Stock - University of Nebraska, Department of Horticulture.

Certified Seed/Stock - Available.

Preparer/Additional Information - S.A. Westerholt, 377 Plant Science, Univ. of Nebraska, Lincoln, NE 68583-0742, (402) 472-1142.

NE 80-88

Developed by the University of Nebraska.

Source - Hybrid cross from selections found in a Falls City park and an Alma municipal golf course.

Method of Breeding - Parents selected on the basis of stress tolerance, color, and quality.

Intended Use - Cool-season turfgrass adapted for the midwest; however, does well in many other areas where Kentucky bluegrass is well adapted. Rapid spread and early greenup is ideal for golf course and home-lawn use.

Description - Cool-season grass with moderately fine leaves, but not as fine as common types. Normal growth habit; up to 45 cm if left unmowed. Very aggressive plant growth; very attractive under high maintenance.

Adapted to - PHZ 3, 4, 5 are areas of best adaptation and will perform well under heat stress as well as ideal management.

Released - Pending.

Breeder Seed/Stock - University of Nebraska, Department of Horticulture.

Certified Seed/Stock - Available.

Preparer/Additional Information - S.A. Westerholt, 377 Plant Science, Univ. of Nebraska, Lincoln, NE 68583-0742, (402) 472-1142.

Newport (Reg. No. 2)

Selected at Plant Materials Center, SCS, Pullman, WA - Jens Clausen, Carnegie Institution of Washington, and Stanford University, Stanford, CA.

Source - Maritime race collected from coastal bluffs at Newport, Lincoln County, OR, by W. E. Lawrence. Propagated at Carnegie Institution of Washington, accession CIW 4466-1 and P-13821.

Method of Breeding - Seed of original collection used to establish spaced planting in 1949. Strain found apomictic (2n=81); bulked seed used for increase in 1953. This strain used in Carnegie Institution of Washington bluegrass hybrid studies.

Intended Use - Turf.

Description - Vigorous, highly productive, coastal race of broad climatic tolerance. Wide, dark-green leaves, low-growing, fair to good in seed production, medium-late in seed maturity, and rapid sod forming. Appears to be fairly resistant to rust and leaf spot.

Adapted to - LRR B, D, E; PHZ 5.

Released - 1958, cooperatively by Washington and Oregon Agricultural Experiment Stations at Pullman and Corvallis, respectively, and Plant Materials Center, SCS, Pullman, WA.

Breeder Seed/Stock - Washington State Crop Improvement Association in cooperation with the Washington Agricultural Research Center and the Plant Materials Center, SCS, Pullman, WA.

Certified Seed/Stock - Available.

Preparer/Additional Information - Plant Materials Center, SCS, Rm 104, Hulbert Agricultural Sciences Bldg., WSU, Pullman, WA 99164-6211, (509) 335-7376.

Nimbus

Source - Originated as a single, highly apomictic plant selected in The Netherlands in 1973. It produces a uniformly dense turf with excellent disease resistance, especially to leaf spot and powdery mildew.

Method of Breeding - A descendent from a single seed; breeder seed was first bulked in the F_3 generation.

Intended Use - Turfgrass areas.

Description - Leaves are medium green and broad, with pubescence present on sheath. Ligules are fairly long: collar is fringed with short hairs. Panicles are open and of medium length with little or no anthocyanin pigment. Produces a very dense, uniform turf which is of a medium green color.

Adapted to - Useful in turfgrass applications throughout Canada and the northern U.S. and with limited use as far south as the transition zone.

Released - Zelder, b.v., Gennep, The Netherlands.

Breeder Seed/Stock - Zelder, B.V., Gennep, The Netherlands.

Certified Seed/Stock - Cascade International Seed Co.

Preparer/Additional Information - Irvin H. Jacob, Cascade International Seed Company, 8483 West Stayton Road, Aumsville, OR 97325, (503) 749-1822.

NuBlue (J-229)

Source - Originated as a single, highly apomictic plant selected from the F_1 progeny of the cross NJE P-59 x Baron.

Method of Breeding - Single-plant selection from the progeny of the cross NJE P-59 x Baron.

Intended Use - Turf.

Description - Moderately low-growing with a medium-dark green color, and a fine-textured leaf blade which produces a dense turf.

Adapted to - Northern and central U.S. and southern Canada.

Released - Jacklin Seed Co., Rutgers University, and Medalist America.

Breeder Seed/Stock - Jacklin Seed Co.

Certified Seed/Stock - Available.

Preparer/Additional Information - Kim Peterson, Jacklin Seed Co., W. 5300 Riverbend Ave., Post Falls, ID 83854, (208) 773-7581.

Nugget

Selected at the Alaska AES, ARS cooperating - H.J. Hodgson, R.L. Taylor, L.J. Klebesadel, and A.C. Wilton.

Source - Single-plant collection made in 1957 at Hope, Cook Inlet, south-central Alaska.

Method of Breeding - Spaced-plant evaluation of bluegrasses collected in Alaska, named varieties and commercial sources. Evaluation of promising selections under turf conditions.

Intended Use - Lawns and revegetation.

Description - Outstanding winter survival. Very dense, dark green, tolerant to powdery mildew and leaf spot, and good seed yields. Comparatively narrow leaves, erect seedheads, good rhizome development, rapid germination and growth. Snow mold reaction comparable with other varieties. Highly apomictic.

Adapted to - LRR W, X, Y; PHZ 2, 3, 4.

Released - 1966, cooperatively by Alaska AES and Plant Science Research Division, ARS.

Breeder Seed/Stock - Alaska AES.

Certified Seed/Stock - Available.

Preparer/Additional Information - Alaska Plant Materials Center, HC 02 Box 7440, Palmer, AK 99645, (907) 745-4469.

NuStar

Source - A single-plant, ecotype selection from an apomictic bluegrass found growing near Ritzville, WA.

Method of Breeding - Transferred to the greenhouse at Jacklin Seed Co. and later replicated in seed-yield trial near Post Falls, ID. Tested for seed yielding ability for five years.

Intended Use - Turf.

Description - A dense, moderately dark green Kentucky bluegrass with improved turf quality performance.

Adapted to - Northern and central U.S. and southern Canada.

Released - 1990, Jacklin Seed Co.

Breeder Seed/Stock - Jacklin Seed Co.

Certified Seed/Stock - Available.

Preparer/Additional Information - Kim Peterson, Jacklin Seed Co., W. 5300 Riverbend Ave., Post Falls, ID 83854, (208) 773-7581.

P-104 (Reg. No. 31)

Developed by Princeton Turf Nurseries in 1974. Lofts Seed, Inc., has world-wide marketing rights. Carried as accession no. 509068, Princeton, 104.

Source - Germplasm obtained from the New Jersey AES.

Method of Breeding - A cross between Warren's A-25 Kentucky Bluegrass x Anheuser Dwarf Kentucky Bluegrass.

Intended Use - Lawns, parks, and sports fields. P-104 grows well in full sun to moderate shade. Ideal for sod production.

Description - A vigorous, very aggressive cultivar with medium green color and excellent resistance to leaf spot, melting-out, stripe smut, and leaf rust.

Adapted to - LRR A, B, C, E, F, G, H, K, L, M, N, R, S; PHZ 3, 4, 5, 6, 7.

Released - 1980 by Princeton Turf Nurseries.

Breeder Seed/Stock - Lofts Seed, Inc.

Certified Seed/Stock - Available.

Preparer/Additional Information - Lofts Seed Inc., Chimney Rock Road, P.O. Box 146, Bound Brook, NJ 08805, (908) 560-1590.

Parade

Source - Southern Netherlands.

Method of Breeding - Plants collected in The Netherlands. Increased by seed and grown in turf plots for four years. Top performers had progeny row-planted to check level of apomixis. Performed well on clay and sandy soil.

Intended Use - Athletic fields, golf courses, lawns, parks, and playgrounds.

Description - Widely adapted, good density, good drought recovery, good treading resistance, and good divot recovery.

Adapted to - PHZ 2, 3, 4, 5, 6.

Released - D.J. Van der Have, The Netherlands.

Breeder Seed/Stock - D.J. Van der Have, The Netherlands.

Certified Seed/Stock - Available from Advanta Seeds West, Inc.

Preparer/Additional Information - Kenneth Hignight, Advanta Seeds West, Inc., 33725 Columbus St. S.E., P.O. Box 1496, Albany, OR 97321-0452, (503) 967-8923.

Park (Reg. No. 4)

Selected at Minnesota AES, St. Paul - H.L. Thomas, Herman Shultz, A.R. Schmid, and H.K. Hayes.

Source - Vegetative material collected from 60 old pastures and waste places throughout Minnesota in 1937.

Method of Breeding - Collections separated into 281 vigorous individual plants; carried through extensive selection and testing program until 1947. Eighteen clones selected for further testing; the 1953 mixture of 15 best apomictic clones increased for testing as Minnesota 95.

Intended Use - Low-maintenance turf.

Description - In Minnesota described as being superior to Merion in seedling and plant vigor, resistance to rust, and sod formation.

Adapted to - Non-humid areas in northern U.S., especially the central Midwest.

Released - 1957, by Minnesota AES.

Breeder Seed/Stock - Minnesota AES.

Certified Seed/Stock - Available.

Preparer/Additional Information - Nancy Ehlke, Minnesota AES, Univ. of Minnesota, 220 Coffy Hall, 1420 Eckles Ave., St. Paul, MN 55108, (612) 625-8761.

Plush

Selected at Farmers Forage Research Cooperative, Lafayette, IN - R.J. Buker. Carried as accession no. FFR 9.031.

Source - Plants obtained from C.R. Funk, Rutgers, N.J., who had screened population for Helminthosporium leaf spot and stripe smut.

Method of Breeding - Superior clone selected on basis of turf trials at Lafayette, IN, and seed production tests at Spokane, WA.

Intended Use - Turf.

Description - Dwarf to semidwarf, resistant to leaf spot and stripe smut, produces good seed yield. Highly apomictic.

Adapted to - Any area where bluegrass is commonly grown; PHZ 3.

Released - By Farmers Forage Research Cooperative, Lafayette, IN.

Breeder Seed/Stock - Farmers Forage Research Cooperative.

Certified Seed/Stock - Available.

Preparer/Additional Information - FFR Cooperative, 4112 E. State Road 225, West Lafayette, IN 47906, (317) 567-2115.

Ram I

Discovered by Ernest W. Brown and Alexander M. Radko. Developed with the cooperative efforts of U.S. Golf Association Green Section, Lofts Pedigreed Seed, and the New Jersey AES. Carried as accession no. Ram I.

Source - Ecotype selection by A.M. Radko on a putting green of the Webhannet Golf Club, Kennebunk Beach, ME. A single, highly apomictic plant.

Method of Breeding - Selection evaluated under spaced-plant progeny tests.

Intended Use - Well suited for golf courses, sod operations, athletic fields, parks, and home lawns.

Description - Moderately low-growing, leafy, turf-type cultivar of medium texture and a rich, dark green color. Good disease resistance, above average ability to resist invasion by annual bluegrass (*Poa annua* L.). Ninety-five percent apomictic and tolerant to close mowing. Adaptable to high or low maintenance.

Adapted to - LRR A, B, C, E, F, G, H, K, L, M, N, R, S; PHZ 3, 4, 5, 6, 7.

Released - 1972 by Lofts Seed, Inc. and Jacklin Seed Co.

Breeder Seed/Stock - Jacklin Seed Company.

Certified Seed/Stock - Available.

Preparer/Additional Information - Kim Peterson, Jacklin Seed Co., W. 5300 Riverbend Ave., Post Falls, ID 83854, (208) 773-7581; and Lofts Seed Inc., Chimney Rock Road, P.O. Box 146, Bound Brook, NJ 08805, (201) 560-1590.

S-21

Jacklin Seed Company, Post Falls, ID - A.W. Jacklin. Carried as accession no. S-21.

Source - Aberrants from Delta.

Method of Breeding - Fifty similar aberrant plants were selected from a row planting of Delta on the basis of rhizome vigor, color, height, and seed production.

Intended Use - Turfgrass, where demand exists for a source of certified common-type Kentucky bluegrass.

Description - Medium-light green color, narrow erect leaf, matures one to two days later than South Dakota.

Adapted to - LRR A, B, E, G, F, K, L, M, R, S; PHZ 5.

Released - 1978 by Jacklin Seed Company, Post Falls, ID.

Breeder Seed/Stock - Jacklin Seed Company, Post Falls, ID.

Certified Seed/Stock - Available.

Preparer/Additional Information - Kim Peterson, Jacklin Seed Co., W. 5300 Riverbend Ave., Post Falls, ID 83854, (208) 773-7581.

Shamrock

Rutgers University - Dr. C. Reed Funk.

Source - Hybridization and selection was by Dr. C. Reed Funk, with additional yield trials and evaluation by R.H. Bailey.

Method of Breeding - Originated as a single, highly apomictic plant selected from the progeny of a hybrid of A25 x Touchdown crossed with Sydsport.

Intended Use - Home lawns; athletic fields; parks; sod production; golf course fairways, tees, and roughs; and institutional and commercial turf. Recommended as a component of bluegrass mixtures as well as bluegrass-ryegrass mixtures. Maximum durability if turf is allowed to establish before heavy use.

Description - Relatively low-growing variety with a dark green genetic color and medium-fine textured leaf blade. Shamrock exhibits early spring green up and provides an aggressive turf of medium-fine texture and good density. Shamrock has exhibited good resistance to leaf spot, dollar sot, stripe smut, and leaf rust; good resistance to melting out and powdery mildew; and very good drought tolerance.

Adapted to - PHZ 3, 4, 5, 6, 7.

Released - 1991, Dr. C. Reed Funk of Rutgers University.

Breeder Seed/Stock - Dr. C. Reed Funk of Rutgers University.

Certified Seed/Stock - Available.

Preparer/Additional Information - Art Wick, LESCO, Inc., 20005 Lake Road, Rocky River, OH 44116, (216) 333-9250; or R.H. Bailey, R.H. Bailey Seed, Inc., P.O. Box 13517, Salem, OR 97309, (503) 362-9700.

SR 2000

Seed Research of Oregon, Pure Seed Testing, and Rutgers Univ. Crop Science Dept. - M.F. Robinson and Leah A. Brilman, W.A. Meyer, D.C. Funk and C.R. Funk. Carried as experimental no. H75-1688.

Source - Apomictic F_1 hybrid from NJEP-154 x Baron cross. NJEP-154 was selected from old lawn located near Museum of Natural History, Washington, D.C., during 1963. Baron is a cultivar selected from an old turf in The Netherlands.

Method of Breeding - Progeny of a single, highly apomictic plant selected from the cross NJEP-154 x Baron made during 1974. Hybrid plant selected from spaced planting in 1975.

Intended Use - Lawn-type turfs in full sun to moderate shade in areas where Kentucky bluegrass is adapted. Should be blended with other adapted Kentucky bluegrasses, turf-type perennial ryegrasses, and/or strong creeping red fescues. Well suited to mixtures with turf-type tall fescues.

Description - Moderately low-growing with medium-dark green color, extensive rhizomes, wide leaves, large seed, and a high percentage of fertile florets. Can produce an aggressive, persistent turf with medium-high density and excellent heat and drought tolerance. Good resistance to leaf rust, stripe smut, leaf spot, and melting-out diseases.

Adapted to - LRR A, B, C, D, E, F, G, H, K, L, M, N, P, R, S, W; PHZ 2, 3, 4, 5, 6, 7.

Released - 1992 by Seed Research of Oregon.

Breeder Seed/Stock - Pure Seed Testing, Hubbard, OR, and Seed Research of Oregon, Corvallis, OR.

Certified Seed/Stock - Available.

Preparer/Additional Information - Seed Research of Oregon, Inc., P.O. Box 1416, Corvallis, OR 97339, (800) 253-5766.

SR 2100

Source - Open-pollinated progeny of A80-336 with Baron and Julia as possible pollen donors. A80-336 is a progeny of Warren's A-25 x Touchdown.

Method of Breeding - Hybridization in greenhouse under conditions favorable to increase sexual reproduction.

Intended Use - Cool-season turfgrass.

Description - Aggressive, vigorous variety with good seed yields. Has shown good drought tolerance, extensive rhizomes, and good turf performance.

Adapted to - PHZ 3, 4, 5, 6, 7. Northeastern, midwest, and western U.S.

Released - 1993, Seed Research of Oregon, Inc. and Rutgers University.

Breeder Seed/Stock - Seed Research of Oregon, Inc.

Certified Seed/Stock - Available.

Preparer/Additional Information - Dr. Leah A. Brilman, Seed Research of Oregon, Inc., P.O. Box 1416, Corvallis, OR 97339, (503) 757-2663.

Suffolk

Jacklin Seed Company, Post Falls, ID, and Rutgers University, New Brunwick, NJ - L.A. Brilman, A.D. Brede, and C.R. Funk. Carried as accession nos. H74-239; 239; 239-S;239-RS.

Source - NJE P-59 x Baron.

Method of Breeding - Originated from a single plant of F_1 progeny, from greenhouse hybridization between NJE P-59 x Baron. The single plant was vegetatively propagated, then multiplied, tested, evaluated, and released.

Intended Use - General turf use.

Description - Medium-dark green leaves, with exceptional cool-weather color retention and early spring green-up; produces turf which is moderately aggressive, with medium texture and medium density; excellent resistance to leafspot, melting out, and red thread; good tolerance to frost and drought.

Adapted to - LRR A, B, E, G, F, K, L, M, R, S; PHZ 5.

Released - 1988.

Breeder Seed/Stock - Jacklin Seed Company, Post Falls, ID.

Certified Seed/Stock - Available.

Preparer/Additional Information - Kim Peterson, Jacklin Seed Co., W. 5300 Riverbend Ave., Post Falls, ID 83854, (208) 773-7581.

Sydsport

Source - Originated from plant material collected at an athletic field in southern Sweden.

Method of Breeding - The original selection was made from wild material of *Poa pratensis* through isolation of plants with turfgrass habits.

Intended Use - Turf. Can be seeded alone or in blends or mixtures.

Description - Very dense and durable. Good for high-wear turf situations on athletic fields, golf courses, parks, and home sites. Good disease resistance, generally attractive, and bright color.

Adapted to - Wherever cold-season grasses are adapted.

Released - Weibulls in mid-1970s. Lofts has the North American breeding and production rights.

Breeder Seed/Stock - Lofts Seed, Inc. by authorization of Weibull.

Certified Seed/Stock - Available.

Preparer/Additional Information - Marie Pompei, Lofts Seed Inc., P.O. Box 146, Bound Brook, NJ 08805, (908) 560-1590.

Tendos

KWS-AG, Einbeck, Germany.

Source - Collection of clones in turf areas in central Germany.

Method of Breeding - Selected as single apomictic clone from numerous collections on basis of turf performance in Germany.

Intended Use - Ornamental, recreational, and athletic turf.

Description - Moderately dark, very leafy, producing an attractive turf and retaining excellent green color during fairly mild winters.

Adapted to - The northern cool, humid region; PHZ 3, 4, 5 (6, 7 limited to low humidity regions).

Released - 1981, KWS-AG.

Breeder Seed/Stock - KWS-AG.

Certified Seed/Stock - Distributed by TMI, Tangent, OR.

Preparer/Additional Information - TMI, 22068 Case Road NE, Aurora, OR 97002, (503) 678-2597.

Touchdown

Source - National Golf Links of America, Southhampton, NY.

Method of Breeding - Single ecotype was collected and subsequently progeny tested to yield the uniform and stable cultivar.

Intended Use - Cool-season turfgrass.

Description - Wear-resistant, highly drought-tolerant, and spreads aggressively when compared to other tall bluegrass varieties. Resists invasion of annual bluegrass and tolerates low cutting height.

Adapted to - PHZ 5, 6, 7, and parts of 8.

Released - 1974 by Pickseed West Inc.

Breeder Seed/Stock - Pickseed West Inc.

Certified Seed/Stock - Available.

Preparer/Additional Information - Pickseed West Inc., P.O. Box 888, Tangent, OR 97389, (503) 926-8886.

Troy

Selected at Montana AES, Bozeman, ARS cooperating - R.E. Stitt.

Source - Increase of PI 119684. Introduced from Turkey by Westover-Wellman expedition in 1936.

Method of Breeding - The original introduction from Turkey was reselected for powdery mildew resistance and released to Jacklin Seed Company for exclusive production and marketing.

Intended Use - Forage-type Kentucky bluegrass. Excellent pasture cultivar.

Description - Vigorous pasture strain. Released for use in irrigated pastures in Montana. Tall, erect-growth habit, good recovery, open sod. Not outstanding with respect to disease resistance. Adapted to cooler parts of Kentucky bluegrass region. Early maturing; ready to graze at Bozeman 10-14 days before other strains.

Adapted to - LRR A, B, E, F, G, K, M, L, R, S; PHZ 5.

Released - Jacklin Seed Company, Post Falls, ID.

Breeder Seed/Stock - Maintained by Montana State Univ., Bozeman, MT.

Certified Seed/Stock - Available.

Preparer/Additional Information - Kim Peterson, Jacklin Seed Co., W. 5300 Riverbend Ave., Post Falls, ID 83854, (208) 773-7581.

Unique

Source - Collection from Rhode Island.

Method of Breeding - Collection of a highly apomictic, single ecotype from an old-turf area.

Intended Use - Cool-season turf for sod, athletic fields, and golf courses.

Description - Early spring greenup, excellent turf quality, and good resistance to stem rust, stripe rust and leaf spot.

Adapted to - PHZ 2, 3, 4, 5, 6, 7, 8, 10.

Released - Turf-Seed, Inc.

Breeder Seed/Stock - Pure Seed Testing, Inc.

Certified Seed/Stock - Available. PVP No. 9200129.

Preparer/Additional Information - Turf-Seed, Inc., P.O. Box 250, Hubbard, OR 97032, (503) 651-2130.

Victa

Selected at Research Division, O.M. Scott & Sons Co., Marysville, OH - T. Fuchigami and E. Mayer.

Source - Collection of plants made in 1962 from lawns in San Fernando area, CA.

Method of Breeding - Single-plant selection made from original collection; progeny tested to assess apomictic stability. Seed from clonally propagated plants used for turf testing at Long Beach, CA; Marysville, OH; Accokeek, MD; and St. Louis, MO. Tested as Ba 62-54.

Intended Use - Permanent turf alone or in mixtures with other cool-season turf species.

Description - Low-growing variety. Moderate to high level of resistance to dollar spot, leaf spot, crown rust, head scarp, and smut. Blue-green leaf color. Seed ca. 900,000/lb. High seed yield combined with moderately good turf performance.

Adapted to - PHZ 2, 3, 4, 5, 6, 7.

Released - 1972 by O.M. Scott & Sons Co.

Breeder Seed/Stock - O.M. Scott & Sons Co.

Certified Seed/Stock - Available.

Preparer/Additional Information - Virgil Meier, O.M. Scott & Sons, Marysville, OH 43041 (513) 644-0011.

Viva

Source - Unknown.

Method of Breeding - Ecotype collection.

Intended Use - Turf for golf courses, playgrounds, athletic fields, and home lawns.

Description - High seed-yielding variety. Dark green color and good density.

Adapted to - PHZ 2, 3, 4, 5, 6.

Released - O.M. Scott & Sons, Marysville, OH.

Breeder Seed/Stock - O.M. Scott & Sons, Marysville, OH.

Certified Seed/Stock - Available.

Preparer/Additional Information - Kenneth Hignight, Advanta Seeds West, Inc., 33725 Columbus Street S.E., P.O. Box 1496, Albany, OR 97321-0452, (503) 967-8923.

Voyager

Source - Tennessee, Alabama, California, and New Jersey.

Method of Breeding - A composite variety comprised of a blend of the progenies of 11 highly apomictic plants.

Intended Use - Cool-season turf for sod, athletic fields, and golf courses.

Description - Good summer performance, with improved insect resistance. Good low-maintenance performance.

Adapted to - PHZ 2, 3, 4, 5, 6, 7, 8, 10.

Released - Turf-Seed, Inc.

Breeder Seed/Stock - Pure Seed Testing, Inc.

Certified Seed/Stock - Available.

Preparer/Additional Information - Turf-Seed, Inc., P.O. Box 250, Hubbard, OR 97032, (503) 651-2130.

Washington

Source - Isolated by Dr. C. Reed Funk and facilities of the New Jersey AES, Rutgers University. H86-526 originated as a single, highly apomictic plant selected from the open-pollinated progeny of AL-17 and Julia, the most likely pollen parent.

Method of Breeding - AL-17 was pollinated by Julia and other selected Kentucky bluegrasses under greenhouse conditions favorable to increased sexual reproduction of highly apomictic Kentucky bluegrass during the late winter of 1985. Seedlings from this cross were transferred to a spaced-plant field nursery at Adelphia, NJ, in September 1985.

Intended Use - Seed produced will be used to produce improved turfs including commercially grown sod in all areas where cool-season grasses are adapted.

Description - An attractive, vigorous plant selected from this nursery on June 11, 1986. Seed harvested from the Adelphia nursery was used to establish additional turf trials and to plant an experimental foundation seed field near Madras, OR, in August 1988. H86-526 is a facultative apomict with approximately 90% of its progeny appearing genetically identical to the maternal parent.

Adapted to - Useful in blends of turfgrass throughout the area where cool-season grasses are used.

Released - 1990.

Breeder Seed/Stock - Cascade International Seed Co.

Certified Seed/Stock - Available.

Preparer/Additional Information - Irvin H. Jacob, Cascade International Seed Company, 8483 West Stayton Road, Aumsville, OR 97325, (503) 749-1822.

Poa secunda J. Presl - big or sandburg bluegrass

***Poa ampla* Merr.**
***Poa juncifolia* Scribn.**

Cool-season, native bunchgrass. Valuable in the Pacific northwest and throughout northern part of intermountain region. Very early, relatively coarse, and palatable. Damaged by overgrazing; drought resistant.

9005460

Selected at Plant Materials Center, SCS, Bridger, MT - M.E. Majerus and J.G. Scheetz. Carried as accession no. WY-738.

Source - Collected from a black alkali site, 20 km southwest of Laramie, WY, by M.E. Majerus.

Method of Breeding - Direct increase of field collection.

Intended Use - Mixtures on saline-alkaline affected soils for grazing or cover and stabilization.

Description - Good seedling vigor, mildly rhizomatous, and has ability to establish excellent stands in saline-alkaline affected soils.

Adapted to - LRR B, D, E, F, G; PHZ 3.

Released - Distributed for limited field testing.

Breeder Seed/Stock - Plant Materials Center, SCS, Bridger, MT.

Certified Seed/Stock - Not available.

Preparer/Additional Information - Plant Materials Center, SCS, RR l, Box 1189, Bridger, MT 59014, (406) 662-3579.

Canbar

Selected at Plant Materials Center, SCS, Pullman, WA - J.L. Schwendiman. Tested as P-851.

Source - Collection from Blue Mountains, WA.

Method of Breeding - Selected as most vigorous accession from large group of collections of this polymorphous bluegrass species.

Intended Use - Understory species in rangeland reseeding.

Description - Vigorous, robust, leafy; characterized by excellent early spring growth, abundant basal leaves, numerous stems 40-45 cm in height, and abundant seedheads. Seed matures in mid-June, when plants become dormant until onset of fall rains. Apomictic and has been used in the Carnegie Institution of Washington bluegrass hybrid studies.

Adapted to - LRR B, E; PHZ 5.

Released - 1979 cooperatively by the Idaho and Oregon Experiment Stations and the Plant Materials Center, SCS, Pullman, WA.

Breeder Seed/Stock - Plant Materials Center, SCS, Pullman, WA.

Certified Seed/Stock - Available.

Preparer/Additional Information - Plant Materials Center, SCS, Room 104, Hulbert Agricultural Sciences Bldg., WSU, Pullman, WA 99164-6211, (509) 335-7376.

Service

Source - Mass selection from a six- meter row of PI 387931, originally received from the U.S. NPGS in 1979. Original materials from a single- plant collection from along the Alaska Highway east of Whitehorse, Yukon Territory, Canada.

Method of Breeding - Open-pollinated increase from original selection.

Intended Use - Erosion control, reclamation, and habitat restoration. Possible value in decorative landscaping.

Description - Erect, perennial bunchgrass, 45-80 cm tall, with distinct blue coloration. In Alaskan trials, surpassed Sherman in vigor and winter-hardiness. Performs best on dry, gravelly, or rocky soils.

Adapted to - LRR W, X, Y (southern half); PHZ 1, 2, 3, 4.

Released - 1989, Alaska Department of Natural Resources.

Breeder Seed/Stock - Alaska Plant Materials Center.

Certified Seed/Stock - Available.

Preparer/Additional Information - Stoney J. Wright, Manager, Alaska Plant Materials Center, State of Alaska, HC 02 Box 7440, Palmer, AK 99645, (907) 745-4469.

Sherman (Reg. No. 6)

Selected at Plant Materials Center, SCS, Pullman, WA - V.B. Hawk, J.L. Schwendiman, and A.L. Hafenrichter.

Source - Collected from native vegetation near Moro, Sherman County, OR, by D.E. Stephens, superintendent of Sherman Branch Experiment Station, Moro, in 1932. Re-collected by SCS in 1935.

Method of Breeding - Comparisons among 178 accessions. Selected accession subjected to mass selection. Tested as P-2716.

Intended Use - Range reseeding and revegetating disturbed lands.

Description - Starts growth very early in spring. Productive, early maturing, 900-965 mm tall, erect growing, fine stemmed. Long-lived perennial bunchgrass; high in seed, forage, and root production. Distinct blue, moderately abundant leaves; large, compact seedhead. Plants apomictic (2n=63). Adapted to conservation seedings alone or with alfalfa in dryland areas in wheat-fallow farmland on light-textured soils. Successfully used for reseeding burned-over forest lands in pine zones of the western states. Used by the Carnegie Institution of Washington in bluegrass hybrid studies.

Adapted to - LRR B, D; PHZ 5.

Released - 1945, cooperatively by Washington, Idaho, and Oregon Agricultural Experiment Stations at Pullman, Moscow, and Corvallis, respectively, and the Plant Materials Center, SCS, Pullman, WA. Distributed for field tests in 1938.

Breeder Seed/Stock - Plant Materials Center, SCS, Pullman, WA.

Certified Seed/Stock - Available.

Preparer/Additional Information - Plant Materials Center, SCS, Room 104, Hulbert Agricultural Sciences Bldg, WSU, Pullman, WA 99164-6211, (509) 335-7376.

Poa supina Schrader - supina bluegrass

Somewhat caespitose, stout, creeping perennial; found in pastures and damp places; from central and southwest Europe and northeast Russia.

Supranova

Source - Initial breeding material was collected in 1976 as an ecotype near Innsbruck, Austria, by Saatzucht Steinach Gmbh.

Method of Breeding - From stock material, several clones were selected in 1982 to form a synthetic for seed production. Nine clones were finally selected to form the variety.

Intended Use - Turf and groundcover.

Description - Extremely wear tolerant, uncommonly shade tolerant, extraordinarily aggressive, and will overcome contaminating grasses in turf. Very low-growing as a mature plant.

Adapted to - PHZ 2, 3, 4, 5, 6, 7 - Pacific northwest, transitional zone, intermountain area, and north-central states.

Released - Saatzuct Steinach GmbH, Austria.

Breeder Seed/Stock - Phillip Berner, Saatzucht Steinach GmbH, Austria.

Certified Seed/Stock - Available from Fine Lawn Research, Inc.

Preparer/Additional Information - David Lundell, Fine Lawn Research, Inc., P.O. Box 1051, Lake Oswego, OR 97034, (503) 636-2600.

Poa trivialis L. - rough stalk bluegrass

Cool-season, sod-forming grass from Europe. Distributed in northern U.S.. Limited use in lawn mixtures for shady areas. Does best in cool, moist environment; dormant during mid-summer; not adapted on dry sites.

Colt

Source - Parental stock originated from New Jersey and Pennsylvania.

Method of Breeding - Synthetic cultivar development via polycross and subsequent progeny testing.

Intended Use - For southern U.S. overseeding and for use as a component in shade seed mixtures.

Description - One of only a few commercially available *Poa trivialis* cultivars. It has performed well in trials for overseeding dormant bermudagrass in southern U.S.

Adapted to - For shade and overseeding use in PHZ 5, 6, 7, 8, 9.

Released - 1985 by Pure Seed Testing and Pickseed West Inc.

Breeder Seed/Stock - Pure Seed Testing.

Certified Seed/Stock - Available.

Preparer/Additional Information - Pickseed West Inc., P.O. Box 888, Tangent, OR 97389, (503) 926-8886.

Cypress

Source - From collection made in northern Europe and from accessions in western Oregon.

Method of Breeding - Extensive clonal evaluation for color, disease resistance, texture and persistence in hot weather, selection of desired types, phenotypic matched mating over three generations.

Intended Use - For overseeding of southern golf turf for winter use.

Description - Excellent dark green color for *Poa trivialis* and very fine texture and performance on overseeded putting greens.

Adapted to - As a perennial in shady, wet locations of all colder locations; as winter annual throughout the southern states of the U.S.

Released - 1991 by KWS-AG, Einbeck, Germany

Breeder Seed/Stock - KWS Seeds, Inc. and TMI.

Certified Seed/Stock - Available.

Preparer/Additional Information - Fred B. Ledeboer, Ph.D., TMI, 22068 Case Road NE, Aurora, OR 97002, (503) 678-2597.

Polder

Source - Mommersteeg International B.V., P.O. Box 1, 5250 AA Vlijmen, The Netherlands.

Method of Breeding - Ecotypes were collected at several places in The Netherlands. Observations were made from plants in a selection field. Selections were clones (synthetic varieties). The synthetics were tested in turf trials. One of the synthetics later became Polder.

Intended Use - Turf.

Description - Prostrate growth habit, narrow leaf width, medium to late seed head.

Adapted to - PHZ 2, 3, 4, 5, 6 - Cool-season regions, transitional zones.

Released - 1992, Mommersteeg International.

Breeder Seed/Stock - Mommersteeg International.

Certified Seed/Stock - Not available.

Preparer/Additional Information - David Lundell, Fine Lawn Research, Inc., P.O. Box 1051, Lake Oswego, OR 97034, (503) 636-2600.

Sabre (Reg. No. 18)

Developed cooperatively by International Seeds, Inc. and the New Jersey AES. Carried as experimental no. PT 4.

Source - Close-cut lawns, tennis courts, and golf course putting greens and fairways in the northeastern U.S.

Method of Breeding - Bi-parental and polycross progenies of 10 select clones were subjected to three cycles of phenotypic mass selection in spaced-plant nurseries.

Intended Use - Turf production under cool, moist shade in the north and for the winter overseeding of dormant warm-season grasses in the sun; combines well with perennial ryegrass and fine fescues in overseeding mixtures.

Description - Early flowering and very fine bladed. Has the ability to produce a denser turf with a darker green color and slower rate of vertical growth than other sources of *Poa trivialis*. Sabre is probably the most widely used variety of Poa trivialis in the U.S.

Adapted to - LRR A, M, N, R, S - damp, shaded areas. C, D, I, J, P, T, U, V - overseeding of dormant warm-season turf; PHZ 5.

Released - 1977 by International Seeds, Inc., Halsey, OR.

Breeder Seed/Stock - Maintained by International Seeds Inc.

Certified Seed/Stock - Available in quantity from International Seeds Inc. U.S. PVP no. 7700104.

Preparer/Additional Information - International Seed Inc, P.O. Box 168, 820 W. First St., Halsey, OR 97348, (503) 369-2251.

***Psathyrostachys juncea* (Fisch.) Nevski - Russian wildrye**

***Elymus junceus* Fisch.**

Cool-season bunchgrass native to central Asia, introduced in 1927 from the former USSR. Used primarily for pasture over general area where crested wheatgrass is adapted. Performed well in northern Great Plains and parts of intermountain region. Starts growth early. Leafy, nutritious, with dense basal leaves. Cures well, used extensively for late summer and fall grazing. Relatively low seedling vigor, deep rooted, drought resistant, salt tolerant. Exacting conditions required for successful seed production.

Bozoisky-Select

Source - Developed from PI 406468 obtained from the former USSR.

Method of Breeding - The breeding population was subjected to two cycles of selection for improved vigor, leafiness, seed yield, coleoptile length, and seedling vigor. Breeder's seed was obtained by bulking the open-pollination seed of 23 clones selected from a nursery consisting of 2,100 second-cycle plants.

Intended Use - Range and pasture plantings.

Description - Bozoisky-Select has been significantly more vigorous and productive than Vinall in range seedings. At eight semiarid range locations, it yielded 23% more forage than Vinall during the first two production years. Stand establishment of the new cultivar has been equal to or superior to Vinall in over 20 trials representing the sagebrush, juniper, shadscale, greasewood, and indian ricegrass ecosystems. Better seedling vigor and larger seeds than Vinall or Swift in laboratory trials. Coleoptile length, a character associated with better seedling emergence from deep plantings, was significantly greater in Bozoisky-Select than in Vinall or Swift. Grazing trials indicate that the strain is equally as palatable to grazing cattle as Vinall.

Adapted to - LRR B, G, F; PHZ 3, 4, 5.

Released - Yes.

Breeder Seed/Stock - ARS Logan, UT.

Certified Seed/Stock - Available.

Preparer/Additional Information - ARS, Utah AES, Logan, UT 84322, (801) 750-3067.

Cabree (Reg. No. 45)

Selected at Agriculture Canada Research Station, Lethbridge, Alberta - S. Smoliak. Carried as accession no. LRS 6757.

Source - The six clones included in Cabree trace to a field of Russian wildrye established in 1952 at Manyberries, Alberta, with commercial seed of unknown origin.

Method of Breeding - Ten plants with excellent seed retention were identified from among selections from a commercial field. The 10 selections were selfed, and 17 S1 plants with superior seed retention were selected. Six of the S1 plants, none which was related, were included in a synthetic after evaluation for seed retention, seedling vigor, forage and seed yield, and culm strength in replicated clonal and polycross progeny tests.

Intended Use - Cabree is recommended for dryland pasture, particularly in dry areas of the Canadian prairie region and the northern Great Plains region of the U.S. that average less than 380 mm of annual precipitation.

Description - Improved seed retention is the main attribute distinguishing Cabree from other Russian wildrye cultivars. It is a hardy perennial with a moderate level of drought tolerance. Averages taller culms and longer leaves than most current cultivars. Exhibits resistance to powdery mildew, spot blotch, and leaf rust under southern Alberta conditions. Cabree is the Blackfoot Indian word for antelope.

Adapted to - LRR F, G; PHZ 2, 3, 4.

Released - 1976 by Agriculture Canada.

Breeder Seed/Stock - Agriculture Canada Research Station, Lethbridge, Alberta T1J 4B1.

Certified Seed/Stock - Available.

Preparer/Additional Information - Surya N. Acharya, Agriculture Canada, Research Station, P.O. Box 3000, Lethbridge, Alberta T1J 4B1, (403) 327-4561.

Mankota (Reg. No. CV-149)

Selected at ARS Northern Great Plains Research Laboratory, Mandan, ND - John D. Berdahl. Carried as accession no. R1808.

Source - Of the six clones included in Mankota, three trace to PI314675, an introduction collected by Q. Jones and W. Keller near Alma-Ata in the former USSR and received in 1966; two to Mandan 34, a bulk population; and one to PI 272136, an introduction received in 1961 from the Alma-Ata Botanic Garden.

Method of Breeding - The six parent clones of the synthetic, Mandan R1808, were selected from a space-plant source nursery, and open-pollinated progenies of these clones were evaluated for seedling vigor in greenhouse and field tests. Replicated clonal plots and open-pollinated progenies of the parent clones were evaluated in field tests for forage and seed yield, spring vigor, resistance to leaf spot diseases, and general appearance.

Intended Use - Mankota is adapted for grazing in the northern Great Plains region of the U.S. and the Canadian prairie region. Primary use is expected to be fall and early winter grazing to complement existing native range or other seeded pastures.

Description - Moderate to good resistance to leaf spot diseases; heading date is approximately two days later than other current cultivars. Has no visual characteristics that distinguish it from other Russian wildrye cultivars. Forage yields have been high at high-yielding sites.

Adapted to - LRR F, G; PHZ 2, 3, 4.

Released - 1991 by ARS, SCS, and the North Dakota AES.

Breeder Seed/Stock - ARS, Northern Great Plains Research Laboratory, Mandan, ND 58554.

Certified Seed/Stock - Available.

Preparer/Additional Information - John D. Berdahl, ARS, Northern Great Plains Research Laboratory, P.O. Box 459, Mandan, ND 58554, (701) 663-6445.

Mayak (Reg. No. 43)

Selected at Agriculture Canada Research Station, Swift Current, Saskatchewan - T. Lawrence. Carried as accession no. ScR3631.

Source - The 20 clones included in Mayak trace to selections from a population derived from strains 1546, 1495, 2355, and D-19 (PI 75737) from the Northern Great Plains Research Laboratory, Mandan, ND, and from Swift Current breeding material.

Method of Breeding - Parent clones included in the synthetic cultivar were identified through several cycles of recurrent selection followed by evaluation in open-pollinated and polycross progeny tests. Primary selection criteria were high forage and seed yield and resistance to spot diseases.

Intended Use - Mayak is adapted for grazing in the Canadian prairie region and the northern Great Plains region of the U.S. Mayak provides excellent pasture throughout the grazing season and has good curing qualities that make it especially useful for late-summer, fall, and early-winter grazing.

Description - Selected for high forage and seed yield and resistance to leaf spot diseases, primarily. Lacks other distinguishing characteristics.

Adapted to - LRR F, G; PHZ 2, 3, 4.

Released - 1971 by Canada Department of Agriculture.

Breeder Seed/Stock - Canada Department of Agriculture Research Station, Swift Current.

Certified Seed/Stock - Available.

Preparer/Additional Information - Paul G. Jefferson, Agriculture Canada, Research Station, P.O. Box 1030, Swift Current, Saskatchewan S9H 3X2, (306) 773-4621.

Swift (Reg. No. 65)

Selected at Agriculture Canada Research Station, Swift Current, Saskatchewan - T. Lawrence. Carried as accession no. ScR3711.

Source - The 26 clones included in Swift trace to progenies of five clones of the cultivar Sawki developed at Swift Current and to Mandan 1546 from the Northern Great Plains Research Laboratory, Mandan, ND.

Method of Breeding - To improve seedling vigor, four cycles of recurrent selection were made for ability to emerge from a 5 cm depth in soil in a greenhouse. Recurrent selection was practiced in spaced-planted field nurseries for spring vigor, resistance to leaf spot diseases, leafiness, seed production, freedom from lodging, and general appearance. Seedling establishment and dry-matter yields in the establishment year were measured in standard performance tests for Swift and check cultivars.

Intended Use - Adapted for grazing in the Canadian prairie region and the northern Great Plains region of the U.S. Good curing qualities that make it especially useful for late-summer, fall, and early-winter grazing.

Description - Emerges from deeper plantings and develops larger plants in the establishment year than other commonly grown cultivars of Russian wildrye. Good resistance to leaf spot diseases. Lacks other distinguishing characteristics. Improved seedling vigor has been a major factor contributing to the popularity of Swift in its area of adaptation.

Adapted to - LRR F, G; PHZ 2, 3, 4.

Released - 1978 by Agriculture Canada.

Breeder Seed/Stock - Agriculture Canada, Research Station, Swift Current, Saskatchewan S9H 3X2.

Certified Seed/Stock - Certified seed is handled by SeCan Association, Suite 512, 885 Meadowlands Dr., Ottawa, Ontario K2C 3M2.

Preparer/Additional Information - Paul G. Jefferson, Agriculture Canada, Research Station, P.O. Box 1030, Swift Current, Saskatchewan S9H 3X2, (306) 773-4621.

Tetracan

Selected at Agriculture Canada Research Station, Swift Current, Saskatchewan - T. Lawrence. Carried as accession no. Sc RN3761.

Source - Tetracan traces to tetraploid Russian wildrye germplasm produced by chromosome doubling with colchicine by Dr. A.E. Slinkard while he was employed at the University of Idaho, Moscow.

Method of Breeding - Recurrent selection was practiced for seedling emergence from a 5-cm depth in soil in a greenhouse. Recurrent selection was practiced in spaced-planted field nurseries for spring vigor, resistance to leaf-spot diseases, leafiness, seed production, freedom from lodging, and general appearance. Grazing data were obtained from pasture tests using yearling steers.

Intended Use - Adapted for grazing in the Canadian prairie region and the northern Great Plains region of the U.S. Tetracan has good curing qualities that make it especially useful for late-summer, fall, and early-winter grazing.

Description - The main attribute of Tetracan, the first tetraploid Russian wildrye cultivar developed in North America, is excellent establishment vigor. Has significantly larger seed than cultivars with a diploid chromosome complement. Seedlings have the ability to emerge from greater depths in the soil. Plant regrowth and seed yields are similar to diploid cultivars. Animal performance from grazing trials is similar to diploid cultivars.

Adapted to - LRR F, G; PHZ 2, 3, 4.

Released - 1988 by Agriculture Canada.

Breeder Seed/Stock - Agriculture Canada, Research Station, Swift Current, Saskatchewan S9H 3X2.

Certified Seed/Stock - Certified seed is handled by SeCan Association, Suite 512, 885 Meadowlands Dr., Ottawa, Ontario K2C 3M2.

Preparer/Additional Information - Paul G. Jefferson, Agriculture Canada, Research Station, P.O. Box 1030, Swift Current, Saskatchewan S9H 3X2, (306) 773-4621.

Vinall (Reg. No. 5)

Selected at ARS Northern Great Plains Research Laboratory, Mandan, ND - G.A. Rogler and H.M. Schaaf. Carried as accession no. Mandan 2355.

Source - Three parental clones derived from PI 75737 and one each from PI 108493 and PI 111549.

Method of Breeding - Synthetic of five clones. At least four generations of selection on single-plant basis in open-pollinated progenies and inbred lines preceded choice of each parent. Progeny tests in yield plots and as spaced plants used to measure effects of outcrossing with other four plants for each plant in synthetic. All five clones were good seed producers; three of them produce exceptionally large seed. Distributed for testing as Mandan 2355.

Intended Use - Vinall is most often stockpiled and used for fall grazing to complement native pastures by extending the grazing season. Vinall is adapted to clay soils and is moderately tolerant to stress cause by saline and sodic soils.

Description - Vinall has no visual characteristics that distinguish it from other Russian wildrye cultivars. Seed yields at Mandan, ND, have been equal to more recently released cultivars, but seedling vigor is inferior. Longevity of established stands has been excellent. It is anticipated that recently developed cultivars of Russian wildrye will replace Vinall. Breeders seed will be maintained because of the widespread use of the cultivar as a standard check in experimental tests.

Adapted to - LRR B, G, F; PHZ 2, 3, 4.

Released - 1960 by ARS, in cooperation with the Agricultural Experiment Stations at North Dakota, South Dakota, Montana, Wyoming, Nebraska, Colorado, and New Mexico.

Breeder Seed/Stock - ARS, Northern Great Plains Research Laboratory, Mandan, ND 58554.

Certified Seed/Stock - Available.

Preparer/Additional Information - John D. Berdahl, ARS, Northern Great Plains Research Laboratory, P.O. Box 459, Mandan, ND 58554, (701) 663-6445.

Pseudoroegneria spicata **(Pursh) A. Love ssp. *inermis* (Scribn. & J.G. Sm.) A. Love - beardless wheatgrass**

Agropyron inerme **(Scribn. & J.G. Sm.) Rydb.**

Similar to bluebunch wheatgrass in appearance except for absence or near absence of awns. Distributed over much the same area, but less abundant than bluebunch.

Whitmar (Reg. No. 4)

Selected at Plant Materials Center, SCS, Pullman, WA - J.L. Schwendiman.

Source - Collected from native Palouse prairie grassland climax near Colton, Whitman County, WA, by L.A. Mullen in area of 500 mm of annual precipitation and elevation of 855 m on palouse silt-loam soil.

Method of Breeding - First observed as outstanding accession, P-3537, in observational tests among more than 500 beardless and bluebunch wheatgrass collections, which represented six ecotypes, from Pacific northwest. Developed by selection from space-planted nursery.

Intended Use - Range reseeding.

Description - Long-lived, native, perennial bunchgrass; drought resistant. Intermediate type, with moderately abundant, erect to semi-erect, medium-coarse stems. Leaves abundant, soft, lax, flat basal, and cauline. Seeds awnless, but short-awned seeds occur occasionally. Seedling vigor good. Seed and forage production high. Good spring and fall recovery and retains feed value and palatability late into summer. (2n=14).

Adapted to - LRR B; PHZ 5.

Released - 1946, by Washington, Idaho, and Oregon Agricultural Experiment Stations at Pullman, Moscow, and Corvallis, respectively, and SCS, Plant Materials Centers, Aberdeen, ID, and Pullman, WA.

Breeder Seed/Stock - Plant Materials Center, SCS, Pullman, WA.

Certified Seed/Stock - Available.

Preparer/Additional Information - Plant Materials Center, SCS, Rm 104, Hulbert Agricultural Sciences Bldg., WSU, Pullman, WA 99164-6211, (509) 335-7376.

Pseudoroegneria spicata ssp. *spicata* (Pursh) A. Love - bluebunch wheatgrass

Agropyron spicatum **(Pursh)**
Elytrigia spicata **(Pursh) Dewey**

Important cool-season bunchgrass in intermountain region from western Montana to central Washington and south into Nevada and Utah. Valuable species in native range where prized for palatability and drought resistance. Spreads slowly from short rhizomes. Leaves abundant, erect to semi-erect, soft, lax, flat, and primarily basal. Stems moderately abundant, erect to semi-erect, and medium coarse. Seeds large, heavy, and awned; must be processed to permit satisfactory seeding. Best adapted at higher elevations where available moisture exceeds 200 mm; good spring and fall recovery. Retains feed value and palatability late into summer and fall. (2n=28).

Goldar

Selected by Plant Materials Center, SCS, Pullman, WA - J.L. Schwendiman. Moved to Aberdeen, ID. Carried as accession no. PI 539873 and SCS-9002950.

Source - Selected from a native plant collection made on Malley Ridge, Umatilla National Forest, Asotin, WA, between 310-475 m elevation, in open ponderosa pine woodland, in 1934.

Method of Breeding - First observed as promising accession in Pullman nursery, which contained more than 500 Pacific northwest beardless and bluebunch wheatgrass collections, representing six ecotypes. Developed by mass selection from spaced plantings. Natural selection from rod rows of native collections and rod rows of progeny. Some crossing is assumed to have occurred during early testing.

Intended Use - Rangeland seedings, critical area stabilization, in mixes for re-establishment of native plant communities, weed control, vegetative firebreaks, and minespoil reclamation.

Description - Goldar is typical of the species *P. spicata,* with strongly divergent awns. The only other closely related variety is Whitmar, *P. spicata* ssp. *inermis,* which is essentially awnless.

Adapted to - LRR B, D, E; PHZ 5, 6 in areas above 310 m elevation and with median annual precipitation above 25 cm.

Released - 1989 by Plant Materials Center, SCS, Aberdeen, ID, Idaho AES, Utah AES, and ARS.

Breeder Seed/Stock - Maintained at Plant Materials Center, SCS, Aberdeen, ID.

Certified Seed/Stock - Available.

Preparer/Additional Information - Plant Materials Center, SCS, P.O. Box 296, Aberdeen, ID 83210-0296, (208) 397-4133.

Secar

Selected at Plant Materials Center, SCS, Pullman, WA - J.L. Schwendiman.

Source - A native plant collection on the Lewiston Grade in the Snake River gorge, near Lewiston, ID, on July 5, 1938.

Method of Breeding - First observed as outstanding accession, P-6409, in observational tests among more than 500 beardless and bluebunch wheatgrass collections, representing six ecotypes from the Pacific northwest. Developed by mass selection from spaced plantings.

Intended Use - Range reseeding.

Description - A low-elevation, dryland ecotype. A densely tufted bunchgrass with abundant, narrow leaves, numerous fine stems, small seeds, and divergent awns; early maturing. Slower to establish and less productive than larger types but more drought tolerant and persistent under adverse conditions. Adapted alone or with an understory grass at low elevations in the 200-300 mm precipitation zones of the Pacific northwest.

Adapted to - LRR B, D, E; PHZ 5.

Released - 1980, cooperatively by the Washington Agricultural Research Center, the Idaho, Oregon, Montana, and Wyoming Agricultural Experiment Stations, and the Plant Materials Center, SCS, Pullman, WA.

Breeder Seed/Stock - Plant Materials Center, SCS, Pullman, WA.

Certified Seed/Stock - Available.

Preparer/Additional Information - Plant Materials Center, SCS, Room 104, Hulbert Agricultural Sciences Bldg., WSU, Pullman, WA 99164-6211, (509) 335-7376.

Schizachyrium scoparium (Michx.) Nash - little bluestem

Important warm-season bunchgrass. Widely distributed throughout eastern and central U.S.. Valuable rangegrass in Flint Hills of east-central Kansas and Oklahoma. More drought resistant and found in more westerly and drier parts of Great Plains than big bluestem.

Aldous

Selected at Kansas AES, Manhattan - A.E. Aldous.

Source - Accessions collected in 1935 from Flint Hill native grasslands south of Manhattan.

Method of Breeding - Composite of progeny of these accessions made after several generations of selection. Distributed for testing as KG-1580.

Intended Use - Range/pasture reseeding.

Description - Tall, leafy, vigorous, medium-late in maturity, and more uniform than field-run accessions. Produces abundant forage and, under favorable conditions, good seed yield. Possesses some resistance to rust.

Adapted to - LRR F, G, H, M, N, O, P, T; PHZ 4.

Released - 1966, cooperatively by Kansas AES; Plant Science Research Division, ARS; and Plant Sciences Division, SCS.

Breeder Seed/Stock - Kansas AES.

Certified Seed/Stock - Available.

Preparer/Additional Information - SCS, 760 S. Broadway, Salina, KS 67401, (913) 823-4541.

Blaze (Reg. No. 3)

Selected at Nebraska AES, Lincoln, ARS cooperating - L.C. Newell.

Source - Domestic collections in 1953 from natural prairies in Nebraska and Kansas.

Method of Breeding - Fifteen clones, selected for late maturity, leafiness, and seed production in space-planted nurseries polycrossed in isolation. Progenies selected for seedling vigor and a green-leaf character. Second generation of synthesis provided breeder seed.

Intended Use - Range/pasture reseeding.

Description - Leafy, mid-tall, and late maturing in central latitude; foliage bright to dull green, turning red in fall. In limited comparisons, produced better stands and was more productive than other selections and native ecotypes. Recommended for conservation plantings and permanent pasture mixtures of warm-season prairie grasses in central and eastern Nebraska and adjacent areas in bordering states. The area of reliable seed production is centered in southeast Nebraska.

Adapted to - LRR G, H, M; PHZ 4, 5.

Released - 1967, cooperatively by Nebraska AES and USDA-ARS.

Breeder Seed/Stock - Nebraska AES and USDA-ARS, Univ. of Nebraska, Lincoln.

Certified Seed/Stock - Available.

Preparer/Additional Information - K.P. Vogel, ARS-USDA, 344 Keim Hall, Univ. of Nebraska or Jeff Pederson, Dept. of Agronomy, Univ. of Nebraska, Lincoln, NE 68583, (402) 472-1564, (402) 472-2811.

Camper

Selected at Nebraska AES, Lincoln, ARS cooperating - L.C. Newell.

Source - Domestic collections in 1953 from natural prairies in Nebraska and Kansas.

Method of Breeding - Developed by synthesis of two unrelated lines, designated for leaf color as "Late Gray" and "Early Blue." Three generations of spaced-plant selection for their dominant leaf color phenotypes. Evaluated in Nebraska Outstate Tests (1964-68). Seed of selected progeny from an intercrossing isolation of parent lines was increased as breeder seed. Result is an improvement upon single-source and wild harvests of little bluestem in broad genetic composition, adaptation, seed production, and stand establishment.

Intended Use - In mixtures of other warm-season prairie grasses for range, non-irrigated pasture, and roadside plantings.

Description - A long-lived perennial forage cultivar with maximum vegetative growth produced in midsummer. Bunch-type, spreading by short rhizomes; leaves are long and narrow. Leaf color is predominantly gray to glaucus blue-green in early growth and under dry conditions, becoming darker with approaching maturity. Seed set results from pollination over an extended period with moderately late maturity for seed harvest; its area of adapation and use is thus extended throughout those parts of Nebraska where little bluestem is grown.

Adapted to - East-central and central Nebraska.

Released - 1973, cooperatively by Nebraska AES and USDA-ARS.

Breeder Seed/Stock - Nebraska AES.

Certified Seed/Stock - Available.

Preparer/Additional Information - K.P. Vogel, ARS-USDA, 344 Keim Hall, Univ. of Nebraska or Jeff Pederson, Dept. of Agronomy, Univ. of Nebraska, Lincoln, NE 68583, (402) 472-1564, (402) 472-2811.

Cimarron

Increased at Plant Materials Center, SCS, Manhattan, KS - R.D. Lippert. Carried as accession no. PM-K-152.

Source - Collected from many sites in southwest Kansas and panhandle of Oklahoma by SCS personnel in 1959 at approximate elevation of 764-1,070 m and precipitation of 380-500 mm.

Method of Breeding - Collections bulked to establish increase field. Natural selection under combine harvesting for more uniform maturity.

Intended Use - Range reseeding and revegetation of disturbed areas.

Description - Cimarron proved significantly better in forage production and disease resistance than other strains or cultivars in extensive trials over its range of adaptation. It has been grown where annual precipitation is as low as 300 mm. Performance has been good on all types of soils, except where high salinity, alkalinity, or excessive weed competition is a problem.

Adapted to - LRR E, G, H, N, P; PHZ 4b.

Released - 1979, cooperatively by Plant Materials Center, SCS, Manhattan, KS, and Kansas AES.

Breeder Seed/Stock - Plant Materials Center, SCS, Manhattan, KS.

Certified Seed/Stock - Available.

Preparer/Additional Information - SCS, 760 S. Broadway, Salina, KS 67401, (913) 823-4541.

Pastura

Selected at Plant Materials Center, SCS, Los Lunas, NM, and New Mexico AES, University Park, cooperating - J.A. Downs, G.C. Niner, and J.E. Anderson. Carried as accession no. PI 476998 & PM NM 272.

Source - Collections made in 1956 and 1957 near Rowe and Pecos, NM, at elevation of 2,000-2,100 m and average annual precipitation of 360 mm.

Method of Breeding - Increase of original collection. Field tested as PM-NM-272.

Intended Use - Range reseeding and revegetation of disturbed areas.

Description - True green, fairly uniform growth, excellent seedling vigor, and resistance to climatic adversities. Herbage production only average, but seed production high, with processing less of problem than with many strains because of less villous inflorescence. Widely scattered field tests show strain is well adapted for sites where species is recommended in central and eastern New Mexico and in eastern Colorado.

Adapted to - LRR G, H; PHZ 4.

Released - 1963, cooperatively by SCS - Los Lunas and New Mexico AES.

Breeder Seed/Stock - Plant Materials Center, SCS, Los Lunas, NM.

Certified Seed/Stock - Not available.

Preparer/Additional Information - Plant Materials Center, SCS, 1036 Miller St. SW, Los Lunas, NM 87031, (505) 865-4684.

Secale cereale L. - cereal rye

Widely domesticated cereal grain adapted to temperate and cold temperate regions of the world. Conservation uses include winter cover crop for erosion control and green manure, annual forage, and temporary stabilization of disturbed sites. Cultivars developed specifically for conservation and forage are listed here.

Aroostook

Selected at the Big Flats Plant Materials Center, SCS, Corning, NY, after extensive field testing in Aroostook County, ME. Carried as accession no. NY-6030 PI 464583.

Source - A selection of common rye received from Stanford Seed Co., Buffalo, NY. Parentage traced to Balbo variety, grown in Missouri.

Method of Breeding - Selection from all available cereal rye lines at Corning, NY, and Presque Isle, ME, for late fall germination and growth, and early and rapid spring growth. Direct increase of original selection.

Intended Use - As winter cover crop, following potatoes or corn; for early spring forage production.

Description - Aroostook is very winter-hardy and initiates spring growth 7-14 days before Musketeer, Frontier, Von Lockon, or Rymin. Head emergence is nine days earlier than

Von Lockon and maturity occurs five days earlier. Leaf area index and leaf area per plant are greater for Aroostook than for Balbo after 220 growing degree-days (base 0°C).

Adapted to - LRR K, L, M, N, northern P, R, S; PHZ 3.

Released - 1981.

Breeder Seed/Stock - Maintained at Big Flats Plant Materials Center, SCS, Corning, NY.

Certified Seed/Stock - Available.

Preparer/Additional Information - John Dickerson, Big Flats Plant Materials Center, SCS, RD l, Rte 352, Box 360A, Corning, NY 14830, (607) 562-8404.

Secale montanum Guss. - mountain brome

A short-lived perennial rye from southwest Asia. It is distributed from Morocco and Spain eastward to Italy, Yugoslavia, Turkey, Iran, and Iraq. Originally tested in the 1930s as a green manure or cover crop on dryland farms. In the 1940s, the USFS looked at its potential as a forage plant. Tested by USDA-FS for use as a short-lived cover crop in range rehabilitation seedings. Adapted to a wide range of semi-arid conditions across the intermountain region of the U.S.. Grows well on all types of soil, elevations of 150-2,140 m, temperatures of -5°C to +30°C, and total annual precipitation of 250-1,000 mm.

9036554

Increased at the Aberdeen Plant Materials Center in Aberdeen, ID, Pullman Plant Materials Center in Pullman, WA, USDA-FS Shrub Sciences Lab in Provo, UT, and the Utah Division of Wildlife Resources - USDA-FS Great Basin Field Station in Ephraim, UT. Carried as accession no 9036554.

Source - Introduced by the ARS, Plant Introduction Center to the Pullman Plant Materials Center from collections made in the former Soviet Union in the 1930s. Parent material was given the accession number P4888, later changed to 9017596, and also given NSSL# 93622.

Method of Breeding - Increase of original accession for evaluation purposes.

Intended Use - Annual weed competition, quick cover for erosion control. Early spring forage, upland bird feed and cover.

Description - Short-lived perennial rye that establishes well in disturbed areas in 250-1,000 mm precipitation zones. It can survive on all types of soils and will not invade into sites that have perennial grass cover. It can become a serious weed in dryland or irrigated winter wheat crops. Extensively evaluated by the plant materials centers and found to have potentially weedy characteristics.

Adapted to - LRR B, D, E; PHZ 4, 5.

Released - Has not been released.

Breeder Seed/Stock - USDA-FS Intermountain Forest and Range Experiment Station, Ephraim, UT.

Certified Seed/Stock - Not available.

Preparer/Additional Information - E.D. McArthur, USDA-FS, Shrub Science Lab, 735 N. 500 E., Provo, UT 84606, (801) 377-5717; or Richard Stevens, Utah Division of Wildlife Resources, 15 S. Main, Ephraim, UT 84627, (801) 283-4441.

Setaria italica (L.) Beauv. - foxtail millet

Warm-season annual from Asia. Used for late sown hay in Great Plains and to limited extent for birdseed and pasture. Grows best under warm conditions; not drought resistant. Largely replaced by sudangrass and early-maturing sorghums for temporary pasture and emergency forage. Many varieties developed when crop was popular, including common, German, Golden Wonder, Goldmine, Hungarian, Siberian, Kursk, and Turkestan.

Sno-fox

Developed at the University of Nebraska, Panhandle Station.

Source - Tested under the experimental number ISc474 which was included in a collection of *Setaria* millets obtained from the Rockefeller Foundation.

Method of Breeding - The original line contained off-type plants for both height and seed color. These were rogued to purify the line prior to release.

Intended Use - Grain production.

Description - Evaluated 1969-79 under dryland conditions for both grain and forage production. Has had good grain yields, generally being the top yielding *Setaria*. Especially well suited to higher elevations where the growing season is short or in late plantings, since it is earlier to mature than other *Setaria* species. Forage yields have been comparable to other *Setaria* varieties and in some cases have been comparable to sudangrass and sorghum-sudangrass hybrids. Widely used for finch food in the bird seed trade and the forage is used for livestock feed.

Adapted to - Nebraska.

Released - University of Nebraska, Panhandle Station, Scottsbluff, NE.

Breeder Seed/Stock - University of Nebraska, Panhandle Station, Scottsbluff, NE.

Certified Seed/Stock - Available.

Preparer/Additional Information - David D. Baltensperger, Panhandle Research & Extension Center, Univ. of Nebraska, 4502 Avenue I, Scottsbluff, NE 69361, (308) 632-1261.

***Sorghastrum nutans* (L.) Nash - indiangrass**

Warm-season, native bunchgrass. Distributed throughout eastern U.S. and west to North Dakota, Colorado, and Arizona. Valuable rangegrass in central and southern Great Plains. Does best on fertile bottom lands, but also grows on sandy soils. Palatable early in season, but only fair for winter use; damaged by overgrazing.

Cheyenne

Selected at SCS Nursery, Woodward, OK - J.E. Smith, Jr., and G.L. Powers.

Source - Native rangeland near Supply, OK, 1942.

Method of Breeding - Increase of bulk collection at Cheyenne, OK, until 1951, then at Texline, TX. Only one seed crop grown by SCS at this location.

Intended Use - Range reseeding, revegetation of disturbed areas, pasture.

Description - Heterogeneous plant makeup. Good forage type, good seed producer. Adapted for range and pasture in western Oklahoma and in Texas.

Adapted to - LRR H; PHZ 5b.

Released - Informally by SCS in 1945, but field established by Max Bower in 1954 at Morton, TX, believed to be first planting made for commercial seed production.

Breeder Seed/Stock - Not available.

Certified Seed/Stock - Not available. Common in commercial production.

Preparer/Additional Information - SCS, 760 S. Broadway, Salina, KS 67401, (913) 823-4541.

Holt (Reg. No. 16)

Developed at Nebraska AES, Lincoln, ARS cooperating - E.C. Conard and L.C. Newell.

Source - Collections in Elkhorn Valley of Holt County in northeastern Nebraska.

Method of Breeding - Mass selection for type. Grown at Lincoln since 1942.

Intended Use - Range/pasture reseeding.

Description - Moderately early maturing; superior in leafiness and yield to early-maturing strains from northern and western sandhill region of Nebraska. Relatively finer leaves and stems than later maturing varieties from more southerly sources, which may produce more total forage. Well suited for revegetation either in pure stands or in mixture with other relatively early maturing, tall, warm-season grasses, such as Nebraska 28 switchgrass and sandhill bluestems. May also be grown in diverse mixtures containing such mid-tall grasses as Butte sideoats grama and Nebraska 27 sand lovegrass. Should be grown in irrigated rows for seed production and will mature seed crops in such relatively short seasons as those of western or northern Nebraska.

Adapted to - LRR F, H, M, G; PHZ 4.

Released - 1960, cooperatively by Nebraska AES and USDA-ARS.

Breeder Seed/Stock - Nebraska AES and USDA-ARS, Univ. of Nebraska, Lincoln.

Certified Seed/Stock - Available.

Preparer/Additional Information - K.P. Vogel, ARS-USDA, 344 Keim Hall, Univ. of Nebraska or Jeff Pederson, Dept. of Agronomy, Univ. of Nebraska, Lincoln, NE 68583, (402) 472-1564, (402) 472-2811.

Llano

Selected at Plant Materials Center, SCS, Los Lunas, NM, and New Mexico AES, University Park, cooperating - J.A. Downs, G.C. Niner, and J.E. Anderson. Carried as accession nos. PI 476999 & PM-NM-275.

Source - Collected in 1956 and 1957 from sandy plains sites in eastern New Mexico (one near Hudson and one near Portales) at elevation of 1,222-1,255 m and average annual precipitation of 400-435 mm.

Method of Breeding - Increased separately, but combined in equal parts to produce heterogeneous second generation. In comparison-row planting and in initial increase, growth, flowering, and seeding characteristics of two collections were similar. Increased and tested as PM-NM-275.

Intended Use - Range reseeding, dryland and irrigated pasture, and revegetation of disturbed areas.

Description - Fairly uniform, with good leaf production extending well up culms. Under irrigation and fertilization, plants mature at 1.5-1.8 m, yielding high-quality seed in excess of 180 kg per acre. Adapted to range and pasture seeding on sandy sites in southwestern Great Plains, where rainfall is erratic and relatively low. Performed well in eastern New Mexico and east-central and southeastern Colorado, but needs adequate testing farther north and east.

Adapted to - LRR G, H; PHZ 4.

Released - 1963, by New Mexico AES and Plant Materials Center, SCS, SCS, Los Lunas, NM.

Breeder Seed/Stock - Plant Materials Center, SCS, Los Lunas, NM.

Certified Seed/Stock - Not available.

Preparer/Additional Information - Plant Materials Center, SCS, 1036 Miller St. SW, Los Lunas, NM 87031, (505) 865-4684.

Lometa (Reg. No. 79)

Selected in 1975 at the James E. "Bud" Smith Plant Materials Center, Knox City, TX, and increased for field testing as PMT-802. Carried as accession no. T-802.

Source - Original seed collected in 1964 near Lometa, TX, by Harry Schofield, an SCS employee.

Method of Breeding - Selected from 28 similar accessions for better forage production in central and south Texas.

Intended Use - Range reseeding, revegetation, pasture planting, and hay production.

Description - Flowers two to four weeks later than Cheyenne or Tejas. Produces twice the amount of forage of Cheyenne and extends range into south Texas. Grows well on most soil textures where precipitation is above 560 mm or on sites with extra runoff. Late seed producer, greens up early. Responds well to fertilizer and/or fertility. Excellent forage quality and cures well.

Adapted to - LRR H, I, J, M, N, O, P, T, U; PHZ 7.

Released - 1981 by Plant Materials Center, SCS, Knox City, TX, and Texas AES.

Breeder Seed/Stock - Plant Materials Center, SCS, Knox City, TX.

Certified Seed/Stock - Available.

Preparer/Additional Information - Plant Materials Center, SCS, Rte 1 Box 155, Knox City, TX 79529, (817) 658-3922.

Nebraska 54

Increased by private seed producer, Harold Hummel, Fairbury, NE.

Source - Seed collected in 1954 by H. Hummel from selected native plants growing along railroad right-of-way and in native ranges in southern part of Jefferson County, NE.

Method of Breeding - Placed directly in seed production.

Intended Use - Range/pasture reseeding.

Description - Tall, leafy, moderately late maturing indiangrass. Can produce high seed yields. Good seedling vigor. Adapted in southern and eastern Nebraska and in northern Kansas.

Adapted to - LRR H, M, N; PHZ 4, 5.

Released - 1957, certified as to origin by Nebraska Crop Improvement Association.

Breeder Seed/Stock - Nebraska AES and USDA-ARS, Univ. of Nebraska.

Certified Seed/Stock - Available.

Preparer/Additional Information - K.P. Vogel, ARS-USDA, 344 Keim Hall, Univ. of Nebraska or Jeff Pederson, Dept. of Agronomy, Univ. of Nebraska, Lincoln, NE 68583, (402) 472-1564, (402) 472-2811.

Osage

Selected at Kansas AES, SCS and ARS cooperating - F.L. Barnett.

Source - Seed collected from eastern and central Kansas and Oklahoma in 1953.

Method of Breeding - Recurrent selection for leafiness, vigor, freedom from rust, and since species tends to mature late, for earliness of maturity. Eight clones selected for inclusion in synthetic. Distributed for testing as Kansas Experimental Strain 3.

Intended Use - Range/pasture reseeding.

Description - Tall, vigorous, leafy, late maturing, and rust resistant. Good seed yields. Adapted in eastern Kansas, western Missouri, and northeastern Oklahoma. Best seed yields south of Kaw and Smoky Hill Rivers.

Adapted to - LRR M, N, H; PHZ 4.

Released - 1966, cooperatively by Kansas AES, Plant Sciences Division, SCS, and Plant Science Research Division, ARS.

Breeder Seed/Stock - Kansas AES.

Certified Seed/Stock - Available.

Preparer/Additional Information - Kansas State Univ., AES, Manhattan, KS 66506, (913) 532-6101.

Oto

Developed at Nebraska AES, Lincoln, ARS cooperating - L.C. Newell.

Source - Collections from natural grasslands of Nebraska and Kansas in 1953-54.

Method of Breeding - Fifteen accessions exhibited bright-green leaves, brown panicles, and late maturity. One group of 100 clones were isolated and progeny bred true for brown-glumed seed. Nebraska tests showed marked superiority in establishment and yield where soil moisture and fertility not limiting.

Intended Use - Range/pasture reseeding.

Description - Long-season variety; plants robust and erect, attaining spread of 0.6 m and height of 1.8 m; leaves long and bright green; panicles broad in anthesis, contracting into golden to dark brown, compact heads. Matures late in the season, with seed harvests at early frost dates in southern Nebraska. Recommended primarily in mixed stands of warm-season prairie grasses in which it improves late summer grazing. Area of recommended use centers in eastern and southern Nebraska, but extending into adjacent areas.

Adapted to - LRR H, M; PHZ 5.

Released - 1970, cooperatively by the Nebraska AES and USDA-ARS.

Breeder Seed/Stock - Nebraska AES and USDA-ARS, Univ. of Nebraska, Lincoln.

Certified Seed/Stock - Available.

Preparer/Additional Information - K.P. Vogel, ARS-USDA, 344 Keim Hall, Univ. of Nebraska or Jeff Pederson, Dept. of Agronomy, Univ. of Nebraska, Lincoln, NE 68583, (402) 472-1564, (402) 472-2811.

PI 514673

Plant Materials Center, SCS, Americus, GA, University of Georgia. Carried as accession no. PI 514673.

Source - Plant Materials Center, SCS, Americus, GA; four lines from Terrell Co., GA, Barbour Co., AL, Sumter Co., GA, and Houston Co., AL.

Method of Breeding - Synthetic from cross of four lines.

Intended Use - Hay, pasture production.

Description - Produces more forage in piedmont/coastal plain of southeastern U.S. than other tested varieties. Currently being tested.

Adapted to - LRR P, T; PHZ 8.

Released - Anticipated release in 1994-96 by SCS and University of Georgia.

Breeder Seed/Stock - Plant Materials Center, SCS, Americus, GA.

Certified Seed/Stock - Not available.

Preparer/Additional Information - SCS, State Office, Robert G. Stephens Federal Bldg., Box 13, 355 E. Hancock Ave., Athens, GA 30601, (912) 706-2115.

Rumsey

Plant Materials Center, SCS, in cooperation with the Missouri AES - Billy Rountree. Carried as accession no. PI 315747.

Source - Collected from a native stand in Jefferson County, IL; evaluated, selected, and increased at the Plant Materials Center, SCS, Elsberry, MO.

Method of Breeding - Increase of field selection.

Intended Use - Pasture, hay production, and forage reseedings.

Description - Increased seedling growth rate, superior forage production, increased resistance to lodging. The flowering date of Rumsey is 10-20 days later than those cultivars evaluated. This late maturity will maintain forage quality later into the growing season.

Adapted to - LRR H, M, N, O; PHZ 4.

Released - 1983.

Breeder Seed/Stock - Plant Materials Center, SCS, Elsberry, MO.

Certified Seed/Stock - Available.

Preparer/Additional Information - Jimmy Henry, Plant Materials Center, SCS, RR 1, Box 9, Elsberry, MO 63343, (314) 898-2012.

Tejas

Selected at Texas Research Foundation, Renner, TX - E.O. Gangstad.

Source - Collections from Texas, Oklahoma, and New Mexico in 1954.

Method of Breeding - Selected for leafiness, forage yield, and seed yield in polycross nurseries planted in 1955, 1957, and 1959. In 1962, 10 highest yielding clones bulk planted and increased as experimental synthetic variety Tejas.

Intended Use - Range/pasture reseeding.

Description - Relatively uniform plant size; robust, vigorous, dark blue-green plant. Lemon-yellow seedhead, with dark awned seeds. Flowers in mid-October at Renner.

Adapted to - LRR H, J; PHZ 6, 7.

Released - No.

Breeder Seed/Stock - Texas Research Foundation.

Certified Seed/Stock - None.

Preparer/Additional Information - Texas A&M Univ., AES, College Station, TX 77843, (409) 845-3041.

Tomahawk (Reg. No. CU-131)

Selected at the Plant Materials Center, SCS, Bismarck, ND, and was developed and evaluated in cooperation with the ARS, Northern Great Plains Research Laboratory, Mandan, ND - John McDermand, Erling T. Jacobson, and Russell J. Haas. Carried as accession nos. PI 478006, PM-ND-444.

Source - Composite of three seed collections from native stands near Ludden, Dickey County, ND, and Britton, Marshall County, and Hecla, Brown County, SD.

Method of Breeding - Initial evaluation of the collections were made in comparison with 16 other accessions at the Bismarck Plant Materials Center. The three accessions were selected based on high seed yield and winter survival. Phenology, forage yield, and persistence have been extensively evaluated in comparative field evaluation studies and field plantings in North Dakota, South Dakota, and Minnesota.

Intended Use - Range and pasture seedings, wildlife habitat, natural areas, surface mine revegetation, and critical area seedings.

Description - Earlier maturity and superior winter-hardiness and persistence. At northern latitudes, forage production is similar to Holt and exceeds Oto and Rumsey. Matures 33 days earlier than Holt, 71 days earlier than Oto, and 82 days earlier than Osage and Rumsey. Chromosome number 2n=40.

Adapted to - LRR F, G, K; PHZ 3a - 4a.

Released - Cooperatively in 1988 by the SCS, Plant Materials Center, Bismarck, ND; ARS; and the North Dakota, South Dakota, and Minnesota Agricultural Experiment Stations.

Breeder Seed/Stock - ARS Northern Great Plains Research Laboratory, Mandan, ND.

Certified Seed/Stock - Available in limited quantity.

Preparer/Additional Information - Russell J. Haas, SCS, P.O. Box 1458, Bismarck, ND 58502, (701) 250-4425.

Spartina alterniflora **Loisel. - smooth cordgrass**

Smooth cordgrass is the dominant emergent macrophyte in many coastal salt marshes in the U.S. It occurs from Newfoundland to Texas, and is found almost without exception growing in the intertidal zone. It has been introduced on the Pacific coast in Washington through oyster cultures and appears to be spreading; it is also present along the eastern coast of South America and in Europe along the coasts of England and France. It is a perennial grass, spreading primarily by vegetative rhizomal reproduction, and producing a sparse crop of seed. Growth height varies with the species (0.5-2.5 m). Although forming gregarious stands that cover extensive areas, it has little value as forage because of its tough foliage. Its use in recent years had been to re-establish intertidal marshes along the Atlantic and Gulf coasts.

Smooth cordgrass is defined as a facultative halophyte; it is tolerant but not dependent on salt as a physiological requirement for growth and reproduction. Smooth cordgrass grows on a wide range of substrate textures from coarse sands to high organics. It is well adapted to anaerobic substrates because of its well developed aerenchyma tissue.

Bayshore

Source - Several culms from a single clone were collected along the Chesapeake Bay shore of Dorchester County, MD, by W. Curtis Sharp.

Method of Breeding - Vegetative reproduction of original planting stock.

Intended Use - Revegetate and stabilize tidal shorelines and restoration of saline and brackish wetlands.

Description - A salt-tolerant perennial grass that grows to height of 1.2 m. Creeping rhizomes with short internodes that produce extensive colonies. Bayshore is typical of the species but has superior rhizome growth, stem density, and foliage abundance.

Adapted to - LRR R, S, T; PHZ 6a - 9b.

Released - 1992, jointly by USDA-SCS, NJ AES, and Virginia Dept. of Conservation and Recreation.

Breeder Seed/Stock - SCS, Plant Materials Center, Cape May Court House, NJ, or its designee is solely responsible for breeders stock. This cultivar is maintained by vegetative means.

Certified Seed/Stock - Plants can be reproduced by commercial nurseries according to state seed and plant certification rules.

Preparer/Additional Information - D.W. Hamer, Cape May Plant Materials Center, SCS, 1536 Rte 9 N, Cape May Court House, NJ 08210, (609) 465-5901.

Vermilion

Golden Meadow Plant Materials Center, SCS, Golden Meadow, LA. Carried as accession no. 9054025.

Source - Accession material collected in Vermilion Parish, LA.

Method of Breeding - Vermilion was selected from an assembly of 90 accessions, thus representing an ecotypic selection. Vermilion is a vegetative release.

Intended Use - It has been shown to be an important plant in maintaining the stability of saltwater marshes, shorelines, canal banks, dunes. and other marsh-water interfaces, thus promoting the re-establishment of emergent wetland vegetation.

Description - Morphologically, it is fairly typical of the taxa *Spartina alterniflora* Loisel. There are some slight variations within the variety, such as stem color, but these conceivably can be attributed to ecotypic influence. Selection was based primarily on phenotypic expression of non-morphological characteristics, e.g., transplant survival, disease resistance, and rate and number of tiller production.

Adapted to - LRR P, T; PHZ 8, 9, 10.

Released - 1989, SCS, and the Louisiana State Univ. Agricultural Center.

Breeder Seed/Stock - A vegetative release. Foundation stock will be maintained by the Golden Meadow Plant Materials Center, Golden Meadow, LA.

Certified Seed/Stock - Not available.

Preparer/Additional Information - Michael Materne, SCS, P.O. Box 16030, University Station, Baton Rouge, LA 70893, (504) 389-0335; or Gary Fine, SCS, Plant Materials Center, P.O. Box 2202, Galliano, LA 70354, (504) 475-5280.

Spartina patens (Ait.) Muhl. - saltmeadow cordgrass

Saltmeadow cordgrass occurs along the coast of North America from Quebec to Florida and Texas and in some inland saline marshes in New York and Michigan. It is fine stem, and can reach a height of one meter; however, there is much growth variation within the species. Reproduction is primarily from strong rhizomes, although limited seed is produced. Its normal occurrence along the eastern and Gulf coasts is just above the normal high tide area, and occasionally occurs on dryer sand dunes. Its use is primarily with smooth cordgrass in coastal marsh restoration.

Avalon

SCS, and Cook College, Rutgers University - D.W. Hamer, C.R. Belcher, R.W. Duell. Carried as accession no. PI 421237.

Source - Two plants randomly selected by W. Curtis Sharp near Avalon, NJ, September 1965.

Method of Breeding - Vegetative reproduction of original planting stock.

Intended Use - The initial vegetation and restabilization of tidal shorelines above mean high tide.

Description - A perennial salt-tolerant grass that grows to 0.9 m tall. It has long slender rhizomes that account for most spreading. Avalon is typical for the species with medium-textured slender culms.

Adapted to - LRR R, S, T; PHZ 6a - 9b.

Released - 1986, jointly by SCS and NJ AES.

Breeder Seed/Stock - The cultivar is maintained only by vegetative means. The Plant Materials Center, SCS, Cape May Court House, NJ, or its designee is solely responsible for breeders stock.

Certified Seed/Stock - Plants can be reproduced by commercial nurseries according to State Seed and Plant Certification rules.

Preparer/Additional Information - D.W. Hamer, Plant Materials Center, SCS, 1536 Rte 9N, Cape May Court House, NJ 08210, (609) 465-5901.

Flageo

SCS, Plant Materials Centers, Americus, GA, and Brooksville, FL. Carried as accession no. PI 421238.

Source - SCS accession collected in sand-dune area near Maneto, NC, by Karl Graetz, SCS, in 1971.

Method of Breeding - Selected as one of 10 superior accessions from a 79-accession collection established at Plant Materials Center, SCS, Cape May Court House, NJ. Chosen as one of two superior lines and test planted along the Atlantic and Gulf coasts.

Intended Use - For soil erosion prevention on low dunes/high salt marsh along coastal areas.

Description - Ability to persist and spread at greater rates than other tested germplasm. Course stem, upright , more open sod developed than with Avalon. Vegetatively reproduced at present. Grows well with good survival in southern coastal areas, spreads well and forms a good rhizome network. Sister selection (PI 421237) released as cultivar Avalon, which is better adapted north of the Carolinas.

Adapted to - LRR U, T, P; PHZ 8.

Released - 1990, SCS, Plant Materials Centers, Americus, GA, and Brooksville, FL, and Fort Valley College, Fort Valley, GA.

Breeder Seed/Stock - Plant Materials Center, SCS, Americus, GA.

Certified Seed/Stock - Not available. Please contact: Donald Surrency, SCS, State Office, Federal Bldg., Box 13, 355 E. Hancock Ave., Athens, GA 30601.

Preparer/Additional Information - Charles M. Owsley, Plant Materials Center, SCS, 295 Morris Dr., Americus, GA 31709, (912) 924-2286.

PI 415141

Plant Materials Center, SCS, Brooksville, FL. Carried as accession no. 415141.

Source - SCS Accession collected at Abbeville, Vermillion Parish, LA 1972.

Method of Breeding - Selected as a superior accession from collection established at Americus, GA. Transferred to Brooksville, where this accession grew better than at Americus.

Intended Use - Vegetation of coastal salt-marsh areas.

Description - Vegetatively reproduced at present. Grows well with good survival both in coastal sand areas and inland sites. It spreads better than other accessions in these areas, and has a more open leaf blade than is typical of most other *S. patens.* May have potential for inland use on saline areas and oil exploration sites. More tolerant of extended periods of shallow standing water than some other *S. patens.* Responds well to foliar application of nutrient solutions.

Adapted to - LRR T, U; PHZ 8, 9, 10.

Released - May be released in future.

Breeder Seed/Stock - Plant Materials Center, SCS, Brooksville, FL.

Certified Seed/Stock - Not available.

Preparer/Additional Information - Plant Materials Center, SCS, 14119 Broad St., Brooksville, FL 34601, (904) 796-9600.

Sporobolus airoides (Torr.) Torr. - alkali sacaton

Native bunchgrass found from Washington and South Dakota south into Mexico. Abundant in parts of southwestern U.S. on lower alkaline flats; also grows on rocky soils and open plains; tillers extensively on moist sites. Grazes well during growing season; unpalatable when mature and not good as winter forage.

Salado

Selected at Plant Materials Center, SCS, Los Lunas, NM, and New Mexico AES, University Park, cooperating - G.C. Niner, W.R. Oaks, and J.E. Anderson. Carried as accession no. PI 15617.

Source - Collected from shallow upland range site in 1958, 12 miles south of Claunch, NM, at elevation of 1,770 m and precipitation about 300 mm annually.

Method of Breeding - Increase of original collection as NM-184.

Intended Use - Range reseeding and revegetation of disturbed areas.

Description - Similar in growth form to bottomland types. Seedling vigor and forage production good. Seed production average.

Adapted to - LRR D, G, H, E; PHZ 5.

Released - 1982, cooperatively by SCS - Los Lunas and New Mexico AES.

Breeder Seed/Stock - Plant Materials Center, SCS, Los Lunas, NM.

Certified Seed/Stock - Available.

Preparer/Additional Information - Plant Materials Center, SCS, 1036 Miller St. SW, Los Lunas, NM 87031, (505) 865-4684.

Saltalk

Selected in 1973 at the James E. "Bud" Smith Plant Materials Center, Knox City, TX, and increased for field planting evaluation as PMT-1733. Carried as accession no. PI 434445.

Source - Original seed collected from native stand near Erick, OK, in 1967 by SCS employees, Orvis Lowry and Burton Connally.

Method of Breeding - Selected from among 32 similar accessions for seed production, leafiness, and seedling vigor.

Intended Use - Erosion control, revegetation of saline and alkaline sites, and range reseeding.

Description - Germinates and establishes well in highly saline and alkaline soils of fine to moderately fine texture. Requires little maintenance except grazing management; highly palatable. Withstands flooding and considerable silt deposition.

Adapted to - LRR H, J; PHZ 6.

Released - 1981, Plant Materials Center, SCS, Knox City, TX, Texas AES, and Oklahoma AES.

Breeder Seed/Stock - Plant Materials Center, SCS, Knox City, TX.

Certified Seed/Stock - Not available.

Preparer/Additional Information - Plant Materials Center, SCS, Rte 1 Box 155, Knox City, TX 79529, (817) 658-3922.

Stenotaphrum secundatum (Walt.) Kuntze - St. Augustinegrass

Warm-season, sod-forming grass indigenous in West Indies and common in tropical Africa, Mexico, and Australia. Used as lawn grass from Florida to eastern Texas and for pasture on muck soils in southern Florida. Subject to winterkilling north of Augusta, GA, and Birmingham, AL. Grows best on relatively fertile, well-drained soils; tolerates shade. Subject to chinch bug damage. Propagated vegetatively. Texas common is used extensively on lawns in eastern half of Texas. Subject to severe damage by chinch bugs and brownpatch. A virus known as St. Augustinegrass decline is a major problem in lower Gulf coast area.

DelMar

Source - O.M. Scott & Sons breeding program.

Method of Breeding - Single plant F_1 progeny of Seville and a cold-tolerant selection from Tennessee.

Intended Use - Turf.

Description - Excellent cold tolerance; good resistance to St. Augustine decline virus and gray leaf spot; and good shade tolerance. Attractive turf. Diploid (2n=18).

Adapted to - Southern portion of PHZ 8 and warmer regions.

Released - 1986, O.M. Scott & Sons.

Breeder Seed/Stock - O.M. Scott & Sons Co.

Certified Seed/Stock - Not available.

Preparer/Additional Information - Virgil Meier, O.M. Scott & Sons Co., Marysville, OH 43041, (513) 644-0011.

Floralawn

Developed at University of Florida - A.E. Dudeck, J.A. Reinert, and P. Busey.

Source - FA-108, a probable seedling descendant of Roselawn. Male parent unknown.

Method of Breeding - Selected from seedlings based on resistance to the St. Augustine decline strain of panicum mosaic virus and resistance to the southern chinch bug, the latter a serious problem in Florida. Floralawn is propagated asexually by stolon cuttings.

Intended Use - Lawns, conservation areas, and other irrigated turf areas. Floralawn has poor shade tolerance, thus it may be inappropriate for mature landscapes.

Description - Leaf color blue-green, often with a pinkish color in the collars. Stolons relatively thick (>2.9 mm). Spikelets are long (>5.2 mm), thus Floralawn can usually be distinguished from Bitterblue. Characteristics are very similar to Floratam, except that Floralawn has a distinctive banding pattern for alcohol dehydrogenase. More resistant to atrazine and leaf spot disease than Bitterblue. The level of resistance to the southern chinch bug is often very strong, and insecticidal controls are usually not needed; however, a race of southern chinch bug is adapted to Floratam and Floralawn and has damaged both cultivars in the field. Floralawn is equal to Floratam in most respects; however, it has been observed to have better fall color retention in Florida.

Adapted to - LRR C, D, I, O, P, T, U, V; PHZ 9, 10; microenvironment and management can greatly affect zone of hardiness.

Released - 1985, Florida Foundation Seed, Inc.

Breeder Seed/Stock - Environmental Horticulture Department, University of Florida, Gainesville.

Certified Seed/Stock - Not available.

Preparer/Additional Information - Philip Busey, Fort Lauderdale Research and Education Center, Univ. of Florida, 3205 College Ave., Fort Lauderdale, FL 33314, (305) 475-8990.

Floratam

Developed at University of Florida and Texas A&M University, cooperating - G.C. Horn, A.E. Dudeck, and R.W. Toler.

Source - FA-110, a probable seedling descendant of Roselawn. Male parent unknown.

Method of Breeding - Selected from seedlings based on resistance to the St. Augustine decline strain of panicum mosaic virus. PMV-SAD had been observed in 1966 and 1967 in southern Texas, and was becoming a serious problem for Texas sod producers, who were thus anxious to produce a resistant grass. Floratam was also discovered to be resistant to the southern chinch bug, which had long been a serious problem in Florida. Floratam is propagated asexually by stolon cuttings.

Intended Use - Lawns, conservation areas, and other irrigated turf areas. Because of its poor shade tolerance, Floratam thins out and often performs poorly in mature landscapes.

Description - Leaf color blue-green, often with a pinkish color in the collars. Stolons relatively thick (>2.9 mm).

Spikelets are long (>5.2 mm), thus Floratam can usually be distinguished from Bitterblue, with which it is sometimes confused. More resistant to atrazine and gray leaf spot disease than Bitterblue. The level of resistance to the southern chinch bug is often very strong, and insecticidal controls are usually not needed; however, a race of southern chinch bug has overcome Floratam and damaged it in most counties of Florida. It was grown on more than 12,000 hectares for sod in Florida in the 1980s.

Adapted to - LRR C, D, I, O, P, T, U, V; PHZ 9, 10; microenvironment and management can greatly affect zone of hardiness.

Released - 1973, Florida Foundation Seed, Inc., and Texas A&M University.

Breeder Seed/Stock - Fort Lauderdale Research and Education Center, University of Florida.

Certified Seed/Stock - Not available.

Preparer/Additional Information - Philip Busey, Fort Lauderdale Research and Education Center, Univ. of Florida, 3205 College Avenue, Fort Lauderdale, FL 33314, (305) 475-8990.

Floratine

Selected at Florida AES, Gainesville - G.C. Nutter and R.J. Allen, Jr.

Source - Selected originally in 1948 and maintained at Belle Glade, FL; vegetative material moved to Gainesville in 1953 for evaluation as turfgrass.

Method of Breeding - Natural selection (probably a seedling from Bitter Blue).

Intended Use - Turf.

Description - Low-growing, fine textured, and attractive blue green. Stolons branch prolifically, producing dense turf of short (25-40 mm) and narrow (8 mm) leaves; average internode length 200 mm as compared with 50 mm for Bitter Blue and 75 mm for common. Average maximum unmowed height in test plots was 90 mm compared with 130 mm for Bitter Blue; tolerates close mowing and has survived at 12 mm. Rate of coverage and other characteristics similar to those of Bitterblue.

Adapted to - LRR U, T; PHZ 8, 9, 10.

Released - 1959, by Florida AES.

Breeder Seed/Stock - Florida AES.

Certified Seed/Stock - Not available. Commercially available.

Preparer/Additional Information - Environmental Horticultural Dept., 1545 Fifield Hall, Univ. of Florida, Gainesville, FL 22611, (904) 392-7938.

FX-10

Developed at Fort Lauderdale Research and Education Center, University of Florida - Philip Busey.

Source - F_2 hybrid produced in 1983 by B.J. Center; derived from four polyploid (2n=30) accessions from Africa. Parents had been introduced in 1963 and 1964 by W.W. Huffine and A.J. Oakes.

Method of Breeding - Over 100 F_1 and F_2 hybrids were produced through controlled crossing of polyploid African introductions. Hybrids evaluated in field plots for turf quality, disease resistance, and drought resistance. Irrigation was permanently curtailed in established plots, during the dry season of 1988. During the first six weeks, FX-10 suffered no loss of canopy, while other genotypes, such as Floratam, were severely damaged. Antibiosis screening of FX-10 revealed moderate resistance to the PDP (Polyploid Damaging Population) chinch bug, based on reduced oviposition rate and shortened longevity of bugs confined on FX-10. FX-10 displayed less damage from gray leaf spot than Bitterblue or Floratam. Distinctive esthetic and morphological features were also involved in the discovery of FX-10. Propagated asexually by stolon cuttings.

Intended Use - Lawns, conservation areas, and other turf areas receiving minimal or no irrigation. Appropriate for low-lying, humid coastal areas subject to seasonal drought, but where subsurface moisture is present.

Description - Superior drought resistance compared with Floratam and Floralawn. Roots extend to at least 1.4 m deep in sand soil in south Florida. Moderately resistant, based on laboratory antibiosis tests, to the PDP southern chinch bug, which destroys Floratam. FX-10 has many (about 20) small hairs (about 1 mm long) on the upper surface of the youngest leaf blades, concentrated toward the base of the leaf blade; sparser hairs are present on the lower leaf blade surface. FX-10 can usually be spotted in the field because its leaf color is more bluish gray than other St. Augustinegrass varieties. Several other characteristics, such as chromosome number (2n=30), spikelet length (about 4.5 mm), and length of the flowering region on the inflorescence (about 75 mm) are diagnostic for FX-10 when used in combination. The unmown height of FX-10 is considerably shorter than Floratam. More susceptible to atrazine in sand soil than Floratam. Less susceptible to leaf spot disease than Floratam, Bitterblue, or FX-33. Generally slower in growth rate than Floratam. Performs poorly in dense (>=95%) shade.

Adapted to - LRR C, D, I, O, P, T, U, V; PHZ 8 (in part), 9, 10; microenvironment and management can greatly affect zone of hardiness.

Released - 1990 and licensed exclusively to the Florida Sod Growers Cooperative, Box 874, LaBelle, FL 33935. U.S. Plant Patent 7852 has been issued for FX-10. Crop Reg. CV-153.

Breeder Seed/Stock - Fort Lauderdale Research and Education Center, University of Florida. Subject to proprietary restrictions.

Certified Seed/Stock - Not available.

Preparer/Additional Information - Philip Busey, Fort Lauderdale Research and Education Center, Univ. of Florida, 3205 College Ave., Fort Lauderdale, FL 33314, (305) 475-8990.

FX-33

Selected at Fort Lauderdale Research and Education Center, University of Florida - Philip Busey.

Source - F_2 hybrid produced in 1983, B.J. Center; derived from four polyploid (2n=30) accessions from Africa. Parents had been introduced in 1963 and 1964 by W.W. Huffine and A.J. Oakes.

Method of Breeding - Over 100 F_1 and F_2 hybrids were produced through controlled crossing of polyploid African introductions. Hybrids were evaluated in field plots for turf quality, disease resistance, and drought resistance. Irrigation was permanently curtailed in established plots during the dry season of 1988. FX-33 suffered no loss of canopy, while other genotypes such as Floratam were severely damaged. Antibiosis screening of FX-33 revealed resistance to two southern chinch bug populations, based on reduced oviposition rate and shortened longevity of bugs confined on FX-33.

Intended Use - Lawns, conservation areas, and other turf areas receiving minimal or no irrigation. Appropriate for low-lying, humid coastal areas subject to seasonal drought, but where subsurface moisture is present.

Description - Superior drought resistance compared with Floratam and Floralawn. Roots extend to at least 1.4 m deep in sand soil in south Florida. Resistant, based on laboratory antibiosis tests, to the PDP southern chinch bug, which destroys Floratam. Abundant hairs present on the upper and lower surfaces of the blades of young leaves, by which FX-33 can always be distinguished in the field from all other cultivars of St. Augustinegrass. Chromosome number 2n=30. More susceptible to atrazine in sand soil than Floratam or Bitterblue. Prone to an unidentified leaf spot disease, typified by small, pin-head-sized spots.

Adapted to - PHZ 9.

Released - 1990, licensed exclusively to the Florida Sod Growers Cooperative, Box 874, LaBelle, FL 33935. U.S. Plant Patent 7699 has been issued for FX-33.

Breeder Seed/Stock - Fort Lauderdale Research and Education Center, University of Florida.

Certified Seed/Stock - Not available.

Preparer/Additional Information - Philip Busey, Fort Lauderdale Research and Education Center, Univ. of Florida, 3205 College Ave., Fort Lauderdale, FL 33314, (305) 475-8990.

FX-261

Selected at Fort Lauderdale Research and Education Center, University of Florida - Philip Busey. Carried as Experimental No. FX-261.

Source - Pedigreed hybrid produced in 1984 by B.J. Center from diploid (2n=18) accessions of the Breviflorus Race (Gulf Coast and Dwarf).

Method of Breeding - Thousands of hybrids and selfed progeny were created and evaluated in field plots from 1978 through 1986. Parents were recurrently selected for high turfgrass quality (density, color, fine texture, and dwarf habit) and coverage under conditions of weed competition and minimal pesticide use. FX-261 was recognized as the most well-adapted dwarf genotype. It was tested as an asexually propagated clone.

Intended Use - High-visibility lawns and other turf areas receiving supplemental irrigation. The coarse texture and tall habit of most St. Augustinegrass genotypes is often considered aesthetically objectionable. Because it is lower growing and finer textured, FX-261 conveys a more uniform appearance than Floratam and some other cultivars. When left unmown for several weeks, FX-261 reaches only 41% of the height of Floratam.

Description - Finer texture and shorter leaves than Bitterblue, Floratam, Florida Common, and Seville. Reduced seedhead abundance and higher quality ratings than Seville when maintained under low (10 g N m-2 y-1) fertilization. Distinctive because of short (<50 mm) floral regions and mauve anthers. More prone to wilt during periods of irrigation curtailment than Floratam or Bitterblue. Although FX-261 can be damaged by the southern chinch bug, the incidence of damage is less than for Florida Common and Bitterblue.

Adapted to - PHZ 9.

Released - Proposed for commercial release. Limited experimental distribution to sod growers beginning in 1987.

Breeder Seed/Stock - Fort Lauderdale Research and Education Center, University of Florida.

Certified Seed/Stock - Not available.

Preparer/Additional Information - Philip Busey, Fort Lauderdale Research and Education Center, Univ. of Florida, 3205 College Ave., Fort Lauderdale, FL 33314, (305) 475-8990

Jade

Source - O.M. Scott & Sons breeding program.

Method of Breeding - Single-plant F_1 progeny of cold-tolerant selection from Tennessee and Seville.

Intended Use - Turf.

Description - Very good turf performance, good cold tolerance, short internodes. Diploid (2n=18).

Adapted to - PHZ 9, 10, 11.

Released - 1988 by O.M. Scott & Sons Co.

Breeder Seed/Stock - O.M. Scott & Sons Co.

Certified Seed/Stock - Not available.

Preparer/Additional Information - Virgil Meier, O.M. Scott & Sons Co., Marysville, OH 43041, (513) 644-0011.

Raleigh

Source - Yard in Raleigh, NC.

Method of Breeding - Five years of individual plant selection for winter-hardiness under minimum culture in field plots. In sixth year combined 10 best plants for increase.

Intended Use - Lawn turf.

Description - Cold-hardy to -12°C in field. Resistant to SAD virus. Quite shade tolerant.

Released - 1983 by North Carolina State Univ. Experiment Station.

Breeder Seed/Stock - North Carolina State Univ. Turf Program.

Certified Seed/Stock - Available as sod.

Preparer/Additional Information - Dr. Arthur H. Bruneau, Box 7620, North Carolina State Univ., Raleigh, NC 27695-7103, (919) 515-5854.

Seville

Source - O.M. Scott & Sons breeding program.

Method of Breeding - Single-plant progeny of 1081.

Intended Use - Turf.

Description - Excellent turf performance and semi-dwarf growth habit. Diploid (2n=18).

Adapted to - PHZ 9, 10, 11.

Released - 1978 by Pursley, Inc.

Breeder Seed/Stock - O.M. Scott & Sons Co.

Certified Seed/Stock - Not available.

Preparer/Additional Information - Virgil Meier, O.M. Scott & Sons Co., Marysville, OH 43041, (513) 644-0011.

Sunclipse

Source - O.M. Scott & Sons breeding program.

Method of Breeding - Single-plant F_1 progeny of an unreleased plant from Florida and a cold-tolerant selection from Tennessee.

Intended Use - Turf.

Description - Good turf performance, especially well adapted to southern California climate, short internodes. Diploid (2n=18).

Adapted to - PHZ 9, 10, 11.

Released - 1988, Pacific Sod Co.

Breeder Seed/Stock - O.M. Scott & Sons Co.

Certified Seed/Stock - Not available.

Preparer/Additional Information - Virgil Meier, O.M. Scott & Sons Co., Marysville, OH 43041, (513) 644-0011.

Stipa nelsonii **Scribn.**

Stipa columbiana **auct. non Macoun - Columbia needlegrass**

Culms mostly 30-60 cm tall, sometimes as much as 1 m. Dry plains, meadows, and open woods, at medium and high altitudes, South Dakota to Yukon Territory, south to Texas and California.

9040039

Selected for advanced testing by the Upper Colorado Environmental Plant Center (UCEPC), Meeker, CO. Carried as accession no. 9040039.

Source - Collected by Dave Bearden, August 1980, Dry Fork, Rio Blanco County, CO.

Method of Breeding - Increased from original collection and rogued by the UCEPC. Very likely this accession will be blended with 9040047 for the release. Good winter-hardiness; close grazing and drought may reduce stand.

Intended Use - Revegetation of mined land, transmission corridors, roadsides, and as a component in rangeland seedings.

Description - Good forage and seed production. Good regrowth and fair drought tolerance. Good seedling vigor and fair establish-ability.

Adapted to - LRR D, E; PHZ 5.

Released - No. (Distributed for field testing.)

Breeder Seed/Stock - Breeders (original) seed will be maintained at the Upper Colorado Environmental Plant Center.

Certified Seed/Stock - Not available.

Preparer/Additional Information - Randy Mandel, Upper Colorado Environmental Plant Center, P.O. Box 448, Meeker, CO 81641, (303) 878-5003.

9040047

Selected for advance testing by the Upper Colorado Environmental Plant Center, Meeker, CO. Carried as accession no. 9040047.

Source - Collected by Charles Holcomb and H. Hodgkinson, August 1980, northwest of Gateway, Mesa County, CO.

Method of Breeding - Increased from original collection and rogued by the UCEPC. Very likely this accession will be blended with 9040039 for the release. Good winter-hardiness; close grazing and drought may reduce stands.

Intended Use - Revegetation of mined land, transmission corridors, roadsides, and as a component in a rangeland seeding.

Description - Good forage and seed production. Good regrowth and fair drought tolerance. Good seedling vigor and fair establish-ability.

Adapted to - LRR D, E; PHZ 5.

Released - No. (Distributed for field testing.)

Breeder Seed/Stock - Breeders (original) seed will be maintained at the Upper Colorado Environmental Plant Center, Meeker, CO.

Certified Seed/Stock - Not available.

Preparer/Additional Information - Randy Mandel, Upper Colorado Environmental Plant Center, P.O. Box 448, Meeker, CO 81641, (303) 878-5003.

9040137

Selected for advance testing by the Upper Colorado Environmental Plant Center, Meeker, CO. Carried as accession no. 9040137.

Source - Collected by Glenn Niner, August 1975, in the San Luis Valley area, Rio Grande County, CO.

Method of Breeding - Increased from original collection and rogued by the UCEPC.

Intended Use - Revegetation of mined land, transmission corridors, roadsides, and as a component in rangeland seeding.

Description - Drought tolerance, fair seed with excellent seed conditioning characteristics. Fair forage and regrowth.

Adapted to - LRR D, E; PHZ 5.

Released - No. (Distributed for field testing.)

Breeder Seed/Stock - Original seed will be maintained at the Upper Colorado Environmental Plant Center.

Certified Seed/Stock - Not available. Good winter-hardiness; close grazing and drought may reduce stand.

Preparer/Additional Information - Randy Mandel, Upper Colorado Environmental Plant Center, P.O. Box 448, Meeker, CO 81641, (303) 878-5003.

X Stiporyzopsis caduca (Beol) B.L. Johnson & Rogler *[Stipa viridula X Oryzopsis hymenoides]* - mandan ricegrass

Mandan

Selected at ARS, Northern Great Plains Research Laboratory, Mandan, ND - G.A. Rogler.

Source - Amphidiploid of natural cross of *Stipa viridula* Trin. and *Oryzopsis hymenoides* (Roemer & J.A. Schultes) Richer ex Piper. F_1 hybrid occurred in nursery at Mandan in 1941; fertile F_2 plant found in 1945.

Method of Breeding - Natural intergeneric hybridization followed by spontaneous chromosome doubling.

Intended Use - Conservation grass for semiarid sites.

Description - Morphologically intermediate between two parent species, with growth habit more closely approaching that of *Oryzopsis hymenoides.*

Adapted to - LRR B, E, G; PHZ 3, 4.

Released - No. Distributed for testing.

Breeder Seed/Stock - USDA-ARS Northern Great Plains Research Laboratory, Mandan, ND.

Certified Seed/Stock - Not available.

Preparer/Additional Information - John D. Berdahl, ARS, Northern Great Plains Research Laboratory, P.O. Box 459, Mandan, ND 58554, (701) 663-6445.

Tripsacum dactyloides (L.) L. - eastern gamagrass

Warm-season, native sod-forming grass with thick, scaly rhizomes. Found in natural grassland prairies of central and eastern U.S. along streambanks and other lowland sites. Grows in large clumps; must be properly managed for hay or pasture production. Highly nutritious forage.

Pete

KSU and SCS, Manhattan, KS - K.L. Anderson, and R.D. Lippert. Carried as accession no. PM-K-24.

Source - Collections in 1958 from Kansas and Oklahoma.

Method of Breeding - Seed from 70 collections bulked and used to establish seed-increase field in 1960. Seed from this field advanced two generations by harvesting and replanting. Natural selection under combine harvesting for more uniform maturity.

Intended Use - Hay production, warm-season pasture, wildlife food and cover, and re-establishment of native prairie.

Description - Well-adapted, leafy ecotype; typical in general appearance to species.

Adapted to - LRR H, M, N, O, P, R; PHZ 5b.

Released - 1988, Kansas AES, Plant Materials Center, SCS, Manhattan, KS, Southern Plains Range Research Station, ARS, Woodward, OK AES, Stillwater, OK.

Breeder Seed/Stock - Plant Materials Center, SCS, Manhattan, KS.

Certified Seed/Stock - Available.

Preparer/Additional Information - SCS, 760 S. Broadway, Salina, KS 67401, (913) 823-4541.

Vulpia myuros (L.) K.C. Gmel - foxtail fescue

***Festuca megalura* Nutt.**
***Festuca myuros* L.**
***Vulpia megalura* (Nutt.) Rydb.**

Cool-season, low-growing annual grass. Ubiquitous in arid, mild-winter areas of Europe and North and South America. Valuable as an erosion control plant in desert and Mediterranean-like climate zones. Relatively low fire hazard and low competition with associated species in seeding mixtures.

Zorro

Selected at Plant Materials Center, SCS, Lockeford, CA - G. Edmonson, R. Clary, K. Croeni and R. Slayback. Carried as accession nos. PI 421020, PL-109-71.

Source - Collected from native plants on Columbia fine sandy loam, Plant Materials Center, SCS, San Joaquin County, CA, on June 8, 1971.

Method of Breeding - Original seed increased and tested with other annual grasses for critical area stabilization.

Intended Use - Cover crop, revegetation of disturbed areas, acid mine spoils, and revegetation of wildfire burns.

Description - Aggressive, early-maturing, winter-growing annual grass. Excellent seedling vigor; has shown superior seedling establishment to *Bromus hordeaceus* L. ssp. *hordeaceus* [synonym = *B. mollis* auct. non L.] on infertile, shallow, or droughty soils. Matures seed about two weeks earlier than Blando brome and assures perpetuation in low rainfall areas.

Adapted to - LRR A, C, D; PHZ 8b, 9.

Released - 1977, cooperatively by California AES, Davis and Plant Materials Center, SCS, Lockeford, CA.

Breeder Seed/Stock - Plant Materials Center, SCS, Lockeford, CA.

Certified Seed/Stock - Available.

Preparer/Additional Information - Robert Slayback, SCS, 2121-C, 2nd St., Davis, CA 95616, (916) 757-8257.

Zoysia japonica Steud. - Japanese lawngrass

Warm-season, sod-forming grass from Asia. Used for general-purpose turf and erosion control in southeastern U.S. Relatively winter-hardy, but does not thrive or compete well where summers are short or cool. Grows best on heavy soils; not drought resistant. Relatively coarse and tough. Produces seed, but generally planted vegetatively.

Belair (Reg. No. 104)

ARS, Beltsville, MD - J.J. Murray and N.R. O'Neill. Carried as accession no. R52-25.

Source - Unknown, but most likely North Korea.

Method of Breeding - Single-plant selection from F_1 polycross progeny of a promising vegetative selection from an old turfgrass nursery at the Beltsville Agricultural Research Center.

Intended Use - Turf use on lawns, parks, athletic fields, and golf courses.

Description - Established turf is medium coarse in texture, less dense than Meyer and other fine-leafed zoysiagrasses, and has a medium-dark green color during the growing season. In several locations, Belair has been superior to Meyer in rate-of-spread, zoysia rust resistance, drought tolerance, fall color retention, and early spring growth. It is easier to mow with homeowners' rotary-type lawnmowers than other zoysiagrass cultivars and is less prone to thatching. At present, Belair is the only commercially available zoysiagrass cultivar that exhibits a green- or white-colored rhizome as opposed to red- or purple-colored rhizomes on other cultivars.

Adapted to - LRR J, M, N, P; PHZ 6.

Released - 1987 by ARS, Beltsville, MD.

Breeder Seed/Stock - Vegetative breeders material available only from ARS, Beltsville, MD. Belair is highly self-sterile and therefore not released as seeded cultivar.

Certified Seed/Stock - Not available.

Preparer/Additional Information - Kevin N. Morris, National Turfgrass Evaluation Program BARC-West, Bldg. 001 Room 3233, Beltsville, MD 20702, (301) 504-9125.

Cashmere

Developed by Pursley Turf Farms.

Source - Coastal Florida.

Method of Breeding - Discovered variety.

Intended Use - Turf.

Description - Fine leaved, rapid growth laterally. Tolerates low mowing height and has fair shade tolerance.

Adapted to - LRR T, U; PHZ 9, 10, 11.

Released - Pursley Turf Farms.

Breeder Seed/Stock - Pursley Turf Farms.

Certified Seed/Stock - Not available.

Preparer/Additional Information - Pursley Turf Farms, P.O. Box 1448, Palmetto, FL 34220, (813) 822-4919.

FLR-800

Source - Germplasm from Belair, Meyer, Illiasis, and the Republic of South Korea.

Method of Breeding - First-generation seed produced by random crossing of two vegetatively propagated clones of *Zoysia japonica* Steud.

Intended Use - Low to moderately maintained turfgrass areas.

Description - Two-clone synthetic producing plants with purple stems and flowers. Medium leaf texture. Faster rate of spread than Meyer or Belair with less density and thatch. Rust resistance is good and better cold-hardiness than Meyer. Better adapted to seeding with or overseeding with cool-season grasses, especially tall fescues, than Meyer.

Adapted to - PHZ 4, 6, 7, and upper part 8.

Released - Expected in 1994.

Breeder Seed/Stock - Fine Lawn Research, Inc.

Certified Seed/Stock - Not available.

Preparer/Additional Information - David Lundell, Fine Lawn Research, Inc., P.O. Box 1051, Lake Oswego, OR 97034, (503) 636-2600.

FLR-900

Source - Germplasm from Belair, Meyer, Illiasis, and the Republic of South Korea.

Method of Breeding - First-generation seed produced by random crossing of two vegetatively propagated clones of Zoysia japonica Steud.

Intended Use - Low to moderately maintained turfgrass areas.

Description - Two-clone synthetic producing plants with yellow-green stems and flowers. Medium leaf texture (wider than Meyer and thinner than Belair). Less density and thatch, and easier to mow than Meyer, but with similar cold-hardiness. Good rust resistance. Requires lower maintenance than Meyer to provide acceptable turf quality for golf course fairways, homelawns, and institutional grounds in most areas.

Adapted to - PHZ 4, 6, 7, and upper part 8.

Released - Expected in 1994.

Breeder Seed/Stock - Fine Lawn Research, Inc.

Certified Seed/Stock - Not available.

Preparer/Additional Information - David Lundell, Fine Lawn Research, Inc., P.O. Box 1051, Lake Oswego, OR 97034, (503) 636-2600.

FZ-26

Selected at Fort Lauderdale Research and Education Center, University of Florida - Philip Busey.

Source - Vegetative clone presented in 1916 by J.C. Koningsberger, Bogor Botanic Gardens, Java, Indonesia. Introduced as PI 42839.

Method of Breeding - Two cycles of clonal selection were performed in field plots which were severely infested with the sting nematode. FZ-26 was selected based on superior adaptive ratings compared to Emerald and Meyer in replicated plots in Broward, Collier, and Palm Beach Counties, Florida. It also persisted and expanded from an abandoned plot which had irrigation and fertilization withheld for five years.

Intended Use - Lawns and other turf areas. May be suitable for use in subtropical areas such as Florida, where other cultivars are poorly adapted.

Description - Leaf blades flat, sparsely pubescent on the upper surface, and relatively soft compared with other zoysiagrass genotypes, 3.5-4.5 mm wide and about 60 mm long. Faint reddish pigmentation in rhizome internodes and prophylls. Second glume mucronate, anthers purple and stigmas white. Similar to Meyer in overall appearance. Open-pollinated progeny vary in stem color, thus FZ-26 must be propagated asexually.

Adapted to - LRR N, P, M, J; PHZ 7.

Released - Available for experimental use.

Breeder Seed/Stock - Florida Research and Education Center, University of Florida.

Certified Seed/Stock - Not available.

Preparer/Additional Information - Philip Busey, Fort Lauderdale Research and Education Center, Univ. of Florida, 3205 College Ave., Ft Lauderdale, FL 33314, (305) 475-8990.

Meyer (Reg. No. 12)

Selected at Arlington Farms, VA, and Plant Industry Station, Beltsville, MD, in cooperation with U.S. Golf Association Green Section - Ian Forbes, M.H. Ferguson, and F.V. Grau.

Source - Japanese lawngrass seed introduced in 1930 from northern Korea. *Z. japonica* known to have been present in U.S. in 1895.

Method of Breeding - Promising individual plant selected at Arlington Farms in 1940. Vegetative material moved to Beltsville in 1941; increased for testing in 1947-48 as Z-52.

Intended Use - Lawns and golf courses.

Description - Develops tough, wear-resistant turf. Leaf width intermediate between that of *Z. matrella* (L.) Merr. and common *Z. japonica.* Drought resistant, but will turn brown during long, dry periods. Grows and persists on relatively poor soils. Rate of spread and color improved by applications of fertilizer and irrigation. Competes very satisfactorily with weeds and other grasses in areas where adapted. Competition from other species increases time required to attain complete coverage. Winter-hardy, but in general only recommended in areas with long, warm growing season. Warm-season grass; becomes dormant and brown with first frost.

Adapted to - PHZ 6, 7, 8, 9, 10, 11 (mid-Atlantic and southeastern states, central to lower Midwest and the southwestern U.S.).

Released - 1951, cooperatively by Plant Science Research Division, ARS, and U.S. Golf Association Green Section. The name, Meyer, honors memory of Frank N. Meyer, USDA plant explorer.

Breeder Seed/Stock - Plant Industry Station, Beltsville.

Certified Seed/Stock - Available.

Preparer/Additional Information - Kevin Morris, National Turfgrass Evaluation Program, USDA Turf, BARC-West, Bldg. 001, Rm. 333, Beltsville, MD 20705, (301) 504-5125.

SR 9000

Developed by Turfgrass Germplasm Services - Jack Murray.

Source - Zoysia japonica collection from old turf.

Method of Breeding - Two clones identified with similar growth form, seed color, high seed production, and uniform progeny. These are vegetatively multiplied into production fields, allowed to intercross, and the harvested seed represents the variety.

Intended Use - Home lawns, golf course fairways and roughs, sports fields. Can be blended with turf-type tall fescues.

Description - Seeded zoysiagrass that is more uniform than Korean common and with texture and density making it compatible with turf-type tall fescues. Capable of rapidly forming a uniform, low-maintenance turf.

Adapted to - LRR G, H, I, J, N, O, P, S, T, U; PHZ 6, 7, 8, 9, 10.

Released - 1994 by Seed Research of Oregon, Inc., Corvallis.

Breeder Seed/Stock - Jack Murray, Turfgrass Germplam Services and Seed Research of Oregon.

Certified Seed/Stock - Available.

Preparer/Additional Information - Dr. Leah A. Brilman, Seed Research of Oregon, Inc., P.O. Box 1416, Corvallis, OR 97339, (503) 757-2663.

SR 9100

Developed by Turfgrass Germplasm Services - Jack Murray.

Source - Zoysia japonica collections from old turf sites.

Method of Breeding - Two clones identified with growth form and color similar to Meyer and the same seed color, high number of seedheads, and uniform progeny. These are vegetatively multiplied in production fields, allowed to intercross, and the harvested seed represents the variety.

Intended Use - Home lawns, golf course fairways and tees, sports fields.

Description - Seeded zoysia that is similar to the vegetative variety Meyer, but can be established more quickly by seed. Can form a dense, fairly low-growing, medium textured turf.

Adapted to - LRR G, H, I, J, N, O, P, S, T, U; PHZ 6, 7, 8, 9, 10.

Released - 1994 by Seed Research of Oregon, Inc., Corvallis.

Breeder Seed/Stock - Jack Murray, Turfgrass Germplasm Service and Seed Research of Oregon.

Certified Seed/Stock - Available.

Preparer/Additional Information - Dr. Leah A. Brilman, Seed Research of Oregon, Inc., P.O. Box 1416, Corvallis, OR 97339, (503) 757-2663.

Sunrise Brand

Source - Chinese common.

Method of Breeding - None.

Intended Use - Turf.

Description - Common type characteristics. Treated for improved germination.

Adapted to - Entire U.S.

Released - 1991, Jacklin Seed Co. and International Seeds, Inc.

Breeder Seed/Stock - None.

Certified Seed/Stock - Not available.

Preparer/Additional Information - Kim Peterson, Jacklin Seed Co., W. 5300 Riverbend Ave., Post Falls, ID 83854, (208) 773-7581.

Zoysia japonica **Steud. *X Zoysia matrella*** **(L.) Merr. var. *tenuifolia*** **(Willd. ex Thiele) Sasaki - Zoysia hybrids**

Emerald (Reg. No. 7)

Selected at Plant Industry Station, Beltsville, MD - Ian Forbes.

Source - Selected from several F_1 hybrids between *Zoysia matrella* (L.) Merr. var. *tenuifolia* (Willd. ex Thiele) Sasaki and *Z. japonica* Steud. *Z. japonica* parent introduced from Korea and *Z. matrella* var. *tenuifolia* parent from AES at Guam.

Method of Breeding - Hybrids made in all possible combinations between *Z. japonica* and *Z. matrella* var. *tenuifolia.* Selection in F_1 based on turf quality (leaf width, density, color, and growth habit) and winter-hardiness. Tested as experimental 34-35.

Intended Use - Lawns and golf courses.

Description - Vegetatively propagated F_1 hybrid *(Z. matrella* var. *japonica X Z. matrella* var. *tenuifolia).* In comparison with *Z. japonica, Z. matrella, Z.* var. *tenuifolia,* and Meyer zoysia at Beltsville and Tifton, GA, Emerald had the best total turf-quality score at both locations for three years. Combined to varying degrees of greater winter-hardiness, nonfluffy growth habit, and faster rate of spread of its *Z. japonica* parent with finer leaves, denser turf, and dark-green color of its *Z. matrella* var. *tenuifolia* parent. Exhibited hybrid vigor in rate of spread, browning, and density ratings. Considerably more shade and frost tolerant than bermudagrass.

Adapted to - PHZ 7, 8.

Released - 1955, cooperatively by Georgia AES, Tifton; Plant Science Research Division, ARS: and U. S. Golf Association Green Section.

Breeder Seed/Stock - Not available.

Certified Seed/Stock - Available in quantity.

Preparer/Additional Information - ARS Coastal Plain Experiment Station, P.O. Box 748, Tifton, GA 31793, (912) 386-3353.

Zoysia matrella **(L.) Merr. - manilagrass**

Warm-season, sod-forming grass introduced from Asia. Used as lawngrass in southeastern U.S.. Finer, denser sod, but less winter-hardy than *Zoysia japonica* Steud.

Matrella

Increased at Alabama AES, Auburn.

Source - Received from H.N. Vinall in 1927. F.C. 13521 obtained originally from J.B. Norton, Hartsville, SC; probably selection from PI 48574 obtained from the ARS, Southern Plant Introduction Station.

Method of Breeding - Increase of F.C. 13521.

Intended Use - Turf.

Description - Fine, dark green leaf blades usually 75-125 mm long when not mowed. Grows very dense. Produces creeping stolons that root profusely; ends of stolons cling to ground and thus grow under competing plants. Stands considerable shade. Produces seedheads and some seed in spring. Rather free from diseases and insects. Susceptible to drought, but recovers rapidly when moisture becomes available.

Released - Alabama AES.

Adapted to - PHZ 7, 8, 9, 10.

Breeder Seed/Stock - Vegetatively propagated.

Certified Seed/Stock - Available.

Preparer/Additional Information - Dr. Ray Dickens, AL AES, Auburn Univ., Auburn, AL 36849, (205) 844-4000.

Bibliography

Hanson, A.A. 1972. Grass varieties of the United States. Agriculture Handbook No. 170. USDA, Agricultural Research Service, Beltsville, MD.

Holliday, P. 1990. A Dictionary of Plant Pathology. Cambridge Univ. Press, London, UK

Horst, R.K. 1979. Westcott Plant Diseases Handbook, 4th ed. Van Nostrand Reinhold Co., New York.

Soil Conservation Service 1992. Plant List of Accepted Nomenclature, Taxonomy, & Symbols (PLANTS). USDA, SCS, National Plant Materials Center, Beltsville, MD.

Appendix A: USDA Plant Hardiness Zone Map

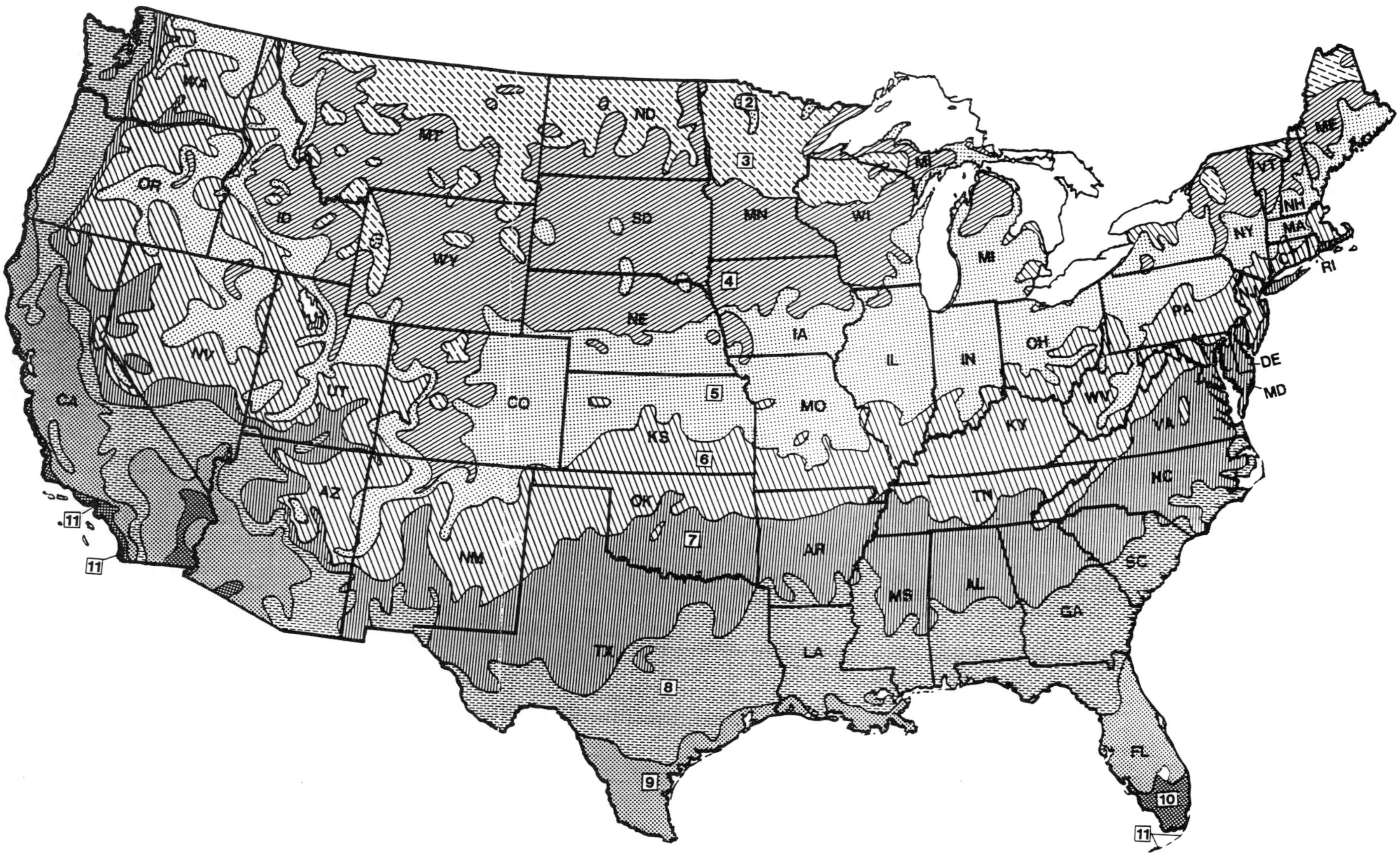

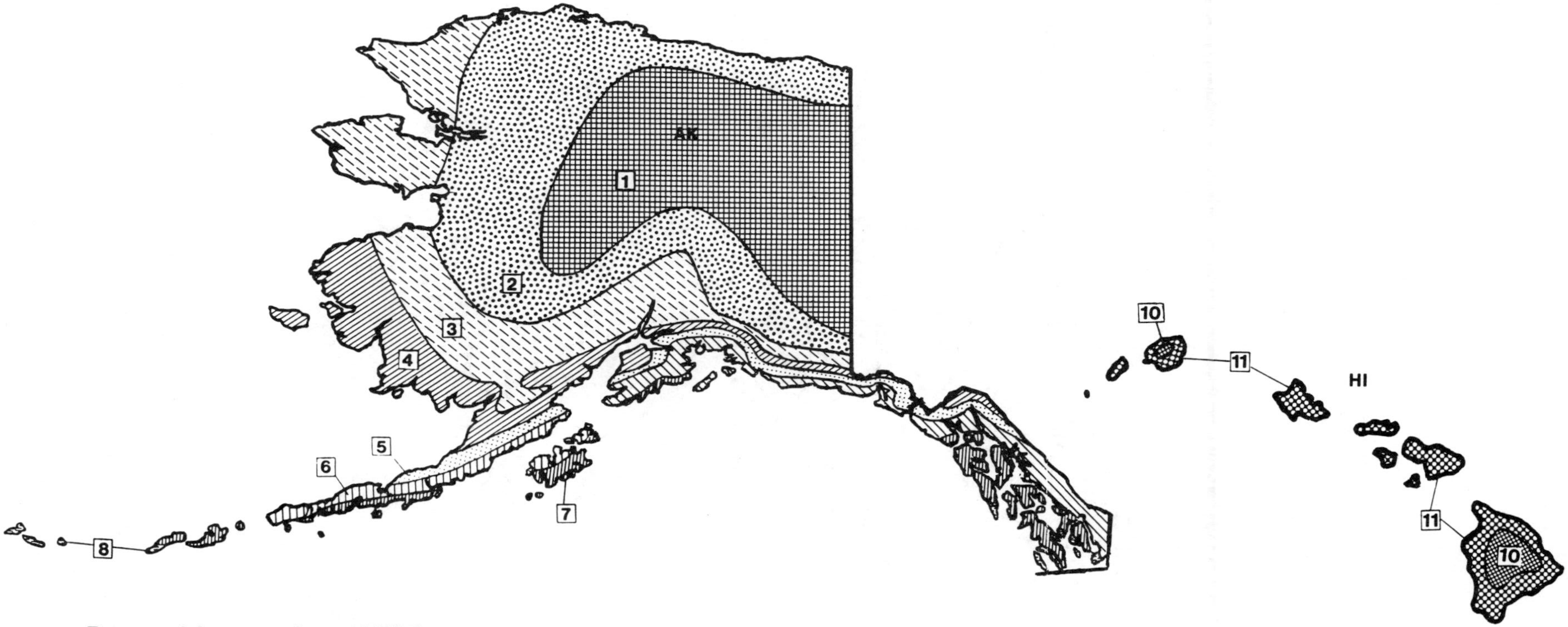

Range of Average Annual Minimum Temperature for Each Zone

Zone 1	Below -50° F
Zone 2	-50° to -40°
Zone 3	-40° to -30°
Zone 4	-30° to -20°
Zone 5	-20° to -10°
Zone 6	-10° to 0°
Zone 7	0° to 10°
Zone 8	10° to 20°
Zone 9	20° to 30°
Zone 10	30° to 40°
Zone 11	Above 40°

Appendix A: Land Resource Regions of the United States

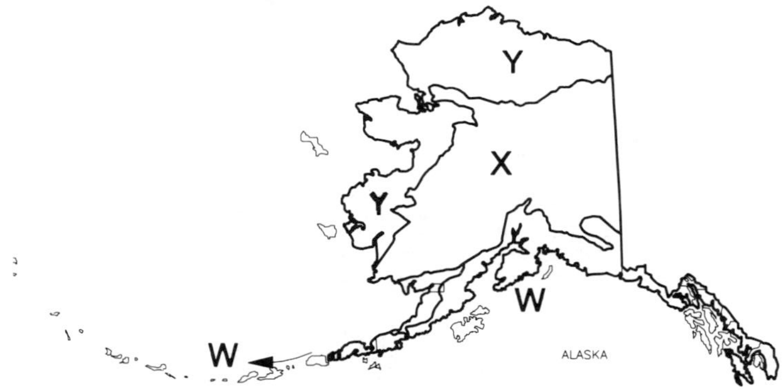

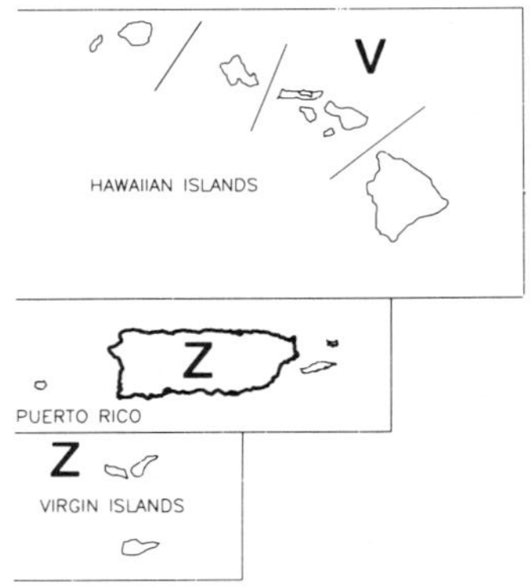

Legend for Map of Land Resource Regions

A NW Forest, Forage and Specialty Crops
B NW Wheat and Range
C California Subtropical Fruit, Truck, and Speciality Crops
D Western Range and Irrigated
E Rocky Mountain Range and Forest
F Northern Great Plains Spring Wheat
G Western Great Plains Range and Irrigated
H Central Great Plains Winter Wheat and Range
I SW Plateaus and Plains Range and Cotton
J SW Prairies Cotton and Forage
K Northern Lake States Forest and Forage
L Lake States Fruit, Truck, and Dairy
M Central Feed Grains and Livestock
N East and Central Farming and Forest
O Mississippi Delta Cotton and Feed Grains
P S. Atlantic and Gulf Slope Cash Crops, Forest, and Livestock
R Northeastern Forage and Forest
S Northern Atlantic Slope Diversified Farming
T Atlantic and Gulf Coast Lowland Forest and Crop
U Florida Subtropical Fruit, Truck Crop, and Range
V Hawaii
W Southern Alaska
X Interior Alaska
Y Arctic and Western Alaska
Z Caribbean Area

Appendix B: Abbreviations Used

The following abbreviations are used in *Grass Varieties in the United States:*

Abbreviation	Full Name
AES	Agriculture Experiment Station
ARS	Agricultural Research Service
BARC	Beltsville Agricultural Research Center
NTEP	National Turfgrass Evaluation Program
PI	Plant Introduction
PVP	Plant Variety Protection
SCS	Soil Conservation Service
U.S.	United States
USDA-ARS	Agricultural Research Service
USDA-FS	Forest Service
USDA-SCS	Soil Conservation Service
U.S. NPGS	National Plant Germplasm System
USSR	Union of Soviet Socialist Republics

Appendix C: Obsolete Grass Varieties and Experimental Lines and/or Varieties No Longer Available in the U.S.

Name, Variety or Number	Descriptive or Scientific Name	AH-170 Handbook Edition
A-10	Kentucky bluegrass	1972
A-10675	Siberian wheatgrass	1965
A-11527	sand lovegrass	1959,1965
A-12445	*Bromus coloratus*	1959
A-12496	intermediate wheatgrass	1972
A-12638	kleingrass	1972
A-12752	Boer lovegrass	1959,1965
A-14156	kleingrass	1959
A-1770	fairway wheatgrass	1959
A-20	Kentucky bluegrass	1972
A-2514	Russian wildrye	1959
Algerian	blue panicgrass	1959
Alta 4-36	tall fescue	1959,1965
Alta 144	tall fescue	1959
Angleton	bluestem	1972
Arboretum	Kentucky bluegrass	1972
Argentine	bahiagrass	1972
Arlington	creeping bentgrass	1972
Artex	Texas millet	1972
Asheville	tall fescue	1972
Astor	annual ryegrass	1972
Astra	timothy	1972
Aubade	Italian ryegrass	1972
Avon	orchardgrass	1972
B-230	dallisgrass	1972
B-430	dallisgrass	1972
B. in. 12	smooth brome	1959,1965
Barbarossa	meadow fescue	1972
Barenza	perennial ryegrass	1972
Barfalla	chewings fescue	1972
Bargenta	Kentucky bluegrass	1972
Bariton	timothy	1972
Barkas	meadow fescue	1972
Bayshore (Gene Tift)	bermudagrasses	1972
Beltsville 117-27(6)	Kentucky bluegrass	1972
Beltsville Selections	Japanese lawngrass	1965
Beltsville Synthetic 4	sudangrass	1959,1965
Belturf	Kentucky bluegrass	1972
Bitter Blue	St. Augustinegrass	1972
Blair	smooth brome	1972
Boone	orchardgrass	1972
Boutleoua eriopoda	blue grama	1972
Brage	orchardgrass	1972
Bromus tectorum	cheatgrass	1972
C-52	creeping bentgrass	1972

Name, Variety or Number	Descriptive or Scientific Name	AH-170 Handbook Edition
California 23	sudangrass	1972
Campus	Kentucky bluegrass	1972
Carson	sand bluestem	1972
Chapel Hill	rescuegrass	1972
Cherry	sand bluestem	1972
Clatsop	red fescue	1972
Cohansey	creeping bentgrass	1972
Cold Hardy	Lehmann lovegrass	1959,1965
Collins	creeping bentgrass	1972
Common (Garawi)	sudangrass	1972
Congressional	creeping bentgrass	1972
Cornell 1777	timothy	1959
Cornell 4059	timothy	1959
Coronado	sideoats grama	1972
Cougar	Kentucky bluegrass	1972
Dahlgren	creeping bentgrass	1959,1965
Dayton	orchardgrass	1972
De Soto	sorgrass	1972
Dural	timothy	1972
Duraturf	red fescue	1972
Durlawn	red fescue	1972
Ea 611081	St. Augustinegrass	1972
Empire	foxtail millet	1972
Engmo	timothy	1972
Essex	timothy	1972
Evansville	creeping bentgrass	1972
Everglades	bermudagrass	1972
Evergreen	timothy	1972
Festina	meadow fescue	1972
Fischer	smooth brome	1972
Flagstaff	black grama	1959,1965
Formosa	yellow bluestem	1959,1965
Fortune	tall fescue	1972
Fox	smooth brome	1972
Frontier	reed canarygrass	1972
Gahi 1	pearl millet	1972
Game	perennial ryegrass	1972
Gasel	rescuegrass	1972
Georgia 337	sudangrass	1972
Georgia Inbred Lines	pearl millet	1972
Tift 239DA2		
Tift 239DB2		
Tift 23A1		
Tift 23B1		

Name, Variety or Number	Descriptive or Scientific Name	AH-170 Handbook Edition
Tift 23DA1		
Tift 23DB1		
German 8	foxtail millet	1959
German 8A	foxtail millet	1959
German R	foxtail millet	1959
Golfrood	chewings fescue	1972
Green Stipagrass	green needlegrass	1972
H-1	Italian ryegrass	1959,1965
HA-333	buffelgrass	1972
HA-716	guineagrass	1972
Hardy	Lehmann lovegrass	1959
Heidemij	timothy	1972
Hercules	orchardgrass	1972
Highlight	chewings fescue	1972
Holfior	colonial bentgrass	1972
Homesteader	smooth brome	1972
Hopkins	timothy	1972
Huron	timothy	1972
Hybrid SJ	pearl millet	1959,1965
Idaho 100	Russian wildrye	1972
Idaho 3	intermediate wheatgrass	1959,1965
Idaho 4	intermediate wheatgrass	1959,1965
Iowa M2-10820	intermediate wheatgrass	1959,1965
Iowa synthetics	orchardgrass	1959
Iowa Synthetics	reed canarygrass	1972
Jackson	orchardgrass	1972
Jeanerette	smooth brome	1959,1965
Jolanda	Italian ryegrass	1972
Kalahari	Lehmann lovegrass	1959,1965
Kansas Experimental	Indiangrass	1965
KB-143	Kentucky bluegrass	1959
KB-176	Kentucky bluegrass	1959
Kenmont	tall fesuce	1973
Kentucky 45-50	tall fescue	1965
Kentucky Select	orchardgrass	1965
Kenwell	tall fescue	1972
Kernwood	velvet bentgrass	1972
King	timothy	1972
Kuhl	smooth brome	1965
Ky. Experimental	Lolium-Festuca derivative	1972
La Estanzuela 284	Italian ryegrass	1972
Lahoma	sudangrass	1972
Liso	smooth brome	1972
Lomas	blue wildrye	1959,1965
Lorain	timothy	1972
M2-10302	indiangrass	1959,1965
M2-10820	intermediate wheatgrass	1965
M2-11108	Canada wildrye	1959,1965
M2-11142	orchardgrass	1959,1965
Mandan 1274	intermediate wheatgrass	1972
Mandan 2194B	crested wheatgrass	1959,1965
Mandan 2359	crested wheatgrass	1959,1965
Mandan 315	reed canarygrass	1959,1965
Mandan D-19	Russian wildrye	1972
Marash	Old World bluestem	1972
Marfa	blue grama	1972
Marietta	timothy	1972
Martin	smooth brome	1972
Massa	Italian ryegrass	1972
Masshardy	orchardgrass	1972
Medon	timothy	1972
Menuet	chewings fescue	1972
Mesa	buffalograss	1972
Metropolitan	creeping bentgrass	1972
Michigan B-2	smooth brome	1959
Midwest	Japanese lawngrass	1972
Milton	timothy	1972
Mississippi Fine Stem	johnsongrass	1972
Mississippi ISJ	sorgrass	1972
Mississippi SJ-1	sorgrass	1959,1965
Mississippi SJ-2	sorgrass	1972
N-7035	sudangrass	1972
NB-280S	sorghum X sudangrass	1972
NCS-511	tall fescue	1959
Nebraska 10	crested wheatgrass	1959,1965
New York Synthetic B	timothy	1959,1965
New York Synthetic D	orchardgrass	1959
New York Synthetic E	orchardgrass	1959,1965
New York Synthetic L	smooth brome	1959,1965
New Zealand	tall fescue	1959
NK-100	perennial ryegrass	1972
NK-101	perennial ryegrass	1972
NK-37	bermudagrass	1972
NK-Experimental K9-124	perennial ryegrass	1972
NK-T2	Italian ryegrass	1972
NK-T3	Italian ryegrass	1972
NK-T4	Italian ryegrass	1972
No Mow	bermudagrass	1972
Norbeck	creeping bentgrass	1959,1965
Nordstern	orchardgrass	1972
Nu Dwarf	Kentucky bluegrass	1972
Oklahoma 1	smooth brome	1959,1965

Name, Variety or Number	Descriptive or Scientific Name	AH-170 Handbook Edition
Oklahoma 8	sudangrass	1959,1965
Oklahoma 4880	Boer lovegrass	1959
Oklahoma	smooth brome	1959,1965
Olds	red fescue	1972
Ormond	bermudagrass	1972
P-106	Kentucky bluegrass	1972
P-14762	creeping foxtail	1972
P-14893	hardinggrass	1972
P-14897	western wheatgrass	1972
P-14943	thickspike wheatgrass	1965
P-15589	prairie sandreed	1972
P-15626	yellow bluestem	1972
P-15638	green needlegrass	1972
P-15656	foxtail millet	1972
P-1822	thickspike wheatgrass	1959,1965
P-2369	reed canarygrass	1972
P-2447	bunchgrass	1965
P-2575	Indian ricegrass	1972
P-2662	blue wildrye	1972
P-312	perennial ryegrass	1972
P-4874	bulbous bluegrass	1972
P-57	Kentucky bluegrass	1972
P-6435	tall fescue	1972
P-69	Kentucky bluegrass	1972
P-727	western wheatgrass	1972
P-84	Kentucky bluegrass	1972
P-851	canby bluegrass	1972
Palestine	orchardgrass	1972
Paraguay	bahiagrass	1972
Parkland	smooth brome	1972
Pasture laboratory Synthetics 1 - 7	orchardgrass	1959,1965
Pelo	perennial ryegrass	1972
Pennington 63-70	tall fescue	1972
Pennington 69-10	Italian ryegrass	1972
Pennlate	orchardgrass	1972
Pennlu	creeping bentgrass	1972
Pennmead	orchardgrass	1972
Pennpar	creeping bentgrass	1972
Pennstar	Kentucky bluegrass	1972
Pennsylvania synthetics	orchardgrass	1959
Pensacola x Common	bahiagrass	1972

Name, Variety or Number	Descriptive or Scientific Name	AH-170 Handbook Edition
Perennial Sweet	sorgrass	1972
Petra	perennial ryegrass	1972
Phalaris coerulescens	sunolgrass	1972
PI 111994	sunolgrass	1972
PI 155084	buffelgrass	1972
PM-C-14	alkali sacaton	1965
PM-C-29	bluebunch wheatgrass	1965
PM-C-30	western wheatgrass	1965
PM-M-161	bluebunch wheatgrass	1965
PM-ND-264	alkali sacaton	1972
PMT-389	Arizona cottontop	1972
Polis	rough stalk bluegrass	1972
Prairie	rescuegrass	1972
Prato	Kentucky bluegrass	1972
Pretoria 90	Diaz bluestem	1972
Primo	Kentucky bluegrass	1972
Promenade	Italian ryegrass	1972
Raritan	velvet bentgrass	1972
Redpatch	smooth brome	1972
Ree	intermediate wheatgrass	1972
Reno	plains bluegrass	1959,1965
Rescue 440	rescuegrass	1959,1965
Reveille	perennial ryegrass	1972
Rhode Island 6	red fescue	1965
Rideau	orchardgrass	1972
Rise	reed canarygrass	1972
Roselawn	St. Augustinegrass	1972
Ryegrass 12	Italian ryegrass	1959,1965
S-101	perennial ryegrass	1972
S-143	orchardgrass	1972
S-215	meadow fescue	1972
S-23	perennial ryegrass	1972
S-24	perennial ryegrass	1972
S-37	orchardgrass	1972
S-50	timothy	1972
S-5573	reed canarygrass	1972
Sandburg	smooth brome	1972
Sand strain	western wheatgrass	1959
Sawki	Russian wildrye	1972
SC 20-672	brunswickgrass	1972
Seaside	creeping bentgrass	1972
Selection 7	pearl millet	1959,1965
Shelby	timothy	1959,1965
Sodco	Kentucky bluegrass	1972

Name, Variety or Number	Descriptive or Scientific Name	AH-170 Handbook Edition
Sonora	black grama	1972
South Dakota 15	crested wheatgrass	1959,1965
South Dakota 20	intermediate wheatgrass	1959
South Dakota 4305	Kentucky bluegrass	1972
Starr	pearl millet	1972
Stoneville rust-resistant strains	Italian ryegrass	1959,1965
Stoneville Selection	sudangrass	1959,1965
Stoneville Synthetic 1	sudangrass	1959,1965
Stoneville Synthetic 2	sudangrass	1965
Sturdy	meadow fescue	1972
Suhi 1	sudangrass hybrid	1972
Sunturf	bermudagrass	1972
Superior	reed canarygrass	1972
Swallow	timothy	1972
Sweet 372	sudangrass	1972
Sweet 372 (S1)	sudangrass	1972
Synthetic 1	kleingrass	1959
T-15327	blue panicum	1972
T-20258	*Stipa hyalina*	1959,1965
Taptoe	perennial ryegrass	1972
Tardus II	orchardgrass	1972
Tennessee strains	tall fescue	1959,1965
Tetrone	Italian ryegrass	1972
Texas 46	rescuegrass	1972
Texas Synthetic 63-1	kleingrass	1965
Texas Synthetic 63-4	blue panicgrass	1965
Texturf 10	bermudagrass	1972
Texturf 1F	bermudagrass	1972
Tiffine (Reg. No. 3)	bermudagrass	1972
Tifhi 1	bahiagrass	1972
Tifhi 2	bahiagrass	1972
Tifleaf 1	pearl millet	
Tifton	Italiain ryegrass	1972
Toronto	creeping bentgrass	1972
Trigo	pubescent wheatgrass	1972
Trogdon	orchardgrass	1959,1965
Tualatin	tall oatgrass	1972
Tucson	sideoats grama	1972
Turkish	fairway wheatgrass	1959,1965

Name, Variety or Number	Descriptive or Scientific Name	AH-170 Handbook Edition
U-3	bermudagrass	1972
Uganda	bermudagrass	1972
Utah 109	pubescent wheatgrass	1959
Utah 42-1	crested wheatgrass	1959,1965
Utah Synthetic	orchardgrass	1959,1965
Uvalde	sideoats grama	1972
Virginia 70	orchardgrass	1972
Washington	creeping bentgrass	1972
Washington 88	orchardgrass	1959,1965
Washington H-2	orchardgrass	1959,1965
Western	Indiangrass	1972
Wheeler	sudangrass	1972
Windsor	Kentucky bluegrass	1972
Wintergreen	red fescue	1972
Wisconsin B-55	smooth brome	1959,1965
Wisconsin B-63	smooth brome	1959,1965
Wisconsin F-52A	orchardgrass	1965
Woodward Strain W-4	sideoats grama	1959,1965
Woodward Strains W1, W2	switchgrass	1959
Woodward Strains W1 W2 W3 W4 W5	sand lovegrass	1959,1965
Woodward Strains Block E Hope Temple W1 W2 W3	sideoats grama	1972
Z-73	Japanese lawngrass	1972

Appendix D: Common and Scientific Names of Diseases and Insects

Common Name	Scientific Name
	Diseases
anthracnose	*Colletotrichum graminicola* (Ces.) G. Wilson
brown patch	*Rhizoctonia solani* Kuhn
brown spot	*Pyrenophora bromi* (Died.) Drechs.
brown stripe	*Scoledcotrichum graminis* (Fuckel) Deighton *Schlerophthora macrospora* (Sacc.) Thirum.
candii stripe rust	*Puccinia crandilii* Pommel & H. Hume in H. Hume
coryneum blight	*Cladosporium phlei* (C.T. Gregory) G.A. De Vries
covered smut	*Ustilago hordei* (Pers.) Lagerh.
crown rust	*Puccinia coronata* Corda
dollar spot	*Sclerotina homoeocarpa* F.T. Bennett
ergot	*Claviceps purpurea* (Fr.) Tul.
endophyte	*Acremonium coenophialium* Morgan-Jones & W. Gams *Acremonium* Link *Ovularia lolii* Volkart
fusarium blight	*Fusarium* Link:Fr.
fusarium patch	*Fusarium culmorum* (Wm.G. SM.) Sacc.
head scrap	*Fusarium roseum* Link:Fr.
head smut	*Ustilago bullata* Berk. *Helminthosporium* Link
leaf blotch	*Salenophonia* Maire.
leaf rust	*Uromyces dactylidis* G. Otth
leaf scald	*Rhynchosporium* Heinsen ex A.B. Frank.
leaf spot	*Ascochyta brachypodii* (Syd.) R. Sprague & A.G. Johnson *Bipolaris* Shoemaker *Bipolaris setariae* (Swada) Shoemaker *Drechslera dactylidis* Shoemaker *Drechslera siccans* (Drechs.) Shoemaker *Mastigosporium* spp. Riess in Fresen. *Pyricularia grisea* (Cooke) Sacc. *Septoria spraguei* Uecker & J.M. Krupinsky
leaf spotting	*Bipolaris* Shoemaker
leaf streak	*Scolicotrichum* Kunze in Kunze & J.C. Schmidt:Fr. *Helminthosporium* Link
melting out	*Drechslera poae* (Baudys) Shoemaker
net blotch	*Pyrenophora dictyoides* Paul & Parbery
pink patch	*Laetisaria fuciformis* (McAlpine) Burdsall
pink snow mold	*Microdochium nivale* (Fr.) Samuels & I.C. Hallet
powdery mildew	*Erysiphe graminis* DC.
purple eye spot	*Mastigosporium rubricosm* (Dearn. & Barth.) Nannf.
pythium blight	*Pythium ultimum* Trow
red leaf spot	*Drechslera erythropila* (Drechs.) Shoemaker
red thread	*Laetisaria fuciformis* (McAlpine) Burdsall *Phythium aphanidermatum* (Edson) Fitzp.
root rot	*Helminthosporium* Link *Pythium* Pringsh.
rust	*Puccinia* Pers.:Pers. *Puccinia* cynodontis Lacr. ex Desmaz. *Puccinia substriata* Ellis & Barth.
scald	*Rhynchosporium secalis* (Oudem.) J.J. Davis
silver top	*Fusarium poae* (Peck) Wollenweb.
smut	*Ustilago* (Pers.) Roussel. *Ustilago striiformis* (Westend.) Niessl
snow mold	*Microdochium nivale* (Fr.) Samuels & I.C. Hallett *Typhula incarnata* Fr. *Acremonium boreale* Smith & Davidson
stem rust	*Puccinia graminis* Pers.:Pers. ssp. *graminicola* Z. Urban
stripe rust	*Puccinia striiformis* Westend.

Common Name	Scientific Name
summer patch	*Leptosphaeria korrae* J. Walker & A.M. Smith *Phialophora radicicola* Cain
typula blight	*Typhula* (Pers.) Fr.
yellow stripe rust	*Puccinia striiformis* Westend.
	Insects
banks grass mite	*Oligonychus pratensis* Banks
bluegrass billbug	*Sphenophorus parvulus* Cyllenhal
lygus bug	*Lygus lineolaris* P. deB
mite	*Eriophyes* spp.
	Bryobia praetiosa Kotch
southern chinch bug	*Blissus insularis* Barber
sod webworm	*Parapedistia teterrella* Zinckins
	Pediasia trisecta Walker
wheat curl mite	*Eriophyes* tulipae Keifer

Index of Scientific Plant Names, Common Names, Varieties, and Strains